AF830860

Nanotechnology in the Food Industry

Unlock the mysteries of nanotechnology's transformative role in the food industry with *Nanotechnology in the Food Industry: Applications, Recent Trends, and Future Perspectives*. Embark on a journey through the latest research and developments in nanomaterials synthesis, characterization, and manipulation techniques aimed at aligning with consumer expectations. Discuss the fundamental principles underlying nanotechnology and nanomaterials, illuminating their pivotal significance in shaping the future of food production and consumption, as well as applications of nanotechnology in the food industry, from revolutionizing packaging and ensuring food safety to enhancing consumer perception and extending shelf life.

Key Features

- Provides comprehensive information on different aspects of nanotechnology within the food industry
- Presents a wealth of new facts on utilizing the potential of nanotechnology for food materials
- Reflects the contemporary landscape of nanotechnology in the food sector

With a focus on recent advances and future prospects, this book provides detailed discussions on nanosensors, nanoparticles in food formulations, and strategies for shelf-life enhancement. It is an indispensable resource for students, researchers, and scientists seeking to deepen their understanding of nanotechnology's role in shaping the future of food.

Nanotechnology in the Food Industry

Applications, Recent Trends, and Future Perspectives

Edited by

Sukhvinder Singh Purewal and Avneet Kaur

CRC Press
Taylor & Francis Group
Boca Raton London New York

CRC Press is an imprint of the
Taylor & Francis Group, an **informa** business

Cover image: ustas7777777/Shutterstock.com Stock ID 1170993091

First edition published 2025
by CRC Press
2385 NW Executive Center Drive, Suite 320, Boca Raton, FL 33431

and by CRC Press
4 Park Square, Milton Park, Abingdon, Oxon, OX14 4RN

CRC Press is an imprint of Taylor & Francis Group, LLC

©2025 selection and editorial matter, Sukhvinder Singh Purewal and Avneet Kaur; individual chapters, the contributors

Reasonable efforts have been made to publish reliable data and information, but the author and publisher cannot assume responsibility for the validity of all materials or the consequences of their use. The authors and publishers have attempted to trace the copyright holders of all material reproduced in this publication and apologize to copyright holders if permission to publish in this form has not been obtained. If any copyright material has not been acknowledged please write and let us know so we may rectify in any future reprint.

Except as permitted under U.S. Copyright Law, no part of this book may be reprinted, reproduced, transmitted, or utilized in any form by any electronic, mechanical, or other means, now known or hereafter invented, including photocopying, microfilming, and recording, or in any information storage or retrieval system, without written permission from the publishers.

For permission to photocopy or use material electronically from this work, access www.copyright.com or contact the Copyright Clearance Center, Inc. (CCC), 222 Rosewood Drive, Danvers, MA 01923, 978-750-8400. For works that are not available on CCC please contact mpkbookspermissions@tandf.co.uk

Trademark notice:Product or corporate names may be trademarks or registered trademarks and are used only for identification and explanation without intent to infringe.

ISBN: 978-1-032-55922-3 (hbk)
ISBN: 978-1-032-57169-0 (pbk)
ISBN: 978-1-003-43816-8 (ebk)

DOI: 10.1201/9781003438168

Typeset in Times
by Newgen Publishing UK

Contents

About the Editors vii
List of Contributors ix
Preface xiii
List of Abbreviations xv

Chapter 1 Nanoparticles and Methods Involved in Their Synthesis 1

Zahra Emam-Djomeh, Mohammad Ekrami, Nima Mobahi, Rezvan Esmaeily, and Salma Samiei Tousi

Chapter 2 Nanotechnology and Food-Grade Enzymes 20

Leila Ziaeifar, Arash Hosseini, Maryam Salami, Zahra Emam-Djomeh, and Golam Reza Askari

Chapter 3 Nanotechnology for Encapsulation and Delivery of Functional Food Ingredients 55

Sovan Samanta, Jhimli Banerjee, Rubai Ahmed, Swarnali Das, Sandeep Kumar Dash, Avneet Kaur, and Sukhvinder Singh Purewal

Chapter 4 Role of Nanotechnology in Food Industries 93

Zahra Emam-Djomeh, Nima Mobahi, Mohammad Ekrami, and Elaheh Pourmohammad

Chapter 5 Nanotechnology in Food Packaging and Preservation 111

Anna Rafaela Cavalcante Braga, Monica Masako Nakamoto, Monize Bürck, Letícia Guerreiro da Trindade, and Marcelo Assis

Chapter 6 Nanosensors to Detect Food Contaminants 132

Selenay Sadak, İpek Küçük, Didem Nur Ünal, Cem Erkmen, and Bengi Uslu

Chapter 7 Role of Nanotechnology in Detection of Heavy Metals in Food Products 157

Gurveer Kaur, Chirasmita Panigrahi, and Swati Agrawal

Chapter 8 Nanosensors for Pathogen Detection in Food Products 183

Ruslan Mehadi Galib and Mahabub Alam

Chapter 9 Impact of Nanomaterials on Human Health through/via Food Chain......................203

Fahima Siddikey, Md. Istiakh Jahan, Syeda Sabrina Akter, A S M Sayem, and Mahabub Alam

Chapter 10 Nanotechnology Current Scenario and Future Prospects..225

Ruslan Mehadi Galib, Tanjilla Akter Sumi, and Mahabub Alam

Chapter 11 Market Potential of Innovations Based on Food Nanotechnology...........................243

Mahabub Alam, Tanjilla Akter Sumi, Raju Mia, and Md. Jamil Imtiaj

Index...259

About the Editors

Sukhvinder Singh Purewal, PhD, is presently working as an assistant professor, University Centre for Research & Development (UCRD), Chandigarh University, Mohali, Punjab, India. In 2018, he was awarded a major research project spanning 2018–2022 from the SEED Division, Department of Science and Technology (DST), New Delhi. His research interests encompass: solid-state fermentation, extraction of bioactive compounds from natural sources, antioxidants, and the processing of fruits. Having authored more than 45 research papers in internationally recognized journals, Dr. Purewal has been honored with the Best Poster Awards at both national and international conferences. He is an active life member of the Association of Microbiologists of India (AMI), the Mycological Society of India (MSI), the Association of Food Scientists and Technologists (AFSTI) in Mysore, India, and the Indian Science Congress Association (ISCA). In 2023, Dr. Purewal was honored with the prestigious Young Scientist Award by AMI in recognition of his outstanding contributions to the field of food microbiology. Furthermore, Dr. Purewal has demonstrated his expertise by authoring, *Millets: Properties, Processing, and Health Benefits*, and editing, *Maize Nutritional Composition, Processing, and Industrial Uses* and *Chickpea and Cowpea: Nutritional Profile, Processing, Health Prospects, and Commercial Uses* with CRC Press/Taylor & Francis.

Avneet Kaur, PhD, is presently working as an assistant professor in the Department of Chemistry, University Institute of Sciences, Chandigarh University, Mohali, Punjab, India. During her PhD tenure she has published more than five research papers in journals of high repute. Her research during her PhD focused primarily on two distinct areas. Firstly, she explored thermo-physical and theoretical examination of binary mixtures comprising choline-based ionic liquids and hydrogen bond donor solvents such as water, straight-chain alcohols, and alkane diols, with potential industrial applications. Secondly, she explored the development of choline-based ionic liquid-assisted nanostructures for electrochemical sensing of pollutants, showcasing their potential as intelligent designs for various engineering and industrial applications. In the last 2 years, she has been actively working on biochemistry of food products. In 2023, she was honored with the Best Poster Award at a national conference. Dr. Kaur is an active life member of Indian Science Congress Association (ISCA).

Contributors

Swati Agrawal
Agricultural and Food Engineering Department
Indian Institute of Technology Kharagpur
Kharagpur, West Bengal, India

Rubai Ahmed
Department of Physiology
University of Gour Banga
Malda, West Bengal, India

Syeda Sabrina Akter
Department of Food Engineering and Technology
Sylhet Agricultural University, Sylhet, Bangladesh

Tanjilla Akter Sumi
Department of Clothing and Textile
Bangladesh Home Economics College, Dhaka, Bangladesh

Mahabub Alam
Department of Food Engineering and Tea Technology
Shahjalal University of Science and Technology
Sylhet, Bangladesh

Golam Reza Askari
Department of Food Science and Engineering
College of Agriculture and Natural Resource
University of Tehran
Karaj, Iran
Functional Food Research Core
University of Tehran
Tehran, Iran

Marcelo Assis
Department of Analytical and Physical Chemistry
University Jaume I (UJI)
Castelló, Spain

Jhimli Banerjee
Department of Physiology
University of Gour Banga
Malda, West Bengal, India

Anna Rafaela Cavalcante Braga
Department of Biosciences
Department of Physiology
Department of Chemical Engineering
Universidade Federal de São Paulo (UNIFESP)
São Paulo, Brazil

Monize Bürck
Department of Biosciences
Universidade Federal de São Paulo (UNIFESP)
Santos, Brazil
Department of Physiology
Universidade Federal de São Paulo (UNIFESP)
São Paulo, Brazil

Swarnali Das
Department of Physiology
University of Gour Banga
Malda, West Bengal, India

Sandeep Kumar Dash
Department of Physiology
University of Gour Banga
Malda, West Bengal, India

Mohammad Ekrami
Transfer Phenomena Laboratory (TPL)
Department of Food Science
Technology and Engineering
Faculty of Agricultural Engineering and Technology
University of Tehran
Karaj, Iran

Zahra Emam-Djomeh
Transfer Phenomena Laboratory (TPL)
Department of Food Science
Technology and Engineering
Faculty of Agricultural Engineering and Technology
University of Tehran
Karaj, Iran

Cem Erkmen
Department of Analytical Chemistry
Faculty of Pharmacy
Istanbul Aydin University
Istanbul, Turkey

Rezvan Esmaeily
Department of Food Science and Technology
Faculty of Food Science
Bu-Ali Sina University
Hamedan, Iran

Ruslan Mehadi Galib
Department of Food Engineering and Tea Technology
Shahjalal University of Science and Technology
Sylhet, Bangladesh

Letícia Guerreiro da Trindade
Department of Chemical Engineering
Universidade Federal de São Paulo (UNIFESP)
Campus Diadema
Diadema, São Paulo, Brazil

Arash Hosseini
Department of Food Science and Engineering
College of Agriculture and Natural Resource
University of Tehran
Karaj, Iran

Md. Jamil Imtiaj
Department of Food Engineering and Tea Technology
Shahjalal University of Science and Technology
Sylhet, Bangladesh

Md. Istiakh Jahan
Department of Food Engineering and Tea Technology
Shahjalal University of Science and Technology
Sylhet, Bangladesh

Avneet Kaur
Department of Chemistry
University Institute of Sciences
Chandigarh University
Gharuan, Mohali, Punjab, India

Gurveer Kaur
Department of Processing and Food Engineering
Punjab Agricultural University
Ludhiana, Punjab

İpek Küçük
Department of Analytical Chemistry
Faculty of Pharmacy
Ankara University
Ankara, Turkey
The Graduate School of Health Sciences
Ankara University
Ankara, Turkey

Raju Mia
Department of Management Studies
Rabindra University
Bangladesh

Nima Mobahi
Transfer Phenomena Laboratory (TPL)
Department of Food Science
Technology and Engineering
Faculty of Agricultural Engineering and Technology
University of Tehran
Karaj, Iran

Monica Masako Nakamoto
Department of Biosciences
Universidade Federal de São Paulo (UNIFESP)
Santos, Brazil
Department of Physiology
Universidade Federal de São Paulo (UNIFESP)
São Paulo, SP, Brazil

Chirasmita Panigrahi
Chemical Engineering Department
Indian Institute of Technology Madras
Chennai, Tamil Nadu

Elaheh Pourmohammad
Department of Food Science and Technology
Faculty of Agricultural
Urmia University
Urmia, Iran

Sukhvinder Singh Purewal
University Centre for Research & Development (UCRD)
Chandigarh University
Gharuan, Mohali, Punjab, India

Selenay Sadak
Ankara University
Faculty of Pharmacy
Department of Analytical Chemistry
Ankara, Turkey
Ankara University
The Graduate School of Health Sciences
Ankara, Turkey

Maryam Salami
Department of Food Science and Engineering
College of Agriculture and Natural Resource
University of Tehran
Karaj, Iran
Functional Food Research Core
University of Tehran
Tehran, Iran

Sovan Samanta
Department of Physiology
University of GourBanga
Malda, West Bengal, India

A S M Sayem
Department of Food Engineering and Tea Technology
Shahjalal University of Science and Technology
Sylhet, Bangladesh

Fahima Siddikey
Department of Food Engineering and Tea Technology
Shahjalal University of Science and Technology
Sylhet, Bangladesh

Salma Samiei Tousi
Transfer Phenomena Laboratory (TPL)
Department of Food Science
Technology and Engineering
Faculty of Agricultural Engineering and Technology
University of Tehran
Karaj, Iran

Didem Nur Ünal
Faculty of Pharmacy, Department of Analytical Chemistry
Ankara University
Ankara, Turkey
The Graduate School of Health Sciences
Ankara University
Ankara, Turkey

Bengi Uslu
Department of Analytical Chemistry
Faculty of Pharmacy
Ankara University
Ankara, Turkey

Leila Ziaeifar
Department of Food Science and Engineering
College of Agriculture and Natural Resource
University of Tehran
Karaj, Iran

Preface

Chapter 1 discusses the synthesis methods and characteristics of nanoparticles. **Chapter 2** extensively covers the intersection of nanotechnology and food-grade enzymes. In **Chapter 3**, the utilization of nanotechnology for encapsulating and delivering functional food ingredients is elaborated upon. The role of nanotechnology in the food industry is thoroughly discussed in **Chapter 4**. **Chapter 5** places a significant emphasis on the utilization of nanotechnology for food packaging and preservation. **Chapter 6** explores the application of nanosensors for detecting food contaminants. The detection of heavy metals in food products through nanotechnology is discussed in **Chapter 7**. **Chapter 8** elaborates the workings of nanosensors for pathogen detection in food products. In **Chapter 9**, the detailed impact of nanomaterials on human health via the food chain is discussed. The current status and future prospects of nanotechnology are outlined in **Chapter 10**. Lastly, **Chapter 11** provides an in-depth analysis of the market potential of innovations stemming from food nanotechnology.

Abbreviations

AAS	atomic absorption spectroscopy
AChE	acetylcholinesterase
AES	atomic emission spectroscopy
AgNC	aptamer-modified silver nanocluster
AgNP	silver nanoparticle
Al^{3+}	aluminum
AO	acridine orange
Apt	aptamer
APTES	amino propyl triethoxysilane
ASP	anti solvent precipitation
ASV	anodic stripping voltammetry
Au/Cu–MOF	AuNPs-functionalized Cu–metal organic framework
AuE	gold electrode
AuNC	gold nanocluster
AuNP	gold nanoparticle
Au-SPE	gold screen-printed electrode
BBA	bivalent binding aptamer
BHQ3	Black Hole Quencher-3
BOMC	boron-doped ordered meso-porous carbon
BSA	bovine serum albumin
B–Ser–His-GQD	serine and histidine-functionalized grapheme quantum dot
BTPA	3-(3-(benzo[d]thiazol-2-yl)-2-hydroxyphenyl)-2-(4′-(diphenylamino)-[1,1′-biphenyl]-4-yl)acrylonitrile
BTSIXO	3′,6′-bis(diethylamino)-2-((3,4,5-trimethylbenzylidene) amino)spiro isoindoline-1,9′-xanthen]-3-one
C	carbon
CA	cellulose acetate
CAB	cellulose acetate butyrate
CAC	critical aggregation concentration
CB	cell-biomass
CC	carbon cloth
CD	carbon dot
Cd^{2+}	cadmium ion
cDNA	complementary DNA
CDs-BP	carbondots-black phosphorus nanohybrid
CdTe	cadmium telluride
CdTeQD	cadmium telluride quantum dot
Ce	cerium
Ch	chitosan
CHD	coronary heart disease
CN^-	cyanideion
CNC	cellulose nanocrystal
CNF	cellulose nanofibers
CNT	carbon nanotube
COD	covalent organic frameworks
Co-MOF	cobalt metal–organic framework

COOH–GO	carboxylated graphene oxide
CQD	carbon quantum dot
Cr	chromium
CTAB	cetyl trimethyl ammonium bromide
Cu	copper
Cu^-	cupric cation
Cu-MOF	functionalized copper–metal organic framework
CuNP	copper nanoparticle
CuS	copper sulfide
CV	cyclic voltammetry
CVD	chemical vapor deposition
CVS	chemical vapor synthesis
Cys-AgNP	cysteine-coated silver nanoparticle
DBF	4,5-bis-dihydroxyboron fluorescein
DC	dendritic cell
DMZ	dimetridazole
DPV	differential pulse voltammetry
DTN	DNA tetrahedron nanostructure
ECL	electro chemi-luminescence
EFA	essential fatty acid
EIS	electrochemical impedance spectroscopy
EO	essential oil
EU	European Union
Eu-In-BTEC	In-codoped Europium-based MOFs
EμPAD	electrochemical micro paper analytical device
FA	folic acid
FAM	DNA probe
FDA	Food and Drug Administration
Fe^{3+}	iron(III) ion
Fe_3O_4	ironoxide (II, III) magnetic nanoparticles
FGP	fluorinated graphene
FITC	fluorescein isothiocyanate
FMNB	fluorescent magnetic nanobead
FNS	fumed silica nanoparticle
FQICTS	fluorescence quenching immune chromatographic test strip
FRET	Forster resonance energy transfer
FSP	flame spray pyrolysis
g-C_3N_4	graphitic-carbon nitride
GCE	glassy carbon electrode
GC-MS	gas chromatography spectrometry
GL-VS_2	grass-like vanadium disulfide
GNS	gold nanosphere
GO	graphene oxide
GQD	graphene oxide quantum dot
H_2O_2	hydrogen peroxide
HCR	hybridization chain reaction
h-DNA	hairpin-DNA
HEBM	high-energy ball milling
Hg^{2+}	mercuryion

HMI	heavy metal ion
hNF	hybrid nanoflower
HPH	high-pressure homogenization
ICP-MS	inductively coupled plasma mass spectrometry
IGC	inert gas condensation
IL	ionic liquid
IMS	immune magnetic separation
ISO	International Organization for Standardization
ITO	indium tin oxide
IUPAC	International Union of Pure and Applied Chemistry
LbL	layer-by-layer
LCN	liquid crystalline nanoparticle
LIBS	laser-induced breakdown spectroscopy
LOC	lab-on-chip
LOD	limit of detection
LSG	laser scribed graphene
LSPR	localized surface plasmon resonance
LSV	linear sweep voltammetry
MagQDNP	magnetic quantum dot nanoparticle
MAPK	mitogen-activated protein kinase
MCH	6-Mercapto-1-hexanol
MCT	medium chain triglyceride
MFDS	Ministry of Food and Drug Safety
MF-MIP	magnetic and fluorescent molecularly imprinted polymer
MIL-101	Materials Institute Lavoisier
MIOPPy	molecularly imprinted over oxidized polypyrrole
MMIP	magnetic molecularly imprinted polymer
MnO_2	manganese dioxide
MNP	magnetic nanoparticle
Mn-SnO_2	manganese-doped tin oxide
MOF	metal–organic framework
MPBA	4-mercaptophenylboronic acid
MSNP	mesoporous silica nanoparticle
MWCNT	multi-walled carbon nanotube
N-GQD	nitrogen-doped grapheme quantum dot
NHAP	nano-sized hydroxyl apatite
NLC	nanostructured lipid carrier
NM	nanomaterial
NMR	nuclear magnetic resonance
NP	nanoparticle
NP-1	novel pyrene modified nanocrystalline cellulose
NPG	nanoporous gold
NRF	non-covalent ratio metric fluorophore
NSM	nanostructured material
OCS	optical chemical sensor
OECD	Organization for Economic Cooperation and Development
OPDA	o-phenylene diamine
OSA	octenyl succinic anhydride
PAA	Polyacrylic acid

PACVD	plasma-aided chemical vapor deposition
p-ATP	p-aminothiophenol
Pb^{2+}	lead
PBE	paper-based electrode
PBS	polybutylene succinate
PCN 222	zirconium–porphyrin MOF
PCR	polymerase chain reaction
Pd	palladium
PDA	polydopamine
Pd NP	palladium nanoparticle
PECVD	plasma-enhanced chemical vapor deposition
PGE	pencil graphite electrode
PHB	poly hydroxy butyrate
POC	point-of-care
Poly-NR	poly neutral red
POPDGNS	polymer–nanoparticles composite
PP6	phosphate pillar [6]arene
PPy	polypyrrole
PQ	paraquat
PtNP	platinum nanoparticle
PVA	polyvinyl alcohol
PVC	polyvinylchloride
PVDF	poly(vinylidene fluoride)
QD	quantum dot
qPCR	quantitative-polymerase chain reaction
RFN	ratio fluorescence nanoprobe
rGO	reduced grapheme oxide
rGO-CNT	reduced graphene oxide-carbon nanotubes
RhB	rhodamine B
RSSP	rhodamine 6G silica particle
SDBS	sodium dodecyl benzene sulfonate
SERS	surface-enhanced Raman scattering
SF	silk fibroin
S-GQD	sulfur-doped graphene quantum dot
SiO_2	silicon dioxide
SLN	solid lipid nanoparticle
SnO_2	tin oxide
SP	signal probe
SPCE	screen-printed carbon electrode
SPGrE	screen-printed graphene paste electrode
SPION	super paramagnetic press oxide nanoparticle
SPR	surface plasmon resonance
SWV	square wave voltammetry
Tb^{3+}	Terbium (III)
Tbdcpcpt	$[Me_2NH_2][Tb_3(dcpcpt)_3(HCOO)]$
Tb-MOF	$[Tb_2(H_2btec)(btec)(H_2O)]\cdot 4H_2O$
TiO_2	titanium oxide
TPS	thermoplastic starch
UCNP	up conversion nanoparticle
UiO-66	zirconium metal–organic framework

WE	working electrode
WPC	whey protein concentrate
ZIS-HNC	hollow $ZnIn_2S_4$nanocage
[Zn • (BA) • (BBI)]	a two-dimensional zinc(II)-based metal–organic framework
ZnO	zinc oxide
$ZnWO_4$	zinc tungstate
Zr	zirconium
ZV-Mn NP	zero valent manganese nanoparticle

1 Nanoparticles and Methods Involved in Their Synthesis

Zahra Emam-Djomeh, Mohammad Ekrami, Nima Mobahi, Rezvan Esmaeily, and Salma Samiei Tousi

1.1 INTRODUCTION

The word "nanoparticle"(NP) often excludes individual molecules and is mostly used to describe ultrafine particles (Ekrami et al., 2022b). The NPs may display distinct size-dependent characteristics markedly distinct from tiny particles and bulk materials. Despite being often linked with current science, NPs have a lengthy historical background. The intriguing and unexpected characteristics of NPs are ascribed to their substantial surface area, which surpasses the influence of their relatively modest volume (Ijaz et al., 2020). By manipulating the size, shape, and composition of particles, it becomes feasible to regulate solar absorption (Emam-Djomeh et al., 2023a). The NPs have been affixed to textile strands to produce intelligent and utilitarian garments. Nanoscale material changes, which display wave-like properties as a result of quantum mechanics, impact the behavior of electrons inside matter and atomic interactions (Izci et al., 2021). Nanoscale structure fabrication makes it possible to control material attributes like magnetic susceptibility, charge capacity, color, and melting point without changing their chemical composition (Shrestha et al., 2020). Harnessing this potential will result in the development of novel, cutting-edge goods, and previously unattainable technologies. Now, let's quickly examine the techniques used to create NPs (Mittal et al., 2020).

1.2 METHODS OF NP SYNTHESIS

There is a strong need for high-capacity NPs that possess excellent quality and precise control asboth properties are crucial for their successful implementation in various industries. Typically, the production techniques for NPs may be categorized into three groups: (1) chemical approaches, (2) physical methods, and (3) bio-assisted methods.

1.2.1 Chemical Methods

1.2.1.1 Sol–Gel Method

Two primary processes, hydrolysis and condensation, make up the sol–gel process. The first stage in creating the gel phase is hydrolysis, which entails dissolving the precursor bonds with water (Ekrami et al., 2022d). Subsequently, condensation occurs, resulting in the creation of nanomaterials. Subsequent removal of surplus water determines the ultimate structure of the material (Bokov et al., 2021).

The experimental configuration is indeedsimple. Sol is derived by hydrolysis or polymerization, achieved by introducing appropriate reagents into the precursor solution. The sol may be applied onto desired substrates as thin films utilizing spin coating and dip coating (Eddy et al., 2023).

DOI: 10.1201/9781003438168-1

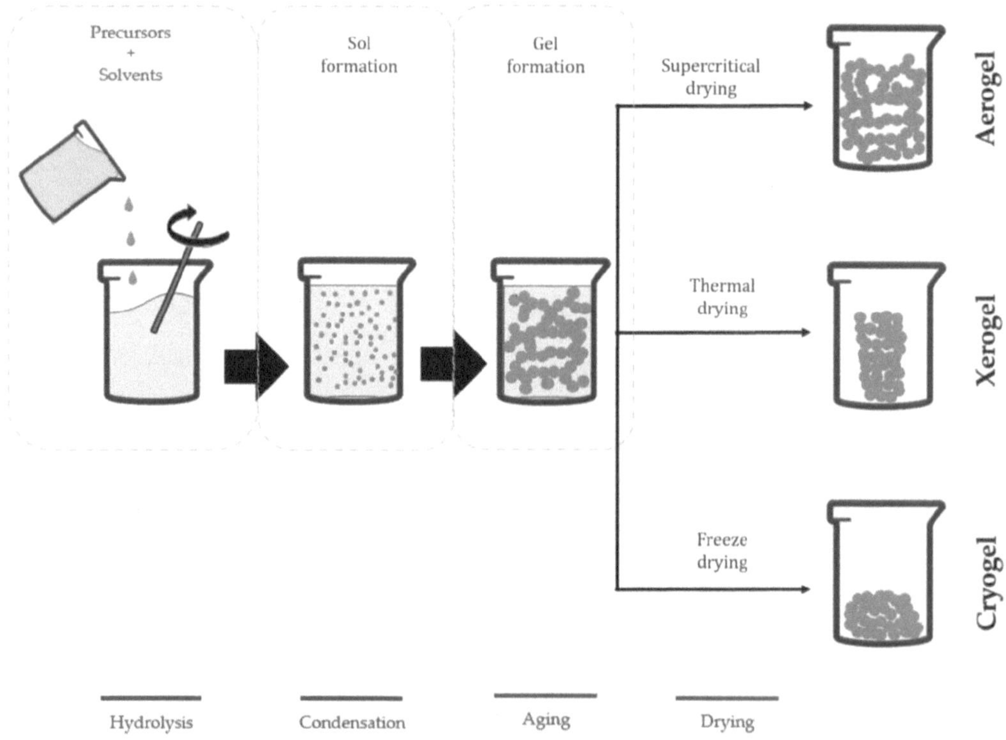

FIGURE 1.1 Schematic diagram of sol–gel processing.

The sol–gel process can produce several types of materials, allowing for precise control over the resulting materials' phase, shape, and size (Dhand et al., 2015). The particles created inside a gel matrix have a consistent shape and size, improving the materials' inherent properties. The approach is mostly used by researchers in the biomedical and biotechnical fields for biomedical applications because of its high purity, high-quality yield, and favorable physical features (Phromma et al., 2020). Figure 1.1 depicts the schematic diagram of the sol–gel technique.

1.2.1.2 Microemulsion Technique

In a microemulsion, there are at least three parts: a polar phase (often water), a hydrocarbon liquid or oil as a non-polar phase, and a surfactant. All of these mixtures need to be isotropic, stable, homogeneous, and see-through (Salvador et al., 2021). The interface tension between the microemulsion and the surplus phase is reduced when surfactant molecules create a layer between the organic and aqueous phases. In addition, they block the droplets from combining (Pineda-Reyes and Olvera, 2018). A microemulsion system is composed of uniformly sized spherical droplets (with diameters varying from 600 nm to 8000 nm) that may be either water-in-oil (w/o) or oil-in-water (o/w) depending on the kind of surfactant used (Tianimoghadam and Salabat, 2018). For the best results for synthesizing NPs, use a reverse micellar system devoid of water. One kind of microemulsion is called a reverse micelle. The organic tails of the surfactant molecules are pointed outward in this micelle form, while the polar head groups are located inside the water-filled center (Ekrami et al., 2022d). The microemulsion technique is commonly used to synthesize various types of inorganic nanomaterials, such as metal salt NPs (Ghadami Jadval Ghadam and Idrees, 2013; He et al., 2018; Zhang et al., 2017), metal NPs (Maslova et al., 2021; Sánchez et al., 2022; Sandhu et al., 2021), magnetic NPs (Baig et al., 2019; Fatima et al., 2023; Rodríguez et al., 2017), semiconducting metal sulfide NPs (Angelescu et al., 2013; Chakraborty and Moulik, 2005; Khomane et al., 2002), metal oxide NPs (Bozkurt, 2020; Li et al., 2014; Parejas et al., 2017), and composite NPs (Lavand et al., 2015; Thangavelu, 2022; Zhang et al., 2020). Metal NPs were readily synthesized utilizing the microemulsion process using the reduction approach (Richard et al., 2017).

1.2.1.3 Hydrothermal Synthesis

Producing NPs of metal oxide, while precisely controlling their characteristics, is achieved by subjecting the water to different pressure and temperature settings near or at supercritical water (Gan et al., 2020). The NPs are generated in a chemical solution by combining two or more phases of matter (such as solids, liquids, or gases) in a colloidal system (Hayashi and Hakuta, 2010). This procedure, which may include combining gels and foams, is carried out with meticulous temperature and pressure control. This method has the potential to mass-produce inexpensive NPs with ideal dimensions, geometries, chemical makeup, and surface chemistry (Yang and Park, 2019).

1.2.1.4 Polyol Synthesis

A metal-containing molecule may be synthesized using the polyol technique by reacting it with poly(ethylene glycol)s. This procedure makes use of poly(ethylene glycol), which dissolves stabilizing and protective compounds, complexing agents, a solvent, and a reducing agent all at once (Favier et al., 2019; Kotoulas et al., 2019). A diverse array of metal-based NPs (Lee et al., 2020; Sharma et al., 2019b; Zeroual et al., 2020), metal oxide NPs (Lahane et al., 2022; Mahamuni et al., 2019; Setiawan et al., 2022), magnetic NPs (Iacovita et al., 2019; Iacovita et al., 2020), and metal hybrid NPs was synthesized using this chemical technique. The polyol technique created low-temperature methods for synthesizing magnetic NPs such as Fe_3O_4 (Cheah et al., 2021). The morphological development of NPs in two distinct polyol processes is shown in Figure 1.2.

1.2.1.5 Chemical Vapor Techniques

Chemical vapor deposition (CVD) often produces solid films by inducing chemical reactions in a high-temperature vapor phase. Thin films generated by the CVD method under certain circumstances also include ultrafine particles (Gleason, 2020). Therefore, the production of NPs may also be achieved using this approach, provided the CVD system is adjusted or improved according to the specified conditions (Dhand et al., 2015). Therefore, when the process parameters are altered to generate NPs instead of thin solid films in the CVD process, the changed method is called chemical vapor synthesis (CVS) (Jo et al., 2021). This synthesis technique involves generating precursor materials in vapor inside a reactor, where nucleation processes occur under certain conditions, including solid, liquid, and gaseous stages.

Furthermore, CVS synthesizes multicomponent or doped NPs using various precursors (Jia et al., 2021). Composite NPs are created by encasing two or more materials within each other; for example, silicon tetrachloride and sodium chloride may be combined to make silicon particles encased in sodium chloride. Many different materials have been utilized to synthesize NPs using the CVS technique (Levish et al., 2023).

1.2.1.6 Plasma-Enhanced Chemical Vapor Deposition (PECVD)

One popular CVD technique for depositing thin films is plasma-augmented CVD, which is sometimes called plasma-aided CVD (PACVD). It is also possible to create NPs using the PECVD method. Plasma, as its name suggests, speeds up the chemical reactions that produce NPs and thin films (Tanemura et al., 2001). Unlike the typical CVD method, producing thin films and NPs using PECVD occurs at much lower temperatures. Asplasma is a gas that is partly ionized, forming thin films and NPs involves the participation of ionized species and radicals. Producing NPs from various materials was shown using PECVD (Yanase et al., 2019). When the input power is minimal, the reactivity of the plasma stays low, leading to a sluggish pace of chemical reactions and

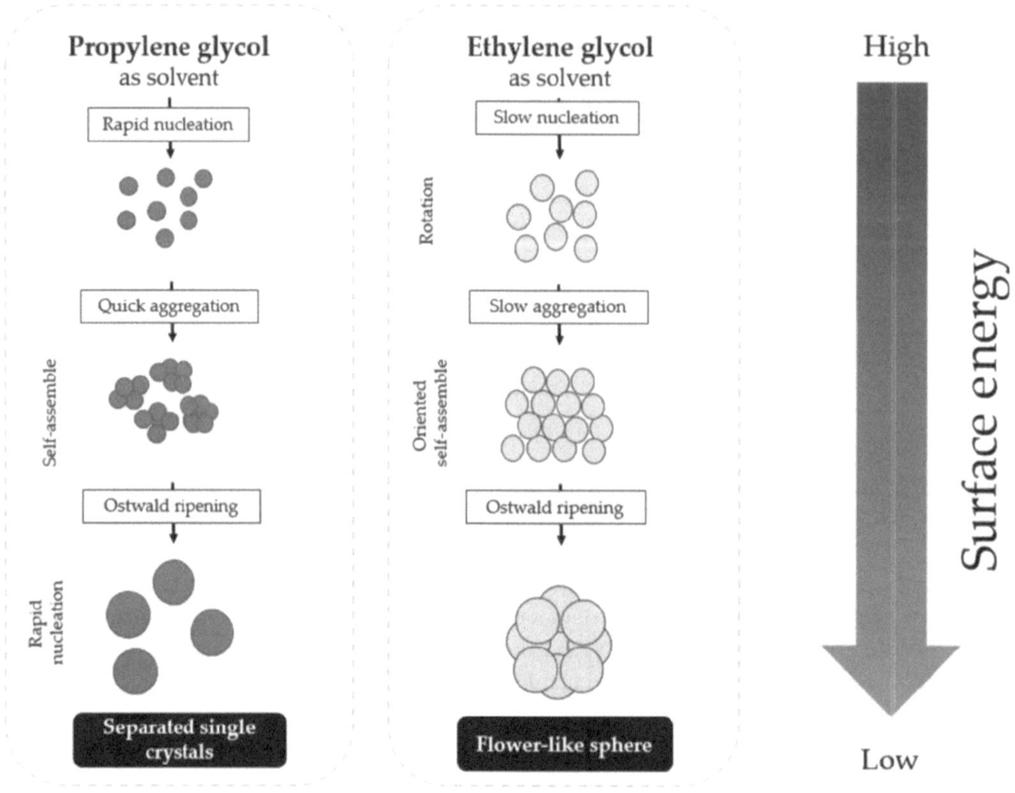

FIGURE 1.2 The morphology evolution of NPs in polyol processes using two different polyols.

the consequent synthesis of monomers. Due to the sluggish reaction rate, only a limited number of nuclei are produced, while a substantial quantity of precursor molecules remains unreacted. When a significant quantity of unreacted substances condenses and reacts with a relatively small number of nuclei, the resultant particles increase in size (Mishin et al., 2015).

1.2.2 Biological Methods

1.2.2.1 Biogenic Synthesis

Bioreactors for NP production commonly use prokaryotic bacteria, actinomycetes, fungi, algae, and yeast. This method for synthesizing a variety of NPs was the subject of much scientific study (Ekrami et al., 2023). Microorganisms extract certain ions from their surroundings and then convert them into metallic elements using enzymes produced by cellular processes (Ekrami et al., 2022c; Mughal et al., 2021). The production of NPs may be categorized as intracellular or extracellular, depending on where the synthesis occurs. The intracellular approach entails the transportation of metal ions into the microbial cell, where they undergo enzymatic processes to create NPs (Raj et al., 2021). Producing NPs outside of cells requires first immobilizing metal ions onto cell surfaces and then reducing these ions using enzymes. To reduce the concentration of metal ions, bacteria interact with proteins, enzymes, sugars, and a variety of anionic functional groups found in their biomass (Sadhasivam et al., 2020). Anionic functional groups such as peptidoglycan, teichoic acids, lipoteichoic acids, proteins, and polysaccharides are abundant in the cell walls of Gram-positive bacteria (Ekrami et al., 2022c; Shakouri et al., 2023). Biosorption and reduction of silver cations may take place at these functional groups (Idris and Roy, 2023). Spyridopoulouet al. used the probiotic strain *Lactobacillus casei* ATCC 393, previously shown to inhibit colon cancer cell growth, to synthesize biogenic selenium NPs (Spyridopoulou et al., 2021). A potential nanofactory for the synthesis of ZnO NPs was investigated by Mohd Yusof et al. (2020) using the cell-biomass (CB) and supernatant (CFS) of zinc-tolerant *Lactobacillus plantarum* TA4.

Using fungi in green chemistry to produceNPs has many benefits, including enhanced bioaccumulation, economic feasibility, and a straightforward synthesis technique that may be readily scaled up owing to simplified downstream processing and biomass handling (Ekrami et al., 2021; Mughal et al., 2021). Microscopical examination confirms that enzymes found on mycelial cell walls enable NP synthesis on the mycelia surface (Guilger-Casagrande and Lima, 2019). The use of several fungus species in the synthesis of silver nanoparticles (Ag NPs) has been the subject of many studies. Bagur et al. (2022) investigated the possibility of biogenic Ag NPs mediated by an endophytic fungal extract from *Tinosporacordifolia* for therapeutic applications. Gupta and Saxena (2023) designed to formulate Ag NPs from *Aspergillus oryzae* MTCC No. 3107.

It has been shown that actinomycetes and yeast may be beneficial for the bio-inspired production of NPs. A member of the kingdom fungi, yeast is a kind of eukaryotic microorganism (Bachheti et al., 2021). Lian et al. (2019) reported a facile one-pot and effective green process for biogenic selenium NPs was obtained using the cell-free extracts of a novel yeast *Magnusiomycesingens* LH-F1. Zhang et al. (2016) achieved the biogenic synthesis of gold NPs (AuNPs) using the yeast cells of *Magnusiomyces ingens* LH-F1.

A diverse range of microorganisms exhibits distinct reactions when exposed to metal precursors, producingNPs (Bhakya et al., 2016). For instance, bacteria and fungi can produce substances inside or outside their cells, and these processes vary in various microorganisms. Metal ion transport through the cell wall is an integral part of intracellular production. In particular, positively charged metal ions engage in an interaction with the negatively charged wall. Enzymes inside the cell lower the ions to the level of metal NPs (Sharma et al., 2019a). Figure 1.3 depicts the process of synthesizing AuNPs using microorganisms.

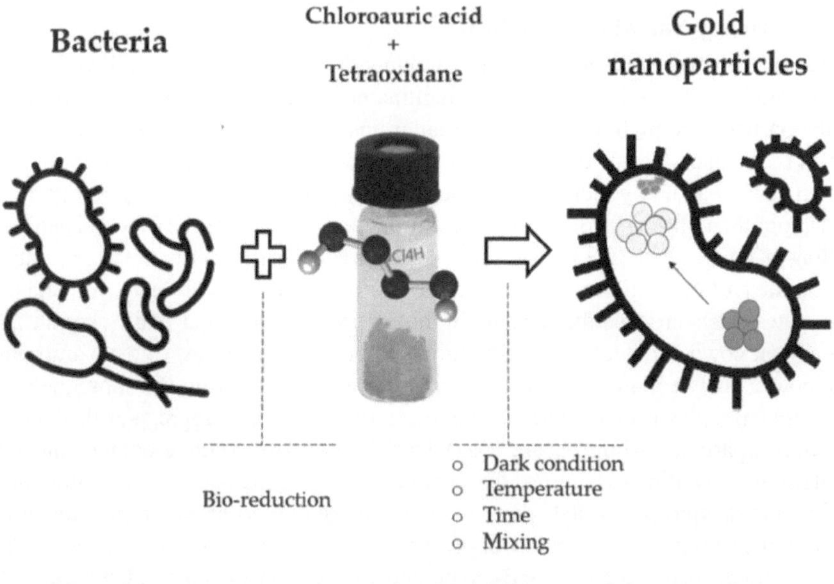

FIGURE 1.3 Mechanism of the intracellular-cell-bound synthesis of AuNPs using bacteria.

1.2.2.2 Templates Based on Biomolecules

Nucleic acids, membranes, viruses, and diatoms serve as templates for the synthesis of NPs. DNA is well-known for its remarkable ability to serve as a biomolecular template, demonstrating a strong attraction to transition metal ions (Baron et al., 2007). There has been some research suggesting that DNA cross-linked hydrogel might be a useful scaffold for the synthesis of AuNPs. DNA's strong affinity for gold and other transition metals makes it possible to build up Au precursor inside a hydrogel. The DNA hydrogel further reduces the $HAuCl_4$, leading to the creation of spherical AuNPs that are 2–3 nm in size, evenly distributed, and free of aggregates (Zinchenko et al., 2014).

Previous studies have shown the fast synthesis of electroless, electrically conductive nanowires of Au, Pd, and CdS using ultraviolet-irradiation and microwaves (MWs), with DNA serving as both a reducing and capping agent (Liu et al., 2014). Some possible uses for these nanowires include building blocks for miniature computers, sensors, optoelectronic devices, and functioning nanodevices (Nyamjav and Ivanisevic, 2005). The fact that biological membranes have such minute gaps in their structure also allowed them to be used as templates to generate NPs (Wang and Ding, 2014).

1.2.2.3 Plant Biometabolites

Using plant extracts or biomass for NP biosynthesis is a highly efficient, swift, ecologically friendly, non-toxic, and clean technique (Ekrami et al., 2022a; Ekrami et al., 2022c; Ekrami et al., 2022d; Lee et al., 2011). Created with this method have been NPs made of noble metals, metal oxides, bimetallic alloys, and similar materials. Because of their useful properties, plant biometabolites may be used as reducing and capping agents in NP manufacturing (Mittal et al., 2013). The rates of NP photosynthesis are much higher in comparison to other biosynthesis methods and sometimes similar to the rate of chemical routes. Glutathione, a kind of antioxidant made up of three amino acids, is present in many organisms such as plants, mammals, fungi, and archaea. It has been shown that glutathione aids in the creation of structured clusters of AuNPs by enhancing the contacts between the particles (Jadoun et al., 2021).

1.2.3 Physical Methods

1.2.3.1 High-Energy Ball Milling (HEBM)

HEBM technique involves subjecting specific amounts of powders to the repetitive impact of balls strategically fired by ball milling equipment (Jamima et al., 2020). Generally, the milling process decreases the particle size by fracturing the initial grains into smaller ones. This behavior becomes evident when the surface area of the powder is increased at a macroscopic level. Particle fragmentation is often accompanied by increased powder reactivity (Chen et al., 2020). For particles with a diameter of more than 100 μm, compression, attrition, and impact are the key mechanisms at play during milling (Kumari et al., 2023). The surface area gradually increases due to the fragmentation of polycrystalline particles into smaller grains. The size of the crystallites decreases, and the milling activity generates fresh surfaces that engage in mutual interaction (Avar and Logoglu, 2022). In a single-component powder, defects gradually build up, and various phases (mostly amorphous) may form at the borders between grains (Chagas and Ferreira, 2023). During the subsequent phase, the particle mixture has a plastic characteristic due to creating secondary aggregates that counteract the newly created tiny particles (Muñoz et al., 2007). During this phase, a single-component powder may either maintain its crystalline structure and achieve nanosized dimensions of crystallite domains, or it can transition to an amorphous state due to the buildup of defects at the grain boundaries (Zheng et al., 2012). When a multicomponent powder combination undergoes a reaction, new products form at the points where the original compounds come into contactwith each other. Ultimately, during the

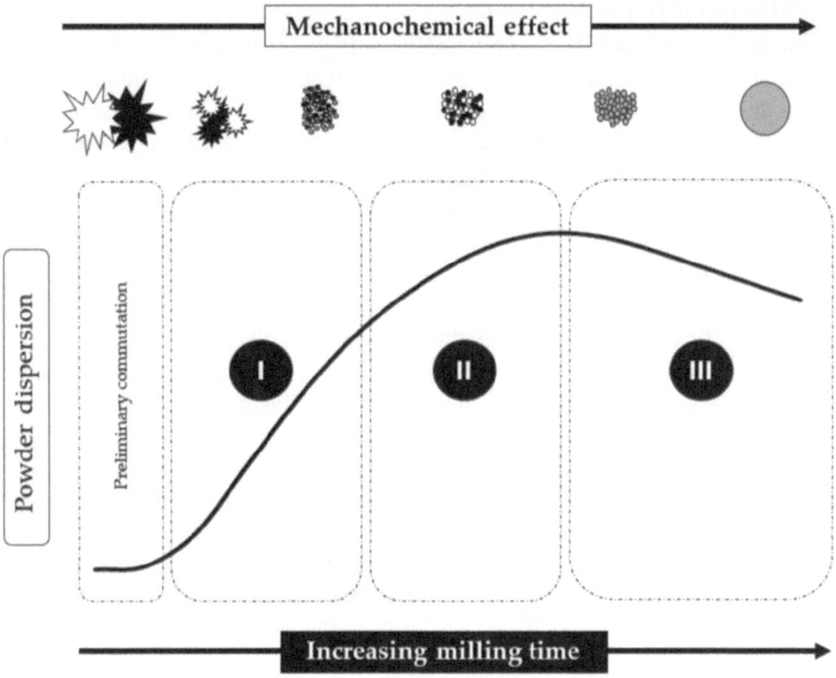

FIGURE 1.4 The activation of powders using high-energy ball milling.

third phase, there is an expansion of previously formed products, accompanied by a little decrease in surface area (Salah et al., 2011). At different stages of the milling process, HEBM has varied impacts on a powder combination, as shown in the diagram in Figure 1.4. Preliminary fragmentation and mixing occur during the first stage (when the particle size is more than 100 μm). During stage I (10–100 μm), the surface area increases gradually as the particles shrink in size. Novel pristine interacting surfaces are formed, followed by the buildup of imperfections in organized regions. Various stages of germs may initiate nucleation, and chemical reactions can occur at the interfaces between grains. During stage II, the emergence of secondary aggregates offsets the presence of recently produced tiny particles. The powder combination has a malleable characteristic, and the surface area stays somewhat consistent. Emerging phases and chemicals mostly manifest during this period. During stage III, there is an increase in particle development, and the process of achieving a uniform mixing of the powder is facilitated. A little reduction in surface area is noticed (Dhand et al., 2015).

1.2.3.2 Inert Gas Condensation (IGC)

One simple way to make NPs is by using IGC. To make the NPs, this method calls for a substrate holder that has been cooled with liquid nitrogen and inert gases like argon or helium (Krasnochtchekov et al., 2022). The volatilized substances are conveyed via inert gases and then condensed onto the substrate, which is cooled using liquid nitrogen (Pandya and Kordesch, 2015). The average particle size exhibited a positive correlation with both the evaporation temperature and helium gas pressure. Particle sizes measured varied between 10 nm and 30 nm. One step of DC sputtering was to create a vapor cloud of metal atoms close to the surface of interest. These vapors then condensed into nanoclusters as they moved through the aggregation zone. The fabricated hybrid NPs have great potential as suitable candidates for magneto-optic applications in biotechnology (Pandya and Kordesch, 2015; Zhao et al., 2013).

1.2.3.3 Physical Vapor Deposition (PVD)

A class of techniques known as PVD is often used for the production of NPs and the deposition of very thin layers of materials (Deng et al., 2020). These layers generally range in thickness from a few nanometers to several micrometers (Rane et al., 2018). PVD is a vacuum deposition technology that is environmentally benign. It involves three essential steps: (1) converting a solid substance into vapor, (2) transporting the vaporized material, and (3) causing the vapor to condense and form thin films and NPs by nucleation and growth (Dhand et al., 2015).

1.2.3.4 Laser Pyrolysis (LAP)

Vapor phase synthesis may be accomplished using the CO_2LAP process. Many different types of NPs may be manufactured using this method (Dhand et al., 2015). At least one reactant or precursor has to be able to absorb infrared (IR) CO_2 laser light via resonant vibrational modes for the reaction system to achieve energy coupling. The CO_2 laser's strong intensity causes multiple IR photons to be absorbed by a single molecule in a specific order. The average gas temperature quickly increases as a result of energy transfer between vibrations and translations in collision-assisted energy pooling, which follows (Reau et al., 2007). This process often results in the formation of a flame in the area where the laser interacts with the gas. When molecules are energized beyond the point of dissociation, they undergo molecular decomposition, which may lead to chemical reactions. This process results in the creation of condensable and volatile substances (Dhand et al., 2015).

In contrast to conventional procedures using the vapor phase, LAP enables precise and accelerated heating, resulting in rapid nucleation, as well as quicker suppression of particle development, often occurring within a few milliseconds (Galvez et al., 2002). Therefore, this approach shows potential in producing NPs with an average diameter spanning 5–60 nm and a tightly controlled size distribution in the high-temperature region (Dhand et al., 2015).

1.2.3.5 Flame Spray Pyrolysis (FSP)

FSP is the most recent advancement among all flame aerosol technologies. The combustion process is a one-step process in which the precursor is liquid. The combustion enthalpy is much larger, accounting for more than 50% of the energy released during burning (Meierhofer et al., 2020). For this reason, an organic solvent is often used. Process technical requirements include a self-sustaining flame, liquid feeds, less volatile precursors, clear scalability, high-temperature flames, and significant temperature gradients (Wallace et al., 2013). Making sophisticated and useful NPs is a common application of this technology. The gas-to-particle approach yields particles with more uniform sizes and shapes compared to the droplet-to-particle procedure; however, both methods may be used to convert liquid precursors into particles (Neto et al., 2020).

1.2.3.6 Electrospraying Technique

Although it has certain similarities with electrospinning, electrospraying generates different materials. The former is used in the creation of NPs, while the latter is put to use in the manufacturing of nanofibers (Farahmand et al., 2023). The electrospraying approach utilizes an electromechanical device that involves a syringe carrying a solution consisting of a specific polymer and solvent. By applying a high voltage to the capillary tip, charged droplets are generated (Pawar et al., 2018). The electrospraying approach offers a combination of versatility and precise manipulation of surface characteristics (Esmaili et al., 2018). The electrospraying technology has many advantages, such as the capacity to produce NPs of consistent size, a rapid preparation process, and the potential for large-scale assembly of NPs (Yanilmaz et al., 2014). Nevertheless, this approach might potentially deteriorate some macromolecules due to heat or shear stresses during the drying and syringe processes (Jayaraman et al., 2015).

1.2.3.7 Melt Mixing

In this method, modified nanofillers are mechanically mixed with a polymer employing kneading, extrusion, or injection molding. Utilizing NPs as fillers in polymer composites is a long-standing technique for achieving certain material properties. The mechanical approach described is widely used due to its environmentally benign nature and compatibility with contemporary industrial methods (Dhand et al., 2015).

1.2.4 Assistive Methods

1.2.4.1 Ultrasound-Assisted

Sonochemistry pertains to chemical processes that are facilitated by ultrasound. Acoustic cavitation is responsible for the sonochemical phenomena seen in liquids. Acoustic cavitation involves the expansion of pre-existing bubbles and their subsequent implosion in a liquid. Intense heating of the bubbles happens when acoustic cavitation collapses (Karthik et al., 2019; Montaño-Priede et al., 2017).

Furthermore, the shock waves resulting from cavitation in mixtures of liquid and solid substances generate collisions between particles at high velocities (Sathya et al., 2017). Ultrasound has been successfully used in several applications, because of its significant physical and chemical effects (Mirzakhani et al., 2018; Montaño-Priede et al., 2017). Furthermore, there have been recent advancements in the use of ultrasound in the precise creation of NPs and porous materials, as well as in the breakdown of organic contaminants for environmental remediation purposes, namely via sonochemical oxidation in wastewater treatment (Asfaram et al., 2016; Barbhuiya et al., 2022). The phenomenon of cavitation during sonolysis generates both oxidative and reductive radicals, which can trigger chemical processes (Behbahani et al., 2017; Wu et al., 2014). Sonochemical methods may create various physical and chemical effects that are advantageous for creating or modifyingNPs. The primary benefit of sonochemical synthesis is its minimal chemical requirements for the process. Acoustic cavitation is the physical phenomenon that causes the sonochemical process (Behbahani et al., 2017). Generally, NPs produced using ultrasound-assisted methods have superior catalytic performance compared to NPs made using classical chemical synthesis. Using ultrasound in synthesis significantly benefitsthe production of high-quality catalyst materials (Biswas et al., 2019; Wu et al., 2014).

1.2.4.2 Microwave-Assisted

One practical alternative to traditional methods for producing organic compounds, polymers, inorganic materials, and NPs is MW heating technology, which has been dubbed the Bunsen burner of the modern day (Hasanpoor et al., 2015). This method offers a higher accuracy level than traditional heating techniques (Qiao et al., 2022). Researchers may now proceed to the next phase of designing and developing nanomaterials by precisely regulating the exact parameters of temperature, pressure, and temperature ramping and selecting suitable solvents. MW-assisted chemical reactions have become widely accepted in laboratory settings; however, there is still considerable debate over how MW irradiation might improve or impact the results of chemical processes (Kahrilas et al., 2014; Zhao et al., 2014). Alternatively, whether these observations can be attributed to "nonthermal" or "specific MW" effects is considered. Recently, there have been notable and groundbreaking advancements in developing MW hardware, namely using silicon carbide reaction vessels and precisely monitoring temperature by fiber-optic temperature probes (Kooti and Naghdi Sedeh, 2013). These developments have greatly contributed to the understanding and analysis of MW effects. SiC reactors are considered favorable substitutes for MW clear borosilicate glass due to their exceptional MW absorptivity (Zhao et al., 2014).

Consequently, they prove to be effective instruments in debunking the purported mystical effects of MWs. This allows for assessing the impact of the electromagnetic field on certain chemical processes while ensuring that the heating conditions remain the same. The temperature is monitored correctly using a fiber-optic probe (Hong et al., 2016). This account provides an overview of the present state of MW-assisted synthesis, focusing on developing different equipment prototypes, exploring organic reactions using nanomaterials, and synthesizing distinct and versatile nanomaterials. The resulting nanomaterials exhibit a range of dimensional shapes (Kahrilas et al., 2014). MW-assisted techniques have produced precisely defined metallic NPs with noble and transition core–shell structures conveniently and consistently, allowing for adjustable shell thicknesses (Kannan et al., 2022). The combination of sustainable nanomaterials and their use in environmentally friendly substances is an excellent combination for the advancement of more environmentally friendly methods in organic synthesis. MW heating greatly enhances the overall sustainability of the process by increasing its efficiency, particularly in flow systems (Hong et al., 2016).

1.2.4.3 Phyto-Assisted

In recent decades, the increasing fascination with green chemistry and nanotechnology has driven the adoption of environmentally friendly methods for producing nanomaterials using plants, microorganisms, and other biological systems. Using polyphenols derived from plant extracts enables the environmentally friendly production of NPs, resulting in the creation of consistent NPssustainably (Jobitha et al., 2012). The size and form of the NPs are contingent upon the pH and other factors of the fluid. Plant extract-mediated NPs have gained significant research interest because of their cost, non-invasive nature, accessibility, and eco-friendliness (Navada et al., 2020). These NPs also show promise for many applications across several industries. Phytochemicals function as capping agents, facilitating the practical production of the NP. It substantially influences the environment toward sustainability and future progress in nanoscience. The rapid expansion of NP applications is expected due to their environmentally friendly production method. Nevertheless, it is important to use prudence when considering the possible enduring effects on both people and animals, as well as the buildup of these particles in the environment (Sivri et al., 2023). Due to the efficient production of NPs using a few phytochemicals derived from plants, the synthesis of NPs by plant-based methods has become a prominent study focus in metallic NPs. Nevertheless, more endeavors are required to enhance the efficiency and output of the reactions to facilitate commercialization and large-scale manufacturing (Jobitha et al., 2012; Santhoshkumar et al., 2017). Using plants to synthesize metallic NPs has many benefits, including safety, non-toxicity, cost-effectiveness, environmental friendliness, and increased sustainability. Nevertheless, several restrictions also exist, including the extraction of natural chemicals, the duration of the reaction, and the uniformity of the resulting NPs (Sivri et al., 2023; Santhoshkumar et al., 2017). The research on the toxicity of NPs in aquatic systems is challenging and intricate due to the difficulties in sampling, assessing stability, and understanding the fate of NPs in various environments (Firdhouse and Lalitha, 2012). According to a recent research, NPs pose a greater risk to fish compared to other microorganisms in aquatic environments (Malhotra et al., 2023). Another challenge in the bioengineering of NPs is constructing an unresolved reaction mechanism (Taha and Da'na, 2022). The concentration of the phytochemical in the plant extract fluctuates, especially during a dry season, when the plants may experience a loss of the phytochemical, resulting in a very low concentration or even leaf shedding. Significant amounts of plant foliage or other components may be required to compensate for the phytochemical concentration, which might have negative effects on the ecological equilibrium (Firdhouse and Lalitha, 2012). Many studies have proposed a reaction mechanism for the synthesis of NP; however, these proposals lack empirical evidence. Phytochemicals in different plants may cause variations in the hypothesis, requiring more investigation to fully understand the response process (Santhoshkumar et al., 2017). Hence, without the discovery of a coherent approach, the intricate chemical process that drives photosynthesis will persist as a substantial and challenging subject of investigation. Ensuring optimal conditions for the parameters is difficult due to variations in the production of the NPs across different plant species (Vignesh et al., 2013).

1.3 OUTLOOK

1.3.1 Integrated DesignBy Three-Dimensional (3D) Printing (3DP)

3DP, or additive manufacturing, creates items or structures in three dimensions using a computer-assisted design model. 3DP encompasses a range of techniques, including constructing things by depositing, solidifying, or melting material layer by layer, all under computer control (Xu et al., 2021). Figure 1.5 illustrates the fundamental idea of 3DP.

NPs include many materials in many forms, with at least one dimension between 1 nm and 100 nm (Tao et al., 2019). Researchers need to use efficient deagglomeration techniques to obtain optimal enhancement efficiency of anisotropic NPs and align the particles in the required direction. The alignment of NPs in 3DP may be achieved either using inherent mechanical processes, such as liquid flow, or by applying external fields, such as electrical or magnetic forces (Nishiguchi et al., 2020; Xu et al., 2021). Effective aggregate breaking and preferred NP alignment in response to external fields have been achieved using external field-assisted 3DP. Research on 4DP—a term for structures that can respond to stimuli—has focused on electrically and magnetically assisted printing processes. The incorporation of an external field into 3DP, however, is complex for several reasons. As a first step, these 3DP methods need highly reactive NPs that can engage in robust electrical or magnetic field interactions with surrounding fields (Xu et al., 2021). This limitation substantially limits the variety of NPs that may be used in composites based on NPs or 3D-printed NPs. Also, the current external fields produced by vibrational generators, electricity, and magnets are not compatible with all 3DP platforms, so you may only use them with particular ones (Tan et al., 2019). Lastly, creating spatial or time-relevant fields limits the number of NPs per printing layer, and low NP concentrations are required to provide satisfactory printing resolutions. By contrast, intrinsic forces, such as mechanical shear produced as the substrate moves or in the extrusion channel, may aid in aligning NPs. These forces can be harnessed by making simple changes to

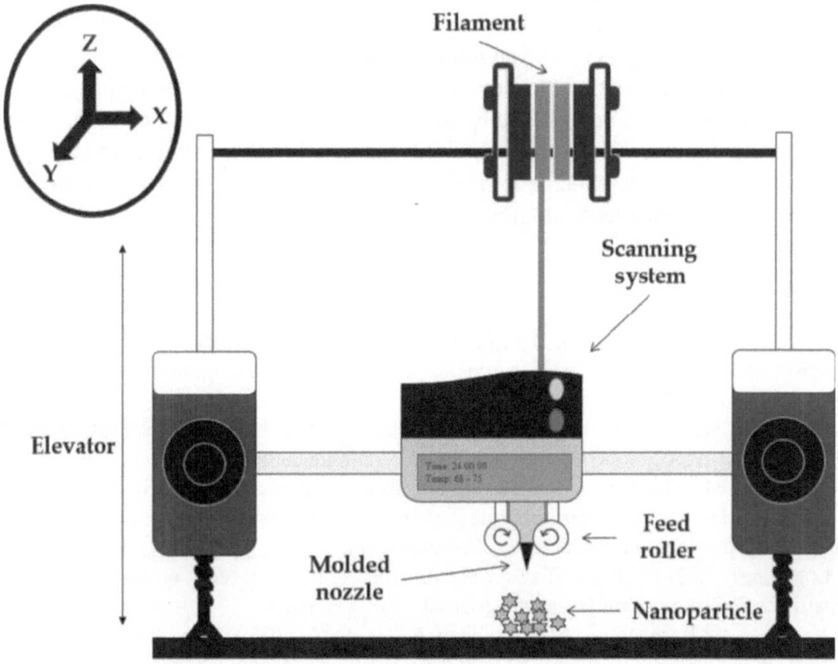

FIGURE 1.5 3DP principle.

current 3DP equipment. Thus, this research primarily focuses on the alignment of NPs produced by shear forces (Podstawczyk et al., 2020; Tao et al., 2019).

1.3.2 Controllable Hydrogel

The development of hydrogel NPs is a new strategy for improving the efficacy of chemotherapeutic drugs by targeted delivery to tumor sites (Emam-Djomeh et al., 2023b; Guo et al., 2020). An ideal design of the NPs would enable the integration of a substantial quantity of the medication, ensuring effective release specifically in the tumor area, ideally regulated, while minimizing release in other regions. This is one of the most difficult elements (De Koker et al., 2016). The design of the NP matrix to improve and facilitate the regulated release of the medication is often dependent on environmental cues, such as temperature, pH, light, glucose, antigen, and reducing agents like glutathione (Silva et al., 2023). Furthermore, external heating methods such as acoustic, magnetic field, or light-mediated heating might create further temperature variations in the tumor site, specifically targeting the NPs (Gao et al., 2016). The temperature sensitivity of the polymer interactions inside these NPs allows for the manipulation of NP size and polymer density. As a result, the effectiveness of drug release is modified (Appel et al., 2015; Emam-Djomeh et al., 2023a; Nejat et al., 2022).

1.4 CONCLUSION AND FUTURE SCOPE

Various physical, chemical, and biological technologies are already accessible to create and produce NPs with the desired size and shape, according to their specific application or usage. This study thoroughly examines the frequently used and investigated methods for synthesizing NPs. It discusses their underlying principles, techniques for creating NPs, and related literature. This study is expected to serve as a comprehensive reference for researchers in selecting an appropriate synthesis technique to develop specific NPs with the required material type, size, surface qualities, and intended application. NP research is now leading the way due to its significant technical capabilities. When synthesizing NPs for targeted drug administration and nanomedicines, achieving accurate size and specified surface features is crucial asthese are the main desired qualities of the leading nanomaterials. Electrospraying is seeing significant growth as a dry method for creating polymeric NPs without the need for any harmful chemicals. Moreover, the use of electrospraying for synthesizing metal and ceramic NPs remains largely uncharted. Plasmonics, optoelectronics, information, and data storage, among others, are important fields that need further exploration and optimization of synthesis methodologies for scalable manufacture of NPs. This may contribute to the development of commercially feasible technologies.

REFERENCES

Angelescu, D.G., Munteanu, G., Anghel, D.F., Peretz, S., Maraloiu, A.V., Teodorescu, V.S. 2013. Formation mechanism of CdS nanoparticles with tunable luminescence via a non-ionic microemulsion route. Journal of Nanoparticle Research. 15, 1–15.

Appel, E.A., Tibbitt, M.W., Webber, M.J., Mattix, B.A., Veiseh, O., Langer, R. 2015. Self-assembled hydrogels utilizing polymer–nanoparticle interactions. Nature Communications. 6, 6295.

Asfaram, A., Ghaedi, M., Goudarzi, A. 2016. Optimization of ultrasound-assisted dispersive solid-phase microextraction based on nanoparticles followed by spectrophotometry for the simultaneous determination of dyes using experimental design. Ultrasonics Sonochemistry. 32, 407–417.

Avar, E.C., Logoglu, E. 2022. Milling effect on the structural and antibacterial properties of ZnO and CuO nanoparticles. Pharmaceutical Chemistry Journal. 56, 1004–1010.

Bachheti, R.K., Abate, L., Bachheti, A., Madhusudhan, A., Husen, A. 2021. Algae-, fungi-, and yeast-mediated biological synthesis of nanoparticles and their various biomedical applications. In: Handbook of Greener Synthesis of Nanomaterials and Compounds, pp. 701–734, Elsevier.

Bagur, H., Medidi, R.S., Somu, P., Choudhury, P.J., Karua, C.S., Guttula, P.K., Melappa, G., Poojari, C.C. 2022. Endophyte fungal isolate mediated biogenic synthesis and evaluation of biomedical applications of silver nanoparticles. Materials Technology. 37, 167–178.

Baig, M.M., Yousuf, M.A., Agboola, P.O., Khan, M.A., Shakir, I., Warsi, M.F. 2019. Optimization of different wet chemical routes and phase evolution studies of $MnFe_2O_4$ nanoparticles. Ceramics International. 45, 12682–12690.

Barbhuiya, R.I., Singha, P., Asaithambi, N., Singh, S.K. 2022. Ultrasound-assisted rapid biological synthesis and characterization of silver nanoparticles using pomelo peel waste. Food Chemistry. 385, 132602.

Baron, R., Willner, B., Willner, I. 2007. Biomolecule–nanoparticle hybrids as functional units for nanobiotechnology. Chemical Communications. 323–332.

Behbahani, E.S., Ghaedi, M., Abbaspour, M., Rostamizadeh, K. 2017. Optimization and characterization of ultrasound assisted preparation of curcumin-loaded solid lipid nanoparticles: application of central composite design, thermal analysis and X-ray diffraction techniques. Ultrasonics Sonochemistry. 38, 271–280.

Bhakya, S., Muthukrishnan, S., Sukumaran, M., Muthukumar, M. 2016. Biogenic synthesis of silver nanoparticles and their antioxidant and antibacterial activity. Applied Nanoscience. 6, 755–766.

Biswas, A., Saha, S., Jana, N.R. 2019. $ZnSnO_3$ nanoparticle-based piezocatalysts for ultrasound-assisted degradation of organic pollutants. ACS Applied Nano Materials. 2, 1120–1128.

Bokov, D., Turki Jalil, A., Chupradit, S., Suksatan, W., Javed Ansari, M., Shewael, I.H., Valiev, G.H., Kianfar, E. 2021. Nanomaterial by sol–gel method: synthesis and application. Advances in Materials Science and Engineering. 2021, 1–21.

Bozkurt, G. 2020. Synthesis and characterization of α-Fe_2O_3nanoparticles by microemulsion method. Erzincan University Journal of Science and Technology. 13, 890–897.

Chagas, E.F., Ferreira, E.S. 2023. Mechanical milling of ferrite nanoparticles. In: Ferrite Nanostructured Magnetic Materials, pp. 85–102, Elsevier.

Chakraborty, I., Moulik, S.P. 2005. On PbS nanoparticles formed in the compartments of water/AOT/n-heptane microemulsion. Journal of Nanoparticle Research. 7, 237–247.

Cheah, P., Qu, J., Li, Y., Cao, D., Zhu, X., Zhao, Y. 2021. The key role of reaction temperature on a polyol synthesis of water-dispersible iron oxide nanoparticles. Journal of Magnetism and Magnetic Materials. 540, 168481.

Chen, B., Kondoh, K., Li, J., Qian, M. 2020. Extraordinary reinforcing effect of carbon nanotubes in aluminium matrix composites assisted by in-situ alumina nanoparticles. Composites Part B: Engineering. 183, 107691.

De Koker, S., Cui, J., Vanparijs, N., Albertazzi, L., Grooten, J., Caruso, F., De Geest, B.G. 2016. Engineering polymer hydrogel nanoparticles for lymph node-targeted delivery. Angewandte Chemie. 128, 1356–1361.

Deng, Y., Chen, W., Li, B., Wang, C., Kuang, T., Li, Y. 2020. Physical vapor deposition technology for coated cutting tools: a review. Ceramics International. 46, 18373–18390.

Dhand, C., Dwivedi, N., Loh, X.J., Ying, A.N.J., Verma, N.K., Beuerman, R.W., Lakshminarayanan, R., Ramakrishna, S. 2015. Methods and strategies for the synthesis of diverse nanoparticles and their applications: a comprehensive overview. RSC Advances. 5, 105003–105037.

Eddy, N.O., Garg, R., Garg, R., Eze, S.I., Ogoko, E.C., Kelle, H.I., Ukpe, R.A., Ogbodo, R., Chijoke, F. 2023. Sol–gel synthesis, computational chemistry, and applications of CaO nanoparticles for the remediation of methyl orange contaminated water. Advances in Nano Research. 15, 35–48.

Ekrami, A., Ghadermazi, M., Ekrami, M., Hosseini, M.A., Emam-Djomeh, Z., Hamidi-Moghadam, R. 2022a. Development and evaluation of *Zhumeria majdae* essential oil-loaded nanoliposome against multidrug-resistant clinical pathogens causing nosocomial infection. Journal of Drug Delivery Science and Technology. 69, 103148.

Ekrami, M., Ekrami, A., Esmaeily, R., Emam-Djomeh, Z. 2022b. Nanotechnology-based formulation for alternative medicines and natural products: an introduction with clinical studies. In: Biopolymers in Nutraceuticals and Functional Foods, pp. 545–580. Royal Society of Chemistry, Polymer Chemistry Series, Edited by: Sreerag Gopi; Preetha Balakrishnan; Matej Bračič, ISBN: 978-1-83916-804-8 https://doi.org/10.1039/9781839168048-00545

Ekrami, M., Ekrami, A., Hosseini, M.A., Emam-Djomeh, Z. 2022c. Characterization and optimization of Salep mucilage bionanocomposite films containing *Allium jesdianum* Boiss. Nanoliposomes for antibacterial foodpackaging utilization. Molecules. 27, 7032.

Ekrami, M., Ekrami, A., Moghadam, R.H., Joolaei-Ahranjani, P., Emam-Djomeh, Z. 2022d. Food-based polymers for encapsulation and delivery of bioactive compounds. In: Biopolymers in Nutraceuticals and Functional Foods, pp. 488–544, The Royal Society of Chemistry.

Ekrami, M., Emam-Djomeh, Z., Joolaei-Ahranjani, P., Mahmoodi, S., Khaleghi, S. 2021. Eco-friendly UV protective bionanocomposite based on Salep-mucilage/flower-like ZnO nanostructures to control photo-oxidation of Kilka fish oil. International Journal of Biological Macromolecules. 168, 591–600.

Ekrami, M., Shakouri, M., Nikkhou, S., Emam-Djomeh, Z. 2023. Extraction and physicochemical characterization of gum. In: Handbook of Natural Polymers, Volume 1, pp. 597–630, Elsevier.

Emam-Djomeh, Z., Ekrami, M., Ekrami, A. 2023a. Overview of types of materials used for food component encapsulation. In: Materials Science and Engineering in Food Product Development, pp. 73–92, Wiley.

Emam-Djomeh, Z., Ekrami, M., Ekrami, A. 2023b. Design and use of hydrogels for food component encapsulation. In: Materials Science and Engineering in Food Product Development, pp. 211–226, Wiley.

Esmaili, Z., Bayrami, S., Dorkoosh, F.A., Akbari Javar, H., Seyedjafari, E., Zargarian, S.S., Haddadi-Asl, V. 2018. Development and characterization of electrosprayed nanoparticles for encapsulation of Curcumin. Journal of Biomedical Materials Research Part A. 106, 285–292.

Farahmand, E., Emam-Djomeh, Z., Ekrami, M., Razavi, S.H. 2023. Polymethacrylate coated electrospun chitosan/PEO nanofibers loaded with thyme essential oil: a newfound potential for antimicrobial food packaging. Journal of Food and Bioprocess Engineering. 6, 8–16.

Fatima, G., Bibi, I., Majid, F., Kamal, S., Nouren, S., Ghafoor, A., Raza, Q., Al-Mijalli, S.H., Alnafisi, N.M., Iqbal, M. 2023. Mn-doped $BaFe_{12}O_{19}$ nanoparticles synthesis via micro-emulsion route: solar light-driven photo-catalytic degradation of CV, MG and RhB dyes and antibacterial activity. Materials Research Bulletin. 168, 112491.

Favier, I., Pla, D., Gómez, M. 2019. Palladium nanoparticles in polyols: synthesis, catalytic couplings, and hydrogenations. Chemical Reviews. 120, 1146–1183.

Firdhouse, M.J., Lalitha, P. 2012. Phyto-assisted synthesis and characterization of silver nanoparticles from *Amaranthus dubius*. International Journal of Applied Biology and Pharmaceutical Technology. 3, 96–101.

Galvez, A., Herlin-Boime, N., Reynaud, C., Clinard, C., Rouzaud, J.-N. 2002. Carbon nanoparticles from laser pyrolysis. Carbon. 40, 2775–2789.

Gan, Y.X., Jayatissa, A.H., Yu, Z., Chen, X., Li, M. 2020. Hydrothermal synthesis of nanomaterials. Journal of Nanomaterials. 2020, 1–3.

Gao, W., Zhang, Y., Zhang, Q., Zhang, L. 2016. Nanoparticle-hydrogel: a hybrid biomaterial system for localized drug delivery. Annals of Biomedical Engineering. 44, 2049–2061.

Ghadami Jadval Ghadam, A., Idrees, M. 2013. Characterization of $CaCO_3$ nanoparticles synthesized by reverse microemulsion technique in different concentrations of surfactants. Iranian Journal of Chemistry and Chemical Engineering (IJCCE). 32, 27–35.

Gleason, K.K. 2020. Nanoscale control by chemically vapour-deposited polymers. Nature Reviews Physics. 2, 347–364.

Guilger-Casagrande, M., Lima, R.D. 2019. Synthesis of silver nanoparticles mediated by fungi: a review. Frontiers in Bioengineering and Biotechnology. 7, 287.

Guo, B., Hoshino, Y., Gao, F., Hayashi, K., Miura, Y., Kimizuka, N., Yamada, T. 2020. Thermocells driven by phase transition of hydrogel nanoparticles. Journal of the American Chemical Society. 142, 17318–17322.

Gupta, T., Saxena, J. 2023. Biogenic synthesis of silver nanoparticles from *Aspergillus oryzae* MTCC 3107 against plant pathogenic fungi *Sclerotinia sclerotiorum*MTCC 8785. Journal of Microbiology, Biotechnology and Food Sciences. 12, e9387–e9387.

Hasanpoor, M., Aliofkhazraei, M., Delavari, H. 2015. Microwave-assisted synthesis of zinc oxide nanoparticles. Procedia Materials Science. 11, 320–325.

Hayashi, H., Hakuta, Y. 2010. Hydrothermal synthesis of metal oxide nanoparticles in supercritical water. Materials. 3, 3794–3817.

He, X., Ran, J., Zhang, W., Fu, H. 2018. Microemulsion-based interfacial diffusion synthesis of uniform $BaCO_3$ nanorods. In: 2018 7th International Conference on Energy, Environment and Sustainable Development (ICEESD 2018), pp. 200–203, Atlantis Press.

Hong, X., Wen, J., Xiong, X., Hu, Y. 2016. Shape effect on the antibacterial activity of silver nanoparticles synthesized via a microwave-assisted method. Environmental Science and Pollution Research. 23, 4489–4497.

Iacovita, C., Fizeşan, I., Pop, A., Scorus, L., Dudric, R., Stiufiuc, G., Vedeanu, N., Tetean, R., Loghin, F., Stiufiuc, R. 2020. In vitro intracellular hyperthermia of iron oxide magnetic nanoparticles, synthesized at high temperature by a polyol process. Pharmaceutics. 12, 424.

Iacovita, C., Florea, A., Scorus, L., Pall, E., Dudric, R., Moldovan, A.I., Stiufiuc, R., Tetean, R., Lucaciu, C.M. 2019. Hyperthermia, cytotoxicity, and cellular uptake properties of manganese and zinc ferrite magnetic nanoparticles synthesized by a polyol-mediated process. Nanomaterials. 9, 1489.

Idris, D.S., Roy, A. 2023. Biogenic synthesis of Ag–CuO nanoparticles and its antibacterial, antioxidant, and catalytic activity. Journal of Inorganic and Organometallic Polymers and Materials. 1–13.

Ijaz, I., Gilani, E., Nazir, A., Bukhari, A. 2020. Detail review on chemical, physical and green synthesis, classification, characterizations and applications of nanoparticles. Green Chemistry Letters and Reviews. 13, 223–245.

Izci, M., Maksoudian, C., Manshian, B.B., Soenen, S.J. 2021. The use of alternative strategies for enhanced nanoparticle delivery to solid tumors. Chemical Reviews. 121, 1746–1803.

Jadoun, S., Arif, R., Jangid, N.K., Meena, R.K. 2021. Green synthesis of nanoparticles using plant extracts: a review. Environmental Chemistry Letters. 19, 355–374.

Jamima, J., Veeramani, P., Kanagaraju, P., Kumanan, K. 2020. Synthesis and characterization of selenium nano particles by high energy ball milling (HEBM). Indian Journal of Veterinary and Animal Sciences Research. 49, 45–51.

Jayaraman, P., Gandhimathi, C., Venugopal, J.R., Becker, D.L., Ramakrishna, S., Srinivasan, D.K. 2015. Controlled release of drugs in electrosprayed nanoparticles for bone tissue engineering. Advanced Drug Delivery Reviews. 94, 77–95.

Jia, F., Shih, Y.-L., Pui, D.Y., Li, Z.-Y., Tsai, C.-J. 2021. Generation of ZnO nanoparticles by chemical vapor synthesis using quenching air. Journal of Nanoparticle Research. 23, 1–15.

Jo, Y.-S., Lee, H.-J., Park, H.-M., Na, T.-W., Jung, J.-S., Min, S.-H., Kim, Y.K., Yang, S.-M. 2021. Chemical vapor synthesis of nonagglomerated nickel nanoparticles by in-flight coating. ACS Omega. 6, 27842–27850.

Jobitha, G.D.G., Kannan, C., Annadurai, G. 2012. A facile phyto-assisted synthesis of silver nanoparticles using the flower extract of *Cassia auriculata* and assessment of its antimicrobial activity. Drug Invention Today. 4, 579–584.

Kahrilas, G.A., Wally, L.M., Fredrick, S.J., Hiskey, M., Prieto, A.L., Owens, J.E. 2014. Microwave-assisted green synthesis of silver nanoparticles using orange peel extract. ACS Sustainable Chemistry & Engineering. 2, 367–376.

Kannan, K., Radhika, D., Vijayalakshmi, S., Sadasivuni, K.K., Ojiaku, A., Verma, U. 2022. Facile fabrication of CuO nanoparticles via microwave-assisted method: photocatalytic, antimicrobial and anticancer enhancing performance. International Journal of Environmental Analytical Chemistry. 102, 1095–1108.

Karthik, K., Nikolova, M.P., Phuruangrat, A., Pushpa, S., Revathi, V., Subbulakshmi, M. 2019. Ultrasound-assisted synthesis of V_2O_5 nanoparticles for photocatalytic and antibacterial studies. Materials Research Innovations. 24, 229–234.

Khomane, R., Manna, A., Mandale, A., Kulkarni, B. 2002. Synthesis and characterization of dodecanethiol-capped cadmium sulfide nanoparticles in a Winsor II microemulsion of diethyl ether/AOT/water. Langmuir. 18, 8237–8240.

Kooti, M., Naghdi Sedeh, A. 2013. Microwave-assisted combustion synthesis of ZnO nanoparticles. Journal of Chemistry. 2013, 562028–562031.

Kotoulas, A., Dendrinou-Samara, C., Angelakeris, M., Kalogirou, O. 2019. The effect of polyol composition on the structural and magnetic properties of magnetite nanoparticles for magnetic particle hyperthermia. Materials. 12, 2663.

Krasnochtchekov, P., Albe, K., Averback, R. 2022. Simulations of the inert gas condensation processes. International Journal of Materials Research. 94, 1098–1105.

Kumari, S., Raturi, S., Kulshrestha, S., Chauhan, K., Dhingra, S., András, K., Thu, K., Khargotra, R., Singh, T. 2023. A comprehensive review on various techniques used for synthesizing nanoparticles. Journal of Materials Research and Technology. 27, 1739–1763.

Lahane, T., Agrawal, J., Singh, V. 2022. Optimization of polyol synthesized silver nanowires for transparent conducting electrodes application. Materials Today: Proceedings. 59, 257–263.

Lavand, A.B., Malghe, Y.S., Singh, S.H. 2015. Synthesis, characterization, and investigation of visible light photocatalytic activity of C doped TiO_2/CdS core–shell nanocomposite. Indian Journal of Materials Science. 2015.

Lee, H.-J., Lee, G., Jang, N.R., Yun, J.H., Song, J.Y., Kim, B.S. 2011. Biological synthesis of copper nanoparticles using plant extract. Nanotechnology. 1, 371–374.

Lee, Y.-J., Kim, K., Shin, I.-S., Shin, K.S. 2020. Antioxidative metallic copper nanoparticles prepared by modified polyol method and their catalytic activities. Journal of Nanoparticle Research. 22, 1–8.

Levish, A., Joshi, S., Winterer, M. 2023. Chemical vapor synthesis of nanocrystalline iron oxides. Applications in Energy and Combustion Science. 15, 100177.

Li, X., Zheng, W., He, G., Zhao, R., Liu, D. 2014. Morphology control of TiO_2 nanoparticle in microemulsion and its photocatalytic property. ACS Sustainable Chemistry & Engineering. 2, 288–295.

Lian, S., Diko, C.S., Yan, Y., Li, Z., Zhang, H., Ma, Q., Qu, Y. 2019. Characterization of biogenic selenium nanoparticles derived from cell-free extracts of a novel yeast *Magnusiomyces ingens*. 3 Biotech. 9, 1–8.

Liu, Q., Song, C., Wang, Z.-G., Li, N., Ding, B. 2014. Precise organization of metal nanoparticles on DNA origami template. Methods. 67, 205–214.

Mahamuni, P.P., Patil, P.M., Dhanavade, M.J., Badiger, M.V., Shadija, P.G., Lokhande, A.C., Bohara, R.A. 2019. Synthesis and characterization of zinc oxide nanoparticles by using polyol chemistry for their antimicrobial and antibiofilm activity. Biochemistry and Biophysics Reports. 17, 71–80.

Malhotra, A., Chauhan, S.R., Rahaman, M., Tripathi, R., Khanuja, M., Chauhan, A. 2023. Phyto-assisted synthesis of zinc oxide nanoparticles for developing antibiofilm surface coatings on central venous catheters. Frontiers in Chemistry. 11, 1138333.

Maslova, V., Quadrelli, E.A., Gaval, P., Fasolini, A., Albonetti, S., Basile, F. 2021. Highly-dispersed ultrafine Pt nanoparticles on microemulsion-mediated TiO_2 for production of hydrogen and valuable chemicals via oxidative photo-dehydrogenation of glycerol. Journal of Environmental Chemical Engineering. 9, 105070.

Meierhofer, F., Mädler, L., Fritsching, U. 2020. Nanoparticle evolution in flame spray pyrolysis—process design via experimental and computational analysis. AIChE Journal. 66, e16885.

Mirzakhani, M., Ekrami, M., Moini, S. 2018. Chemical composition, total phenolic content and antimicrobial activities of *Zhumeria majdae*. Journal of Food and Bioprocess Engineering (JFBE). 1, 8.

Mishin, M.V., Zamotin, K.Y., Protopopova, V.S., Alexandrov, S.E. 2015. Chain assemblies from nanoparticles synthesized by atmospheric pressure plasma enhanced chemical vapor deposition: the computational view. Journal of Nanoscience and Nanotechnology. 15, 9966–9974.

Mittal, A.K., Chisti, Y., Banerjee, U.C. 2013. Synthesis of metallic nanoparticles using plant extracts. Biotechnology Advances. 31, 346–356.

Mittal, D., Kaur, G., Singh, P., Yadav, K., Ali, S.A. 2020. Nanoparticle-based sustainable agriculture and food science: recent advances and future outlook. Frontiers in Nanotechnology. 2, 579954.

Mohd Yusof, H., Abdul Rahman, N.A., Mohamad, R., Zaidan, U.H., Samsudin, A.A. 2020. Biosynthesis of zinc oxide nanoparticles by cell-biomass and supernatant of *Lactobacillus plantarum* TA4 and its antibacterial and biocompatibility properties. Scientific Reports. 10, 19996.

Montaño-Priede, J.L., Coelho, J.O.P., Guerrero-Martínez, A.S., Peña-Rodríguez, O., Pal, U. 2017. Fabrication of monodispersed Au@SiO_2 nanoparticles with highly stable silica layers by ultrasound-assisted Stober method. The Journal of Physical Chemistry C. 121, 9543–9551.

Mughal, B., Zaidi, S.Z.J., Zhang, X., Hassan, S.U. 2021. Biogenic nanoparticles: synthesis, characterisation and applications. Applied Sciences. 11, 2598.

Muñoz, J.E., Cervantes, J., Esparza, R., Rosas, G. 2007. Iron nanoparticles produced by high-energy ball milling. Journal of Nanoparticle Research. 9, 945–950.

Navada, K.M., Nagaraja, G., D'Souza, J.N., Kouser, S., Ranjitha, R., Manasa, D. 2020. Phyto assisted synthesis and characterization of *Scoparia dulsis* L. leaf extract mediated porous nano CuO photocatalysts and its anticancer behavior. Applied Nanoscience. 10, 4221–4240.

Nejat, M.S., Ekrami, M., Emam-Djomeh, Z. 2022. Microencapsulation liposomal technologies in bioactive functional foods and nutraceuticals. In: Biopolymers in Nutraceuticals and Functional Foods, pp. 232–263, The Royal Society of Chemistry.

Neto, P.B., Meierhofer, F., Meier, H.F., Fritsching, U., Noriler, D. 2020. Modelling polydisperse nanoparticle size distributions as produced via flame spray pyrolysis. Powder Technology. 370, 116–128.

Nishiguchi, A., Shima, F., Singh, S., Akashi, M., Moeller, M. 2020. 3D-printing of structure-controlled antigen nanoparticles for vaccine delivery. Biomacromolecules. 21, 2043–2048.

Nyamjav, D., Ivanisevic, A. 2005. Templates for DNA-templated Fe_3O_4 nanoparticles. Biomaterials. 26, 2749–2757.

Pandya, S.G., Kordesch, M.E. 2015. Characterization of InSb nanoparticles synthesized using inert gas condensation. Nanoscale Research Letters. 10, 1–7.

Parejas, A., Montes, V., Hidalgo-Carrillo, J., Sánchez-López, E., Marinas, A., Urbano, F.J. 2017. Microemulsion and sol–gel synthesized ZrO_2-MgO catalysts for the liquid-phase dehydration of xylose to furfural. Molecules. 22, 2257.

Pawar, A., Thakkar, S., Misra, M. 2018. A bird's eye view of nanoparticles prepared by electrospraying: advancements in drug delivery field. Journal of Controlled Release. 286, 179–200.

Phromma, S., Wutikhun, T., Kasamechonchung, P., Eksangsri, T., Sapcharoenkun, C. 2020. Effect of calcination temperature on photocatalytic activity of synthesized TiO_2 nanoparticles via wet ball milling sol–gel method. Applied Sciences. 10, 993.

Pineda-Reyes, A.M., Olvera, M.D.L.L. 2018. Synthesis of ZnO nanoparticles from water-in-oil (w/o) microemulsions. Materials Chemistry and Physics. 203, 141–147.

Podstawczyk, D., Skrzypczak, D., Połomska, X., Stargała, A., Witek-Krowiak, A., Guiseppi-Elie, A., Galewski, Z. 2020. Preparation of antimicrobial 3D printing filament: in situ thermal formation of silver nanoparticles during the material extrusion. Polymer Composites. 41, 4692–4705.

Qiao, Y., Xu, Y., Liu, X., Zheng, Y., Li, B., Han, Y., Li, Z., Yeung, K.W.K., Liang, Y., Zhu, S. 2022. Microwave assisted antibacterial action of Garcinia nanoparticles on Gram-negative bacteria. Nature Communications. 13, 2461.

Raj, S., Trivedi, R., Soni, V. 2021. Biogenic synthesis of silver nanoparticles, characterization and their applications—a review. Surfaces. 5, 67–90.

Rane, A.V., Kanny, K., Abitha, V., Thomas, S. 2018. Methods for synthesis of nanoparticles and fabrication of nanocomposites. In: Synthesis of Inorganic Nanomaterials, pp. 121–139, Elsevier.

Reau, A., Guizard, B., Mengeot, C., Boulanger, L., Ténégal, F. 2007. Large scale production of nanoparticles by laser pyrolysis. In: Materials Science Forum, pp. 85–88, Trans Tech Publications.

Richard, B., Lemyre, J.-L., Ritcey, A.M. 2017. Nanoparticle size control in microemulsion synthesis. Langmuir. 33, 4748–4757.

Rodríguez, A.A.R., Trejo, M.B.M., Zaragoza, M. J. M., Martinez, V.C., Ortiz, A.L., Guerra, E.M., Dorminguez, M.S. 2017. 10.4 Synthesis of MFe_2O_4 nanoparticles by the oil-in-water microemulsion reaction method and its exploration for photocatalytic water splitting. XVII International Congress of the Mexican Hydrogen Society. 519–532.

Sadhasivam, S., Vinayagam, V., Balasubramaniyan, M. 2020. Recent advancement in biogenic synthesis of iron nanoparticles. Journal of Molecular Structure. 1217, 128372.

Salah, N., Habib, S.S., Khan, Z.H., Memic, A., Azam, A., Alarfaj, E., Zahed, N., Al-Hamedi, S. 2011. High-energy ball milling technique for ZnO nanoparticles as antibacterial material. International Journal of Nanomedicine. 863–869.

Salvador, M., Gutiérrez, G., Noriega, S., Moyano, A., Blanco-López, M.C., Matos, M. 2021. Microemulsion synthesis of superparamagnetic nanoparticles for bioapplications. International Journal of Molecular Sciences. 22, 427.

Sánchez, M.D., Falcone, R.D., Ritacco, H.A. 2022. Production of Pd nanoparticles in microemulsions. Effect of reaction rates on the particle size. Physical Chemistry Chemical Physics. 24, 1692–1701.

Sandhu, R.K., Kaur, A., Kaur, P., Rajput, J.K., Khullar, P., Bakshi, M.S. 2021. Solubilization of surfactant stabilized gold nanoparticles in oil-in-water and water-in-oil microemulsions. Journal of Molecular Liquids. 336, 116305.

Santhoshkumar, J., Rajeshkumar, S., Kumar, S.V. 2017. Phyto-assisted synthesis, characterization and applications of gold nanoparticles—a review. Biochemistry and Biophysics Reports. 11, 46–57.

Sathya, K., Saravanathamizhan, R., Baskar, G. 2017. Ultrasound assisted phytosynthesis of iron oxide nanoparticle. Ultrasonics Sonochemistry. 39, 446–451.

Setiawan, H., Triyatna, F., Nurmanjaya, A., Subechi, M., Sarwono, D., Billah, A., Rindiyantono, F. 2022. Synthesis and characterization of gadolinium nanoparticles using polyol method as a candidate for MRI contrast agent. In: Journal of Physics: Conference Series, pp. 012010, IOP Publishing.

Shakouri, M., Salami, M., Lim, L.-T., Ekrami, M., Mohammadian, M., Askari, G., Emam-Djomeh, Z., McClements, D.J. 2023. Development of active and intelligent colorimetric biopolymer indicator: anthocyanin-loaded gelatin-basil seed gum films. Journal of Food Measurement and Characterization. 17, 472–484.

Sharma, D., Kanchi, S., Bisetty, K. 2019a. Biogenic synthesis of nanoparticles: a review. Arabian Journal of Chemistry. 12, 3576–3600.
Sharma, R., Wang, Y., Li, F., Chamier, J., Andersen, S.M. 2019b. Particle size-controlled growth of carbon-supported platinum nanoparticles (Pt/C) through water-assisted polyol synthesis. ACS Omega. 4, 15711–15720.
Shrestha, S., Wang, B., Dutta, P. 2020. Nanoparticle processing: understanding and controlling aggregation. Advances in Colloid and Interface Science. 279, 102162.
Silva, A.T., Figueiredo, R., Azenha, M., Jorge, P.A., Pereira, C.M., Ribeiro, J.A. 2023. Imprinted hydrogel nanoparticles for protein biosensing: a review. ACS Sensors. 8, 2898–2920.
Sivri, D., Kurel, A., Baytore, D.I., Akdogan, N.G., Akdogan, O. 2023. Recycling of hazelnut husk; from bio-waste to phyto-assisted synthesis of silver nanoparticles. ChemistrySelect. 8, e202302262.
Spyridopoulou, K., Tryfonopoulou, E., Aindelis, G., Ypsilantis, P., Sarafidis, C., Kalogirou, O., Chlichlia, K. 2021. Biogenic selenium nanoparticles produced by *Lactobacillus casei* ATCC 393 inhibit colon cancer cell growth in vitro and in vivo. Nanoscale Advances. 3, 2516–2528.
Taha, A., Da'na, E. 2022. Phyto-assisted assembly of metal nanoparticles in chitosan matrix using *S. argel*leaf extract and its application for catalytic oxidation of benzyl alcohol. Polymers. 14, 766.
Tan, H.W., An, J., Chua, C.K., Tran, T. 2019. Metallic nanoparticle inks for 3D printing of electronics. Advanced Electronic Materials. 5, 1800831.
Tanemura, M., Iwata, K., Takahashi, K., Fujimoto, Y., Okuyama, F., Sugie, H., Filip, V. 2001. Growth of aligned carbon nanotubes by plasma-enhanced chemical vapor deposition: optimization of growth parameters. Journal of Applied Physics. 90, 1529–1533.
Tao, J., Zhang, J., Du, T., Xu, X., Deng, X., Chen, S., Liu, J., Chen, Y., Liu, X., Xiong, M. 2019. Rapid 3D printing of functional nanoparticle-enhanced conduits for effective nerve repair. Acta Biomaterialia. 90, 49–59.
Thangavelu, P. 2022. Tuning the optical property of titania nanotubes array using CdS microemulsion sensitization for enhanced photocatalytic activity. Solid State Communications. 354, 114887.
Tianimoghadam, S., Salabat, A. 2018. A microemulsion method for preparation of thiol-functionalized gold nanoparticles. Particuology. 37, 33–36.
Vignesh, V., Anbarasi, K.F., Karthikeyeni, S., Sathiyanarayanan, G., Subramanian, P., Thirumurugan, R. 2013. A superficial phyto-assisted synthesis of silver nanoparticles and their assessment on hematological and biochemical parameters in *Labeo rohita* (Hamilton, 1822). Colloids and Surfaces A: Physicochemical and Engineering Aspects. 439, 184–192.
Wallace, R., Brown, A., Brydson, R., Wegner, K., Milne, S. 2013. Synthesis of ZnO nanoparticles by flame spray pyrolysis and characterisation protocol. Journal of Materials Science. 48, 6393–6403.
Wang, Z.-G., Ding, B. 2014. Engineering DNA self-assemblies as templates for functional nanostructures. Accounts of Chemical Research. 47, 1654–1662.
Wu, Z., Wang, Y., Sun, L., Mao, Y., Wang, M., Lin, C. 2014. An ultrasound-assisted deposition of NiO nanoparticles on TiO_2 nanotube arrays for enhanced photocatalytic activity. Journal of Materials Chemistry A. 2, 8223–8229.
Xu, W., Jambhulkar, S., Ravichandran, D., Zhu, Y., Kakarla, M., Nian, Q., Azeredo, B., Chen, X., Jin, K., Vernon, B. 2021. 3D printing-enabled nanoparticle alignment: a review of mechanisms and applications. Small. 17, 2100817.
Yanase, T., Miura, T., Shiratori, T., Weng, M., Nagahama, T., Shimada, T. 2019. Synthesis of carbon nanotubes by plasma-enhanced chemical vapor deposition using $Fe_{1-x}Mn_xO$ nanoparticles as catalysts: how does the catalytic activity of graphitization affect the yields and morphology?C—Journal of Carbon Research. 5, 46.
Yang, G., Park, S.-J. 2019. Conventional and microwave hydrothermal synthesis and application of functional materials: a review. Materials. 12, 1177.
Yanilmaz, M., Lu, Y., Dirican, M., Fu, K., Zhang, X. 2014. Nanoparticle-on-nanofiber hybrid membrane separators for lithium-ion batteries via combining electrospraying and electrospinning techniques. Journal of Membrane Science. 456, 57–65.
Zeroual, S., Estellé, P., Cabaleiro, D., Vigolo, B., Emo, M., Halim, W., Ouaskit, S. 2020. Ethylene glycol based silver nanoparticles synthesized by polyol process: characterization and thermophysical profile. Journal of Molecular Liquids. 310, 113229.

Zhang, W., Yu, Y., Yi, Z. 2017. Controllable synthesis of $SrCO_3$ with different morphologies and their co-catalytic activities for photocatalytic oxidation of hydrocarbon gases over TiO_2. Journal of Materials Science. 52, 5106–5116.

Zhang, Z., Lin, Y., Liu, F. 2020. Preparation and characterization of CdS/ZnS core-shell nanoparticles. Journal of Dispersion Science and Technology. 41, 725–732.

Zhao, X., Wang, C., Wang, D., Hahn, H., Fichtner, M. 2013. Ge–Cu nanoparticles produced by inert gas condensation and their application as anode material for lithium ion batteries. Electrochemistry Communications. 35, 116–119.

Zhao, X., Xia, Y., Li, Q., Ma, X., Quan, F., Geng, C., Han, Z. 2014. Microwave-assisted synthesis of silver nanoparticles using sodium alginate and their antibacterial activity. Colloids and Surfaces A: Physicochemical and Engineering Aspects. 444, 180–188.

Zheng, L., Cui, B., Zhao, L., Li, W., Hadjipanayis, G.C. 2012. Sm_2Co_{17} nanoparticles synthesized by surfactant-assisted high energy ball milling. Journal of Alloys and Compounds. 539, 69–73.

Zinchenko, A., Miwa, Y., Lopatina, L.I., Sergeyev, V.G., Murata, S. 2014. DNA hydrogel as a template for synthesis of ultrasmall gold nanoparticles for catalytic applications. ACS Applied Materials & Interfaces. 6, 3226–3232.

2 Nanotechnology and Food-Grade Enzymes

Leila Ziaeifar, Arash Hosseini, Maryam Salami, Zahra Emam-Djomeh, and Golam Reza Askari

2.1 INTRODUCTION

Nanotechnology is a multidisciplinary field that was developed to engineer biological elements like atoms, molecules, and supramolecules at nanoscales of approximately 1–100 nm to overcome current challenges by developing new devices and identifying material structures with special properties to study and comprehend the deadly biological issues, followed by disease diagnosis and treatment (Reza Mozafari et al., 2008). The physical property of size that distinguishes one nanomaterial from another is that it increases surface area, which in turn improves the chemical characteristics of nanoparticles and nanomaterials. Additionally, despite their counterpart macromolecules and bulk materials, nanostructures exhibit extraordinary bioactivity, and optical, electrical, and mechanical properties because of the quantum effect at the nanoscale, which makes them more functional and applicable to use. Numerous industrial sectors, notably agri-food operations, benefit greatly from the diverse physical, chemical, and biological properties of nanoparticles produced through various synthesis methods (Ali et al., 2021). Enzymes have long served as useful instruments in a range of scientific and technical fields. Enzymes as sustainable, green catalysts provide some benefits over chemical catalysts for industrial purposes due to their strong catalytic activity, substrate selectivity, low toxicity, and nonexistent synthesis of unwanted products (Singh et al., 2015; Song et al., 2015); thus, they are popular in the subject areas of medicine, chemistry, pharmaceutical research, biochemistry, food, and textile. However, the use of free enzymes has some disadvantageous such as decomposition and instability in extreme temperatures, pH, high salt concentrations, surfactants, and high cost for large-scale use (Ribeiro et al., 2021; Sirisha et al., 2016). Therefore, enzyme immobilization, a new technique has been introduced to be a promising technique to overcome this limitation of enzymes. Enzyme immobilization is a technique involving physical confining or attaching enzymes in a specific space (Khan, 2021). The enzyme immobilization technique can offer the preservation, protection, and retention activity of enzymes under extreme conditions, improve the enzyme separation from the operation mixture, increase the efficient recovery, and the reusability of enzymes for several successive cycles (Khan, 2021; Kharazmi, Taheri-Kafrani, & Soozanipour, 2020). The biocatalytic activity of enzymes also could improve via immobilization on suitable supporting material because of the inhibition activity against enzyme aggregation and would preserve enzyme conformation (Shariat et al., 2018). Immobilization techniques are commonly divided into four basic categories: physical adsorption, entrapment, cross-linking, and covalent bonding (Ding et al., 2022). Physical adsorption (reversible method) involves weak forces such as hydrophobic interaction, hydrogen bonding, van der Waals, and ionic bonding. Immobilization by physical adsorption method is more convenient and because the enzyme could be easily removed from the surface of support material under mild conditions, it is also very attractive as when the

 DOI: 10.1201/9781003438168-2

enzyme activity is reduced, the support can be reused and reloaded with fresh enzyme (Dascălu et al., 2018; Mohamad et al., 2015). The entrapment is also defined as the physical attachment of enzymes into a finite matrix, which can improve mechanical stability and reduce enzyme leaching (Datta & Christena, 2013). Covalent bonding and cross-linking provide strong bonding, but they can deactivate enzymes because of the conformation changes of enzymes (Behram et al., 2023). The covalent method occurs through interaction between amino acid residues of enzymes (such as aspartic acid, histidine, and arginine) and functional groups (such as indolyl, phenolic hydroxyl, and imidazole) (Datta & Christena, 2013). The conventional immobilization methods, however, have a number of drawbacks, including a lack of effective reusability, difficulties in immobilization, a severe loss of enzymatic activity caused by the blocking of the active site, limited flexibility, and mass transfer restrictions between the enzyme and substrate (Wang et al., 2010). Nowadays, development in nanotechnology offers to produce new functional nanostructures for enzyme immobilization to improve the enzyme activity and stability, shelf life, efficiency, and reusability compared with conventional bulk materials. In this chapter, a summary of the most functional food-grade enzymes and their application in the industry (especially in food processing and preserving areas) are referred followed by their immobilization using nanostructure materials such as nanoparticles, nanofibers, nanoliposomes, nanoflowers, sol–gel, nanotubes, and nanosheets, and most of their applications in the immobilized state have been reviewed. Future perspectives on the nanotechnology role in novel immobilization techniques are discussed at the end of the chapter.

2.2 FOOD-GRADE ENZYMES

Food-grade enzymes are commonly used in food processing, food safety, and formulation in producing some ingredients for specific organoleptic features. They are generally classified into hydrolases (EC 3), an extensive class of enzymes that includes carbohydrases (such as amylases, β-galactosidases, cellulases, and pectinases), lipases, and proteases, which are the main food-grade enzymes with practical relevance. As food is naturally perishable, additional hydrolases such as lysozymes and chitinases are mostly used for safety purposes. Oxidoreductases (such as lactoperoxidases (LPOs) and glucose oxidases (GOXs)), another class of significance, are being sought after for their contributions to food safety. Table 2.1 indicates various food-grade enzymes with optimum activity conditions such as pH, temperature as well as the most producing resources. We reviewed the main food-grade enzymes in terms of major families and their optimum conditions of activity as well as the main applications in the industry.

2.2.1 Carbohydrases

2.2.1.1 Amylase

Amylases are members of glycosyl hydrolase family, which specifically hydrolyzes O-glycoside bonds in starch (Ghollasi, 2018) and are classified into three important enzymes including α-amylase (EC 3.2.1.1), β-amylase (EC 3.2.1.2), and glucoamylase (EC 3.2.1.3). Starch converts to maltose and dextrin by hydrolysis activity of α-amylase which occurs randomly, and β-amylases activity which starts from non-reducing ends. Glucoamylase hydrolyzes $\alpha(1\rightarrow4)$, $\alpha(1\rightarrow3)$, and $\alpha(1\rightarrow6)$ linkages to form glucose (Ramos & Malcata, 2011). Amylases are vastly being used in industries such as the conversion of starch, food, detergent, paper, and textile industries, as well as the generation of fuel alcohol for biofuels (Aydemir et al., 2020). It can be used in fields related to biotechnology such as removing environmental pollutants, conversion of starch to desired substrates by many microorganisms, infiltration of waste containing starch, and production of biochemical material with the help of starch substrate (bio-alcohols, high-fructose corn syrup, maltose, high-molecular-weight branched dextrins, maltotetrose syrups, and oligosaccharides mixtures) (Singh et al., 2015).

TABLE 2.1
Common food-grade enzymes, optimum conditions, major sources, and main applications

Properties	Temp(°C)	pH	Major sources	Applications	Reference
Proteases (EC 3.4)					
Serine proteases	50–70	6–11	*Bacillus, Aspergillus*, animaltissue (gut), *Tritirachium album* (thermostable)	Mashing cereal, brewing, removing beer haze, generating protein hydrolysates, manufacturing cheese (coagulation stage), and baking	(Naveed et al., 2021) (de Souza et al., 2015)
Cysteine or thiol proteases	40–55	2–3	*Aspergillus*, stem of pineapple (*Ananas comorus*), latex of figtree (*Ficus* sp.), papaya (*Carica papaya*), *Streptococcus, Clostridium*		
Metallo proteases	65–85	5–7	*Bacillus, Aspergillus, Penicillium, Pseudomonas, Streptomyces*		
Aspartic or carboxyl proteases	40–55	3–5	*Aspergillus, Mucor, Endothia, Rhizopus, Penicillium, Neurospora,* animal tissue (stomach)		
Carbohydrases					
α-Amylase (EC 3.2.1.1)	30–50	6–8	*A. oryzae, A. niger, B. subtilis, T. aestivum*	Saccharification of starch, antistaling agent in bread	(Farooq et al., 2021) (Fernandez Caresani et al., 2020) (Singh et al., 2015)
β-Amylases (EC 3.2.1.2)	50–65	4.5–6.2	Plant based: Alfalfa, tap roots, *Arabidopsis* sp., *Calystegia sepium* rhizome, soybean, sweet potato, barley, rye seeds, and sweet potato bacterial: *B. cereus, B. megaterium, B. polymyxa, C. thermosulfurogenes, Thermoactinomyces* sp.	Breakdown of starch, seed germination, and fruit ripening by adding sweetness	(Das & Kayastha, 2019)
Glucoamylase (EC 3.2.1.3)	50	4–5	*A. niger, Anguina tritici, A. brasiliensis, R. oryzae*	Food, fermentation, textile, and paper	(Wang et al., 2008) (da Costa Luchiari et al., 2021)
Cellulases (EC 3.2.1.4)	40–50	4.5–5.5	*A. oryzae., Fusarium* spp., *Trichoderma* spp., *Penicillium* spp., *Trichoderma* spp., *Bacillus* spp., *Clostridium* spp., *Cellulomonas* spp.	Production of glucose	(Khoshnevisan et al., 2019) (de Freitas et al., 2023)
Pectinase (EC 3.2.1.15)	30–80	3.5–8.5	Plant-based: wheat bran, apple pomace, corn barn, citrus wastes, coffee pulp, sugar cane baggase, and raw cassava tuber starch Fungal: *A. niger*	Fruit juice production	(Aguilar et al., 2008) (Kharazmi, Taheri-Kafrani, & Soozanipour, 2020)

β-Galactosidase (EC 3.2.1.23)	55–60	3–5	Plant-based: papaya, strawberry, tomato, apple, muskmelon, avocado, kiwifruit, coffee, mango, and Japanese pear Fungal: *Kluyveromyces lactis, K. marxianus, Guehomyces pullulans, K. fragilis, A. niger* and *A. oryzae*	Lactose-free products, synthesis of prebiotics and whey	(Movahedpour et al., 2022) (Ustok et al., 2010)
Lipases (EC 3.1.1.3)	40–50	4–5, 7–8	*Rhizomucor miehei, A. niger, Candida rugosa, P. aeruginosa, Staphylococcus epidermidis, B. subtilis, Anthus cervinus, Cissus antarctica, R. oryzae*	Chocolate, production of maltose and lactose-like sugar fatty acid esters, ice cream, flavor industry	(Mohammadi et al., 2016) (dos Santos et al., 2017) (Ismail & Baek, 2020) (Bencze et al., 2016)
Lysozyme (E.C. 3.2.1.17)	50°C	6.5	Egg white, milk, saliva, tears, mucus	Antibacterial, anti-inflammatory, antineoplastic, clinical diagnosis of diseases, and food safety, inhibitory activity against lactic bacteria in the wine industry	(Wang et al., 2020) (Cappannella et al., 2016)
Chitinase (E.C. 3.2.1.14)			Crops sources barley, onion, rice, and wheat. Bacteria sources *Serratia marcescens, Enterobacter* sp., *Bacillus, Aeromonas hydrophila, A. punctata, Serratia, Vibrio,* and *Streptomyces,* Fungal sources *Rhizoctoniasolani, Sclerotium rolfsii, Fusarium* sp.	Medical use, production of ophthalmic products, the production of single-cell protein, and bioethanol and biofertilizer	(Singh et al., 2020) (Dikbaş et al., 2021)
Lactoperoxidase (E.C. 1.11.1.7)	50°C	6	Naturally exists in milk and other exocrine such as tears, saliva, and lung fluidic lining	Medical and food preservation such as tomato and mango, bleaching of whey, antimicrobial activity in fish and meat processing, iodine detection	(Shariat et al., 2018) (Babadaie Samani et al., 2016)
Glucose oxidase (E.C. 1.1.3.4)	3–60°C	3.5–6.5	Insects' sources: honeybee, locust cuticle, *Spodoptera exigua*, and *Helicoverpa armigera* larvae. Fungal sources: *Aspergillus* sp., *penicillium* sp.	Biosensors, reduced soluble oxygen in probiotic products, limitation of oxygen in food packages	(Tian et al., 2019) (Afjeh et al., 2019) (Wang et al., 2022b)

2.2.1.2 Cellulases (EC 3.2.1.4)

Cellulase is a multi-component enzyme containing endoglucanase, exoglucanase, and β-glucosidase, which hydrolyzes cellulose to glucose by cleaving 1, 4, β-glucosidic bonds of cellulose (Poorakbar et al., 2018; Yassin et al., 2019). Cellulases are widely employed in a variety of industries, including food, pulp and paper, laundry, drinks, textiles, agriculture, pharmaceutics, medicine, and particularly the generation of biofuels (Li et al., 2007a-b; Poorakbar et al., 2018). Cellulases are also used as green biocatalysts in gas or liquid fuel production by converting cellulosic waste to soluble sugars (Li et al., 2019; Mihono et al., 2016).

2.2.1.3 Pectinase (EC 3.2.1.15)

Pectinases are multi-component enzymes composed of polygalacturonase, pectin esterase, and pectate lyase. Each degrades segments of pectin and produces soluble compounds such as galacturonic acid (Nouri & Khodaiyan, 2020). These enzymes are vastly used in the juice and wine industry for clarification. Unfortunately, due to the expensiveness, reusability, and low solubility of pectinase enzymes, their industrial applications are limited, which can be solved through immobilization techniques (Faramarzi-Aghgonbad et al., 2021).

2.2.1.4 β-Galactosidase (EC 3.2.1.23)

β-Galactosidase belongs to the glycoside hydrolases family, which contains an (α-β) 8-barrel catalytic domain among its six overall domains (de Freitas et al., 2023). It has become important in the industry due to its ability to produce lactose-free products and lacto-sucrose because of hydrolytic and transgalactosylation activity, respectively (Alshanberi et al., 2021). Despite its therapeutic application for lactose intolerance patients, it improves the flavor and sweetness of dairy products and suppresses crystallization in frozen products (Khan et al., 2017).

2.2.2 Proteases (EC 3.4)

Proteases are important industrial enzymes with numerous uses in detergent, pharmaceutical, and food production fields, and possess 60% of industrial enzymes on the market (Aydemir et al., 2020). The ability to produce natural products, biodegradability, stereo-specificity, specificity, and activity under mild reaction conditions are the main benefits of utilizing proteases. Many well-known proteases, including keratinases, papain, and bromelain, come from plant sources. Protease biosynthesis in plants is also a time-consuming procedure. The most well-known proteases derived from animals include trypsin, chymotrypsin, pepsin, and rennin. Nowadays, proteases that can be found on the market are produced by microbes. This is the result of several factors, including their fast rate of production, low need for cultivating areas, a wide range of biochemical variety, simplicity of geneticmanipulation, and desirable traits that make them suitable for biotechnological applications. Proteins in food are altered by fungus proteases. In comparison to proteases made from bacterial sources, those from fungal sources have more advantages and are known as genetically generally regarded as safe (GRAS) strains (Naveed et al., 2021). Depending on where their active site is located, proteases can be divided into four classes including serine proteases, cysteine or thiol proteases, metalloproteases, and aspartic or carboxyl proteases. Some proteases such as alkaline protease, which is commonly used in the food and medical area, have been immobilized via various nanostructures which results inimproving their mechanical stability and providing convenient separation from reaction media (Hu et al., 2015). Proteases are used to catalyze proteolytic reactions in waste-activated sludge to convert these wastes to products with higher biological values. In another study, keratinase was immobilized on nanostructures to biodegrade feather wastes (Sarathi et al., 2019).

2.2.3 Lipases (EC 3.1.1.3)

Lipases (triacetylglycerol acylhydrolase) are aqueous soluble enzymes that play catalytic activity in various reactions such as acidolysis, esterification, interesterification, and oil hydrolysis (Jafarian et al., 2020). Lipases are found in a wide variety of creatures, including humans, animals, plants, and microorganisms. However, microbial lipases are commercially the most significant due to the simplicity of their production and the potential for genetic manipulation to increase yield (Shalini et al., 2023). Microbial-origin lipases are widely used in several biotransformations due to their high selectivity, stability, and tolerance to solvents frequently used in reaction mediums (Carvalho et al., 2018). Lipases have been employed in food, biodiesel processing, chiral pharmacy, and other biosynthetic industries (Ke et al., 2016). There are two conformational forms for lipase based on crystallographic studies in terms of the closed form, in which the active site of an enzyme is shielded bya "lid" (a helical oligopeptide unit) and an open form in which the lid has been dislocated creating a totally discrete structure, leading to the exposure ofthe active site of the enzyme. Dislocation of the lid provides extra space in the lipase active site to fit a ligand and shape the catalytic site to a more active conformation. This idea has been used to specifically immobilize lipases from different sources on a variety of hydrophobic supports through open forms (Soni et al., 2020). Utilization of free lipase in cheese ripening could appear off-flavor and have bad texture due to excessive lipolytic activity, which develops encapsulation of lipase in liposomes (Kheadr et al., 2002). Several other limitations such as poor stability, narrow optimum pH, and sensitivity to extreme environmentshave attracted interest in the immobilizationof lipase (Ke et al., 2016).

2.2.4 Hydrolases

2.2.4.1 Lysozyme (EC 3.2.1.17)

Lysozyme is a natural cationic enzyme belonging to the hydrolases, which possess antibacterial activity on Gram-positive microorganisms because of the hydrolysis activity of $\beta(1\rightarrow4)$ glycoside bonds between N-acetyl-D-glucosamine and N- acetylneuraminic acid in peptidoglycans (Wang et al., 2021b). Lysozyme is a small globular protein, which is composed of a single polypeptide chain with a molecular weight of 14.3 kDa. Due to its inherent hydrophobic domain, and good solubility in aqueous environment, it is mainly immobilized on nanostructures such as nanotubes (Lopes et al., 2019; Nie et al., 2011; Noor et al., 2020). Lysozymes are mainly utilized in medical fields to ameliorate abscesses, rheum, inflammation, and stomatitis (Shareghi et al., 2015). The commercial lysozyme purified from egg white indicated antimicrobial effect only on G^+ bacteria; so immobilization of lysozyme has been extended to develop the application range and activity (Niu et al., 2020) such as inactivation of lactic bacteria in the wine industry (Cappannella et al., 2016), reduction in the activity of *Staphylococcus aureus* and *Pseudomonas aeruginosa* (Cerón et al., 2023).

2.2.4.2 Chitinase

Chitinase (EC 3.2.1.14) is a hydrolysis enzyme able to decompose chitin based on the amino acid sequence. Chitinase isclassified into endochitinases, which are able to cleave chitin at internal areas to create multimers of N-Acetylglucosamine (GlcNAc), and exochitinases whichenhance the hydrolysis rate of chitin to generate chitotriose or GlcNAc, chitobiose (Singh et al., 2020). Chitinases commonly exist in organisms that require to reshape their own chitin or dissolve and degrade the chitin of fungi or animals (Cano-Salazar et al., 2011). Chitinases are mostly usedin biological poisons against different insects and fungalpathogens to decrease the use of chemical insecticides and fungicides. So, they have a good potential to be anenvironmentally friendlyalternative to chemical components (Singh et al., 2020).

2.3 OXIDOREDUCTASES

2.3.1 Lactoperoxidase (EC 1.11.1.7)

Lactoperoxidase is a glycoprotein possessing heme moiety, which is mainly purified from bovine milk, saliva, and tears. It plays a protective role in the intestinal tract of newborn infants against pathogenic microorganisms via catalyzing halides and pseudohalides as well as preservation of the lactating mammary gland (Altinkaynak, Yilmaz, et al., 2016b; Babadaie Samani et al., 2016). LPO exhibits antimicrobial characteristics in the mentioned secretions due to its ability to catalyze the conversionof thiocyanate ion (SCN^-) to hypothiocyanite using H_2O_2, which shows antimicrobialproperties by oxidation of sulfhydryl of microbial enzymes (Shariat et al., 2018). The use of anLPO system for improving the shelf lifeof dairy products is a popular way, which is done by the addition of quantities of H_2O_2 tomilk. These antimicrobial systems could have some limitations such as resistance development, selective inhibition, and adsorption of the antimicrobial ingredients by protein and fat. To resolve these shortcomings, the LPO system could be immobilized on nanostructure systems to improve its activity and stability (Bhattacharya et al., 2015; Sheikh et al., 2018).

2.3.2 Glucose Oxidase

Glucose oxidase (EC 1.1.3.4) is a dimeric protease with flavin adenine dinucleotide (FAD), which oxidizes β-D glucose to D-glucono-δ-lactone via oxygen molecule followed by hydrolyzation to H_2O_2 and glucuronic acid usingthe FAD coenzyme as an electron carrier (Pourahmad et al., 2019; Wang et al., 2022a). GOX has been used widelyto determine free glucose in body blood, agriculture, and food products. It has been used to gain stronger dough in the bakery industry, as an alternative optionto oxidant ingredients such as bromate and L-ascorbic acid, elimination of oxygen from food packaging, and removal of D-glucose from egg white to inhibit browning (Netto et al., 2013). GOX also is mainly used to designglucose biosensors for glucose detection (Guan et al., 2019).

2.4 APPLYING NANOSTRUCTURED APPROACH FOR FOOD-GRADE ENZYME IMMOBILIZATION

Nowadays, nanotechnology opens new frontiers in developing enzyme immobilization and has proven the enormous benefits of nanostructure materials compared with conventional immobilization methods (Darwesh et al., 2020). Nanostructures are expected to be favorable carriers for enzymes asthey make it possible for enzymes to be used in a vast range of temperatures, pH, and improve enzyme efficiency and reusability compared with conventional bulk materials (Dascălu et al., 2018; Poorakbar et al., 2018). These advantages are related to their small size and higher surface area to volume ratio (to load large amounts of enzymes), flexible structure, and alterable porosity, enabling catalysis in non-aqueous medium, minimum diffusional limitation, and efficient surface functionalization capacity (El Harrad et al., 2018; Shariat et al., 2018; Temoçin, 2021). Enzyme immobilization methods can be classified aschemical methods (including covalent and cross-linking) and physical adsorption (adsorption and entrapment). Moreover, glutaraldehyde (Barbosa et al., 2013) and carbodiimide as a cross-linker (Noor et al., 2020) are common coupling agents used for the chemical immobilization of enzymes onto nanostructures. The structure and morphology of supports have a significant influence on the efficiency of immobilized enzymes as well as the immobilization methods mentioned above. Thus, the support material properties, functional groups, and immobilization conditions are important factors to obtain an appropriate immobilization system. Various food-grade enzymes have been immobilized in nanoscale such asnanoparticles, nanoflowers, nanoliposomes, nanotubes, nanosheets, and nanofibers. In the following, some important nanostructures and their popular applications are discussed. Some of the examples citing immobilization of various kinds of food-grade enzymes are listed in Table 2.2 where immobilized enzymes exhibited thermal stability which remains active in vast ranges of pH and improved the values of kinetic constants.

2.4.1 Nanoparticles

Nanoparticles are three-dimensional (3D) entities of the order of 100 nm or less andindicate characteristics different from bulk materials such as high surface area, effective enzyme loading, tunable size range, mass transfer resistance, and high dispersibility in an aqueous environment. Recently, nanoparticles have been widely used to immobilize enzymes, but the separation of enzymes from the reaction medium is their main disadvantage. For this reason, magnetic nanoparticles (MNPs) are known as suitable carriers for enzymes compared with other nanomaterials. MNPs are nano-forms of magnetic elements such as Fe, Co, Ni, Mn, Gd, and Cr and their derivatives. They have received more attention as a potential scaffold material for enzyme immobilization due to fast kinetics and better adsorption capacity, high specific surface area, easily modifiable surface by chemical procedures, good stability in a broad range of temperatures and pH values, and easy separation from the reaction mixture by a magnetic field. MPNs can be selectively attached to a functional molecule and transferred to a targeted location by usinga magnetic field (Darwesh et al., 2020). However, there is concern about their safety and toxicity in the food industry. Therefore, in these fields, superparamagnetic Fe_3O_4 nanoparticles are most used due to their better compatibility and lack of toxicity (Netto et al., 2013). The Fe_3O_4 nanoparticles could be synthesized through the chemical co-precipitation process of Fe^{2+} and Fe^{3+}ions in a 1:2 molar ratio followed by the gradual addition of ammonia solution as a precipitation agent. This reaction is carried out for 24 h at 70°C until the formation of black precipitation (MNPs) (Behram et al., 2023). MNPs have been functionalized via different organic and inorganic polymers with various functional groups by grafting or coating. The coated polymer layersprotect them against oxidation and aggregation (Babadaie Samani et al., 2016; Chen et al., 2018a; Ladole et al., 2021; Sarkar et al., 2012). The enzyme immobilization onto MNPs can be reversible physical adsorption (such as hydrogen, van der Waals bonding, ionic linkages, as well as hydrophobic interactions) or an irreversible immobilization via covalent bonds throughvarious chemical bonding methods, such as cross-linking agents, surface reactive functional groups, or multifunctional reagents (Dhavale et al., 2018; Mohamad et al., 2015). Covalent bonding involves the attachment between the enzyme and supports via amino acid residues of the enzyme, which could be introduced through cross-linking agents such as maleic anhydride, genipin, and glutaraldehyde (Hu et al., 2015). For instance, lipase from *Rhizopus oryzae* was successfully immobilized onto MNP Fe_3O_4, which was functionalized with 3-aminopropyltriethoxysilane (APTES) and glutaraldehyde to introduce aldehyde groups. The results showed much higher hydrolysis activity and thermo-stability of immobilized enzyme compared with free enzyme (Zhao et al., 2019). However, covalent bonding can reduce enzyme activity but exhibits more stability in comparison with physical adsorption (Noor et al., 2020). Vargas et al. (2021) immobilized invertase on magnetic composites via both covalent immobilization and physical adsorption that support material activated with coupling agents such as glutaraldehyde improves the retention, stability, and catalytic efficiency of the invertase (Barbosa et al., 2013; Ranjbari et al., 2019). Despite some beneficial properties of glutaraldehyde such as low cost and ease of use, it also poses somedisadvantages including toxicity and poor mechanical stability (Mitra et al., 2014). Dextran-aldehyde has been used as an alternative cross-linking agent to immobilize β-galactosidase on MNPs and improvethe enzyme stability after being incubated at 50°C, whichis higher than the one produced with glutaraldehyde (de Freitas et al., 2023). In another study, poly aldehyde kefiran has been used to immobilize pectinaseonto chitosan magnetic microparticles. Results exhibited that the immobilized pectinase showed better stability and durabilityand retained residual activity of about 50% after 10 cycles and could successfully be utilized for apple juice clarification (Nouri &

TABLE 2.2
The application and performance of food-grade enzymes immobilized on various functionalized nanostructures

Nanostructured approach	Enzymes	Application	Performance	Reference
Chitosan-coated Fe_3O_4 MNPs	α-Amylase	Starch digestion	Retained 66% of its initial activity after 20 days. Retained 79% of its initial activity after 20 cycles. Retained 58.3%and 78% of original enzyme activity at 80°C and pH 10, respectively.	(Dhavale et al., 2018)
Fe_3O_4 MNPs (covalent bonding)	Glucoamylase and α-amylase	Curcumin extraction from turmeric root powder	Retained about 95% of initial activity after 30 days. Retained about 50% of residual activity after 10 cycles. The extraction yield increased by 1.3–1.5 fold compared to the individual approach, which resulted in 54% (w/w) isolation with 91% purity of curcuminoids.	(Patil & Rathod, 2022)
Magnetic nickel nanostructure (covalent bonding)	Cellulase	Lignocellulosic biomass hydrolysis	Retained 84% of initial activity after 10 cycles Revealed the highest residual activity at 40°C and pH 5.	(Rashid et al., 2022)
Chitosan coated MNPs (covalent bonding)	Pectinex (*A. aculeatus* commercial enzyme)	Fructo-oligosaccharide production with two activities of sucrose hydrolysis and fructosyl transfer	Retained 70% and 86% of initial activity after 6 cycles for hydrolytic and transfructosylation activities, respectively. Obtained 101.56 g/l concentration of fructo-oligosaccharides in lab-scale experiments with high concentrations of 1-kestose.	(de Oliveira et al., 2020)
Fe3O4 MNPs (covalent bonding)	Pectinase from *B. subtilis*	Industrial uses	Retained about 90% of initial activity after incubation for 120 h at 50°C. Retained 90% of initial activity after 15 consecutive reactions	(Behram et al., 2023)
Chitosan-coated Fe_3O_4 MNPs (poly-aldehyde kefiran cross-linked)	Pectinase	Clarification of apple juice	Retained about 70% of initial activity after 10 cycles. Retained about 60% of initial activity after 30 days. The values of kinetic constants were calculated as K_m 3.201 mg/ml and V_{max} 0.595 µmol of galacturonic acid/min with free enzyme and 2.680 mg/ml (K_m) and 0.801 µmol of galacturonic acid/min (Vmax) with immobilized enzyme.	(Nouri & Khodaiyan, 2020)
MNPs (covalent bonding)	β-Galactosidase from *A. oryzae*	Production of tagatose from lactose	Immobilized enzyme with glutaraldehyde retained 100% of initial activity after storage at 4°C for 10 weeks. Maintained about 80% of activity after 10 cycles of reuse.	(de Freitas et al., 2023)
Fe_3O_4 MNPs	Lipase from *A. niger*	Production of biodiesel	Immobilized enzymeslost only 15% of activity after 5 cycles of reuse. Retained its activity after storage of 60 days.	(Shalini et al., 2023)

Fe_3O_4 MNPs (covalent bonding with APTES and GA modification)	Lipase from *R. oryzae*	Synthesis of (1,3-DAG)	Maintained 80% of its initial activity after 60 cycles The values of kinetic constants were calculated as K_m 4.88 Mm and V_{max} 10.34 µM $min^{-1}mg^{-1}$, K_{cat}5.52 with immobilized enzyme and 5.56 mM (K_m) and 1.14 mM $min^{-1}mg^{-1}$ (V_{max}), and 0.61S^{-1} (K_{cat}) with free enzyme. K_{cat}/K_m values of the immobilized enzyme were 1.13 $mM^{-1}S^{-1}$ which was much higher than that of the free enzyme (0.11 $mM^{-1}S^{-1}$) and exhibited great accessibility of substrate to the active site	(Zhao et al., 2019)
MNPs (physical and covalent bonding)	Proteas (pepsin)	Hemoglobin hydrolysis	Retained about 95–98% of initial activity after 3 months at 4°C. Retained 58.5% and 46% of initial activity for the physical and covalently immobilized pepsin after 10 cycles, respectively. The values of kinetic constants were calculated as K_m 110 µM and V_{max} 117.4 µmol/min/mg pepsin with physical adsorption and 140 µM (K_m) and 59.88 µmol/min/mg pepsin with covalently immobilized pepsin.	(Dascălu et al., 2018)
MNPs (covalent bonding)	Cellulase and lysozyme	Extraction of lipids from microalgae	Immobilized enzymes indicated a half-life three times higher than those of free enzymes. Retained about 60% of itsinitial activity after 6 cycles of reuse.	(Chen et al., 2018a)
Zinc oxide nanoparticles	Chitinase enzyme from *L. coryniformis*	Investigated on death rate and mean time of death of *S. zeamais*	Immobilized enzyme exhibited the mortality rate of 100% and an average death time of 2.4 days with nanoformulation of 6 mg/l.	(Dikbaş et al., 2021)
Silver nanoparticle AgNPs	Lactoperoxidase	Antibacterial against *Escherichia coli*	Immobilized enzyme on AgNPs did not show bacterial growth as shown by lack of increase in OD beyond 6 h and potentially for longer than 10 h.	(Sheikh et al., 2018)
Fe_3O_4MNPs (silica-coated)	Lactoperoxidase	Bleaching of fluid whey, antimicrobial activity in fish and meat industry, in dairy industry, which can be deactivated with heavy metals such as cadmium	The values of kinetic constants were calculated as K_m 0.34 mM and V_{max} 1.09 mM/min with immobilized enzyme and 0.55 mM (K_m) and 0.59 mmol/min with free enzyme. Immobilized enzyme indicated more stability in the presence of $CdCl_2$.	(Babadaie Samani et al., 2016)
Fe_3O_4MNPs (chitosan-coated)	Glucose oxidase	To improve the viability of probiotic bacteria in drinking yogurt	The yogurt containing immobilized enzyme had the maximum number of *Bifidobacterium lactis* and *L. acidophilus*	(Afjeh et al., 2019)
Liposome (thin film hydration)	Amyloglucosidase from *A. niger*	Glucose production	Retained 0% activity after 3 cycles. Kinetic constants were calculated as K_m 1.15 mg/ml and V_{max} 0.35 mg/ml/min with multilamellar vesicles immobilized enzyme and 1.64 mg/ml and V_{max} 0.56 mg/ml/min with large unilamellar vesicles.	(Li et al., 2007b)
Liposome (freeze–thawing	Cellulase	Glucose production	Immobilized liposome-bound cellulase and immobilized cellulase retained about 85% and 50% of initial activity after 5 cycles, respectively.	(Li et al., 2007a)

(continued)

TABLE 2.2 (Continued)
The application and performance of food-grade enzymes immobilized on various functionalized nanostructures

Nanostructured approach	Enzymes	Application	Performance	Reference
		To improve the catalytic activity of in ionic liquids	The maximum production of glucose (10.8 mM-glucose/(h M-lipid) was achieved at the reaction time of 48 h.	(Mihono et al., 2016)
Liposome (dehydration–rehydration vesicle method)	β-Galactosidases	to overcome the shortcoming ofthe sweet taste of hydrolyzed-lactose milk	The entrapment efficiency of 28% was obtained with Ch:PC and E:L ratios of 0.53 and 13.76, respectively.	(Rodríguez-Nogales & López, 2006)
Liposome (coated by silica)	Lipase from *R. miehei*	Transesterification of triolein with methanol to methyl esters	Immobilized enzyme retained its activity after 5 cycles of reuse.	(Macario et al., 2013)
Liposome (thin film hydration and freeze–thawing	Trypsin		There was a linear relationship between the phospholipid concentration and trypsin encapsulation yield. pH had more effect on the trypsin activity compared with ionic strength.	(Hwang et al., 2012)
Liposome (pectin-coated)	Lysozyme	Inhibitory activity against *L. monocytogenes*	Decreased the population of *L. monocytogenes* by 2 log CFU/ml in milk and 5 log CFU/ml in low-fat milk.	(Lopes et al., 2019)
Liposome (thin film hydration)	Chitinase	To lysis of fungal cell and sclerotial wall	Activity of free chitinase decreased about 15% in the presence ofsurfactant, while the activity in the liposome immobilizes chitinase increased at 50% under thesame conditions.	(Cano-Salazar et al., 2011)
Liposome (thin film hydration, chitosan coated)	Glucose oxidase	Design a biosensor to rapid analysis of glucose	Indicated a wider range of glucose (0.01–10mM/l). K_m value was lower than free enzyme related to better affinity between the immobilized enzyme and glucose.	(Guan et al., 2019)
Nanoflower (using Cu^{2+})	α-Amylase	Provides a sustainable and mild route to industrial scale in biomedicine, biofuel cells, biosensors, and tissue engineering	Retained about 84% of initial activity after 8 cycles. Retained about 90% of initial activity after 27 days of storage. Immobilized enzymes indicated high affinity toward substrates.	(Wang et al., 2013)
Nanoflower (using Cu^{2+})	Three pancreatic digestive enzymes (α-amylase, lipase, and protease)	can be used for the treatment of wastewater, biosensors, biocatalysts, and bio-related devices	Immobilized multienzymes indicated high catalytic activity at higher temperatures (50°C and 70°C) and a broad range of pH (6–9).	(Aydemir et al., 2020)
Nanoflower (using Ca^{2+})	β-galactosidase	Synthesis of galacto-oligosaccharides	Immobilized enzyme indicated a half-life 11 times higher than free enzyme at pH 9. Obtained high maximum recovery of enzyme (82%).	(Tavernini et al., 2021)

Nanoflower (using Cu^{2+})	β-Galactosidase	Lactose decomposition	Maintained about 74% of activity following storage at 25°C for 5 weeks. Retained about 70% of activity after 15 cycles of use. Obtained substrate conversion of 82% in the first reaction.	(Li et al., 2023)
Nanoflowers (using salts of Cu^{2+}, Mn^{2+}, Zn^{2+}, Co^{2+}, and Ca^{2+})	β-Galactosidases	Lactose hydrolysis in milk	Enzyme–Ca^{2+} nanoflowers retained about 100% of initial activity after 3 cycles, and complete hydrolysis of the lactose contained in a glass of milk, in about 10 min.	(Talens-Perales et al., 2020)
Nanoflower (using salts of Ca^{2+})	Lipase	Could be employed in the biofuel synthesis	Maintained more than 75% of initial activity after 15 days of storage. Retained 80% of initial activity after 5 cycles.	(Ke et al., 2016)
Nanoflower (using salts of Ca^{2+})	Lipase from *A. oryzae*	Hydrolysis of p-nitrophenol	Retained 90% of initial activity after storage for 27 days.	(Li et al., 2020)
Nanoflower (using salts of Zn^{2+})	Protease (papain)	Casein hydrolysis	The optimum temperature of the reaction enhanced from 55°C to 70°C. Retained 88.8% of initial activity after 10 cycles.	(Zhang et al., 2016)
Nanoflower (using salts of Cu^{2+} with magnetic properties)	Protease (papain)	Proteolytic of milk allergic protein	Immobilized enzymes showed 1556% higher proteolytic activity than native enzymes.	(Feng et al., 2020)
Nanoflower (using salts of Ca^{2+})	Protease (α-chymotrypsin)	bovine serum albumin (BSA) and human serum albumin (HSA) digestion	The enzyme activity enhanced up to 266%. Retained about 60% of activity after 8 cycles.	(Yin et al., 2015)
Nanoflower (using salts of Ca^{2+})	Lactoperoxidase	Nanosensor to detection of dopamine and epinephrine.	Immobilized enzyme enhanced about 160% and 360% in activities at pH 6 and pH 8, respectively. Retained about 65% of initial activity after 6 cycles of reuse.	(Altinkaynak et al., 2016)
Nanoflower (using salts of Cu^{2+})	Co-immobilizing three recombinant enzymes CBH, EG, and BG	The conversion of cellulosic biomass to biofuels	The K_m and K_{cat}/K_m values for ECG-NFs were 9.33 g/l and 0.0051 l/min g, respectively. Immobilized multienzymes were more active than free multienzymes over a broad range of pH (3–8) and temperature (20–70°C). Retained about 60% of initial activity after 8 cycles.	(Han et al., 2021)
Nanofiber (covalent bonding)	α-Amylase	Suitable matrix for industrial applications	The kinetic constants, V_{max} 0.16 μM/min and K_m were calculated as 2.3 mg/ml with immobilized enzyme, while it was 0.15 μM/min and 0.125 mg/ml with free enzyme. Retained about 54% of initial activity after storage for 9 days.	(Ghollasi, 2018)
Nanofiber (chitosan coated)	α-Amylase	Production of fermentable sugars from cassava fibrous waste	Retained about 90% of initial activity after 6 cycles.	(Gali et al., 2021)
Nanofiber (cellulose nanofiber, covalent bonding)	Cellulase	Biomedical and food packaging applications	Retained 85% of initial activity after 6 cycles. Immobilized enzymes retained approximately 80% of initial activity over a broad range of temperatures (30–70°C).	(Yassin et al., 2019)

(*continued*)

TABLE 2.2 (Continued)
The application and performance of food-grade enzymes immobilized on various functionalized nanostructures

Nanostructured approach	Enzymes	Application	Performance	Reference
Nanofiber (covalent bonding)	β-Galactosidase	Bioconversion of lactose into GOS through β-galactosidase in batch and repeated batch systems.	Immobilized enzyme enhanced catalytic reaction lactose conversion from 41% to 86% and GOS yield from 28% to 40% in comparison with native enzyme.	(Misson, Jin, et al., 2020)
Nanofiber	Lipase from *P. fluorescens*		The immobilized enzyme indicated about 40% of activity at 85°C, retained about 60% of activity after 90 days indicating reusability after 10 cycles of reuse.	(Dwivedee et al., 2019)
Nanofiber (PVA, cross-linked)	Protease (pepsin)		The immobilized enzymatic activity obtained 96% of the crude enzyme and indicated reusability of 4 cycles and maintain for 10 days of storage.	(Loredo-Alejos et al., 2022)
Nanofiber (PVA, cross-linked)	Protease (ficin)		The immobilized ficin extract showed 92% of the enzyme activity of the crude ficin extract. The enzymatic activity of the immobilized ficin extract was conserved after a total of 9 reuse cycles and maintained after being stored for 25 days.	(Rojas-Mercado et al., 2018)
Nanofiber (cross-linked by Millard reaction)	Lysozyme	Antibacterial activity against *S. aureus* and *E. coli*	Retained 49% and 60% of its initial activity after storage at 25°C and 4°C for 40 days, respectively. Maintained 78% of its initial activity after 12 cycles of reuse.	(Wanget al., 2021a)
Nanofibers (covalent bonding)	Glucose oxidase	Bioelectrochemical electrode.	Biochemical electrode indicated long-term storage stability over 63 days and indicated a great linear responseto the glucose concentration ranges with two separate calibration curves, from 2 to 8 mM/l and from 10 to 30 mM/l.	(Temoçin, 2021)
Sol–gel (organosilane-functionalized silicas)	α-Amylase from *B. subtilis*, α-amylase from *A. oryzae*, and barley β-amylase	Saccharification of starch	The biocatalytic activity relies on both the nature of the enzyme and the used organosilane. β-Amylase indicated better catalytic activity after encapsulation.	(Caresani et al., 2021)
Sol–gel (physical bonding)	β-Amylase	Hydrolysis of selected starches	Immobilized enzyme maintained about 70% of its initial activity after incubation at different pH for 2 h. Maintained about80% of activity after incubation at70°C, and remained active in the presence of KCl, Na^+, and EDTA.	(Abu et al., 2022)

Sol–gel	β-Galactosidase	Bioconversion of lactose to galacto-oligosaccharides	Immobilized enzyme obtained the conversion yield of 59%, while this value was 74% for native enzyme. The highest galacto-oligosaccharides were achieved for the immobilized enzyme (28%) compared with the native enzyme (19%).	(Misson et al., 2016)
Sol–gel (covalent bonding)	Lipase from *A. Niger*	Olive oil hydrolysis	The immobilized biocatalyst and free lipase had anoptimum pH of 2 and 5, respectively. Maintained relative activity above 50% until 7 cycles of reuse in olive oil hydrolysis. The values of kinetic constants were calculated as K_m 115 mM and V_{max} 714 U mg^{-1} with immobilized enzyme and 77 mM (K_m) and 1250 Umg^{-1} (V_{max}) with free enzyme. indicated that the immobilization process reducedenzyme-substrate affinity.	(dos Santos et al., 2017)
Sol–gel (physical bonding)	Alcalase from *B. licheniformis*	The synthesis of C-terminal peptide amides	Immobilized enzyme retained 100% of the original activity even after 14 cycles. The activity of the immobilized enzyme was two times greater than native enzyme after incubation at 25°C for 50 days.	(Corîci et al., 2011)
MWCNT (covalent bonding)	α-Glucosidase		Retained 84.16% of initial activity. After storage at 4°C for 30 days.	(Mohiuddin et al., 2014)
MWCNT (covalent bonding)	Cellulase	To producebio alcohol, food, brewing, oilexploration, medicinalingredients	Immobilized enzyme was more active than free enzyme over a broad pH range of 3–7 and indicated activity of about 92% at 50°C. Retained 70% of initial activity after 7 cycles. Maintained 88% of initial activity after storage of 7 days.	(Li et al., 2019)
MWCNT (physical bonding)	Cellulase		Retained 52% of initial activity after 6 cycles.	(Mubarak et al., 2014)
MWCNT (covalent and physical bonding, polyaniline cobalt-coated)	β-Galactosidase from *A. oryzae*	Immobilized β-galactosidase provides a potential therapeutic agent forthe decomposition of lactose from milk products and prevents lactose crystallization	Covalently and physically immobilized enzymes retained 92% and 74% of initial activity after 10 successive cycles, respectively, and maintained 46% and 78% of initial activity after incubation at 60°C for 2 h. Covalent immobilization retained 70% activity in the presence of 5% galactose. The K_m value was calculated as 0.4 mM for adsorbed enzyme and 0.3 mM forcovalently attached enzyme, respectively, indicating higher substrate–enzyme affinity	(Khan et al., 2017)
MWCNT (non-covalent bonding)	Lipase B from *C. antarctica*	Lactones production through the chemo-enzymatic Baeyer-Villiger oxidation of cyclic ketones	Exhibited activity of 60% after 6 cycles of reuse. The immobilized enzyme retained activity over a vast pH and temperature.	(Markiton et al., 2017)

(continued)

TABLE 2.2 (Continued)
The application and performance of food-grade enzymes immobilized on various functionalized nanostructures

Nanostructured approach	Enzymes	Application	Performance	Reference
SWCNT (covalent bonding)	Lipase from *C. antarctica*	Production of biodiesel	Maintained 90% of activity after 10 cycles. Indicated low diffusional limitation in the ethanolysis of the sunflower oil.	(Bencze et al., 2016)
SWCNT and MWCNT (covalent bonding)	Lipase from *R. miehei*		Immobilized enzymes on MWCNT and SWCNT indicated a half-life of 4.2 and 4.6-fold longer than native enzymes at 40°C. Both immobilized enzymes retained more than 90% of initial activity after 10 cycles of reuse.	(Alagöz et al., 2021)
MWCNT	Protease (bromelain)		High concentrations of enzymes induced aggregation between nanostructures and enzymes.	(Jha & Venkatesu, 2016)
SWNT (noncovalent adsorptionand covalent bonding)	Lysozyme	Antimicrobial activity	Retained 100% of the originalactivity even after 14 cycles. The activity of theimmobilized enzyme was two times greater than native enzyme after incubation at 25°C for 50 days. Then on covalently bonded enzyme exhibited a higher activity rate of 0.93 min^{-1} compared to the native LSZ rate of 0.57 min^{-1}. The covalently immobilized enzyme exhibited a much lower rate of 0.014 min^{-1} but retained its activity over a longer period.	(Noor et al., 2020)
Nanosheet (covalent bonding)	α-Amylase	Biosensor for starch detection	Maintained 83% of activity after storage at 4°C for 100 days, and 73% of initial activity after 10 cycles.	(Singh et al., 2015)
Double hydroxides nanosheets@ Fe_3O_4 (covalent bonding)	Cellulase		Maintained maximum activity at 50°C. Maintained about 34% of activity after 6 cycles.	(Pei et al., 2018)
Nanosheet (covalent bonding)	Pectinase		The kinetic constant K_{cat}/K_m for the immobilized enzyme was calculated as 3.7 $mg^{-1}min^{-1}ml^{-1}$, which was higher than the free enzyme. The addition of surfactant improved enzyme activity under an acidic environment.	(Liu et al., 2014)
Nanosheet (covalent bonding)	Alkaline protease	Proteolysis of waste-activated sludge	Immobilized enzyme retained about 90% of its initial activity after 20 days at 4°C	(Su et al., 2013)

Nanosheet (covalent bonding)	Trypsin	Activity with immobilized enzyme	Immobilized trypsin generated polypeptides with high antioxidant activity. Maintained 90% of activity after storage at 4°C for 60 days. Retained 80.5% of initial activity after 10 cycles.	(Siddiqui et al., 2020)
Nanosheet (covalent bonding)	Protease from *P. vannamei*	Pharmaceutics, therapeutics, diagnostics, and supplements in dietsanalyzing proteins and peptides and screening drugs based on enzyme inhibition	Immobilized protease improved thermal stability at extreme pH values. After 24 h of incubation at 90°C, the free enzyme retained less than 10% of the activity, while the immobilized enzyme kept more than 90% of its original activity	(Ranjbari et al., 2019)
Magnetic graphene oxide	Protease (ficin)	Production of biopeptides, pharmaceutical and brewing industries	The activities of immobilized ficin at 60°C and pH 7 resulted in recovery rates of approximately 92% and 97%, respectively. The immobilized ficin on the GO and GO–Fe_3O_4 lost 30% and 26% of their original activity, respectively 4°C (pH 7) over a period of120 days	(Tahsiri et al., 2023)
Graphene oxide nanosheets	Lactoperoxidase	Application in sensors, fuel cells, and nanoelectronics	K_m, V_{max}, for native enzymes, was 178 mM, 0.63 µM/ml/min, respectively, whereas, for immobilized enzymes, K_m, V_{max} values were 104 mM, 0.36 µM/ml/min, respectively. The optimum pH shifted to the alkaline range and the optimum temperature enhanced up to 60°C. Retained about 50% of initial activity after 12 cycles.	(Shariat et al., 2018)
Graphene oxide nanosheets	Glucose oxidase	Glucose biosensor	The excellent sensitivity of about 30 µA/Mm/cm^2 After 47 days, lost only 10% of the response current.	(Karuppiah et al., 2014)

Note: 1-Ethyl-3-(3-dimethylaminopropyl) carbodiimide (EDC), 1,3-Diacyglycerols (1,3-DAG), 3-aminopropyltriethoxysilane (APTES), β-Glucosidase (BG), Bovineserum albumin (BSA), Cellobiohydrolase (CBH), Cholesterol:Phosphatidylcholine ratio (Ch:PC), Endo-glucanase (EG), Enzyme:lipid (E:L), Galacto-oligosaccharides (GOS), Glutaraldehyde (GA), Human serum albumin (HSA), Polyvinyl alcohol (PVA), Multiwall carbon nanotubes (MWCNTs), Polyethyleneimine (PEI), Single-walled carbon nanotubes (SWNTs).

Khodaiyan, 2020). Pectinase extracted from *Bacillus subtilis* was immobilized on magnetic iron oxide nanoparticles using the co-precipitation process. Immobilized pectinase maintained more than 90% of its initial activityafter keeping it at 50°C for 120 h (Behram et al., 2023). There are several studies about immobilization of food-grade enzymes on MNPs and their applications such as GOX on chitosan-MNP for viability of probiotic bacteria in drinking yoghurt (Pourahmad et al., 2019), LPO immobilization on silver nanoparticle to improve its antimicrobial activity and increasing the shelf life of dairy products (Sheikh et al., 2018), chitinase from *Lactobacillus coryniformis* immobilized on zinc oxide to control the maize weevil *Sitophiluz zeamais,* which is known as s warehouse pest (Dikbaş et al., 2021), lysozyme immobilized on MNPs to improve the rate of *Micrococcus lysodeikticus* lyses (Shareghi et al., 2015), lipase immobilized on MNPs in biodiesel production, alkaline protease onto amino-functionalized $Fe_3O_4$to catalyze hydrolysis of oat bran into oat polypeptides (Hu et al., 2015), β-galactosidase on gold nanoparticles to produce lactose-free dairy products (Alshanberi et al., 2021), cellulase on magnetic nickel to degradation of lignocellulosic biomass (Rashid et al., 2022), and α-amylase on chitosan-coated Fe_3O_4MNPs to facilitate starch digestion (Dhavale et al., 2018). Whenchitosan-covered MNPs are used as support for a commercial enzyme preparation from *Aspergillus aculeatus*, they display between 70% and 86% of enzyme activity after 6cycles of reuse (de Oliveira et al., 2020). Immobilized xylanase on MNPs coated with polyethylene glycol (which has specific characteristics including nontoxicity, high hydrophilicity, flexibility, biocompatibility, and non-antigenicity) maintained residual activity of 50% after 9 cycles and could be utilized for apple juice clarification. The juice clarification was investigated by measuring the turbidity reduction. More than 42% reduction in turbidity was achieved for the immobilized enzyme while it was 52.8% for the native enzyme. The results showed a higher V_{max} value compared withthe native enzyme suggesting higher catalytic activity (Kharazmi, Taheri-Kafrani, Soozanipour, et al., 2020). Patil and Rathod (2022) studied co-immobilization of glucoamylase and α-amylase on MNPs using glutaraldehyde as pre-treatment to increase the extraction yield of Curcuma longa powder. A residual activity of 50% was retained after 10 consecutive cycles of operation and maintained more than 95% activity after 30 days of storage. The co-immobilization of two enzymes including pectinase and cellulase on MNPs also improved the lycopene extraction from tomato peel and displayed more than 50% lycopene yield after 6 cycles of reuse (Ladole et al., 2018). Recently, gold and silica-coated MNPs have been utilized due to their unique chemical and optical properties along with chemical inertness thatprovide high surface functionality and improve enzyme stability against uncontrolled oxidation and irreversible aggregation processes (Singh et al., 2016). For instance, trypsin was immobilized on magnetic Fe_3O_4 nanoparticles coated with a layer of gold nanoparticles through strong but reversible Au–S and Au–NH_2 bonds. The use of gold nanoparticles as intermediate ligands for the attachment of trypsin exhibited a significant acceleration of enzyme activity (Cao et al., 2016). Magnetic silica nanocomposites due to high surface space, uniform pore size, easy functionalization, and hydrothermal stability as well as biocompatibility have made them an appropriate choice of support material to immobilize various enzymes (Poorakbar et al., 2018). Bebić et al. (2020) developed fumed silica nanoparticles (FNS), modified with aminopropyltrimethylsilane (APTMS), to enhance amino group on the surface of FNS, and employed as a platform for more stable electrostatic interactions of laccase from *Myceliophthora thermophila.* Along with the high activity of immobilized laccase in the degradation of pesticide lindane, it demonstrated recyclability up to 6 repeated cycles with 70% activity.

2.4.2 Nanoliposomes

Liposomes can be defined as artificial sphere-shaped vesicles that are composed of one or more phospholipid bilayers with an aqueous inner volume reservoir to encapsulate the hydrophilic compounds and the hydrophobic domain entrapped lipophilic compounds. Liposomes can be classified into different categories based on their sizes in the range of 20–100 nm for small unilamellar to more than 1000 nm for multivesicular vesicles (Pasarin et al., 2023). Nanoliposomes are nanoscale liposomes that provide several benefits compared with conventional liposomes including high specific surface area, confining unwanted interactions with proteolytic enzymes, improving solubility and bioavailability, stabilizing against harsh conditions, and allowing for sustained and targeted release (Alonso-Estrada et al., 2022; Malam, 2009). The environment in which the enzyme is immobilized is a main factor to maintain its native and active conformation. Thus, entrapping enzymes within a natural aqueous-containing membrane such as liposomes should be appropriate. Asthe enzymes are entrapped within the liposomes, they will not be able to attach their substrates and remain inactive in the environment, so entrapping enzymes in liposomes makes it possible to control the time, rate, and duration of the release (Reza Mozafari et al., 2008). There are several methods to fabricate nanoliposomes including conventional methods (such as thin film hydration, ether/ethanol injection), reverse phase evaporation (such as detergent depletion, microfluidic channel, heating, membrane extrusion, high shear homogenization, and sonication), and modern methods such aslyophilization, dual asymmetric centrifugation, and supercritical fluid methods (Andra et al., 2022). The fabrication technique and its condition (including phospholipid type, lipid concentration, and enzyme purity) play a critical role in enzyme encapsulation efficiency (Meesters, 2010; Sanchez & Perillo, 2000). Hwang et al. (2012) evaluated the influence of several factors such as ionic strength, phospholipid concentration, and pH on trypsin encapsulation. They proved that there wasa linear relationship between the phospholipid concentration and trypsin encapsulation yield, and they also showed that pH had more effect on the trypsin activity compared with ionic strength. Nanoliposomes are thermodynamically unstable, so represent several disadvantages such as short half-life, susceptibility to oxidation, hydrolysis, and leakage of entrapped enzymes through decomposition of the system (Akbarzadeh et al., 2013). These shortcomings can be solved by coating the surface of liposomes with certain polymers which induce electrostatic repulsion, therefore it canimprove enzyme stability (Pasarin et al., 2023; Reza Mozafari et al., 2008). For instance, Macario et al. (2013) entrapped lipase from *Rhizomucor miehei* in a liposome composed of α-phosphatidylcholine followed bycoating with a porous inorganic silica shell to stabilize the internal liposomal phase. The entrapped lipase is used to produce methyl esters via transesterification of triolein with methanol. Results showed higher total productivity than the native enzyme and maintain its activity after 5 cycles. Most of the food-grade enzymes entrapped into nanoliposomes have been used in the dairy industry, especially cheese processing. For instance, Jahadi et al. (2015) encapsulated flavorzyme in liposomes via heating method without using any chemical solution or detergents. They obtained the highest entrapment efficiency and activity in lecithin concentration of 4.5%, flavorzyme/lecithin ratio of 5%, at 45°C, and mixing for 30 min. In another study, they immobilized flavorzyme on liposomes with a diameter of 189 nm with an entrapment efficiency of 26.5%, which could accelerate the ripening time of cheese from 20 days to 10 days (Jahadi et al., 2020). Rodríguez-Nogales and López (2006) successfully immobilized β-galactosidase on liposomes by dehydration–rehydration vesicle method. They optimized liposomal formulation with cholesterol:phosphatidylcholine and enzyme: lipid ratios of 0.53 and 13.76, respectively, at pH 6 and obtained 28% entrapment efficiency and the stability of β-galactosidase against proteolytic activity was enhanced by the entrapment process. The glucose biosensor was successfully designed by immobilizing GOX on a nanoliposome made up of a glassy carbon electrode and chitosan. The measurement of glucose was based on the detection of H_2O_2 concentration which was produced by GOX activity. The immobilized GOX indicated higher affinity to the substrate and exhibited excellent reproducibility and stability (Guan et al., 2019). Liposome–cellulase complex exhibited higher efficiency, lower deactivity through reusing, and higher stability against ionic liquids (Mihono et al., 2016). Lysozyme encapsulated in phosphatidylcholine liposome coated with pectin decreased the population of *Listeriamono cytogenes* by 2 log CFU/ml in whole milk and 5 log CFU/ml in skim milk (Lopes et al., 2019).

2.5 ORGANIC–INORGANIC HYBRID NANOFLOWERS

Organic–inorganic hybrid nanoflowers (hNFs) are modern nanostructures forimmobilization of enzymes that improve enzyme activity and stability in a wide range of pH levels, temperature, and salt concentrationcomparedwith free enzyme or conventional bulk materials (Altinkaynak et al., 2016a). Constructing enzymes into hNFs provides a high surface area, which improves mass-transfer limitations, active cooperation of immobilized enzymes on each other due to nanoscale entrapped enzymes in hNFs, and suitable enzyme conformation in hNFs (Altinkaynak, Tavlasoglu, et al., 2016a; Wu et al., 2015). Geet al. (2012) discovered the first example of enzyme-inorganic crystal hNFs by adding copper (II) sulfate to phosphate-buffered saline (PBS) solution withbovine serum albumin (BSA). The construction of hNFs, a self-assembly process, contains three stages including nucleation, growth, and completion. During the first step, crystals of copper phosphate are formed followed by nucleation through the intervention of amide groups in the enzyme structure; in the second step, the discrete nanopetals resulted from preliminary protein crystals created due to nucleation acting as seed particles. Finally, the protein molecules play as a template for nanopetals and serve asglue to bind the petals together for the formation of large protein agglomerates. The production and construction of nanoflowers are affected by several factors such as environmental conditions (temperature and pH), biomineralization time, and the reagent concentration (such as metal ions, phosphate, and enzymes) (Tavernini et al., 2021). Employing this method, Altinkaynak et al. (2016b) investigated the influence of various PBS pH levels and temperatures (4°C and 20°C) on synthesizing the LPO–copper phosphate hNFs. They demonstrated that LPO hNFs were formed at pH 9 and 10, which is because of the proximityto pH value of LPO, while at pH of 4 and 5 (acidic conditions), there was an inhibition of nanoflower production due to enhancingthe density of positive charges at the enzyme surface and resulted in strong repulsive forces between cooper ions and enzymes as well as among enzymes. They also found that LPO-hNPs indicated a more compact and spherical structure at 4°C compared with 20°C, which related to tight binding between LPO and $Cu_3(PO_4)_2$ nanocrystal at this temperature. Mostafavi et al. (2023) studied the effect of different metal ion solutions (including Cu^{2+}, Ca^{2+}, Co^{2+} , or Mn^{2+}) on the formation of the protease hNFs. They obtainedexcellent enzyme activity, stability, and reusability, for protease–$Mn_3(PO_4)_2$ hNFs. Ke et al. (2016) designed hNFs using *Burkholderia cepacia* lipase as an organic part and calcium phosphate as an inorganic part to improve the activity and stability of lipase. The results exhibited that under the optimum immobilization conditions, the immobilized enzyme activity enhanced up to 308% of the native enzyme; it alsoretained 75% of initial activity after incubation for 15 days and maintained about 80% of initial activity after 5 cycles. Similarly, a thermophilic lipase nanoflower was constructed by steering $Cu_3(PO_4)_2$ solution with PBS and lipase. The activity of lipase was evaluated by monitoring the degradation of *p*-nitrophenyl caprylate. The immobilized enzyme indicated similar catalytic activity as the native enzyme and improved stability under extreme conditions (Liu et al., 2021). Recent progress in nanotechnology has prompted research interest inthe co-immobilization of two or more various enzymes at the same time in onesingle structure. Co-immobilized triple enzyme nanoflower is synthesized by incubating pancreatin solution including protease, lipase, and α-amylasein, a PBS solution containing Cu_SO4, at 4°C for 3 days. At elevated temperatures, the nanoflower-bound triple enzyme was less labile and less sensitive to different pH levels (Aydemir et al., 2020). Li et al. (2023) synthesized β-galactosidase–$Cu_3(PO_4)_2$–hNFs and evaluated its catalytic activity via digestibility of o-nitrophenyl-β-D-galactopyranoside as the substrate. The immobilized enzyme maintained 74.8% of initial activity after incubation at room temperature for 5 weeks and was reusable after 15 cycles. Finally, the immobilized enzyme successfully hydrolyzed 82% of lactose in the first cycle which proved this nanostructure could be an appropriate option for the industrial production of lactose-free products for people with lactose intolerance disorder. β-Glucosidase, cellobiohydrolase, and endo-glucanase were co-immobilized through the integration of binary tags made of elastin-like polypeptide (ELP) and His-tag to use as a triple enzyme to convert cellulose into glucose. Multienzymatic immobilization showed excellent cascade catalytic activity compared to native enzymes (Han et al., 2021). Wang et al. (2013) designed nanoflowers using calcium phosphate crystals with the same procedure which was applied for the construction of Cu–hNFs. The α-amylase used in this study is an enzyme that indicates the allosteric effect when interacted with Ca^{2+} in aqueous solutions. In the absence of Ca^{2+} in an aqueous solution during the assembling of hNFs with $Ca^{2+}$and α-amylase, Ca^{2+} firmly attached to α-amylase and maintained enzyme activity for a longer time compared with the native enzyme. Many other food-grade enzymes have been immobilized in this novel hNPs with various ions such aspapain hNFs with zinc ions (Zhang et al., 2016), chymotrypsin using calcium ions (Yin et al., 2015), trypsin hNFs using zinc ions, lipase hNFs using univalent metal ions (Ag^{+}), bivalentmetal ions (Zn^{2+}, Mn^{2+}, and Ca^{2+}), and trivalent metal ions (Al^{3+}, and Fe^{3+}) (Wang et al., 2022a).

2.6 NANOFIBER

Nanofibers have obtained growing importance among the support materials because of their specific attributes such as self-assembling performance, high specific surface area, high enzyme binding capacity, homogeneous dispersion, high porosity, interconnectivity, and good mechanical stability (Kamaci & Peksel, 2020; Ribeiro et al., 2021). Nanofibers are used as a support systemthrough two different methods including surface adsorption (attributed to the enzyme immobilization on nanofibers via physical or covalent bonding) and encapsulation (attributed to direct co-electrospinning of enzymes and other materials) (Li et al., 2018). Nanofibers could be constructed via many methods such as template synthesis, self-assembly, phase separation, and electrospinning (Işik et al., 2019). Nanofibers fabricated via electrospinning indicate a uniform diameter, and flexible structure although could contain diverse components, thus being used in many bioreactor designs (de Melo Brites et al., 2020). Some limitations of nanofibers can be solved with somemodifications including modification toward biocompatibility, such as the modification of carboxyl-containing nanofiber with chitosan or gelatin to create bilayer biomimetic surface, modification toward enzyme mobility by adding spacer arms onto the surface such as polyethylene glycol (PEG), and modification toward electrical conductivity with incorporating polymers with high electrical conductivity. Recently, the layer-by-layer self-assembly technique has been used to improve mechanical stability via electrospinning, which can be obtained via consecutivelyattaching multivalent species on the nanofiber via electrostaticinteractions, hydrogen bonding, and covalent bonding, thus improving the nanofiber potentials forenzyme immobilization. Wang et al. (2021b) immobilized lysozyme by applying a layer-by-layer self-assembly comprising alternately coating positively charged lysozyme and negatively charged sodium alginate on negatively charged cellulose acetate nanofiber. The immobilized lysozyme exhibited high stability in extreme environments and maintained over 70% of its initial activity after 4 cycles of reuse. They also indicated that lysozyme could maintain55.6% of its activity by inducing Millard reaction between lysozyme and dextran sulfate sodium salton the surface of cellulose acetate nanofibers (Wang et al., 2021b). Polyvinyl alcoholnanofibers were used for the immobilization of ficin and papain, the most famous plant protease. The immobilized enzymes maintained 92% and 88% of the initial activity of the free enzymes, respectively (Moreno-Cortez et al., 2015; Rojas-Mercado et al., 2018). Sass and Jördening (2020) immobilized β-galactosidase from *Aspergillus oryzae* on electrospun gelatin nanofiber usingtwo techniques. In the first method, enzymes were covalently immobilized on activated gelatin nanofiber mats with hexamethylenediamine as a bifunctional linker, and in the second technique, enzyme was entrapped into nanofibers with simultaneouselectrospinning of gelatin and enzyme. The covalently immobilized enzyme exhibited better thermostability compared to free and entrapment enzymes while it indicated lower V_{max} than entrapment techniques. In recent years, researchers have conducted extensive studies on various materials that could be used in food-grade immobilization through nanofibers, including polystyrene for immobilization of β-galactosidase to produce galacto-oligosaccharides (Misson, Saallah, et al., 2020); poly (ε-caprolactone) for immobilization of trypsin to the digestion of casein and gelatin (Pinto et al., 2015; Song et al., 2015);

polyaniline for immobilization of lipase tobe explored in esterification, aminolysis, alcoholysis, and biofuel synthesis (Soni et al., 2020); chitosan-coated polylactic acid nanofibers for immobilization of α-amylaseto produce fermentable sugars (Gali et al., 2021); cellulose polymers nanofibers for immobilization of protease (Badoei-Dalfard et al., 2022); and polyacrylonitrile nanofibers for immobilization of GOX for glucose detection (Çetin et al., 2023).

2.7 SOL–GEL

The sol–gel process provides a potential technique to build inorganic gel that could be in the shape of films, nanospheres, and fibers. This process is a chemical method to manufacture various nanostructures, especially metal oxide nanoparticles, which have some advantages such as low cost, low operating temperature, and excellent production efficiency (Zhang et al., 2022). The sol–gel process has higher popularity and industrial application compared with other nanoparticle synthesized methods. It consists of three main steps:

1. The conversion of precursors to a homogeneous sol.
2. Conversion of a sol into a gel.
3. The removal of the solvent in the gel structure and drying (Bokov et al., 2021). Oxide compositions of zirconium, alumino silicates, alumina, silica, and titanium are commonly utilized for sol–gel production. Among them, silicon dioxide (SiO_2) is the most interesting for enzyme immobilization (Pierre, 2004). This preference is due to several properties such as nontoxicity, biocompatibility, and controlled porosity (Mujahid et al., 2010). Silica is required to be activated through several modifications because of their chemically inertness, such as treating via aminoalkyl triethoxysilanes to introduce amino groups, and further they are functionalized through diverse methods for immobilization of enzymes (Sirisha et al., 2016). Their functionalities can be altered by modifying the silica surface with free silanols via hydrogen bonding, electrostatic interaction, and chemical derivatization. In a common production method, silica precursors, such as tetramethyl orthosilicate or tetraethyl orthosilicate, are hydrolyzed into a colloidal solution or "sol" and adding enzymes to the "sol," followed by gelation via polycondensation of hydrolyzed precursors which caused encapsulation of enzyme in a silicate matrix. Achieving different pores in the final silicate matrices depends on the drying methods of sol–gel such as xerogels obtained from hydrophilic sol–gel dried at room temperature (pores of 1–10 nm in diameter), aerogels obtained from drying the sol–gel with supercritical carbon dioxide (unaltered porosity), aerogels with a large distribution in pore diameters obtained through gelation at pH 7, followed by drying with supercritical carbondioxide (pores of 2–80 nm in diameter), ambigels obtained by drying hydrophobic gels at room temperature which poses the same porosity and size as wet gels, and aerogels with hydrophobic properties produced through drying under the supercritical condition of hydrophobic gels (Rother & Nidetzky, 2009). Abu et al. (2022) entrapped β-amylase on silica sol–gel and evaluated the effect of temperature, pH, and metal ion on entrapped enzyme activity. They also investigated the effectiveness of the immobilized enzyme in the hydrolysis of selected starches. Entrapped enzyme retained more than 70% and more than 60% of initial activity after 2 h incubation at different pH values and incubation at 70°C, respectively. They also revealed that enzyme activity was improved in the presence of KCl, Na^+ and ethylenediaminetetraacetic acid (EDTA). Also, the immobilized enzyme enhancedthe hydrolysis of selected starches. Fernandez Caresani et al. (2020) evaluated amylase immobilization by sol–gel entrapment from three different sources by measuring the reducing sugars. They demonstrated that the addition time of enzyme to sol–gel structure and the concentration of catalyst could influence the immobilized enzyme activity; they also showed that the catalytic activity of encapsulated amylase related to the amylase source. Recently,

aerogels have been used for the immobilization of several enzymes such as lipase (Barbosa et al., 2016), trypsin, and pepsin, and exhibited better activity than those of corresponding free enzyme solutions (Kato et al., 2014). dos Santos et al. (2017) immobilized lipase onto the sol–gel silica that was subjected to a silanization process followed by activating with 10% glutaraldehyde solution. Results indicated that the optimum pH activity decreased from 5 for free lipase to 2 with immobilized lipase. The residual activity of biocatalysts was above 87% at 55°C, after 240 min storage and it retained about 50% of its initial activity after 7 cycles of hydrolyzing olive oil. However, regarding kinetic parameters, covalent immobilization results inconstructional changes in the enzyme, which causes a reduction in the feasibility of the substrate enzyme binding, which leads toa reductionin the maximum reaction rate compared to the native lipase.

2.8 NANOTUBE

Carbon nanotubes (CNTs) have a one-dimensional hollow tube formed by folding or rolling graphite sheets to create special hollow cylinders. CNTs are garnering enormous attention to serve as a matrix for enzyme immobilization because of their advantageous features such as large specific surface area, high mechanical strength, remarkable electronic conductivity, chemical stability, and effective loading capacity in comparison with the traditional counterparts such as silica and other polymeric materials (Liu & Cai, 2007; Rasheed et al., 2019). CNTs deposed either in a single layer (SWCNTs) or several graphite layers surrounding a central tubule (MWCNTs) (Sobhan et al., 2019). SWCNTs have been used in many studies because of their good dispersion in solution, small diameter, high surface-to-area ratio for enzyme interaction, and low mass-transfer resistance due to their nonporous property (Sobhan et al., 2019; Wang et al., 2010). Howver, MWCNTs are attractive because of their lower cost and easier dispersibility, better physical and chemical stability, ease of preparation, and low toxicity (Xie et al., 2019). The native form of CNTs is insoluble in many solvents, hence they cannot interact with enzymes directly. The solubility and reactivity of CNTs can be improved by covalent or non-covalent functionalization of CNTs. Although non-covalent immobilization could maintain the native structure of both the enzyme and support material, the gradual leak of the enzyme could reduce the reusability of enzymes. The immobilization ofenzymes through covalent bonding provides more stable biocatalysts, although it generally reduces enzyme activity due to the conformational alteration of the protein (Bencze et al., 2016). For instance, Ji et al. (2021) introduced cinnamaldehyde containing a benzene ring into the chitosan to modify CNTs and make strong non-covalent interaction through π–π stacking between CNTs and chitosan. They use cinnamaldehyde–chitosan CNTs conjugated to immobilize porcine pancreatin lipase to improvethe synthesis of ethyl caproate (apple-like flavor). Immobilized lipase indicated good thermostability, which is attributed to π–π stacking, hydrogen noncovalent bonding such as hydrogen bonding, electrostatic interaction, and hydrophobic interaction between lipase and functionalized CNTs. Mukhopadhyay et al. (2015) entrapped psychrophilic pectate lyase with calciumhydroxyapatite nanoparticles (play as a calcium substitute, the cationic activator of this enzyme) in SWNTs. The immobilized pectate lyase exhibited both thermo- and psychrostability at 80°C and 4°C, respectively, for several hours. The retention of enzymatic activity (more than 70%) was observed following 7 repeated freeze–thaw cycles. Alagöz et al. (2021) compared covalently immobilized lipase from *R. miehei* through MWCNTs functionalized with carbodiimide and SWCNTs functionalized with 3-APTES. They showed both nanotubes presented better reusability after 10 cycles and longer half-life value at a pH of 7.5 at 40°C. Singh et al. (2019) immobilized inulinase from *Penicillium oxalicum* on modified MWCNTs using APTES, which generates amino-terminated surfaces for better inulinase attachments. Inulinase activity increased up to 60.7% and retained 28% of its residual activity after 10 consecutive batch cycles (Singh et al., 2019). In one such study, α-glucosidase enzyme immobilized on MWCNby treatment with cross linkers, 1-ethyl-3-(3-dimethylaminopropyl) carbodiimide and

N-hydroxysuccinimide via amide bonding. The immobilized α-glucosidase enzyme retains its catalytic activity performance up to 5 cycles of reuse (Mohiuddin et al., 2014). *Trichoderma* cellulase MWCNTs and sodium alginate maintained about 70% of their catalytic performance up to 7 cycles of operation. However, the increase in leakage observed after repeated use could be because of the hydrophilic characteristics of sodium alginate and gradually increased pore size or the relatively weak noncovalent bonding between enzyme and MWCNTs (Li et al., 2019). Lipase B from *Candida antarctica* noncovalently immobilized on MWCNTs provided a system that is highly enzyme-loaded and stable. Immobilized lipase could oxidize cyclic ketones to their corresponding lactones in yields of 86–95% under a moderateenvironment atappropriate reaction times (Markiton et al., 2017). β-Galactosidase was immobilized on surface-modified cobalt/MWCNTs through both covalent bonding and physical adsorption (both hydrophobic and ion exchange interactions), by which the immobilization yields achieved for adsorbed and cross-linked enzymes were 93% and 97%, respectively. The covalently immobilized enzyme was remarkably more stable at extreme conditions (pH and temperature) compared to physically adsorbed and free enzymes and also retained 92% of initial activity after 10 successive cycles, while this amount was 74% for the noncovalent immobilized enzyme (Khan et al., 2017). Various studies have been performed on the immobilization of food-grade enzymes on nanotubes such as the immobilization of bromelain on carboxylated MWCNTs (Jha & Venkatesu, 2016), lysozyme on MWCNTs (Wang et al., 2022a), β-LPO on SWCNTs (Bhattacharya et al., 2015), and cellulase on MWCNTs (Mubarak et al., 2014).

2.9 NANOSHEET

Along with enhancing the use of carbon materials, different methods of constructing nanomaterials with functional properties have emerged. Graphene is mostly utilized in the immobilization of enzymes because of some advantages such as low cost, biocompatibility, availability, high specific surface area, thermal stability, fast electron transport, excellent mechanical flexibility, and high hydrophilicity which is related to several functional groups containing oxygen such as carboxyl, hydroxyl, and epoxide (Jafarian et al., 2020; Liu et al., 2014). It is composed of a dense layer of carbon atoms intertwined in a two-dimensional compact construction somehow creating a honey comb lattice. The most used procedures of synthesis are mechanical exfoliation, chemical vapor deposition, spin coating, and liquid exfoliation (Ramakrishna et al., 2018). Graphene builds an excellent nanostructure for enzyme immobilization because of the retention of enzyme active conformation. Functionalized graphene synthesized by thermal exfoliation of graphite oxide was used to immobilize α-amylase extract and purified from the germinated seeds of *Triticum aestivum.* Graphite oxide was oxidized, and its temperature was increased under an argon atmosphere immediately, followed by tube heating it to 1050°C and then chilling to 25°C. α-Amylase was immobilized through cross-linking method using cysteamine and glutaraldehyde. The immobilized enzyme on functionalized graphene sheets obtained high loading efficiency of 85.16%. Immobilized enzyme showed a vast range of pH and temperature of operation compared with free enzyme (Singh et al., 2015). LPO, foundin milk and other exocrine secretions, provides antibacterial properties due to containing thiocyanate ion (SCN^-) and H_2O_2, was extracted from bovine milk, and subsequently was bounded covalently on graphene oxide nanosheets. Considering kinetic parameters, there was greater affinity between immobilized enzyme and substrate compared with free enzyme. The immobilized LPO was characterized by an increased pH range of activity as well as an operating temperature up to 60°C (Shariat et al., 2018). Covalent immobilization of lipase from *R. miehei* on carboxylated graphene nanosheets was performed using the Ugi four-component reaction, which is a condensation reaction among an aldehyde, an amine, a carboxylic acid, and an isocyanide. Immobilized enzymes retained 100% of their initial activities after 24 h of incubation in the presence of 10% of the seven organic solvents, while this amount was 68–80% for free enzymes (Mohammadi et al., 2016). Protease purified from *Penaeus vannamei* shrimp was covalently immobilized on graphene oxide nanosheets that are activated with glutaraldehyde and exhibited thermal stability significantly, which remained at 90% of initial activity following incubationat 90°C for 24h (Ranjbari et al., 2019). Ketose 3-epimerases constitutethe feasible process for the production of D-psicose (D-ribo-2-hexulose or D-cellulose) thatis considered as a rare low-calorie sugar and displaysimportant physiological functions. Dedania et al. (2017)immobilized D-psicose 3-epimerase from *Agrobacterium tumefaciens* onto graphene oxide (GO) via physical adsorption. Thermostability half-life improved from 3.99 min for native enzyme to 720 min for immobilized enzyme at 60°C and the bioconversion efficiency increased up to 10 cycles of reusability. Skoronski et al. (2017) immobilized laccase from *A. oryzae* on grapheme nanosheets by both physical and covalent bonding. The immobilized laccase was active over a broad range of temperatures and pH. The covalently bound enzyme maintained approximately 80% of its initial activity after 6 cycles of reuse in spite ofthe physically adsorbed enzyme, which retained its activity only upon the second cycle of reuse.

2.10 FUTURE PERSPECTIVES

Immobilization techniques address concerns related to enzyme activity, stability, and recovery. They are crucial for improving catalytic performance and creating biosensors in industrial biotechnology. However, their applications are limited. Advanced immobilization engineering has successfully achieved co-immobilization of several enzymes, requiring careful selection of support matrix and immobilization method. Co-immobilization requires managing substrate channeling, lowering production costs, and creating biocompatible, safe systems. However, co-immobilization reactions also face challenges in achieving equivalent catalytic activity, reusability, and product removal (Betancor & Luckarift, 2010). However, immobilization of many enzymes on a single support is particularly difficult asit must maintain the catalytic activity of all the enzymes engaged in the system and, ideally, increase stability. The most unstable catalytic component will limit the composite's half-life if the immobilization design does not provide an overall operational stabilization to each of the included enzymes (Arana-Peña et al., 2021). According to recent studies, there will probably be two goals for this area in the near future. The first will involve carrying out exceedingly complicated processes including numerous co-immobilized enzymes and cofactors, where the cost of the biocatalyst is not an issue and the issues with enzyme co-immobilization may be lessened (although they will still exist). The second will involve creating solutions for the shortcomings of the current co-immobilization systems for processes where the cost of the biocatalysts may be prohibitive. The use of various supports, active groups, and immobilization protocols to obtain the best possible biocatalysts from each involved enzyme, while simultaneously maximizing the benefits of co-immobilization, should be permitted by these new co-immobilization strategies to fully exploit the immobilization's potential to enhance enzyme properties. Additionally, even after stabilization via immobilization, the issue of the various stabilities of the enzymes must be avoided. In this manner, fully active immobilized enzymes need not be discarded after one of the enzymes has been inactivated (Rajnish et al., 2021).

2.11 CONCLUSION

The wide range of applications for nanomaterials and the recent interest in them have made it possible to use them as immobilization matrices for enzymes, especially to prevent the loss of enzyme activity and to keep them reusable and recyclable. Nanostructures like MNPs, nanosheets, nanotubes, nanofibers, etc. have become effective matrices for immobilizing enzymes and making them a better choice as biocatalysts over chemical catalysts. These enhanced enzymes and catalysts can be used in food safety, food quality, and food processing. The detection of food pesticides and unhealthy residues, higher reusability, and higher sensitivity are examples of the aforementioned.

REFERENCES

Abu, T. F. A., Enujiugha, V. N., & Odumosu, E. (2022). Entrapment of β-amylase on silica sol-gel for enhanced activity on selected starches. *Annals. Food Science and Technology*, *23*(2), 218–223. www.afst.valahia.ro

Afjeh, M. E. A., Pourahmad, R., Akbari-Adergani, B., & Azin, M. (2019). Use of glucose oxidase immobilized on magnetic chitosan nanoparticles in probiotic drinking yogurt. *Food Science of Animal Resources*, *39*(1), 73–83. https://doi.org/10.5851/kosfa.2019.e5

Aguilar, C. N., Gutiérrez-Sánchez, G., Rado-Barragán, P. A., Rodríguez-Herrera, R., Martínez-Hernandez, J. L., & Contreras-Esquivel, J. C. (2008). Perspectives of solid state fermentation for production of food enzymes. *American Journal of Biochemistry and Biotechnology*, *4*(4), 354–366. https://doi.org/10.3844/ajbbsp.2008.354.366

Akbarzadeh, A., Rezaei-Sadabady, R., Davaran, S., Joo, S. W., Zarghami, N., Hanifehpour, Y., Samiei, M., Kouhi, M., & Nejati-Koshki, K. (2013). Liposome: Classification, preparation, and applications. *Nanoscale Research Letters*, *8*(1). https://doi.org/10.1186/1556-276X-8-102

Alagöz, D., Toprak, A., Yildirim, D., Tükel, S. S., & Fernandez-Lafuente, R. (2021). Modified silicates and carbon nanotubes for immobilization of lipase from *Rhizomucor miehei*: Effect of support and immobilization technique on the catalytic performance of the immobilized biocatalysts. *Enzyme and Microbial Technology*, *144*, 109739. https://doi.org/10.1016/j.enzmictec.2020.109739

Ali, S. S., Al-Tohamy, R., Koutra, E., Moawad, M. S., Kornaros, M., Mustafa, A. M., Mahmoud, Y. A. G., Badr, A., Osman, M. E. H., Elsamahy, T., Jiao, H., & Sun, J. (2021). Nanobiotechnological advancements in agriculture and food industry: Applications, nanotoxicity, and future perspectives. *Science of the Total Environment*, *792*, 148359. https://doi.org/10.1016/J.SCITOTENV.2021.148359

Alonso-Estrada, D., Ochoa-Viñals, N., Pacios-Michelena, S., Ramos-González, R., Núñez-Caraballo, A., Michelena Álvarez, L. G., Martínez-Hernández, J. L., Neira-Vielma, A. A., & Ilyina, A. (2022). No solid colloidal carriers: Aspects thermodynamic the immobilization chitinase and laminarinase in liposome. *Frontiers in Bioengineering and Biotechnology*, *9*, 793340. https://doi.org/10.3389/fbioe.2021.793340

Alshanberi, A. M., Satar, R., & Ansari, S. A. (2021). Stabilization of β-galactosidase on modified gold nanoparticles: A preliminary biochemical study to obtain lactose-free dairy products for lactose-intolerant individuals. *Molecules*, *26*(5), 1226. https://doi.org/10.3390/molecules26051226

Altinkaynak, C., Tavlasoglu, S., Özdemir, N., & Ocsoy, I. (2016a). A new generation approach in enzyme immobilization: Organic-inorganic hybrid nanoflowers with enhanced catalytic activity and stability. *Enzyme and Microbial Technology*, *93*, 105–112. https://doi.org/10.1016/j.enzmictec.2016.06.011

Altinkaynak, C., Yilmaz, I., Koksal, Z., Özdemir, H., Ocsoy, I., & Özdemir, N. (2016b). Preparation of lactoperoxidase incorporated hybrid nanoflower and its excellent activity and stability. *International Journal of Biological Macromolecules*, *84*, 402–409. http://doi.org/10.1016/j.ijbiomac.2015.12.018

Andra, V. V. S. N. L., Pammi, S. V. N., Bhatraju, L. V. K. P., & Ruddaraju, L. K. (2022). A comprehensive review on novel liposomal methodologies, commercial formulations, clinical trials and patents. *Bionanoscience*, *12*(1), 274–291. https://doi.org/10.1007/s12668-022-00941-x

Arana-Peña, S., Carballares, D., Morellon-Sterlling, R., Berenguer-Murcia, Á., Alcántara, A. R., Rodrigues, R. C., & Fernandez-Lafuente, R. (2021). Enzyme co-immobilization: Always the biocatalyst designers' choice…or not? *Biotechnology Advances*, *51*, 107584. https://doi.org/10.1016/J.BIOTECHADV.2020.107584

Aydemir, D., Gecili, F., Özdemir, N., & Ulusu, N. N. (2020). Synthesis and characterization of a triple enzyme-inorganic hybrid nanoflower (TrpE@ihNF) as a combination of three pancreatic digestive enzymes amylase, protease and lipase. *Journal of Bioscience and Bioengineering*, *129*(6), 679–686. https://doi.org/10.1016/j.jbiosc.2020.01.008

Babadaie Samani, N., Nayeri, H., & Amiri, G. A. (2016). Effects of cadmium chloride as inhibitor on stability and kinetics of immobilized Lactoperoxidase (LPO) on silica-coated magnetite nanoparticles versus free LPO. *Nanomedicine Journal*, *3*(4), 230–239. https://doi.org/10.22038/nmj.2016.7579

Badoei-Dalfard, A., Saeed, M., & Karami, Z. (2022). Protease immobilization on activated chitosan/cellulose acetate electrospun nanofibrous polymers: Biochemical characterization and efficient protein waste digestion. *Biocatalysis and Biotransformation*, *41*(4), 279–298. https://doi.org/10.1080/10242422.2022.2056450

Barbosa, A. S., Lisboa, J. A., Silva, M. A., Carvalho, N. B., Pereira, M. M., Fricks, A. T., Mattedid, S., Limaa, A. S., Franceschic, E., & Soares, C. M. F. (2016). The novel Mesoporous silica aerogel modified with protic ionic liquid for lipase immobilization. *Quimica Nova*, *39*, 415–422. https://doi.org/10.5935/0100-4042.20160042

Barbosa, O., Ortiz, C., Berenguer-Murcia, Á., Torres, R., Rodrigues, R. C., & Fernandez-Lafuente, R. (2013). Glutaraldehyde in bio-catalysts design: A useful crosslinker and a versatile tool in enzyme immobilization. *RSC Advances*, *4*(4), 1583–1600. https://doi.org/10.1039/C3RA45991H

Bebić, J., Banjanac, K., Ćorović, M., Milivojević, A., Simović, M., Marinković, A., & Bezbradica, D. (2020). Immobilization of laccase from *Myceliophthora thermophila* on functionalized silica nanoparticles: Optimization and application in lindane degradation. *Chinese Journal of Chemical Engineering*, *28*(4), 1136–1144. https://doi.org/10.1016/j.cjche.2019.12.025

Behram, T., Pervez, S., Nawaz, M. A., Ahmad, S., Jan, A. U., Rehman, H. U., Ahmad, S., Khan, N. M., & Khan, F. A. (2023). Development of pectinase based nanocatalyst by immobilization of pectinase on magnetic iron oxide nanoparticles using glutaraldehyde as crosslinking agent. *Molecules*, *28*(1), 404. https://doi.org/10.3390/molecules28010404

Bencze, L. C., Bartha-Vári, J. H., Katona, G., Toşa, M. I., Paizs, C., & Irimie, F. D. (2016). Nanobioconjugates of *Candida antarctica* lipase B and single-walled carbon nanotubes in biodiesel production. *Bioresource Technology*, *200*, 853–860. https://doi.org/10.1016/j.biortech.2015.10.072

Betancor, L., & Luckarift, H. R. (2010). Co-immobilized coupled enzyme systems in biotechnology. *Biotechnology and Genetic Engineering Reviews*, *27*(1), 95–114. https://doi.org/10.1080/02648725.2010.10648146

Bhattacharya, K., El-Sayed, R., Andón, F. T., Mukherjee, S. P., Gregory, J., Li, H., Zhao, Y., Seo, W., Fornara, A., Brandner, B., Toprak, M. S., Leifer, K., Star, A., & Fadeel, B. (2015). Lactoperoxidase-mediated degradation of single-walled carbon nanotubes in the presence of pulmonary surfactant. *Carbon*, *91*, 506–517. https://doi.org/10.1016/j.carbon.2015.05.022

Bokov, D., Turki Jalil, A., Chupradit, S., Suksatan, W., Javed Ansari, M., Shewael, I. H., Valiev, G. H., & Kianfar, E. (2021). Nanomaterial by sol-gel method: Synthesis and application. *Advances in Materials Science and Engineering*, *2021*, 1–21. https://doi.org/10.1155/2021/5102014

Cano-Salazar, L. F., Juárez-Ordáz, A. J., Gregorio-Jáuregui, K. M., Martínez-Hernández, J. L., Rodríguez-Martínez, J., & Ilyina, A. (2011). Thermodynamics of chitinase partitioning in soy lecithin liposomes and their storage stability. *Applied Biochemistry and Biotechnology*, *165*(7–8), 1611–1627. https://doi.org/10.1007/s12010-011-9381-1

Cao, Y., Wen, L., Svec, F., Tan, T., & Lv, Y. (2016). Magnetic AuNP@Fe_3O_4 nanoparticles as reusable carriers for reversible enzyme immobilization. *Chemical Engineering Journal*, *286*, 272–281. http://doi.org/10.1016/j.cej.2015.10.075

Cappannella, E., Benucci, I., Lombardelli, C., Liburdi, K., Bavaro, T., & Esti, M. (2016). Immobilized lysozyme for the continuous lysis of lactic bacteria in wine: Bench-scale fluidized-bed reactor study. *Food Chemistry*, *210*, 49–55. https://doi.org/10.1016/j.foodchem.2016.04.089

Caresani, J. R. F., Dallegrave, A., & Santos, J. H. Z. Dos. (2021). Amylases encapsulated in organosilane-modified silicas prepared by sol-gel: Evaluation of starch saccharification. *Journal of Sol-Gel Science and Technology*, *97*(2), 340–350. https://doi.org/10.1007/s10971-020-05446-1

Carvalho, N. B., Vidal, B. T., Barbosa, A. S., Pereira, M. M., Mattedi, S., Freitas, L. D. S., Lima, Á. S., & Soares, C. M. F. (2018). Lipase immobilization on silica xerogel treated with protic ionic liquid and its application in biodiesel production from different oils. *International Journal of Molecular Sciences*, *19*(7), 1829. https://doi.org/10.3390/ijms19071829

Cerón, A., Costa, S., Imbernon, R., de Queiroz, R., de Castro, J., Ferraz, H., Oliveira, R., & Costa, S. (2023). Study of stability, kinetic parameters and release of lysozyme immobilized on chitosan microspheres by crosslinking and covalent attachment for cotton fabric functionalization. *Process Biochemistry*, *128*, 116–125. https://doi.org/10.1016/J.PROCBIO.2023.02.023

Çetin, M. Z., Guven, N., Apetrei, R. M., & Camurlu, P. (2023). Highly sensitive detection of glucose via glucose oxidase immobilization onto conducting polymer-coated composite polyacrylonitrile nanofibers. *Enzyme and Microbial Technology*, *164*, 110178. https://doi.org/10.1016/j.enzmictec.2022.110178

Chen, Q., Liu, D., Wu, C., Yao, K., Li, Z., Shi, N., Wen, F., & Gates, I. D. (2018a). Co-immobilization of cellulase and lysozyme on amino-functionalized magnetic nanoparticles: An activity-tunable biocatalyst for extraction of lipids from microalgae. *Bioresource Technology*, *263*, 317–324. https://doi.org/10.1016/J.BIORTECH.2018.04.071

Chen, Z., Wang, X., Chen, Y., Xue, Z., Guo, Q., Ma, Q., & Chen, H. (2018b). Preparation and characterization of a novel nanocomposite with double enzymes immobilized on magnetic Fe3O4-chitosan-sodium tripolyphosphate. *Colloids and Surfaces B: Biointerfaces*, *169*, 280–288. https://doi.org/10.1016/j.colsurfb.2018.04.066

Corîci, L. N., Frissen, A. E., van Zoelen, D. J., Eggen, I. F., Peter, F., Davidescu, C. M., & Boeriu, C. G. (2011). Sol-gel immobilization of Alcalase from *Bacillus licheniformis* for application in the synthesis of C-terminal peptide amides. *Journal of Molecular Catalysis B: Enzymatic*, *73*(1–4), 90–97. https://doi.org/10.1016/j.molcatb.2011.08.004

da Costa Luchiari, I., Cedeno, F. R. P., de Macedo Farias, T. A., Picheli, F. P., de Paula, A. V., Monti, R., & Masarin, F. (2021). Glucoamylase immobilization in corncob powder: Assessment of enzymatic hydrolysis of starch in the production of glucose. *Waste and Biomass Valorization*, *12*, 5491–5504. https://doi.org/10.1007/s12649-021-01379-0

Darwesh, O. M., Ali, S. S., Matter, I. A., Elsamahy, T., & Mahmoud, Y. A. (2020). Enzymes immobilization onto magnetic nanoparticles to improve industrial and environmental applications. *Methods in Enzymology*, *630*, 481–502. https://doi.org/10.1016/bs.mie.2019.11.006

Das, R., & Kayastha, A. M. (2019). β-Amylase: General properties, mechanism and panorama of applications by immobilization on nano-structures. In Q. Husain & M. Ullah (Eds.), *Biocatalysis: Enzymatic Basics and Applications* (pp. 17–38). Springer. https://doi.org/10.1007/978-3-030-25023-2_2

Dascălu, A., Ignat, L., Ignat, M.E., Doroftei, F., Belhacene, K., Froidevaux, R., & Pinteală, M. (2018). Reusable biocatalyst by pepsin immobilization on functionalized magnetite nanoparticles. *Roumanian Journal of Chemistry*, *63*(7–8), 685–695. http://web.icf.ro/rrch/

Datta, S., Christena, L. R., & Rajaram, Y. R. (2013). Enzyme immobilization: an overview on techniques and support materials. *Three Biotech*, *3*, 1–9. https://doi.org/10.1007/s13205-012-0071-7

Dedania, S. R., Patel, M. J., Patel, D. M., Akhani, R. C., & Patel, D. H. (2017). Immobilization on graphene oxide improves the thermal stability and bioconversion efficiency of D-psicose 3-epimerase for rare sugar production. *Enzyme and Microbial Technology*, *107*, 49–56. https://doi.org/10.1016/j.enzmictec.2017.08.003

de Freitas, L. A., de Sousa, M., Ribeiro, L. B., de França, Í. W. L., & Gonçalves, L. R. B. (2023). Magnetic CLEAs of β-galactosidase from *Aspergillus oryzae* as a potential biocatalyst to produce tagatose from lactose. *Catalysts*, *13*(2), 306. https://doi.org/10.3390/catal13020306

de Melo Brites, M., Cerón, A. A., Costa, S. M., Oliveira, R. C., Ferraz, H. G., Catalani, L. H., & Costa, S. A. (2020). Bromelain immobilization in cellulose triacetate nanofiber membranes from sugarcane bagasse by electrospinning technique. *Enzyme and Microbial Technology*, *132*, 109384. https://doi.org/10.1016/j.enzmictec.2019.109384

de Oliveira, R. L., da Silva, M. F., da Silva, S. P., de Araújo, A. C. V., Cavalcanti, J. V. F. L., Converti, A., & Porto, T. S. (2020). Fructo-oligosaccharides production by an *Aspergillus aculeatus* commercial enzyme preparation with fructosyltransferase activity covalently immobilized on Fe3O4-chitosan-magnetic nanoparticles. *International Journal of Biological Macromolecules*, *150*, 922–929. https://doi.org/10.1016/j.ijbiomac.2020.02.152

de Souza, P. M., de Assis Bittencourt, M. L., Caprara, C. C., de Freitas, M., de Almeida, R. P. C., Silveira, D., Fonseca, Y. M., Ferreira Filho, E. X., Pessoa Junior, A., & Magalhães, P. O. (2015). A biotechnology perspective of fungal proteases. *Brazilian Journal of Microbiology*, *46*, 337–346. https://doi.org/10.1590/S1517-838246220140359

Dhavale, R. P., Parit, S. B., Sahoo, S. C., Kollu, P., Patil, P., Patil, P., Chougale, A. D., & Chougale, A. D. (2018). α-Amylase immobilized on magnetic nanoparticles: reusable robust nano-biocatalyst for starch hydrolysis. *Materials Research Express*, *5*(7), 075403. https://doi.org/10.1088/2053-1591/aacef1

Dikbaş, N., Uçar, S., Tozlu, G., Öznülüer Özer, T., & Kotan, R. (2021). Bacterial chitinase biochemical properties, immobilization on zinc oxide (ZnO) nanoparticle and its effect on *Sitophilus zeamais* as a potential insecticide. *World Journal of Microbiology and Biotechnology*, *37*, 1–14. https://doi.org/10.1007/s11274-021-03138-8

Ding, S. S., Zhu, J. P., Wang, Y., Yu, Y., & Zhao, Z. (2022). Recent progress in magnetic nanoparticles and mesoporous materials for enzyme immobilization: An update. *Brazilian Journal of Biology*, *82*:e244496. https://doi.org/10.1590/1519-6984.244496

dos Santos, E. A. L., Lima, Á. S., Soares, C. M. F., & de Aquino Santana, L. C. L. (2017). Lipase from *Aspergillus niger* obtained from mangaba residue fermentation: biochemical characterization of free and immobilized enzymes on a sol-gel matrix. *Acta Scientiarum. Technology*, *39*(1), 1–8. https://doi.org/10.4025/actascitechnol.v39i1.29887

Dwivedee, B. P., Soni, S., Bhimpuria, R., Laha, J. K., & Banerjee, U. C. (2019). Tailoring a robust and recyclable nanobiocatalyst by immobilization of *Pseudomonas fluorescens* lipase on carbon nanofiber and its application in synthesis of enantiopure carboetomidate analogue. *International Journal of Biological Macromolecules*, *133*, 1299–1310. https://doi.org/10.1016/j.ijbiomac.2019.03.231

El Harrad, L., Bourais, I., Mohammadi, H., & Amine, A. (2018). Recent advances in electrochemical biosensors based on enzyme inhibition for clinical and pharmaceutical applications. *Sensors*, *18*(1), 164. https://doi.org/10.3390/s18010164

Faramarzi-Aghgonbad, I., Amirkhani, L., Hedayatzadeh, S. M., & Derakhshanfard, F. (2021). Immobilization of pectinase enzyme on hydrophilic silica aerogel and its magnetic nanocomposite. *Iranian Journal of Chemistry and Chemical Engineering*, *41*(10), 3282–3292. https://doi.org/10.30492/ijcce.2021.532874.4811

Farooq, M. A., Ali, S., Hassan, A., Tahir, H. M., Mumtaz, S., & Mumtaz, S. (2021). Biosynthesis and industrial applications of α-amylase: a review. *Archives of Microbiology*, *203*, 1281–1292. https://doi.org/10.1007/s00203-020-02128-y

Feng, N., Zhang, H., Li, Y., Liu, Y., Xu, L., Wang, Y., Fei, X., & Tian, J. (2020). A novel catalytic material for hydrolyzing cow's milk allergenic proteins: Papain-$Cu_3(PO_4)_2 \cdot 3H_2O$-magnetic nanoflowers. *Food Chemistry*, *311*, 125911. https://doi.org/10.1016/j.foodchem.2019.125911

Fernandez Caresani, J. R., Dallegrave, A., & dos Santos, J. H. Z. (2020). Amylases immobilization by sol-gel entrapment: Application for starch hydrolysis. *Journal of Sol-Gel Science and Technology*, *94*(1), 229–240. https://doi.org/10.1007/s10971-019-05136-7

Gali, K. K., Soundararajan, N., Katiyar, V., & Sivaprakasam, S. (2021). Electrospun chitosan coated polylactic acid nanofiber: A novel immobilization matrix for α-amylase and its application in hydrolysis of cassava fibrous waste. *Journal of Materials Research and Technology*, *13*, 686–699. https://doi.org/10.1016/j.jmrt.2021.05.001

Ge, J., Lei, J., & Zare, R. N. (2012). Protein-inorganic hybrid nanoflowers. *Nature Nanotechnology*, *7*(7), 428–432. https://doi.org/10.1038/nnano.2012.80

Ghollasi, M. (2018). Electrospun polyethersulfone nanofibers: A novel matrix for alpha-amylase immobilization. *Journal of Applied Biotechnology Reports*, *5*(1), 19–25. https://doi.org/10.29252/jabr.01.01.04

Guan, H., Gong, D., Song, Y., Han, B., & Zhang, N. (2019). Biosensor composed of integrated glucose oxidase with liposome microreactors/chitosan nanocomposite for amperometric glucose sensing. *Colloids and Surfaces A: Physicochemical and Engineering Aspects*, *574*, 260–267. https://doi.org/10.1016/j.colsurfa.2019.04.076

Han, J., Feng, H., Wu, J., Li, Y., Zhou, Y., Wang, L., Luo, P., & Wang, Y. (2021). Construction of multienzyme co-immobilized hybrid nanoflowers for an efficient conversion of cellulose into glucose in a cascade reaction. *Journal of Agricultural and Food Chemistry*, *69*(28), 7910–7921. https://doi.org/10.1021/acs.jafc.1c02056

Hu, T. G., Cheng, J. H., Zhang, B. B., Lou, W. Y., & Zong, M. H. (2015). Immobilization of alkaline protease on amino-functionalized magnetic nanoparticles and its efficient use for preparation of oat polypeptides. *Industrial and Engineering Chemistry Research*, *54*(17), 4689–4698. https://doi.org/10.1021/ie504691j

Hwang, S. Y., Kim, H. K., Choo, J., Seong, G. H., Hien, T. B. D., & Lee, E. K. (2012). Effects of operating parameters on the efficiency of liposomal encapsulation of enzymes. *Colloids and Surfaces B: Biointerfaces*, *94*, 296–303. https://doi.org/10.1016/j.colsurfb.2012.02.008

Işik, C., Arabaci, G., Doğaç, Y. I., Deveci, İ., & Teke, M. (2019). Synthesis and characterization of electrospun PVA/Zn^{2+} metal composite nanofibers for lipase immobilization with effective thermal, pH stabilities and reusability. *Materials Science and Engineering: C*, *99*, 1226–1235. https://doi.org/10.1016/j.msec.2019.02.031

Ismail, A. R., & Baek, K. H. (2020). Lipase immobilization with support materials, preparation techniques, and applications: Present and future aspects. *International Journal of Biological Macromolecules*, *163*, 1624–1639. https://doi.org/10.1016/j.ijbiomac.2020.09.021

Jafarian, F., Bordbar, A. K., Razmjou, A., & Zare, A. (2020). The fabrication of a high performance enzymatic hybrid membrane reactor (EHMR) containing immobilized *Candida rugosa* lipase (CRL) onto graphene oxide nanosheets-blended polyethersulfone membrane. *Journal of Membrane Science*, *613*, 118435. https://doi.org/10.1016/j.memsci.2020.118435

Jahadi, M., Khosravi-Darani, K., Ehsani, M. R., Pimentel, T. C., da Cruz, A. G., & Mozafari, M. R. (2020). Accelerating ripening of Iranian white brined cheesesusing liposome-encapsulated and free proteinases. *Biointerface Research in Applied Chemistry*, *10*(1), 4966–4971. https://doi.org/10.33263/BRIAC101.966971

Jahadi, M., Khosravi-Darani, K., Ehsani, M. R., Mozafari, M. R., Saboury, A. A., & Pourhosseini, P. S. (2015). The encapsulation of flavourzyme in nanoliposome by heating method. *Journal of Food Science and Technology*, *52*, 2063–2072. https://doi.org/10.1007/s13197-013-1243-0

Jha, I., & Venkatesu, P. (2016). Deciphering the interactions of bromelain with carbon nanotubes: Role of protein as well as carboxylated multiwalled carbon nanotubes in a complexation mechanism. *Journal of Physical Chemistry C*, *120*(28), 15436–15445. https://doi.org/10.1021/acs.jpcc.6b03547

Ji, S., Liu, W., Su, S., Gan, C., & Jia, C. (2021). Chitosan derivative functionalized carbon nanotubes as carriers for enzyme immobilization to improve synthetic efficiency of ethyl caproate. *LWT*, *149*, 111897. https://doi.org/10.1016/j.lwt.2021.111897

Kamaci, U. D., & Peksel, A. (2020). Fabrication of PVA-chitosan-based nanofibers for phytase immobilization to enhance enzymatic activity. *International Journal of Biological Macromolecules*, *164*, 3315–3322. https://doi.org/10.1016/j.ijbiomac.2020.08.226

Karuppiah, C., Palanisamy, S., Chen, S.M., Veeramani, V., & Periakaruppan, P. (2014). Direct electrochemistry of glucose oxidase and sensing glucose using a screen-printed carbon electrode modified with graphite nanosheets and zinc oxide nanoparticles. *Microchimica Acta*, *181*, 1843–1850. https://doi.org/10.1007/s00604-014-1256-z

Kato, K., Kawachi, Y., & Nakamura, H. (2014). Silica–enzyme–ionic liquid composites for improved enzymatic activity. *Journal of Asian Ceramic Societies*, *2*(1), 33–40. https://doi.org/10.1016/j.jascer.2013.12.004

Ke, C., Fan, Y., Chen, Y., Xu, L., & Yan, Y. (2016). A new lipase-inorganic hybrid nanoflower with enhanced enzyme activity. *RSC Advances*, *6*(23), 19413–19416. https://doi.org/10.1039/C6RA01564F

Khan, M., Husain, Q., & Bushra, R. (2017). Immobilization of β-galactosidase on surface modified cobalt/multiwalled carbon nanotube nanocomposite improves enzyme stability and resistance to inhibitor. *International Journal of Biological Macromolecules*, *105*, 693–701. https://doi.org/10.1016/j.ijbiomac.2017.07.088

Khan, M. R. (2021). Immobilized enzymes: A comprehensive review. *Bulletin of the National Research Centre*, *45*(1), 1–13. https://doi.org/10.1186/s42269-021-00649-0

Kharazmi, S., Taheri-Kafrani, A., & Soozanipour, A. (2020). Efficient immobilization of pectinase on trichlorotriazine-functionalized polyethylene glycol-grafted magnetic nanoparticles: A stable and robust nanobiocatalyst for fruit juice clarification. *Food Chemistry*, *325*. https://doi.org/10.1016/j.foodchem.2020.126890

Kharazmi, S., Taheri-Kafrani, A., Soozanipour, A., Nasrollahzadeh, M., & Varma, R. S. (2020). Xylanase immobilization onto trichlorotriazine-functionalized polyethylene glycol grafted magnetic nanoparticles: A thermostable and robust nanobiocatalyst for fruit juice clarification. *International Journal of Biological Macromolecules*, *163*, 402–413. https://doi.org/10.1016/j.ijbiomac.2020.06.273

Kheadr, E. E., Vuillemard, J. C., & El-Deeb, S. A. (2002). Acceleration of cheddar cheese lipolysis by using liposome-entrapped lipases. *Journal of Food Science*, *67*(2), 485–492. https://doi.org/10.1111/j.1365-2621.2002.tb10624.x

Khoshnevisan, K., Poorakbar, E., Baharifar, H., & Barkhi, M. (2019). Recent advances of cellulase immobilization onto magnetic nanoparticles: An update review. *Magnetochemistry*, *5*(2), 36. https://doi.org/10.3390/magnetochemistry5020036

Ladole, M. R., Nair, R. R., Bhutada, Y. D., Amritkar, V. D., & Pandit, A. B. (2018). Synergistic effect of ultrasonication and co-immobilized enzymes on tomato peels for lycopene extraction. *Ultrasonics Sonochemistry*, *48*, 453–462. https://doi.org/10.1016/j.ultsonch.2018.06.013

Ladole, M. R., Pokale, P. B., Varude, V. R., Belokar, P. G., & Pandit, A. B. (2021). One pot clarification and debittering of grapefruit juice using co-immobilized enzymes@chitosanMNPs. *International Journal of Biological Macromolecules*, *167*, 1297–1307. https://doi.org/10.1016/j.ijbiomac.2020.11.084

Li, C., Yoshimoto, M., Fukunaga, K., & Nakao, K. (2007). Characterization and immobilization of liposome-bound cellulase for hydrolysis of insoluble cellulose. *Bioresource Technology*, *98*(7), 1366–1372. https://doi.org/10.1016/j.biortech.2006.05.028

Li, C., Zhao, J., Zhang, Z., Jiang, Y., Bilal, M., Jiang, Y., Jia, S., & Cui, J. (2020). Self-assembly of activated lipase hybrid nanoflowers with superior activity and enhanced stability. *Biochemical Engineering Journal*, *158*, 107582. https://doi.org/10.1016/j.bej.2020.107582

Li, D., Wang, Q., Huang, F., & Wei, Q. (2018). Electrospun nanofibers for enzyme immobilization. In B. Ding, X. Wang, & J. Yu (Eds.), *Electrospinning: Nanofabrication and Applications* (pp. 765–781). Elsevier. https:// doi.org/ 10.1016/ B978-0-323-51270-1.00026-1

Li, L. J., Xia, W. J., Ma, G. P., Chen, Y. L., & Ma, Y. Y. (2019). A study on the enzymatic properties and reuse of cellulase immobilized with carbon nanotubes and sodium alginate. *AMB Express*, *9*(1), 1–8. https://doi.org/10.1186/s13568-019-0835-0

Li, M., Hanford, M. J., Kim, J. W., & Peeples, T. L. (2007). Amyloglucosidase enzymatic reactivity inside lipid vesicles. *Journal of Biological Engineering*, *1*, 1–9. https://doi.org/10.1186/1754-1611-1-4

Li, S., Zhang, S., Tao, Y., Chen, Y., Yang, Y., Liang, X., & Li, Q. (2023). Construction of β-galactosidase-inorganic hybrid nanoflowers through biomimetic mineralization for lactose degradation. *Biochemical Engineering Journal*, *197*, 108980. https://doi.org/10.1016/j.bej.2023.108980

Liu, S., & Cai, C. (2007). Immobilization and characterization of alcohol dehydrogenase on single-walled carbon nanotubes and its application in sensing ethanol. *Journal of Electroanalytical Chemistry*, *602*(1), 103–114. https://doi.org/10.1016/j.jelechem.2006.12.003

Liu, Y., Li, Q., Feng, Y. Y., Ji, G. S., Li, T. C., Tu, J., & Gu, X. D. (2014). Immobilisation of acid pectinase on graphene oxide nanosheets. *Chemical Papers*, *68*(6), 732–738. https://doi.org/10.2478/s11696-013-0510-x

Liu, Y., Shao, X., Kong, D., Li, G., & Li, Q. (2021). Immobilization of thermophilic lipase in inorganic hybrid nanoflower through biomimetic mineralization. *Colloids and Surfaces B: Biointerfaces*, *197*, 111450. https://doi.org/10.1016/j.colsurfb.2020.111450

Lopes, N. A., Barreto Pinilla, C. M., & Brandelli, A. (2019). Antimicrobial activity of lysozyme-nisin co-encapsulated in liposomes coated with polysaccharides. *Food Hydrocolloids*, *93*, 1–9. https://doi.org/10.1016/j.foodhyd.2019.02.009

Loredo-Alejos, J. M., Lucio-Porto, R., Pavón, L. L., & Moreno-Cortez, I. E. (2022). Pepsin immobilization by electrospinning of poly (vinyl alcohol) nanofibers. *Journal of Applied Polymer Science*, *139*(9), 51700. https://doi.org/10.1002/app.51700

MacArio, A., Verri, F., Diaz, U., Corma, A., & Giordano, G. (2013). Pure silica nanoparticles for liposome/lipase system encapsulation: Application in biodiesel production. *Catalysis Today*, *204*, 148–155. https://doi.org/10.1016/j.cattod.2012.07.014

Malam, Y., Loizidou, M., & Seifalian, A.M. (2009). Liposomes and nanoparticles: Nanosized vehicles for drug delivery in cancer. *Trends in Pharmacological Sciences*, *30*(11), 592–599. https://doi.org/10.1016/j.tips.2009.08.004

Markiton, M., Boncel, S., Janas, D., & Chrobok, A. (2017). Highly active nanobiocatalyst from lipase noncovalently immobilized on multiwalled carbon nanotubes for Baeyer-Villiger synthesis of lactones. *ACS Sustainable Chemistry and Engineering*, *5*(2), 1685–1691. https://doi.org/10.1021/acssuschemeng.6b02433

Meesters, G. M. H. (2010). Encapsulation of enzymes and peptides. In N. Zuidam & V. Nedovic (Eds.), *Encapsulation Technologies for Active Food Ingredients and Food Processing* (pp. 253–268). Springer. https://doi.org/10.1007/978-1-4419-1008-0_9

Mihono, K., Ohtsu, T., Ohtani, M., Yoshimoto, M., & Kamimura, A. (2016). Modulation of cellulase activity by charged lipid bilayers with different acyl chain properties for efficient hydrolysis of ionic liquid-pretreated cellulose. *Colloids and Surfaces B: Biointerfaces*, *146*, 198–203. https://doi.org/10.1016/j.colsurfb.2016.06.005

Misson, M., Jin, B., Dai, S., & Zhang, H. (2020). Interfacial biocatalytic performance of nanofiber-supported β-galactosidase for production of galacto-oligosaccharides. *Catalysts*, *10*(1), 81. https://doi.org/10.3390/catal10010081

Misson, M., Jin, B., & Zhang, H. (2016). Immobilized β-galactosidase on functionalized nanoparticles and nanofibers: A comparative study. *International Proceedings of Chemical, Biological and Environmental Engineering*, *90*, 7763. https://doi.org/10.7763/IPCBEE

Misson, M., Saallah, S., & Zhang, H. (2020). Nanofiber-immobilized β-galactosidase for dairy waste conversion into galacto-oligosaccharides. In A.Yaser (Ed.), *Advances in Waste Processing Technology* (pp. 37–48). Springer. https://doi.org/10.1007/978-981-15-4821-5_3

Mitra, T., Sailakshmi, G., & Gnanamani, A. (2014). Could glutaric acid (GA) replace glutaraldehyde in the preparation of biocompatible biopolymers with high mechanical and thermal properties?*Journal of Chemical Sciences*, *126*, 127–140. https://doi.org/10.1007/s12039-013-0543-2

Mohamad, N. R., Marzuki, N. H. C., Buang, N. A., Huyop, F., & Wahab, R. A. (2015). An overview of technologies for immobilization of enzymes and surface analysis techniques for immobilized enzymes. *Biotechnology & Biotechnological Equipment*, *29*(2), 205–220. https://doi.org/10.1080/13102818.2015.1008192

Mohammadi, M., Ashjari, M., Garmroodi, M., Yousefi, M., & Karkhane, A. A. (2016). The use of isocyanide-based multicomponent reaction for covalent immobilization of *Rhizomucor miehei* lipase on multiwall carbon nanotubes and graphene nanosheets. *RSC Advances*, *6*(76), 72275–72285. https://doi.org/10.1039/c6ra14142k

Mohiuddin, M., Arbain, D., Shafiqul Islam, A. K. M., Rahman, M., Ahmad, M. S., & Ahmad, M. N. (2014). Covalent immobilization of α-glucosidase enzyme onto amine functionalized multi-walled carbon nanotubes. *Current Nanoscience*, *10*(5), 730–735. https://doi.org/10.2174/157341371066614011 6213526

Moreno-Cortez, I. E., Romero-García, J., González-González, V., García-Gutierrez, D. I., Garza-Navarro, M. A., & Cruz-Silva, R. (2015). Encapsulation and immobilization of papain in electrospun nanofibrous membranes of PVA cross-linked with glutaraldehyde vapor. *Materials Science and Engineering C*, *52*, 306–314. https://doi.org/10.1016/j.msec.2015.03.049

Mostafavi, M., Mahmoodzadeh, K., Habibi, Z., Yousefi, M., Brask, J., & Mohammadi, M. (2023). Immobilization of *Bacillus amyloliquefaciens* protease "Neutrase" as hybrid enzyme inorganic nanoflower particles: A new biocatalyst for aldol-type and multicomponent reactions. *International Journal of Biological Macromolecules*, *230*, 123140. https://doi.org/10.1016/j.ijbiomac.2023.123140

Movahedpour, A., Ahmadi, N., Ghalamfarsa, F., Ghesmati, Z., Khalifeh, M., Maleksabet, A., Shabaninejad, Z., Taheri-Anganeh, M., & Savardashtaki, A. (2022). β-Galactosidase: From its source and applications to its recombinant form. *Biotechnology and Applied Biochemistry*, *69*(2), 612–628. https://doi.org/10.1002/bab.2137

Mubarak, N. M., Wong, J. R., Tan, K. W., Sahu, J. N., Abdullah, E. C., Jayakumar, N. S., & Ganesan, P. (2014). Immobilization of cellulase enzyme on functionalized multiwall carbon nanotubes. *Journal of Molecular Catalysis B: Enzymatic*, *107*, 124–131. https://doi.org/10.1016/j.molcatb.2014.06.002

Mujahid, A., Lieberzeit, P. A., & Dickert, F. L. (2010). Chemical sensors based on molecularly imprinted sol-gel materials. *Materials*, *3*(4), 2196–2217. https://doi.org/10.3390/ma3042196

Mukhopadhyay, A., Bhattacharyya, T., Dasgupta, A. K., & Chakrabarti, K. (2015). Nanotechnology based activation-immobilization of psychrophilic pectate lyase: A novel approach towards enzyme stabilization and enhanced activity. *Journal of Molecular Catalysis B: Enzymatic*, *119*, 54–63. https://doi.org/10.1016/j.molcatb.2015.05.017

Naveed, M., Nadeem, F., Mehmood, T., Bilal, M., Anwar, Z., & Amjad, F. (2021). Protease—A versatile and ecofriendly biocatalyst with multi-industrial applications: An updated review. *Catalysis Letters*, *151*(2), 307–323. https://doi.org/10.1007/s10562-020-03316-7

Netto, C. G. C. M., Toma, H. E., & Andrade, L. H. (2013). Superparamagnetic nanoparticles as versatile carriers and supporting materials for enzymes. *Journal of Molecular Catalysis B: Enzymatic*, *85*, 71–92. https://doi.org/10.1016/j.molcatb.2012.08.010

Nie, H., Wang, H., Cao, A., Shi, Z., Yang, S. T., Yuan, Y., & Liu, Y. (2011). Diameter-selective dispersion of double-walled carbon nanotubes by lysozyme. *Nanoscale*, *3*(3), 970–973. https://doi.org/10.1039/c0nr00831a

Niu, X., Zhu, L., Xi, L., Guo, L., & Wang, H. (2020). An antimicrobial agent prepared by *N*-succinyl chitosan immobilized lysozyme and its application in strawberry preservation. *Food Control*, *108*, 106829. https://doi.org/10.1016/j.foodcont.2019.106829

Noor, M. M., Goswami, J., & Davis, V. A. (2020). Comparison of attachment and antibacterial activity of covalent and noncovalent lysozyme-functionalized single-walled carbon nanotubes. *ACS Omega*, *5*(5), 2254–2259. https://doi.org/10.1021/acsomega.9b03387

Nouri, M., & Khodaiyan, F. (2020). Magnetic biocatalysts of pectinase: Synthesis by macromolecular cross-linker for application in apple juice clarification. *Food Technology and Biotechnology*, *58*(4), 391–401. https://doi.org/10.17113/ftb.58.04.20.6737

Pasarin, D., Ghizdareanu, A. I., Enascuta, C. E., Matei, C. B., Bilbie, C., Paraschiv-Palada, L., & Veres, P. A. (2023). Coating materials to increase the stability of liposomes. *Polymers*, *15*(3), 782. https://doi.org/10.3390/polym15030782

Patil, S. S., & Rathod, V. K. (2022). Combined effect of enzyme co-immobilized magnetic nanoparticles (MNPs) and ultrasound for effective extraction and purification of curcuminoids from *Curcuma longa*. *Industrial Crops and Products*, *177*, 114385. https://doi.org/10.1016/j.indcrop.2021.114385

Pei, J., Huang, Y., Yang, Y., Yuan, H., Liu, X., & Ni, C. (2018). A novel layered anchoring structure immobilized cellulase via covalent binding of cellulase on MNPs anchored by LDHs. *Journal of Inorganic and Organometallic Polymers and Materials*, *28*, 1624–1635. https://doi.org/10.1007/s10904-018-0838-3

Pierre, A. C. (2004). The sol-gel encapsulation of enzymes. *Biocatalysis and Biotransformation*, *22*(3), 145–170. https://doi.org/10.1080/10242420412331283314

Pinto, S. C., Rodrigues, A. R., Saraiva, J. A., & Lopes-da-Silva, J. A. (2015). Catalytic activity of trypsin entrapped in electrospun poly (ϵ-caprolactone) nanofibers. *Enzyme and Microbial Technology*, *79*, 8–18. https://doi.org/10.1016/j.enzmictec.2015.07.002

Poorakbar, E., Shafiee, A., Saboury, A. A., Rad, B. L., Khoshnevisan, K., Ma'mani, L., Derakhshankhah, H., Ganjali, M. R., & Hosseini, M. (2018). Synthesis of magnetic gold mesoporous silica nanoparticles core shell for cellulase enzyme immobilization: Improvement of enzymatic activity and thermal stability. *Process Biochemistry*, *71*, 92–100. https://doi.org/10.1016/j.procbio.2018.05.012

Pourahmad, R., Ein, M., Afjeh, A., Akbari-Adergani, B., & Azin, M. (2019). Use of glucose oxidase immobilized on magnetic chitosan nanoparticles in probiotic drinking yogurt. *Food Science of Animal Resources*, *39*(1), 73. https://doi.org/10.5851/kosfa.2019.e5

Rajnish, K. N., Samuel, M. S., John J, A., Datta, S., Chandrasekar, N., Balaji, R., Jose, S., & Selvarajan, E. (2021). Immobilization of cellulase enzymes on nano and micro-materials for breakdown of cellulose for biofuel production—a narrative review. *International Journal of Biological Macromolecules*, *182*, 1793–1802. https://doi.org/10.1016/J.IJBIOMAC.2021.05.176

Ramos, O. S., & Malcata, F. X. (2011). 3.48—Food-grade enzymes. In M. Moo-Young (Ed.), *Comprehensive Biotechnology* (2nd ed., pp. 555–569). Academic Press. https://doi.org/10.1016/B978-0-08-088504-9.00213-0

Ranjbari, N., Razzaghi, M., Fernandez-Lafuente, R., Shojaei, F., Satari, M., & Homaei, A. (2019). Improved features of a highly stable protease from *Penaeus vannamei* by immobilization on glutaraldehyde activated graphene oxide nanosheets. *International Journal of Biological Macromolecules*, *130*, 564–572. https://doi.org/10.1016/j.ijbiomac.2019.02.163

Rasheed, T., Nabeel, F., Adeel, M., Rizwan, K., Bilal, M., & Iqbal, H. M. N. (2019). Carbon nanotubes-based cues: A pathway to future sensing and detection of hazardous pollutants. *Journal of Molecular Liquids*, *292*, 111425. https://doi.org/10.1016/j.molliq.2019.111425

Rashid, S. S., Mustafa, A. H., Ab Rahim, M. H., & Gunes, B. (2022). Magnetic nickel nanostructure as cellulase immobilization surface for the hydrolysis of lignocellulosic biomass. *International Journal of Biological Macromolecules*, *209*, 1048–1053. https://doi.org/10.1016/j.ijbiomac.2022.04.072

Reza Mozafari, M., Johnson, C., Hatziantoniou, S., & Demetzos, C. (2008). Nanoliposomes and their applications in food nanotechnology. *Journal of Liposome Research*, *18*(4), 309–327. https://doi.org/10.1080/08982100802465941

Ribeiro, E. S., de Farias, B. S., Sant'Anna Cadaval Junior, T.R., de Almeida Pinto, L. A., & Diaz, P. S. (2021). Chitosan-based nanofibers for enzyme immobilization. *International Journal of Biological Macromolecules*, *183*, 1959–1970. https://doi.org/10.1016/j.ijbiomac.2021.05.214

Rodríguez-Nogales, J. M., & López, A. D. (2006). A novel approach to develop β-galactosidase entrapped in liposomes in order to prevent an immediate hydrolysis of lactose in milk. *International Dairy Journal*, *16*(4), 354–360. https://doi.org/10.1016/j.idairyj.2005.05.007

Rojas-Mercado, A. S., Moreno-Cortez, I. E., Lucio-Porto, R., & Pavón, L. L. (2018). Encapsulation and immobilization of ficin extract in electrospun polymeric nanofibers. *International Journal of Biological Macromolecules*, *118*, 2287–2295. https://doi.org/10.1016/j.ijbiomac.2018.07.113

Rother, C., & Nidetzky, B. (2009). Enzyme immobilization by microencapsulation: Methods, materials, and technological applications. *Encyclopedia of Industrial Biotechnology: Bioprocess, Bioseparation, and Cell Technology*, 1–21. https://doi.org/10.1002/9780470054581.eib275

Sanchez, J. M., & Perillo, M. A. (2000). α-Amylase kinetic parameters modulation by lecithin vesicles: Binding versus entrapment. *Colloids and Surfaces B: Biointerfaces*, *18*(1), 31–40. https://doi.org/10.1016/S0927-7765(99)00128-9

Sarathi, M., Doraiswamy, N., & Pennathur, G. (2019). Enhanced stability of immobilized keratinolytic protease on electrospun nanofibers. *Preparative Biochemistry and Biotechnology*, *49*(7), 695–703. https://doi.org/10.1080/10826068.2019.1605524

Sarkar, S., Guibal, E., Quignard, F., & SenGupta, A. K. (2012). Polymer-supported metals and metal oxide nanoparticles: Synthesis, characterization, and applications. *Journal of Nanoparticle Research*, *14*, 1–24. https://doi.org/10.1007/s11051-011-0715-2

Sass, A. C., & Jördening, H. J. (2020). Immobilization of β-galactosidase from *Aspergillus oryzae* on electrospun gelatin nanofiber mats for the production of galactooligosaccharides. *Applied Biochemistry and Biotechnology*, *191*, 1155–1170. https://doi.org/10.1007/s12010-020-03252-7

Ramakrishna, T.R.B, Nalder, T.D., Yang, W., Marshall, S.N., Barrow, C.J. (2018). Controlling enzyme function through immobilisation on graphene, graphene derivatives and other two dimensional nanomaterials. *Journal of Materials Chemistry B*, *6*(20), 3200–3218. DOI: 10.1039/C8TB00313K

Shalini, P., Deepanraj, B., Vijayalakshmi, S., & Ranjitha, J. (2023). Synthesis and characterisation of lipase immobilised magnetic nanoparticles and its role as a catalyst in biodiesel production. *Materials Today: Proceedings*, *80*, 2725–2730. https://doi.org/10.1016/j.matpr.2021.07.027

Shareghi, B., Farhadian, S., Zamani, N., Salavati-Niasari, M., Moshtaghi, H., & Gholamrezaei, S. (2015). Investigation the activity and stability of lysozyme on presence of magnetic nanoparticles. *Journal of Industrial and Engineering Chemistry*, *21*, 862–867. https://doi.org/10.1016/j.jiec.2014.04.024

Shariat, S. Z. A. S., Borzouee, F., Mofid, M. R., & Varshosaz, J. (2018). Immobilization of lactoperoxidase on graphene oxide nanosheets with improved activity and stability. *Biotechnology Letters*, *40*, 1343–1353. https://doi.org/10.1007/s10529-018-2583-7

Sheikh, I. A., Yasir, M., Khan, I., Khan, S. B., Azum, N., Jiffri, E. H., Kamal, M. A., Ashraf, G. M., & Beg, M. A. (2018). Lactoperoxidase immobilization on silver nanoparticles enhances its antimicrobial activity. *Journal of Dairy Research*, *85*(4), 460–464. https://doi.org/10.1017/S0022029918000730

Siddiqui, I., Husain, Q., & Azam, A. (2020). Exploring the antioxidant effects of peptides from almond proteins using PAni-Ag-GONC conjugated trypsin by improving enzyme stability & applications. *International Journal of Biological Macromolecules*, *158*, 150–158. https://doi.org/10.1016/j.ijbiomac.2020.04.188

Singh, K., Srivastava, G., Talat, M., Srivastava, O. N., & Kayastha, A. M. (2015). α-Amylase immobilization onto functionalized graphene nanosheets as scaffolds: Its characterization, kinetics and potential applications in starch based industries. *Biochemistry and Biophysics Reports*, *3*, 18–25. https://doi.org/10.1016/J.BBREP.2015.07.002

Singh, R., Upadhyay, S. K., Singh, M., Sharma, I., Sharma, P., Kamboj, P., Saini, A., Voraha, R., Sharma, A. K., Upadhyay, T. K., & Khan, F. (2020). Chitin, chitinases and chitin derivatives in biopharmaceutical, agricultural and environmental perspective. *Biointerface Research in Applied Chemistry*, *11*(3), 9985–10005. https://doi.org/10.33263/BRIAC113.998510005

Singh, R. S., Chauhan, K., & Kennedy, J. F. (2019). Fructose production from inulin using fungal inulinase immobilized on 3-aminopropyl-triethoxysilane functionalized multiwalled carbon nanotubes. *International Journal of Biological Macromolecules*, *125*, 41–52. https://doi.org/10.1016/j.ijbiomac.2018.11.281

Singh, V., Rakshit, K., Rathee, S., Angmo, S., Kaushal, S., Garg, P., Chung, J. H., Sandhir, R., Sangwan, R. S., & Singhal, N. (2016). Metallic/bimetallic magnetic nanoparticle functionalization for immobilization of α-amylase for enhanced reusability in bio-catalytic processes. *Bioresource Technology*, *214*, 528–533. https://doi.org/10.1016/j.biortech.2016.05.002

Sirisha, V. L., Jain, A., & Jain, A. (2016). Enzyme immobilization: An overview on methods, support material, and applications of immobilized enzymes. *Advances in Food and Nutrition Research*, *79*, 179–211. https://doi.org/10.1016/bs.afnr.2016.07.004

Skoronski, E., Souza, D. H., Ely, C., Broilo, F., Fernandes, M., Fúrigo, A., Jr., & Ghislandi, M. G. (2017). Immobilization of laccase from *Aspergillus oryzae* on graphene nanosheets. *International Journal of Biological Macromolecules*, *99*, 121–127. https://doi.org/10.1016/j.ijbiomac.2017.02.076

Sobhan, A., Lee, J., Park, M. K., & Oh, J. H. (2019). Rapid detection of *Yersinia enterocolitica* using a single-walled carbon nanotube-based biosensor for Kimchi product. *LWT*, *108*, 48–54. https://doi.org/10.1016/j.lwt.2019.03.037

Song, X., Wei, L., Chen, A., & Shao, Y. (2015). Poly(L-lactide) nanofibers containing trypsin for gelatin digestion. *Fibers and Polymers*, *16*(4), 867–874. https://doi.org/10.1007/s12221-015-0867-2

Soni, S., Dwivedee, B. P., & Banerjee, U. C. (2020). Tailoring a stable and recyclable nanobiocatalyst by immobilization of surfactant treated *Burkholderia cepacia* lipase on polyaniline nanofibers for biocatalytic application. *International Journal of Biological Macromolecules*, *161*, 573–586. https://doi.org/10.1016/j.ijbiomac.2020.06.002

Su, R., Zhang, W., Zhu, M., Xu, S., Yang, M., & Li, D. (2013). Alkaline protease immobilized on graphene oxide: Highly efficient catalysts for the proteolysis of waste-activated sludge. *Polish Journal of Environmental Studies*, *22*(3), 885–891.

Tahsiri, Z., Niakousari, M., & Niakowsari, A. (2023). Magnetic graphene oxide, a suitable support in ficin immobilization. *Heliyon*, *9*(6):E16971. https://doi.org/10.1016/j.heliyon.2023.e16971

Talens-Perales, D., Fabra, M. J., Martínez-Argente, L., Marín-Navarro, J., & Polaina, J. (2020). Recyclable thermophilic hybrid protein-inorganic nanoflowers for the hydrolysis of milk lactose. *International Journal of Biological Macromolecules*, *151*, 602–608. https://doi.org/10.1016/j.ijbiomac.2020.02.115

Tavernini, L., Romero, O., Aburto, C., López-gallego, F., Illanes, A., & Wilson, L. (2021). Development of a hybrid bioinorganic nanobiocatalyst: Remarkable impact of the immobilization conditions on activity and stability of β-galactosidase. *Molecules*, *26*(14), 4152. https://doi.org/10.3390/molecules26144152

Temoçin, Z. (2021). Designing of a stable and selective glucose biosensor by glucose oxidase immobilization on glassy carbon electrode sensitive to H_2O_2 via nanofiber interface. *Journal of Applied Electrochemistry*, *51*, 283–293. https://doi.org/10.1007/s10800-020-01502-4

Tian, K., Liu, H., Dong, Y., Chu, X., & Wang, S. (2019). Amperometric detection of glucose based on immobilizing glucose oxidase on g-C_3N_4 nanosheets. *Physicochemical and Engineering Aspects*, *581*, 123808. https://doi.org/10.1016/j.colsurfa.2019.123808

Ustok, F. I., Tari, C., & Harsa, S. (2010). Biochemical and thermal properties of β-galactosidase enzymes produced by artisanal yoghurt cultures. *Food Chemistry*, *119*(3), 1114–1120. https://doi.org/10.1016/j.foodchem.2009.08.022

Vargas, A. Y., Romanelli, G. P., & Martinez, J. J. (2021). Nanopartículas magnéticas funcionalizadas y modificadas con entrecruzamiento para mejorar la inmovilización de la invertasa. *Ciencia En Desarrollo*, *12*(1), 69–77. https://doi.org/10.19053/01217488.v12.n1.2021.12818

Wang, C., Zhou, X., Wang, G., Wang, D., Fang, C., Ru, Y., Hu, J., & Xie, L. (2022). Preparation of lysozyme/carbon nanotube hybrids and their interactions at the nano-bio interface. *Progress in Organic Coatings*, *163*, 106659. https://doi.org/10.1016/J.PORGCOAT.2021.106659

Wang, F., Chen, X., Wang, Y., Li, X., Wan, M., Zhang, G., Leng, F., & Zhang, H. (2022). Insights into the structures, inhibitors, and improvement strategies of glucose oxidase. *International Journal of Molecular Sciences*, *23*(17), 9841. https://doi.org/10.3390/ijms23179841

Wang, L., Wei, L., Chen, Y., & Jiang, R. (2010). Specific and reversible immobilization of NADH oxidase on functionalized carbon nanotubes. *Journal of Biotechnology*, *150*(1), 57–63. https://doi.org/10.1016/j.jbiotec.2010.07.005

Wang, L. B., Wang, Y. C., He, R., Zhuang, A., Wang, X., Zeng, J., & Hou, J. G. (2013). A new nanobiocatalytic system based on allosteric effect with dramatically enhanced enzymatic performance. *Journal of the American Chemical Society*, *135*(4), 1272–1275. https://doi.org/10.1021/ja3120136

Wang, P., Zhang, C., Zou, Y., Li, Y., & Zhang, H. (2021a). Immobilization of lysozyme on layer-by-layer self-assembled electrospun films: characterization and antibacterial activity in milk. *Food Hydrocolloids*, *113*, 106468. https://doi.org/10.1016/j.foodhyd.2020.106468

Wang, P., Zhang, C., Zou, Y., Li, Y., & Zhang, H. (2021b). Immobilization of lysozyme on layer-by-layer self-assembled electrospun nanofibers treated by post-covalent crosslinking. *Food Hydrocolloids*, *121*, 106999. https://doi.org/10.1016/j.foodhyd.2021.106999

Wang, Q., Wang, X., Wang, X., & Ma, H. (2008). Glucoamylase production from food waste by *Aspergillus niger* under submerged fermentation. *Process Biochemistry*, *43*(3), 280–286. https://doi.org/10.1016/J.PROCBIO.2007.12.010

Wang, Y., Li, S., Jin, M., Han, Q., Liu, S., Chen, X., & Han, Y. (2020). Enhancing the thermo-stability and antibacterium activity of lysozyme by immobilization on chitosan nanoparticles. *International Journal of Molecular Sciences*, *21*(5), 1635. https://doi.org/10.3390/ijms21051635

Wang, Z., Liu, P., & Fang, Z. (2022). Trypsin/Zn_3 $(PO_4)_2$ hybrid nanoflowers: Controlled synthesis and excellent performance as an immobilized enzyme. *International Journal of Molecular Sciences*, *23*(19), 11853. https://doi.org/10.3390/ijms231911853

Wu, X., Hou, M., & Ge, J. (2015). Metal-organic frameworks and inorganic nanoflowers: A type of emerging inorganic crystal nanocarrier for enzyme immobilization. *Catalysis Science & Technology*, *5*(12), 5077–5085. https://doi.org/10.1039/c5cy01181g

Xie, F., Yang, M., Jiang, M., Huang, X. J., Liu, W. Q., & Xie, P. H. (2019). Carbon-based nanomaterials—a promising electrochemical sensor toward persistent toxic substance. *TrAC—Trends in Analytical Chemistry*, *119*, 115624. https://doi.org/10.1016/j.trac.2019.115624

Yassin, M. A., Gad, A. A. M., Ghanem, A. F., & Abdel Rehim, M. H. (2019). Green synthesis of cellulose nanofibers using immobilized cellulase. *Carbohydrate Polymers*, *205*, 255–260. https://doi.org/10.1016/j.carbpol.2018.10.040

Yin, Y., Xiao, Y., Lin, G., Xiao, Q., Lin, Z., & Cai, Z. (2015). Enzyme-inorganic hybrid nanoflower based immobilized enzyme reactor with enhanced enzymatic activity. *Journal of Materials Chemistry B*, *3*(11), 2295–2300. https://doi.org/10.1039/c4tb01697a

Zhang, B., Li, P., Zhang, H., Fan, L., Wang, H., Li, X., Tian, L., Ali, N., Ali, Z., & Zhang, Q. (2016). Papain/$Zn_3(PO_4)_2$ hybrid nanoflower: Preparation, characterization and its enhanced catalytic activity as an immobilized enzyme. *RSC Advances*, *6*(52), 46702–46710. https://doi.org/10.1039/c6ra05308d

Zhang, J., Zhang, J., Sun, Q., Ye, X., Ma, X., & Wang, J. (2022). Sol-gel routes toward ceramic nanofibers for high-performance thermal management. *Chemistry*, *4*(4), 1475–1497. https://doi.org/10.3390/chemistry4040098

Zhao, J. F., Lin, J. P., Yang, L. R., & Wu, M. B. (2019). Enhanced performance of *Rhizopus oryzae*lipase by reasonable immobilization on magnetic nanoparticles and its application in synthesis 1, 3-diacyglycerol. *Applied Biochemistry and Biotechnology*, *188*(3), 677–689. https://doi.org/10.1007/s12010-018-02947-2

3 Nanotechnology for Encapsulation and Delivery of Functional Food Ingredients

Sovan Samanta, Jhimli Banerjee, Rubai Ahmed, Swarnali Das, Sandeep Kumar Dash, Avneet Kaur, and Sukhvinder Singh Purewal

3.1 INTRODUCTION

A wide range of foods can be classified as "functional foods" if they have a consistent influence beyond basic nutrition, as well as exert specific physiological effects to encourage health status and reduce the possibility ofchronic diseases (Su et al., 2022). Regular consumption of foods that providefunctional benefits can help manage or lower the risk of a variety of medical conditions, such as cancer, gastrointestinal health, and cardiovascular disease. "Bioactive compounds" are the dietary ingredients that also have positive health effects, but they are often found in trace amounts in some foods (Singh, 2016). Natural functional foods like some particular fruits, vegetables, and cereals that are high in bioactive components such as phenols, flavonoids, terpenoids, omega-3, polyunsaturated fatty acids, etc. make the food more physiologically beneficial by having anti-inflammatory, anti-cancer, and antioxidant qualities, which lowers the risk of non-communicable diseases (Adefegha, 2018). It is not always practicable to consume enough of these compounds by eating a lot of different fruits, vegetables, legumes, oils, and nuts which contain bioactive compounds in high amounts. Due to this reason, food industry is particularly interested in "fortifying" common meals like bread, milk, and beverages with bioactive molecules to provide necessary amount of these functional ingredients to consumers (Singh, 2016). However, because of the lower stability and solubility, decreased permeability, poor bioavailability, prolonged retention time in the digestive system of those components along with targeted delivery and controlled release at that site continue to be difficult. To overcome those problems, nanotechnologyis one of the most desirable and innovativeways to revolutionize the food industry (Livney, 2015; Nile et al., 2020). An enriched core of functional foods, bioactive compounds, nutraceuticals, such as vitamins, antioxidants or probiotics, bioactive peptides, and shell materials, 1–200 nm in diameter with greater surface area, are known as nano-delivery systemsin the form of nano-micelles, nanoliposomes, nano-emulsions, nano-capsules, and nano-nutrients (Ha et al., 2019; Su et al., 2022). Research is being driven to investigate ways to improve food quality with some essential components that have also been found to be non-toxic while having minimal effect on the nutrientcontent of the food product because of concerns about food quality and health impacts (Singh et al., 2017). This technology use serves as a promising solution that helps encapsulate, protect, and deliver to the targeted sites with greater effectiveness and efficacy along with improved biocompatibility and bioavailability of those encapsulated components (McClements & Öztürk, 2021). Enhancing nutritional content, the quality and safety of

DOI: 10.1201/9781003438168-3

food, and the rate at which bioactive substances are released also equally improve with the food efficiency and effectiveness to targeted sitesusing those technologies. Additionally, this technologyalso lessens the adverse effects that may arise from taking some functional foods in excess (Su et al., 2022). Nano-encapsulation, one of the most efficient methods, involves packing gaseous, liquid, and solidsubstances into various carriers, such as capsules. To nano-encapsulate a bioactive peptide using nanotechnologies for enhancing the bioavailability is extremely difficult. Many nutraceuticals and functional dietary ingredients like protein, vitamins, lipids, and minerals are also encapsulated to exert their useful qualities (Biswas & Sengupta, 2022). Nano-emulsions, micelles, liposomes, and biopolymer complexes are used as encapsuling agents, developed by nanomaterials to improve aspects of bioactive compound protection, targeted delivery, assimilation of food matrix, and masking unpleasant flavours (Sharma et al., 2017). In this chapter, we summarize current technologies along with advancements of the nanoparticle (NP) delivery system to enhance the biological activity and encourage the release of functional food components.

3.2 POSSIBILITIES OF NANOTECHNOLOGY FOR ENCAPSULATED DELIVERY OF FUNCTIONAL FOOD INGREDIENTS

Exploiting the special qualities and numerous uses of materials that are created at the nanoscale via nanotechnology has become increasingly popular in several industries, particularly in the field of medicine (Sim & Wong, 2021). Along with these, using nanotechnology and NPs in the food industry is becoming a more and more popular idea. Research on functional foods and nutraceuticals has continued to show a substantial correlation between consuming bioactives and enhancing human health condition (Konstantinidi & Koutelidakis, 2019). The rising prevalence of chronic diseases is the majorissue that people face globally. One of the most efficient methods for avoiding and treating many chronic diseases isto design and develop functional foods that promote human health and nutrition (Tang, 2021). The majority of the active components are used in functional foods such as phenolic compounds, flavonoids (flavonols, flavanols, flavonoids, etc.) alkaloids, and food pigments (β-carotene, lutein, lycopene, and curcumin), and all of which are pH-, temperature-, light-, and humidity-sensitive antioxidants (Su et al., 2022). Thus, the stability of these health-promoting ingredients is currently difficult to directly apply to food products and regulate their specific targeted site release due to their instability during production and when consumed (Otchere et al., 2023). At the moment, scientists are looking for the development of appropriate analytical methods that is crucial to characterize the encapsulation efficiency, release kinetics, stability, and bioavailability of the encapsulated functional food components for a means of enhancing food quality without sacrificing bioactive components. These methods help in evaluating the performance of the encapsulation system and ensuring quality control aspects (McClements, 2020). The encapsulation process involves enclosing the functional food ingredients within a protective shell or matrix, often composed of biocompatible materials. It offers several advantages such as protection of the ingredients from degradation, improves the control of their release at the active physiological site, becomes more stable, as well as enhances bioavailability and targeted delivery (Pateiro et al., 2021). This system also enables to control the functional food ingredients release, which can be achieved through the selection of appropriate encapsulation materials and techniques, allowing for gradual release in specific regions of the digestive system or at predetermined times (Ruiz Canizales et al., 2018). To enhance stability of functional food and to protect it from degradation brought on by oxidative reactions, pH, and environmental factors like light, temperature, and oxygen, these nanotechnologies actually seem to be one of the most promising techniques; they also ensure that the ingredients remain active and retain their functionality during storage and digestion (Min et al., 2018; Su et al., 2022). Side by side the process also can improve the functional food ingredients' bioavailability by protecting them in the gastrointestinal tract from degradation, facilitating them to be absorbed, and enhancing their transport across biological barriers (Amigo & Hernández-Ledesma,

2020). This can be achieved properly by functionalizing the encapsulation materials with ligands that selectively interact with target sites or by designing the particle size and surface properties for specific interactions (Thalhauser & Breunig, 2020). Overall, the encapsulation of NP-mediated functional food ingredients for delivery requires selection of suitable materials and optimization of encapsulation techniques to ensure the safe and efficient delivery of desired functional properties.

3.3 TYPES OF NANOPARTICLES' DELIVERY SYSTEM

To deliver bioactive substances in food items, a variety of nanocarriers with distinct functionalities and advantages are frequently utilized infood industry to enhance the bioavailability and stability of core particles. There are also many NPs, such as dendrimers, quantum dots, nanocellulose, biopolymeric NPs, etc., that hold significant potential for enhancing drug delivery along with an emerging area of investigation in the growing field of nanomedicine. NPs have sparked significant interest in several fields, except drug delivery, due to their unique properties, and also used in functional food delivery which is a relatively newly emerging area of research with promising potential.

3.3.1 Polymeric Micelles

A core-shaped nanostructure with a diameter of 5–100 nm is formed by self-assembly when the block copolymer concentration rises beyond the critical aggregation concentration or critical micelle concentration in aqueous solutions. The block copolymer formed the inner core of missiles via hydrophobic interactionas well as the hydrophilic copolymer blocks that make the outer shell of these missiles are crucial, particularly for their ability to interact with cells and maintain their steric stability (Sawpari et al., 2023; Xu et al., 2013). Due to core–shell structure, they have great potency in drug delivery system particularly in case of hydrophobic or poorly water-soluble drugs to enhance bioavailability, targeted delivery, and sustained release of drugs (Voets et al., 2007). Although micelles have long been employed as carriers of drugs, their use as carriers for functional food ingredients has just recently gained more attention. These polymeric micelles are used to deliver functional food because of their particles, loading, and physiochemical properties, and they are also used as additives to functional food and beverage systems as they are rich with functional nutrients which are natural polyphenols, saponins, lipids, and vitamins, among others (Su et al., 2022). Encapsulation of alpha-tocopherol in fish oil and incorporation of essential oils to lessen lipid oxidation and in flavoured fizzy beverages have been used (Chen et al., 2006). There are many advantages of using of this: it can help eliminate the side effects of drugs, protect the core molecule, enhance the ability to dissolve into water, and regulate the release rate of hydrophobic drugs (Hussein & Youssry, 2018) (Figure 3.1).

3.3.2 Liposomal Delivery

Liposomes are lipid-based NPs, spherical in structure, formed by phospholipid as membrane material, and surrounding an aqueous core like a closed vesicle (Sawpari et al., 2023). This NP with the size between 1 and 1000 nm, formed via self-assembly of amphiphilic molecule, can deliver both hydrophobic and hydrophilic materials (Reza Mozafari et al., 2008; Singh, 2016). The utilization of charged polar lipids in liposome fabrication allows the charged water-soluble ionic species to enclose. Thus, the liposomal core's pH and ionic strength could be different from the continuous phase in which the liposomes are eventually dispersed (Chen et al., 2006). Researchers have started to use liposomes for the controlled distribution of functional substances including proteins and enzymes, vitamins, and flavours in various food applications based on the previous research studies, using of liposome in drug delivery. They also described about the encapsulation of dairy products,

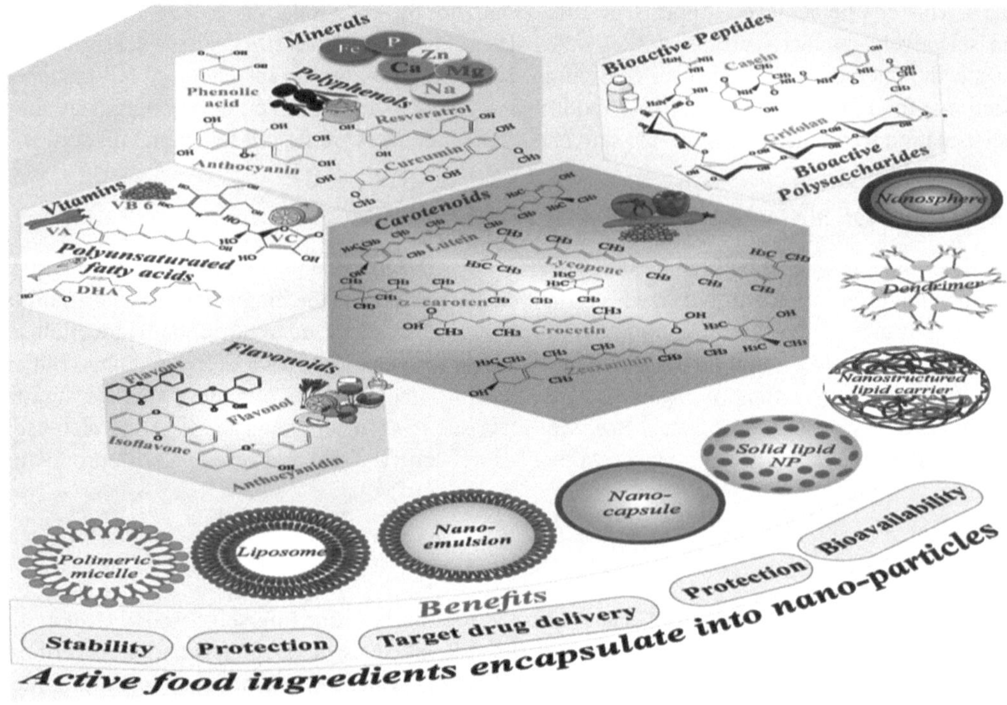

FIGURE 3.1 Nanoparticles are used to encapsulate bioactive food components. Various types of nanoparticles have been used to encapsulate the bioactive food component to enrich the stability and bioavailability as well as to protect the component and deliver to the particular targeted sites with more beneficial role. Parts of this figure weredrawn using pictures from Servier Medical Art. Servier Medical Art by Servier is licensed under a Creative Commons Attribution 3.0 Unported License (https://creativecommons.org/licenses/by/3.0/).

lactoferrin, bacteriostatic glycoproteins, and antimicrobialpolypeptides (Taylor et al., 2005). To increase the bioavailability of the active ingredient in natural foods, decrease toxicity, improve gastrointestinal absorption, and deliver functional component through nanoliposomes, nanometric liposomes, which are structurally, physically, and thermodynamically similar to liposomes are being used. Liposomes also can be used to encapsulate vitamins, proteins, fatty acids, antioxidants, etc. (Reza Mozafari et al., 2008; Singh et al., 2012). The bilayer structure of these entities facilitates drug loading, thus restricting drug absorption and stability, while simultaneously reducing the undesirable consequences associated with site-specific targeted drug administration (Mandpe et al., 2020). There are several advantages associated with using these lipid-based NPs: they can control the drug release and targeting, protect the encapsulated compound from chemical degradation, easily absorbed by epithelial cell, and hydrolysed by particular enzyme in gastrointestinaltract and also in other organs to reduce toxicity (Su et al., 2022) (Figure 3.1).

3.3.3 Nano-Emulsion

A lot of attention is being paid to nano-emulsions in various industries, particularly the food industry (Milani, 2019). Emulsions are thermodynamically unstable nanosize droplet with high surface area commonly made by two immiscible fluids such as oil and water (Choi & McClements, 2020; Islam et al., 2023). The dropletsize less than 100 nm becomes extremely stable between the mean diameter of 20 and 200 nm (Choi & McClements, 2020). The nano-emulsion is significantly more viscous with its effective size of particle or greater number of particle–particle interactions. interactions with stability helps to avoid aggregation (McClements, 2011). Additionally, a low amount of surfactant and uses of organic solvents help enhance its production efficacy (Su et al., 2022). Furthermore, because of their small size, these bioactive macromolecules are easily permeable via gastrointestinal barriers and cellular tissues and are also used to deliver the active substance by enhancing bioavailability and safety along with reducing toxicity (Li et al., 2021; Milani, 2019). It can also boost the solubility and antioxidant activity of the contentsand the release of carotenoids, especially β-carotenes, in the intestine (Schoener et al., 2019). To deliver the nutraceuticals, probiotics, flavours, and colours, nano-emulsions offer significant promise. Food safety, nutrition, and quality are all significantly improved by the use of nano-emulsions containing active substances (antimicrobials) (Aswathanarayan & Vittal, 2019; Islam et al., 2023). It has been also observed that the use of nano-emulsions helped enhance the bioavailability of lipophilic functional compounds and essential oils, solubility of phytochemicals such as curcumin, resveratrol, and carotenoids like lycopene, β-carotene, lutein, and zeaxanthin. Enhancement of bio-accessibility resulted in a larger number of hydrophobic compounds by a nonpolar domain of micelle phase and enhancement of biocompatibility results from the incorporation of isolated bioactive chemicals into emulsion-based delivery methods. Nano-emulsions are widely used to prevent cancer, the treatment of infections of the reticuloendothelial system, liver enzyme replacement therapy, and vaccination (Nile et al., 2020) (Figure 3.1).

3.3.4 Nano-Capsule

Nano-capsules are nanoscale particles, with less than 100 nm in diameter, composed of a shell or matrix that surrounds a core, capable of encapsulatinggas, liquid, or even solid active substances (Assadpour & Jafari, 2019; Jain et al., 2022). When it comes to functional food delivery, nano-capsules offer several advantages due to their small size and versatile properties (Thies, 2012). In case of functional food, the functional components were embedded using nano-capsule technology, which not only minimize the functional ingredient lossduring processing or storage, but also successfully transfer useful components to the human body's gastrointestinal tract or storageand facilitate their absorption (Chopde et al., 2020). Nano-capsules can encapsulate a broad range of bioactive components, including vitamins, minerals, antioxidants, probiotics, and phytochemicals. This encapsulation protects the compounds from degradation, oxidation, and interactions with other food components. It also improves their stability during processing, storage, preserving, and bioavailability by managing the release of functional factors, improving their solubility (Shishir et al., 2018). According to another previous study, the addition of nutritious components (like vitamins) or flavour- or colour-enhancing nano-capsules would remain latent in the meal and would only release when activated by the consumer. Along with that, the addition of nano-capsulated tuna fish oil, which is a source of omega-3 fatty acids, to "Tip-Top" bread also achieved the greatest success (Ameta et al., 2020). Also, it has many advantages: it can alter the encapsulated component distribution and concentrate them in particular target tissues to reduce toxicity and increase efficacy. Additionally, smaller- sized nano-capsules allow passing through biological barriers more easily, thus increasing the likelihood of bioactive compound uptake (De Jong & Borm, 2008; Mitchell et al., 2021). The sensitive ingredient in functional food is protected from the outside environment, such as pH, light, temperature, etc., in the wall material, which is mostly composed of carbohydrates, proteins, plant water-soluble gums, and keeps the ingredient extremely bioactive (de Vos et al., 2010). Another fundamental benefit of utilizing these nanomaterials is that they can also be carriers of flavours and scents, and because of their subcellular size, if food components are scaled down to nanosized crystals, they produce more particles and a larger overall surface area (Kiss, 2020) (Figure 3.1).

3.3.5 Lipid-Based Nanoparticles

Solid lipid nanoparticles (SLNs) are nanoscale particles with 50–1000 nm in diameter and are composed of lipids which solidify at room temperature and incorporate the bioactive chemical into a crystalline or amorphous undercooled matrix. They have gained significant attention in various fields, including pharmaceuticals and food science (Müller et al., 2002). Although SLNs have been utilized to deliver drugs, there is little evidence of their usage in functional food delivery such as vitamins A, D, and K. They may be able to trap labile hydrophobic molecules in a structured solid matrix, increasing their stability, and control the release of these chemicals by creating systems with various melting properties. The potential loss of heat-labile chemical bioactivity at the time of heating process of their synthesis is a significant disadvantage of SLNs (Singh, 2016).

Nanostructured lipid carriers (NLCs) are lipid nano-vehicle partially crystallized particles and have a mean size of 100 nm that are disseminated in an aqueous phase that increase the stability of bioactive food ingredients that are entrapped, allow forenormous loading capacities, and provide long-term release profiles (Livney, 2015). NLCs offer better stability and the potential to encapsulate more bioactive food components compared to SLNs (Reque & Brandelli, 2021).

To achieve several health approaches, numerous elements are required including therapeutic impact, physicochemical stability, and bioavailability, along with restricted intestinal permeability and hepatic first-pass metabolism. To increase the bio-accessibility and bioavailability of nutraceutical substances, several methods have been investigated, like coacervation, inclusion complexation, liposome entrapment, crystallization, and emulsification to improve the solubility and absorption at a particular site (Subramanian, 2021) (Figure 3.1).

3.3.6 Protein-Based Nanocarriers

Due to the capacity to offer special functional features like emulsifying, foaming, gelling, and solubility attributes, protein-based nanocarriers are extensively used in food delivery (Bourbon et al., 2019). Controlled protein aggregation produces hydrogels and nanoparticles that can supply a variety of hydrophilic compounds and give food a nanoscale structure that affects its texture and mouthfeel. NPs that serve as carriers for the encapsulation and dispersion of bioactive substances, such as minerals, vitamins, omega-3 fatty acids, minerals, antioxidants, and probiotic bacteria (Singh, 2016). Mostly the proteins used in delivery are whey, casein, gelatine, silk, albumin, soy, zein, gliadin, and pea (Otchere et al., 2023). Proteins can interact in a variety of ways, including covalently, electrostatically, and hydrophobically and due to polar and non-polar amino acid residues; additionally, functional groups make it easier to influence the release of encapsulated compound (Zhang et al., 2021). Therefore, they have the crucial roleto release their payload to particular environmental triggers, such as ionic strength, temperature, pH, enzyme activity, or redox conditions. This feature can help with the delivery of bioactive molecules in food products as well as in medications. Additionally, protein-based nanocarriers have special functional qualities including emulsification and gelation that make them excellent for encasing bioactive substances (Li et al., 2021; Luykx et al., 2008).

3.3.7 Dendrimers

Dendrimers are highly branched, three-dimensional polymers featuring a well-defined structure. They have attracted significant interest in various fields, including drug delivery, due to their unique properties such as high surface functionality, mono-dispersity, and nanoscale size (Santos et al., 2019). Dendrimers have been the subject of intensive research in the context of drug delivery applications. Their use extends to several disciplines, owing to their capacity to encapsulate molecules of appropriate dimensions within their interior cavities, as well as their capability for covalently grafting molecules onto their periphery. These attributes hold significant promise for diverse applications. (Bacha et al., 2023). Dendrimers can be engineered to provide controlled release of encapsulated compound in response to specific stimuli, such as pH or enzymes existing in the gastrointestinal tract, and also can enhance the bioavailability of poorly soluble or poorly absorbed compound (Chis et al., 2020; Sandoval-Yañez & Castro Rodriguez, 2020). This particle also has the potential to combine therapeutic and diagnostic functions in a single system, known as theranostics. By incorporating both drugs and imaging agents within the dendrimer structure, theranostic dendrimers can enable personalized medicine, allowing simultaneous treatment and monitoring of diseases (Crintea et al., 2023). Side by side of drug delivery, dendrimers also have great potency to deliver bioactive compounds (Tiwari & Takhistov, 2012) (Figure 3.1).

3.3.8 Nanospheres

The formation of nanospheres with a diameter of less than 100 nm occurs when an active core is enveloped by a shell, membrane, or coating. The active chemicals are evenly distributed across nanosphere matrix structures (Rehman et al., 2019). Chitosan (CS) with 100 nm and polyvinyl pyrrolidine with 50–60 nm in diameter have been used to make tiny nanospheres in which hydrophilic compounds can adsorb on the surface and lipophilic drugs can be easily dissolved (Tiwari & Takhistov, 2012). This carrier also has been created into nano-delivery systems to contain bioactive chemicals such nutraceuticals, antimicrobials, antioxidants, and flavours. In addition to these effects, the utilization of micron-sized NPs in food products not only minimizes any negative influence on the sensory properties but also improves the absorption and utilization of encapsulated nutrients (Sampathkumar et al., 2020) (Figure 3.1).

3.4 BIOACTIVE FUNCTIONAL FOOD INGREDIENTS AND THEIR DELIVERY TO ENHANCE THE HEALTH CONDITIONS

Bioactive elements are necessary components of food that have the ability to modify one or more metabolic processes including enhancing the health conditions (Mondal et al., 2021; Teodoro, 2019). As an alternative to traditional treatments for many ailments, functional foods have gained more attention than ever. Functional foods and nutraceuticals have the ability to treat several medical conditions because they contain special functional groups that are created as a result of food metabolism and their molecular variations, in addition toplants naturally producing molecules called phytochemicals that have various biological qualities and medicinal uses (Mondal et al., 2021). According to several research studies, food bioactive components have a beneficial biological influence on human health, acting as a preventative measure against noncommunicable diseases like cardiovascular, cancer, metabolic, and neurological disorders, as well as type II diabetes (Câmara et al., 2020; Mondal et al., 2021; Shaikh, 2022). These advantages have been attributed to secondary metabolites, including flavonoids, fatty acids, plant stanols and sterols, phenolic acids, soy protein, polyols, probiotics and prebiotics phytoestrogens, vitamins, and minerals, glucosinolates, polyphenols, carotenoids, alkaloids, isothiocyanates, terpenoids, saponins, vitamins, and dietary fibres, among others, because of their anti-inflammatory, antithrombotic, antimicrobial, cardioprotective, and vasodilator properties (Câmara et al., 2020; Shaikh, 2022). Fruits, whole grains, and vegetables are the primary sources of the majority of functional plant-based dietary components; however, other components, such as omega-6, -3, and -9 polyunsaturated fatty acids, are also present in some animal products, such as fermented milk products, milk, and some cold-water fish (Shaikh, 2022). Before we can start making dietary recommendations based on scientific study, there is still a lot to be done (Teodoro, 2019).

3.4.1 Carotenoids

Carotenoids are fat-soluble C40 pigmented (Jin & Arroo, 2023) tetraterpenoids, widely distributed in flowers, algae, plants, fruits, fungi, and photosynthetic bacteria. Molds, yeasts, and non-photosynthetic bacteria are another source of carotenoids (Núñez-Gómez et al., 2023; Shaikh, 2022). Lutein, lycopene, α-carotene, β-carotene, γ-carotene, β-cryptoxanthin, astaxanthin, and zeaxanthin are the most common carotenoids found in foods; they either exist in free formulation or have been esterified with fatty acids, allowing the storage of carotenoids (Núñez-Gómez et al., 2023; Shaikh, 2022). Carotenoids are fat-soluble compounds. They can be found in a variety of plant parts, including the leaves, roots, fruit, and seeds, and are found in cells' thylakoids, or photosynthetic tissues, where they coexist with chlorophylls. Some of them contribute to photosynthesis by absorbing light energy (Delgado-Vargas et al., 2000; Núñez-Gómez et al., 2023). Carotenoids are utilized as food colouring agents, as pigments in chicken and fish farms, as antioxidants in nutritional supplements, and as food additives. β-carotene is the principal dietary source of vitamin A. Carotenoids have been found to have protective benefits against major diseases like cardiovascular disease, cancer, and degenerative eye disease. It is also observed that carotenoids play a role in regulating the immune response system and as antioxidants (Bhat et al., 2019; Shaikh, 2022). The human body is unable to produce carotenoids; thus, they must be received through diet. These compounds' positive benefits are typically attributed to their provitamin A activity, which requires an unsubstituted ring structure and an 11-carbon polyene chain. Strong antioxidants known as carotenoids play a role in slowing the consequences of ageing symptoms, which are linked to the demise of cellular activities over time (Núñez-Gómez et al., 2023). Because of their capacity to scavenge free radicals, these substances have been shown to have specific benefits against the ageing of the skin, eyes, and vessels (Núñez-Gómez et al., 2023). Additionally, they enhance cognitive abilities and are effective in neurodegenerative illnesses like Alzheimer's; thus they act as a preventative measure for other oxidative stress-related illnesses like osteoporosis (Núñez-Gómez et al., 2023). They also lower the risk of several cancers, including breast, lung, prostate, and liver cancer, among others. As they can operate as antioxidants and pro-oxidants in certain situations (high oxygen tension and high carotenoid concentration), this anticarcinogenic activity is related to their diverse processes (Rivera-Madrid et al., 2020). It is acknowledged that a wider variety of non-phenolic substances can also be discovered in extracts of high dietary fibre, which is relevant to the concentrations of carotenoids in plant food by-products (Martins et al., 2022) (Figure 3.1).

3.4.2 Dietary Fibres

Dietary fibre is one type of plant-derived food component. The digestive enzymes found in humans are unable to totally fragment them. Fibres are non-starch polysaccharides including mucilages, gums, cellulose, pectins, lignin, hemicellulose, galacto-oligosaccharides, and polyfructose (Dhingra et al., 2012; Shaikh, 2022). Whole grains, beans, citrus peel (34% of cellulose, 10% of hemicellulose), grapefruit wastes (27% of cellulose), broccoli stalks (19% of cellulose), and tomato peel fibre (13% of cellulose), cherry (11% of hemicellulose), brown rice, nuts, bran cereal, baked potato with skin, berries, oatmeal, blackcurrant pomace (constituting 25% of hemicellulose), orange peel (8–53% of pectin), lemon peel (1–17% of pectin), and vegetables are the key sources of dietary fibres. The physicochemical characteristics of dietary fibres, which are directly associated with its physiological effects, have led to substantial research on this topic (Ahmad Khorairi et al., 2023; Núñez-Gómez et al., 2023; Shaikh, 2022; Venkatanagaraju et al., 2020). According to the European Food Safety authority (EFSA), dietary fibre is addressed as "non-digestible" carbohydrates plus lignin and therefore includes: non-starch polysaccharides (hemicelluloses, pectins, cellulose, and hydrocolloids, e.g., mucilages, gums, and glucans), resistant oligosaccharides (galacto-oligosaccharide and fructo-oligosaccharide), resistant starch (includes some types of physically contained starch, unprocessed starch granules, retrograded amylose and starches that have been physically or chemically modified), and lignin associated with the dietary fibre polysaccharides (Núñez-Gómez et al., 2023). Additionally, the Food and Drug Administration (FDA) describes it as non-digestible soluble and insoluble carbohydrates, which have physiological effects that are beneficial to human health (Rezende et al., 2021). The colon ferments fibres, resulting in the production of short-chain fatty acids that have significant health advantages. Functional fibres are manmade fibres that are added to food products to provide comparable health advantages without adding calories, like inulin, cellulose, polydextrose, and maltodextrin. Fibres help avoid colon cancer, prevent obesity and diabetes, lower cholesterol, mitigate irritable bowel syndrome, and boost survivability of breast cancer patients. They also aid to prevent irritable bowel syndrome, coronary and cardiovascular heart disorders (Shaikh, 2022). Additionally, dietary fibres can be categorized not only by their chemical components but also by their characteristics. Each type of fibre exhibits distinct viscosity, solubility, fermentability, properties of hydration, and capacity for absorbing fat, which dictate the positive benefits on human health and the physicochemical relevance. In this regard, fibres are categorized as either soluble (showing favourable effects on serum lipids and metabolism of glucose, both of which are directly related to their capacity to form gels) or insoluble (having a purgative effect on the colon). Despite the fact that not all soluble fibres are viscous, soluble fibres are more typically associated with viscosity than insoluble fibres. These fibres have the ability to bind water, create the satiety-related distension effect, and positively influence biomarkers for heart disease. Moreover, fermentability, or the capacity to be metabolised by colonic microbiota, is a crucial factor to consider when assessing potential physiological impacts because it can change composition in a positive way (Núñez-Gómez et al., 2023). However, consuming too much dietary fibre can have some negative effects, including compact absorption of minerals, proteins, vitamins and calories from the gut, dehydration, increased intestinal gas, intestinal obstruction, which causes flatulence and distention, and dehydration (Bliss et al., 2011; Shaikh, 2022) (Figure 3.1).

3.4.3 Essential Fatty Acids

Asthe human body is unable to generate essential fatty acids (EFAs), it must consume foods that contain them in order to maintain optimum health (Kaur et al., 2014; Shaikh, 2022). Essential fatty acids are long-chain polyunsaturated fatty acids that are referred to be "good fats" because they raise HDL levels and lower LDL levels. The two main necessary fatty acids in the human body are linolenic acid and alpha-linoleic acid (Shaikh, 2022). There are several foods and oils that contain essential fatty acids, including cod liver oil, mackerel, salmon, herring, soybeans, sardines, oysters, anchovies, flax seeds, chia seeds, caviar, walnuts, and canola oils (Shaikh, 2022). Numerous studies have found a beneficial effect of essential fatty acids in several diseased condition-like cardiovascular disease and mortality, helps in baby development, diabetes mellitus, hypertension, cancer prevention (colorectal, breast cancer, and prostate cancer), optimal brain and visual health, arthritis, neuropsychiatric, neurological diseases and improve bodies overall immunity (Kaur et al., 2014). Furthermore, it has been also noted that essential fatty acids are crucial for conditions such as age-related macular degeneration, depression, asthma, psychological disorders (bipolar disorder, schizophrenia), burns, photodermatitis, osteoporosis, skin disease (acne), obesity, insulin sensitivity, and dry-eye syndromes like Sjögren's syndrome (Shaikh, 2022) (Figure 3.1).

3.4.4 Flavonoids

Plants produce pigments known as flavonoids, which are divided into numerous subclasses, each of which contains a wide variety of substances (Shaikh, 2022). Flavonoids have a 15-carbon skeleton with two phenyl (A and B) rings and a heterocyclic (C) ring as part of their chemical structure, where C_6–C_3–C_6 is an acronym for the carbon skeleton. They can be divided into three groups, according to the International Union of Pure and Applied Chemistry (IUPAC) nomenclature: isoflavonoids

(3-phenyl-1,4-benzopyrone), bioflavonoids, and neoflavonoids (4-phenyl-1,2-benzopyrone) (Waheed Janabi et al., 2020). Flavonoids could be further classified into different forms, like isoflavones (biochanin A, daidzein, daidzin, glycitein, genistein, formononetin), flavononols (genistin, astilbin, engeletin, taxifolin), anthocyanidins (delphinidin, cyanidin, malvidin, apigenin, peonidin, petunidin pelargonidin), chalcones (butein okanin), flavonols (quercetin, myricetin isorhamnetin, kaempferol), flavanols (negative-epicatechin, positive-catechin, positive-gallocatechin, negative-epicatechin gallate, negative-epigallocatechin), flavones (chrysin, rutin, apigenin, luteolin), and flavanones (eriodictyol, hesperidin, isosakuranetin, taxifolin, naringenin naringin) (Shaikh, 2022). Different sources of flavonoids are elderberry juice, capers, red onions, sorrel, dried parsley, fresh cranberries, cooked asparagus, goji berries, dried oregano, rocket lettuce, cooked asparagus, blackcurrants, orange juice, grapefruit, lemons, limes, oranges, green tea, artichokes, grapefruit juice, dark chocolate, dried cocoa, black tea, blackberries, pecan nuts, red wine, cooked broad beans, peaches, apples, aronia, bilberries, green pepper, American bilberries, black and red currants, chickpeas, red cabbage, raspberries, and strawberries (Jin & Arroo, 2023; Shaikh, 2022; Waheed Janabi et al., 2020). Plants may produce flavonoids, which are not only responsible for the colour and fragrance of flowers but also act as detoxifying agents, signal molecules, and antimicrobial defence substances to shield plants from a variety of physical and biological threats (Montané et al., 2020; Yuan et al., 2022). There are a lot of health benefits of flavonoids, including antiallergic, antioxidant, antiviral, antifungal, antitoxic, anti-inflammatory, and antibacterial activities (Shaikh, 2022; Ullah et al., 2020). Recent studies have shown several protective functions of flavonoids, such as eye diseases, haemorrhoids, heart diseases, neurodegenerative diseases, diabetes, as well as Alzheimer's or Parkinson's, gout, and periodontal disease (Ullah et al., 2020; Waheed Janabi et al., 2020). Flavonoids are also used to prevent and treat several cancers, including renal cell carcinoma, hepatocellular carcinoma, prostate cancer, pancreatic cancer, ovarian cancer, colon cancer, breast cancer, leukaemia, and oesophageal cancer (Li et al., 2022; Shaikh, 2022) (Figure 3.1).

3.4.5 Phenolic Acids

Phenolic acids contain both carboxyl functional group and phenolic ring. Several examples of phenolic acids are vanillic acid, protocatechuic acid, ferulic acid, p-hydroxybenzoic acid, caffeic acid, p-coumaric acid, sinapinic acid, and syringic acid. Phenolic acids defend against cellular damage caused by free-radical oxidation processes and are absorbed through the intestinal wall (Shaikh, 2022). Different sources of phenolic acids are oilseeds, legumes, fruits, cereals, herbs, vegetables, and beverages. In addition to these sources, they are present in every dietary group (de la Rosa et al., 2019; Kumar & Goel, 2019; Shaikh, 2022). The consumption of phenolic acids has been related to a number of health benefits, like various cancers (prostate cancer, colon cancer), including a reduced risk of type II diabetes, anti-viral properties (human immunodeficiency virus, HIV), neurodegenerative disorders, and cardiovascular disease. These benefits may be attributed to a number of putative mechanisms of actions like glucose regulation, including antioxidation, antiproliferation, anti-inflammation, and microbial modulation (Kumar & Goel, 2019; Shaikh, 2022). Phenolic acid is also crucial for immunoregulation in several diseases, such as asthma, various types of allergic reactions. Additionally, it has been discovered that higher phenolic acid intake significantly lowers diastolic and systolic blood pressure compared to lower phenolic acid intake (Shaikh, 2022) (Figure 3.1).

3.4.6 Vitamins

Vitamins are the organic substances that the body needs to function properly. They are needed in tiny doses and can be obtained through a healthy diet. Vitamins can be divided into two categories: water-soluble vitamins and fat-soluble vitamins. Fat-soluble vitamins are D, A, K, and E and they can be stored in the body. whereas water-soluble vitamins are B-complex and C vitamins, including B12, B6, niacin, biotin, riboflavin, thiamine, pantothenic, and folic acids (FAs). Because anexcess amount of these vitamins is excreted through human fluids like urine and perspiration, they cannot be kept in the body and must be consumed in sufficient amounts daily (Shaikh, 2022). Food reinforcement with vital micronutrients, such as vitamins E, D, and A, promotes human health by guaranteeing adequate ingestion to prevent diseases (including osteoarthritis, cancer, osteoporosis, a suppressed immune system, vision loss, heart disease) and it also increases life quality and survival. The prevalence of malnutrition, particularly in developing nations, has raised the demand for food enriched with lipophilic vitamins A, E, and D (Öztürk, 2017). E, C, and A are among the vitamins that have antioxidant properties. Thus, consuming these vitamin-enriched meals can aid in the treatment of certain chronic conditions (Xiao & Li, 2020). Both food fortification and supplementation are effective at ensuring adequate daily intakes of vitamins and minerals due to their adequatebioavailability (Arshad et al., 2021; Knijnenburg et al., 2019) (Figure 3.1).

3.4.7 Minerals

Minerals are inorganic substances that are found in water and soil and are crucial for maintaining a healthy body. Micro or trace minerals, such as copper, cobalt, chromium, manganese, iodine, iron, fluorine, molybdenum, zinc, and selenium, are needed in very small amounts while macro-minerals, like phosphorus, calcium, magnesium, potassium, sodium, and sulphur, are needed in large amounts (Shaikh, 2022). Mineral deficiencies can lead to a wide range of common diseases and disease symptoms asminerals have numerous potential roles and functions in homeostasis and metabolism. By verifying mineral content details in terms of safe food fortification and processing processes, it is possible to greatly increase mineral absorption and bioavailability. Iron, calcium, zinc, and iodine are the most frequently employed minerals while preparing different food preparations to be fortified. Isotope ratio techniques can be used to precisely determine the bioavailability of dietary minerals. Compared to traditional process parameters, advanced processing methods have less adverse impacts on the uniformity of micro- and macrominerals (Gharibzahedi & Jafari, 2017). The most economical methods to combat global mineral deficiency are food fortification and supplementation. Iodine fortification of salt, which has significantly reduced the prevalence of goitre and other disabilities symptoms in regions where it has been implemented, has been the most promising option. By being incorporated into selenoproteins, the powerful antioxidant selenium (Se) promotes health. As selenoproteins are essential for controlling reactive oxygen species (ROS) and redox conditions in almost all tissue and improve immune system. Additionally, zinc is a crucial trace mineral that is essential for many physiological processes, including the development and functioning of immune cells in the adaptive and innate immune systems. All cells have efficient mechanisms for maintaining zinc homeostasis. The importance of homeostasis disturbances can be seen in a variety of disease models, including infections, autoimmune illnesses, allergies, and cancer (Arshad et al., 2021) (Figure 3.1).

3.5 APPLICATION OF BIOACTIVE COMPOUND ENCAPSULATED NANOPARTICLES AGAINST PATHOPHYSIOLOGICAL DISORDERS

Recently, the advancement of functional foods has increased in response to growing consumer concerns and expectations for products that are natural, nourishing, and nutritious. To sphere the coated component (food or flavour ingredients), nanoencapsulation is a method based on enclosing bioactive agents in solid, liquid, or gaseous states within an inert material and matrix (Pateiro et al., 2021). Nanoencapsulation can expand stability of bioactive compound, regulating their release at physiologically active areas more effectively, enhanced solubility and stability, extended half-life, improved bioavailability and epithelium permeability, boosted tissue targeting, and diminished side effects (Lushchak et al., 2020; Pateiro et al., 2021). However, the primary bioactive ingredients that are employed in foods to promote health contain properties like antimicrobials, antioxidants, vitamins, probiotics, prebiotics, and other elements like minerals, enzymes, etc. Nanotechnology plays a substantial rolein the creation of programmable food, a novel revolutionary idea that promises consumers high-quality food with the necessary nutritional and sensory qualities (Pateiro et al., 2021). Anti-ageing properties such as anti-inflammatory, antioxidant, and anti-tumour (Chavda et al., 2022; Siddiqui & Sanna, 2016), and cardioprotective activities have been reported for various phyto bioactive compounds including quercetin, resveratrol, catechin, curcumin, etc. However, due to their poor bioavailability, stability, and solubility in the gastrointestinal tract, orally taken bioactive chemicals' therapeutic potential is constrained. Recent advances in nanotechnology have improved the bioactivity of naturally occurring bioactive substances, increasing their potential to cure chronic inflammation and combat many illnesses (Lushchak et al., 2020). The capabilities of bioactive compounds in fighting against different disease conditions are described below.

3.5.1 Antidiabetic

Nanoencapsulation can expand stability of bioactive compounds, regulating their release at physiologically active areas more effectively, enhanced solubility and stability, extended half-life, improved bioavailability and epithelium permeability, boosted tissue targeting, and diminished side effects (Lushchak et al., 2020; Pateiro et al., 2021). An investigation compared micellar lutein (control) with lutein-loaded CS–sodium alginate-based nanocarrier systems for lutein bioavailability and pharmacokinetics in diabetic rats. The results showed that these nanocarriers indicated maximal lutein in the plasma, liver, and eye of normal and diabetic rats, and also presented a longer half-life, a longer mean residence time, and a slower plasma clearance rate than the control group, indicating extended circulation. In human retinal pigment epithelial cell line (ARPE-19) cells, pre-treatment with this nanocarriers (10 µM) have drastically reduced H_2O_2 generated cell death, mitochondrial membrane potential, and intracellular ROS compared to control (Toragall & Baskaran, 2021). Additionally, another invivo study claims that lycopene and lutein (extracted from *Spinacia oleracea* L.), both carotenoids, are beneficial for preventing metabolic diseases like diabetes and oxidative stress (Mishra & Kumari, 2021; Mishra et al., 2015). Astaxanthin has a better oral bioavailability when delivered through the use of nano-emulsions, solid lipid NPs, liposomes, CS- and poly (lactic-co-glycolic acid) (PLGA)-based NPs, among other delivery systems. Astaxanthin demonstrated potential biological action in both in vitro and in vivo models, as a powerful antioxidant, anti-lipid peroxidation, gastric and small intestinal circumstances, and as a cardiac disease prevention agent (Abdol Wahab et al., 2022; Liu et al., 2019). The core–shell NPs may be excellent for encasing astaxanthin in functional foods and cosmetics, according to research (Liu et al., 2019). According to invivo research, the MgAl-LDH-hesperidin combination reduces hyperglycaemia and hyperlipidaemia in diabetic rats under treatment by boosting insulin synthesis from beta cells, elevating liver glycogen levels, and boosting lipogenesis. By reducing oxidative stress, increasing the expression of PPARγmRNA, and activating the Nrf2/ARE/antioxidant pathways, as well as by reducing the levels of the pro-inflammatory cytokines interleukin (IL)-17 and tumour necrosis factor-alpha (TNF-α), MgAl-LDH-hesperidin can also improve insulin sensitivity and glucose uptake (El-Shahawy et al., 2021). In diabetic rats, quercetin-conjugated superparamagnetic iron oxide NPs (QCSPIONs) lowered blood glucose levels by reducing miR-29 family expression and consequently improved IGF-1 and GLUT1, 2, 3, 4 expressions (Dini et al., 2021). The protective mechanism of apigenin-SLNs (apigenin-SLNs) results in the reduction of nuclear factor kappa-light-chain-enhancer of activated B cell (NF-κB) expression and the augmentation of nuclear factor erythroid 2-related factor 2 (Nrf2) and haem oxygenase-1 (HO-1) expression in diabetic nephropathy rats. In addition to having anti-inflammatory properties (by inhibiting the release of inflammatory factors), apigenin-SLNs also have antioxidant properties (by reducing the formation of lipid peroxidation)

(Li et al., 2020). Naringenin-loaded PLGA NPs demonstrated 70% encapsulation efficiency in an experimental rat model; the particles appeared to be more efficient than the free drug in ameliorating streptozotocin-induced diabetogenic effects, including hyperglycaemia, hyperlipidaemia, and carbonyl- and iron-mediated oxidative stressors (Maity & Chakraborti, 2020). Current investigation targeted to estimate a formulation of selenium nanoparticles (SeNPs) using Naringenin/Baicalin (NAR/BAI/SeNPs) in streptozotocin-induced diabetic mice and INS-1 (rat insulinoma cell line) pancreatic β-cells. This has a high effectiveness in managing diabetes with good management of fasting blood glucose level by enhancing insulin sensitivity, HbA1c, insulin secretion, glucose tolerance, and HOMA-IR index by reviving the β cells of the pancreas. As a result of improved peripheral glucose utilization, it significantly improves the altered lipid profile, lowers oxidative stress, increases liver glycogen stores, and corrects impaired hepatic glycolysis. This study also indicates how NAR/BAI/SeNPs increase insulin secretion and protect INS-1 pancreatic β cells from glucotoxicity-induced death and dysfunction (Gutiérrez et al., 2023). Additionally, pectin-modified nanomaterial is becoming important for the treatment of T2DM asit can lower insulin production and/or blood glucose levels (Abdulhakeem et al., 2023). For the oral administration of the majority of plant-based chemicals, such as resveratrol, curcumin, quercetin, silymarin, ferulic acid, pelargonidin, crocetin, and oryzanol, PLGA has been proposed as a viable nanocarrier, although it has been asserted that the oral bioavailability and antidiabetic effectiveness of the mentioned plant-derived small molecules are improved by PLGA-loaded and nano-formulations (Dewanjee et al., 2020). Additionally, CShas been recognized as an effective nanocarrier for the oral transport of several bioactive food ingredients, improving the pharmacokinetic properties and compliance of the aforementioned antidiabetic phytochemicals (Dewanjee et al., 2020; Surendran & Palei, 2022) (Figure 3.2). In conclusion, the pharmacokinetic and biopharmaceutical barriers associated with plant-derived antidiabetic compounds can be overcome by the nanoscale formulations, which improve compliance and clinical efficacy. However, more research is needed for clinically efficienttherapeutic nanoformulations of plant-derived antidiabetic compounds to manage diabetes-associated complications.

3.5.2 Anticancer

The world's most common disease and leading cause of morbidity is cancer. To stop the progression of cancer, it is vital to create more efficient and targeted treatment methods (Yuan et al., 2022). According to several studies, natural plant-based bioactive compounds can advance the efficiency of chemotherapy and, in some cases, reduce some of the side effects of medications used as chemotherapeutic agents (Subramaniam et al., 2019). The solubility, membrane permeation, cellular uptake, membrane permeation, bio-accessibility, and stability of carotenoids can be improved through encapsulation within various nanocarriers, which is a remarkable method and creative strategy for the prevention and treatment of cancer (Zare et al., 2021). Crocin-gold nanoparticles (AuNPs) greatly inhibited the proliferation of breast cancer cells, increasing the effectiveness of crocin against cancer (Hoshyar et al., 2016). The most effective crocin nanoliposome formulation was chosen for an in vivo investigation, where the anticancer activity of nanoliposomalcrocin was assessed in BALB/c mice with C26 colon carcinoma. In this study, nanoliposome demonstrates excellent anticancer efficiency. Additionally, crocin-loaded PEG SeNPs showed perfect haemocompatibility and enhanced cytotoxicity to A549 cell lines (human lung cancer cell lines) through mitochondrial apoptosis (Zare et al., 2021). Asastaxanthin is one of the most significant fat-soluble carotenoids, it has positive effects on the biological system (Jafari et al., 2021; Zare et al., 2021). In a research study, tumour-targeted astaxanthin NP, cRGD-PEG2000-DSPE encapsulated astaxanthin (cRGD@AST), was constructed and employed for tumour therapy. Results show that cRGD@AST may enhance astaxanthin's water solubility, can target tumours following cyclic arginylglycylaspartic acid (cRGD) transformation, and can be substantially absorbed by HepG2 cells. Additionally, cRGD@AST can scavenge spare ROS in tumour cells, inhibit cell proliferation, and restore the redox balance in cells (Xie et al., 2023).

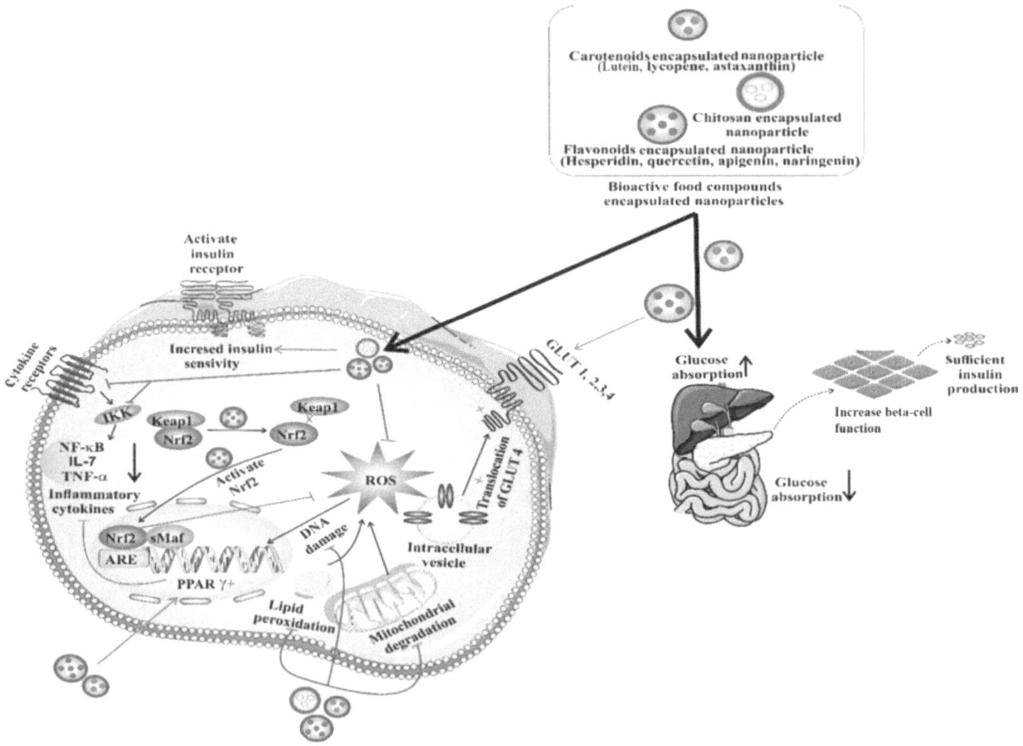

FIGURE 3.2 Schematic representation of antidiabetic targets by carotenoids (lutein, lycopene, astaxanthin), flavonoids (hesperidin, quercetin, apigenin, naringenin), and chitosan encapsulated nanoparticles. Flavonoid nanoparticles activate Nrf2/ARE/antioxidant pathways, increase the expression of PPARγ gene, and enhance absorption of glucose in liver and revive beta-cell function. All mentioned nanoparticles decrease or block diabetes-induced ROS generation, DNA damage, lipid peroxidation, and mitochondrial degradation. Parts of this figure were drawn using pictures from Servier Medical Art. Servier Medical Art by Servier is licensed under a Creative Commons Attribution 3.0 Unported License (https://creativecommons.org/licenses/by/3.0/).

Bharathiraja and his team prepared AuNPs with astaxanthin as a natural reductant, which showed great cytotoxic effect against human breast cancer cells (MDA-MB-231) (Bharathiraja et al., 2016; Zare et al., 2021). Carotenuto and his team investigated the impact of a unique therapeutic approach by combining β-carotene and 5-fluorouracil (5-FU) with NPs as a drug delivery mechanism. Results showed that by inducing apoptosis and cell cycle arrest in the G2/M phase, the combined therapy may be able to overcome 5-FU chemoresistance. Additionally, the combined therapy remarkedly decreased the expression levels of the examined ABC genes. According to this research, treating colorectal cancer cells with low levels of uL3 may be more successful when 5-FU is coupled with carotene (Carotenuto et al., 2023). In another study, the self-assembled nanoprecipitation method was used to create fructose-tethered lipid-nanohybrids that were also loaded with methotrexate (MTX) and beta-carotene. The produced nanocarriers' in vitro cytotoxicity, subcellular localization, and apoptotic activity were assessed against breast cancer cells (MCF-7), where an increase in in vitro cell cytotoxicity was noted (Zare et al., 2021). The epigallocatechin-3-gallate (EGCG) was delivered using a variety of NPs, including gold, liposomes, metallic, polymeric, and carbohydrate-based ones. According to the findings, EGCG NPs encouraged prolonged blood circulation times, enhanced cell internalization in tumour areas, and decreased tumour growth both in vitro and in vivo, mostly in breast and prostate cancer models (Granja et al., 2016). As well as carotenoids, nanocarriers can improve the bioavailability of flavonoids in various cancer treatments.

Researchers developed curcumin–naringenin-loaded dextran-coated magnetic nanoparticles (CUR-NAR-D-MNPs) and examined them in vitro and in vivo models to determine how well they worked in conjunction with radiotherapy to treat tumours. The nanomaterial is biocompatible and has the ability to promote ROS, direct tumour cell apoptosis, and interrupt tumour cell growth. When used in combination with radiotherapy, significant tumour suppression was achieved and the survival ofmice was significantly prolonged through the upregulation of P53, P21, and down-regulation of TNF-α and CD44 (Askar et al., 2021). Researchers discovered that quercetin-loaded PLGA nanoparticles (PLGA-QNPs) can drastically decrease the viability of human cervical and breast cancer cell lines in invivo models. In cells treated with PLGA-QNPs, PI3K/AKT gene expression was downregulated, whereas FoxO1 gene expression was upregulated. These cells also had elevated levels of Caspase-3 and Caspase-7 activity, which are involved in apoptosis (Yadav et al., 2022). Recent studies demonstrated enhanced efficacy of quercetin-loaded chitosan nanoparticles (QCT-CS NPs) in cancer therapy (intravenous treatment of QCT-CS NPs in tumour xenograft mice with A549 and MDA MB 468 cells). This study successfully encapsulated QCT in CS NPs to target the tumour microenvironment. Additionally, researchers discovered that quercetin encapsulated in solid lipid nanoparticles (QT-SLNs) significantly reduced the viability and proliferation of MCF-7 cells by significantly raising ROS levels, malondialdehyde (MDA) concentrations, significantly lowering antioxidant enzyme activity, and decreasing Bcl-2 protein expression, while significantly raising Bax expression (Baksi et al., 2018; Niazvand et al., 2019). Additionally, Kaempferol-loaded nanoparticles (KFP-NPs) demonstrated hepatoprotective and antioxidant benefits in hepatocellular cancer models. KFP-NPs significantly decreasedan mRNA expression of IL-1, IL-6, and TNF-α as well as decreased nuclear factor NF-κB protein expression. It increased the protein and mRNA expression levels of HO-1 and nuclear factor erythroid 2-related factor 2 (Nrf2) (Amjad et al., 2022; Kazmi et al., 2021). Vitamin supplementation (E, C, D3, and B12) during chemotherapy has frequently been linked to lower cancer patient mortality and recurrence rates. Multivitamin (C, D3, and B12)–cisplatin NP complex—NanoCisVital (NCV)—could lessen chemotherapy-induced cancer fatigue. In NCV-treated cells, there are changes in the DDX3X and Ki-67 gene expression observed (Muddineti et al., 2017; Othayoth et al., 2023). In prostate cancer cells, curcumin-loaded cellulose NP (cellulose-CUR) formulation generated the most apoptotic-related ultrastructural alterations (the presence of vacuoles) and had the highest cellular absorption. While cellulose nanocrystals (CNCs) were labelled with fluorescein-5′-isothiocyanate as an imaging agent and coupled to FA as a targeted ligand, both KB and MDA-MB-468 cells showed substantial and folate-receptor-specific binding/uptake of FA-conjugated CNCs. CNCs have a significant potential to be used as innovative nanocarriers for imaging agents and chemotherapeutics in the early diagnosis and treatment of cancer, as shown by the internalization of FA-conjugated CNCs by FR-positive cancer cells and tumours and their extraordinary high affinity for the FR (Bittleman et al., 2018; Muddineti et al., 2017) (Figure 3.3).

3.5.3 Immunomodulation

Bioactive substances are recognized for significant influence on anti-inflammatory response via numerous inflammatory pathways, such as the mitogen-activated protein kinase (MAPK), nuclear factor kappa-light-chain-enhancer of stimulated B cell pathway, Janus kinase/signal transducers and activators of transcription (JAK/STAT), and others. These pathways can regulate the inflammatory rejoinders by producing inflammatory moderators and cytokines, survival of cell, proliferation of cell, dendritic cell (DC) maturation, and T-cell differentiation (Gangwar et al., 2021; Ginwala et al., 2019). The anti-inflammatory potential of naringenin-encumbered liquid crystalline nanoparticles (LCNs) counter to human airway epithelium-derived basal cells (BCi-NS1.1) is being investigated in a research study. Anti-inflammatory ability of Naringenin-LCNs was evaluated by quantitative polymer chain reaction (qPCR), and the results demonstrate that the expression of

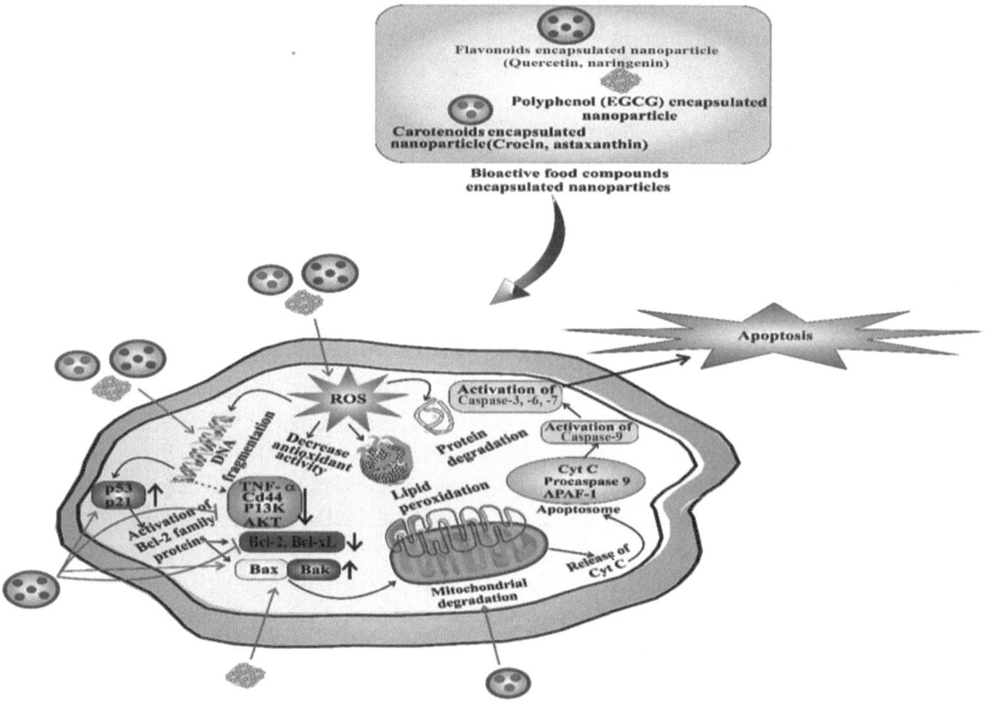

FIGURE 3.3 This schematic representation exhibits anti-cancer efficacy of nanoparticles encapsulated with various bioactive food compound (carotenoids, flavonoids, polyphenols). Here, encapsulated nanoparticles generate ROS in cancer cell, which is the leading cause of decreased antioxidant activity, lipid peroxidation, protein degradation, and DNA damage. Flavonoid encapsuled nanoparticles downregulate the expression of TNF-a, Cd44, PI13, AKT, and Bcl-2 family proteins and upregulate Bax, P53, and P21. In this presentation, green arrows denote upregulation and red lines denote downregulation of cellular signals. Parts of this figure weredrawn using pictures from Servier Medical Art. Servier Medical Art by Servier is licensed under a Creative Commons Attribution 3.0 Unported License (https://creativecommons.org/licenses/by/3.0/).

several interleukins IL-1β, IL-6, IL-8, and TNF-α was reduced in lipopolysaccharide-induced BCi-NS1.1 cells (Wadhwa et al., 2021). Prior studies have demonstrated that PVP-taxifolin liposomes (PVP-TAX-Lips) can cure liver injury (acute) caused by lipopolysaccharide and D-galactosamine (LPS/D-GalN) by stimulating autophagy to reduce the expression of the TLR4 NF-κB/TLR4 signalling pathway and inflammatory markers. Results revealed that PVP-TAX-Lips played a significant role in preventing acute liver damage and also offered a viable drug delivery mechanism for the use of taxifolin (Ding et al., 2022). Diez-Echave and his team assessed the intestinal anti-inflammatory properties of quercetin loaded silk fibroin nanoparticles (QSFNs) in the dextran sulphate sodium experimental model of mouse colitis, which displays some similarities to human inflammatory bowel disease. When compared to the colitic control group, daily injection of QSFNs dramatically decreased disease activity index values. This advantageous effect was supported not only by the histological analysis of the colonic tissues but also by an increase in the colonic expression of several proinflammatory cytokines (Il-1β, TNF-α, Il-6, Icam-1, Mcp-1, Nlrp3, and iNOS) (Diez-Echave et al., 2021). Additionally, using a dextran sulphate sodium (DSS)-induced inflammatory bowel disease model, Khater and his team assessed the antioxidant and anti-inflammatory properties of quercetin nanoparticles (QT-NPs). When compared to the colitic group, the administration of QT-NPs, especially at higher concentrations, dramatically decreased the disease activity index and values of the faecal calprotectin marker. Following treatment with encouraging doses of QT-NPs, the oxidant/antioxidant state of the colon (H_2O_2, ROS, MDA, GPX, SOD, CAT, and TAC) was recovered. Aside from that, QT-NPs at selected dose decreased the expression of the HO-1 and Nrf2 genes, which was consistent with a reduction in the production of COX-2 and iNOS in colonic tissues. Higher QT-NP concentrations significantly reduced pro-inflammatory cytokines, increased the expression of the genes encoding MUC-2, occluded and JAM, and restored the normal topologies of colonic tissues (Khater et al., 2022). Hesperidin-loaded gold nanoparticles (Hsp-AuNPs) were discovered to improve the functional endeavour of macrophages against mice bearing Ehrlich ascites tumour cells. Pro-inflammatory cytokine activation in bone marrow–derived macrophage cells treated with Hsp-AuNPs was also evaluated. The outcomes clearly showed that treatment with Hsp-AuNPs dramatically decreased IL-1, TNF-α, and IL-6 production (Sulaiman et al., 2020). In an invivo investigation, hesperidin also significantly slowed lung damage as seen by decreased inflammatory cytokine amount and inhibited vascular permeability (Jin et al., 2021). In another in vivo study model Freund's complete adjuvant (FCA)-induced arthritis in rats, the well-known flavonoid naringin effectively reduced INF-γ, IL-6, and TNF-α while increasing IL-10 level (Mohanty et al., 2021). Selenium is a trace element that regulates immunological response, enzyme activity, and antioxidant defence through a number of distinct routes. SeNPs were extensively established to demonstrate tremendous efficacy in anti-inflammation, antimicrobial and anticancer treatments by directly cooperating with innate immune cells, like DCs, macrophages, and natural killer cells. The more intriguing fact in that recent research has revealed the possible use of functional SeNPs for immunotherapy, which might be used as a supplement to the present chemotherapy to obtain more effective treatment outcomes. Additionally, SeNPs have the ability to activate and restore a variety of T cells for adaptive immune controls to boost their ability to kill cancer cells, suggesting the promise of SeNPsin the development of immunotherapy strategies (Chen et al., 2022; Moratin et al., 2021; Xia et al., 2022).

3.5.4 Antibacterial

A variety of substances, including phenolics, enzymes/proteins, carotenoids, flavonoids, vitamin C, K, E, bacteriocins, natural polymers, peptides, fatty acids (lipids), organic acids, terpenes, and a number of other bioactive elements aid in the prevention of bacterial development (Fan et al., 2018; Walia et al., 2022). Regarding addressing encapsulation, the term "encapsulation" can refer to the process of integrating bioactive substances into NPs for the purpose of preventing degradation and a reduction in antibacterial action due to stability issues, as well as for the purpose of protecting them or enhancing their qualities. For instance, techniques similar to NLCs, nanoliposomes, biopolymer NPs, nano-emulsions or micelles developed from polysaccharides have been encapsulated as bioactive compounds in liquid, solid, or gaseous states (Pateiro et al., 2021). The nano-emulsion technique, through which the poorly water-soluble bioactive food components are encapsulated, may cause great interaction with bacterial cell membranes to damage it. Cellulose (polysaccharide) encapsulated nano-emulsion with eugenol results in more effective inhibition in growth and development of Gram-positive bacteria than Gram-negative (Wang et al., 2023). As per report of previous study, through using droplet technology a stable nano-emulsion of *T. daenensis* oil was created using lecithin and Tween 80 as emulsifiers had greater antibacterial efficiency against particularly *Acinetobacter baumannii* compared to pure oil (Moghimi et al., 2018). Nano-liposomes contains several beneficial substances that serve the purpose of encapsulating water-soluble, lipid-soluble as well as amphiphilic compounds. A study demonstrated that nisin (small peptide) and crocin (saffron) loaded along with CS-coated nanoliposomes show good bacteriostatic efficacy against *L. monocytogenes* more compared to free nisin and crocin as well as these components containing nanoliposomes without coating (Yousefi et al., 2023). An amphiphilic bioactive component nisin (peptide) nano-emulsion hydroxypropyl methylcellulose (HPMC) film solution shows more antibacterial efficacy than nisin encapsulated HPMC matrix-embedded nanoliposomes against *L. monocytogenes* (Imran et al., 2012). As a chelating agent, CS (polysaccharides) can inhibit the growth of bacteria with its ability to act as a water-binding substance, which prevents a variety of enzymes from working properly, limits the ability of cells to synthesize proteins, and interferes with the creation of mRNA (Yousefi et al., 2023). Some studies reported that electrospun nanofibers with quaternized CSand poly (vinyl alcohol) show good antimicrobial efficacy against *Staphylococcus aureus* and *Escherichia coli*. However, another study also reported that CS-coated metal nanoparticles (AgNPs) fabricated via green (*Carica papaya*) synthesis also show good antibacterial activity against the same bacterial strain (Ignatova et al., 2007; Samanta et al., 2022). Anethole and carvone, both bioactive substances (essential oil), encapsulate PLGA nanospheres. Anethol-loaded NPs show antibacterial activityagainst *Salmonella typhi*, and carvone also shows antibacterial activity against *S. aureus* and *E. coli* bacteria by improving distribution to particular sites through penetration into bacterial cell and being endocytosed by phagocytic cell, as well as enhanced hydrophilicity, sustained release, and greater penetration of encapsulated component (Esfandyari-Manesh et al., 2013). Phenolic compounds are naturally occurring antioxidants noted in various plant-based foods, such as fruits, vegetables, and whole grains (Pandey & Rizvi, 2009). The phenolic bioactive component from the guabiroba fruit loaded within PLGA NPsto increase the functional activity also provides antibacterial efficacy (Pereira et al., 2018). AgNPs decorated quercetin prevented proliferationof *E. coli* by interfering with bacterial metabolism (Sun et al., 2017). A prior study also revealed the antibacterial properties of encapsulated rutin in date and mushroom against *S. aureus* and *E. coli* via interactions of bacterial cell walls and flavonoids, which may cause the breakdown of the lipopolysaccharide layer to inhibit the permeability of the cell membrane (Shah et al., 2022). Vitamin C can be encapsulated into nano-crystal decorated CS to enhance the antibacterial property for food (Baek et al., 2021). Bacteriocins are a class of proteins that are synthesized by bacterial strains through ribosomal processes. These proteins havethe capability to effectively suppress the growth and survival of harmful germs. Additionally, bacteriocins have been recognized for their potential in contributing to the creation of functional foods. An effective strategy for enhancing both the stability and antibacterial efficacy involves the amalgamation of NPs with bacteriocins, resulting in the formation of nano-encapsulated bacteriocins derived from both Gram-negative and Gram-positive bacteria (Shafique et al., 2022). As per a previous study, NPs were fabricated from natural polyphenols and tobramycin antibiotic, which may effectively destroy the biofilm structure and kill bacteria through enhancingthe effectiveness of antibiotics. To eliminate the biofilm, natural polyphenols can compete with autoinducer-2 (AI-2) and suppress quorum sensing in bacteria (Li et al., 2023) (Figure 3.4).

3.5.5 Antiviral

There is a wide range of functional food derived from diverse animal, plant, and fungal sources that have the potential to enhance antiviral immunity. The majority of these items contain naturally occurring vitamins A, C, and D, minerals like Zn, Se, and Fe, and some phenols and phytochemicals (polyphenols, alkaloids, saponins, flavonoids, terpenes, proanthocyanidins, lignans, quinones, tannins, steroids, organosulfur compounds, polysaccharides, and coumarins) that can prevent various RNA viruses from replicating, inhibit viral hemagglutination, increase the secretion of interferon and pro-inflammatory cytokines, and cause the production of specific antibodies against viruses in host cells. Additionally, alkaloids prevent HIV reverse transcriptase and viral protein synthesis, lectins prevent penetration and viral multiplication, and flavonoids prevent reverse transcriptase andsynthesis of viral protein (Goyal et al., 2022; Omer et al., 2022). Despite of having a strong antiviral efficacy of functional food they are also encapsulated into NPs to enhance the solubility and stability as well as to low the degradation rate. Nanoencapsulation of polyphenolic compounds including flavonolignan and flavonoid has a broad range of possible applications with high antiviral activity (Milinčić et al., 2019). The natural flavonoid vitexin, which

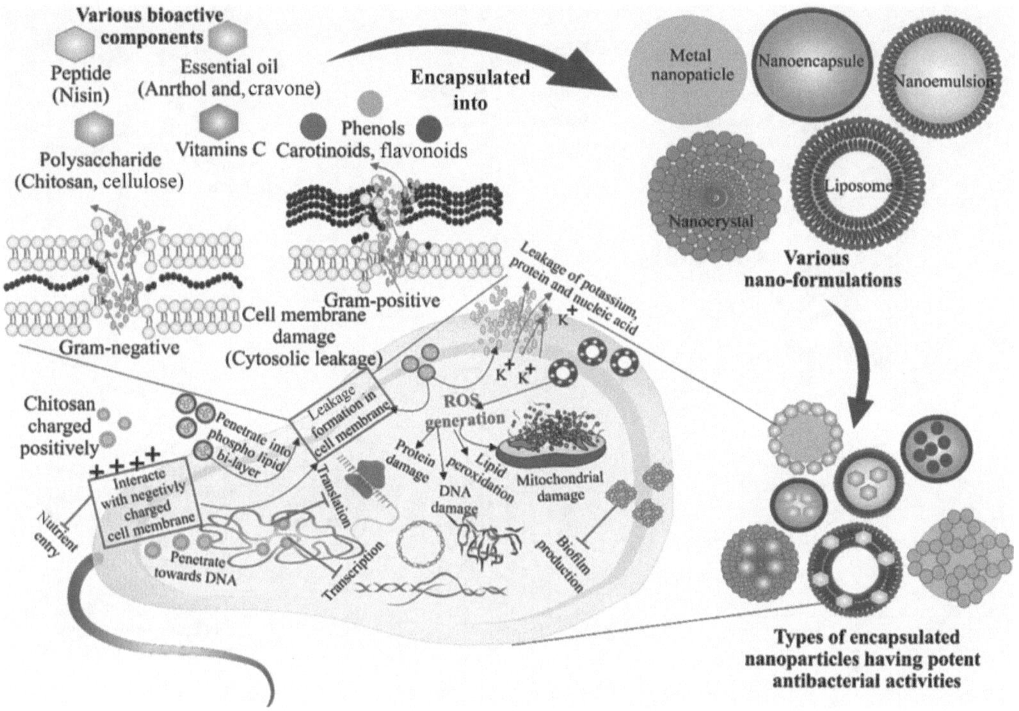

FIGURE 3.4 This schematic diagram represents encapsulation of numerous types of bioactive components into various of nanoparticles to achieve potent antibacterial activities. Through different mechanisms they lead to bacterial cell death such as inhibition of transcription, translation, and biofilm production, also leadingto cell membrane damage, which causes potassium, protein, nucleicacid, and cytosolic leakage. Parts of this figure were drawn using pictures from Servier Medical Art. Servier Medical Art by Servier is licensed under a Creative Commons Attribution 3.0 Unported License (https://creativecommons.org/licenses/by/3.0/).

is present in many medicinal plants, has antiviral properties, and the main mechanism of action is the suppression of viral neuraminidase, DNA/RNA polymerases, proteases, as well as the alteration of numerous viral proteins but with poor water solubility. So, to overcome this limitation and to improve the dissolution rate, the NPs of vitexin were combined the antisolvent precipitation (ASP) and high-pressure homogenization (HPH) methods, followed by lyophilization (Gu et al., 2017; Ninfali et al., 2020). Curcumin's antiviral properties show noticeably strong suppression of various viruses via limiting cell-to-cell transmission by preventing the cell membrane integrity, which prevents attachment and fusion of virus to hepatocytes. The beneficial effects of curcumin are constrained by its poorer solubility, bioavailability, and cellular absorption. Using polymeric NPs such as curcumin is encapsulated into CSNPs to increase the antiviral activity against entry and replication of hepatitis C virus-4 (HCV-4). This could overcome curcumin's limitations by enhancing its sustained release to target diseased cells, improve bioavailability, preventfrom degradation or metabolism, and increasetherapeutic potential (Loutfy et al., 2020). Naringenin, a citrus-derived flavonoid, exhibits antiviral action in addition to being more bioavailable due to formulations like naringenin-loaded NPs (Tutunchi et al., 2020). Carotenoids like lutein, zeaxanthin, and carotene have also drawn interest for their antiviral function. Emulsion-based delivery systems have been employed to distribute carotenoids due to their poor oral bioavailability and stability (Boonlao et al., 2022). Carvacrol, thymol, and γ-terpinene are antiviral compounds found in oregano essential oils. Therefore, micro- and nano-encapsulation techniques were used to increase physicochemical stability, extend shelf life, and accomplish controlled release (Plati & Paraskevopoulou, 2022).

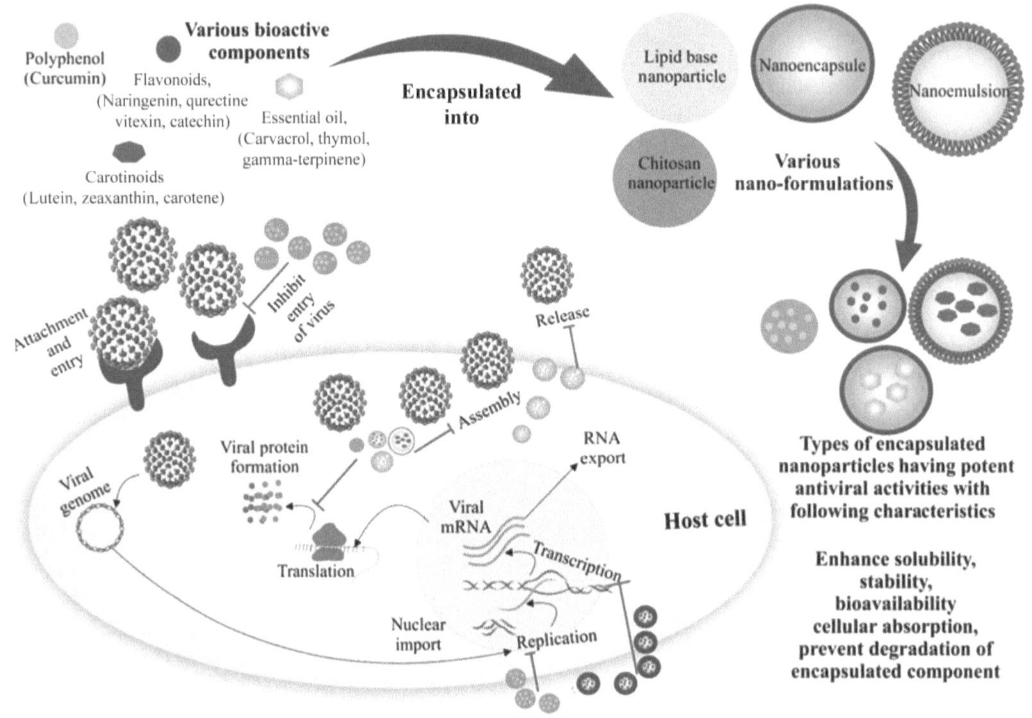

FIGURE 3.5 In this figure, bioactive components are encapsulated into nanoparticle which may enhance the solubility, bioavailability, targeted delivery of those encapsulated component, and also protect them from degradation. Along with those, the encapsulated nanoparticle shows potent antiviral activities by inhibiting the entry of virus into host cell, as well as inhibiting the viral replication, transcription, translation, and cell-to-cell contact. Parts of this figure were drawn using pictures from Servier Medical Art. Servier Medical Art by Servier is licensed under a Creative Commons Attribution 3.0 Unported License (https://creativecommons.org/licenses/by/3.0/).

According to a prior study, nano-formulations are also employed to increase bioavailability and as a carrier molecule of phytoconstituents (Goyal et al., 2022). PLGA-encapsulated quercetin and catechin by oral delivery system also show antiviral activity (Figure 3.5).

3.6 NUMEROUS BIOACTIVE COMPOUNDS IN DISEASE MANAGEMENT REPORTED IN CLINICAL TRIAL STUDIES

A clinical trial (ClinicalTrials.gov: NCT02195960) study evaluates the association between the effect of polyphenol-rich food supplementation and toxicity of radiotherapy for breast cancer. The medication will begin a week before the radiotherapy begins and resume throughout the entire radiotherapy course (3 or 5 weeks) in 300 consecutive breast cancer patients who are radiation candidates. In a literature review study, Neethu and his team described the possible contribution of vitamin C to the treatment of COVID-19 patients and provided data from previous clinical studies. This work has revealed the potential function of vitamin C in combating COVID-19. Several clinical trials included in this study demonstrated that vitamin C has excellent antioxidative and antiviral properties, which opened up a new door for its possible application in therapy and prognosis of COVID-19 (Neethu et al., 2022). Patients with COVID-19 and reduced oxygenation responded favourably to high-dose intravenous vitamin C (HDIVC) therapy, according to phase I and phase II clinical trials (ClinicalTrials.gov: NCT04357782). Case studies indicate that hospital admissions, morbidity (ventilator use, ICU beds, and organ failures), and mortality are influenced by sepsis and the acute respiratory distress syndrome (ARDS). HDIVC may lessen the inflammatory response and the emergence of respiratory failure necessitating intubation. Additionally, to build a mechanistic consideration and hazard stratification of response to HDIVC infusion, researchers document changes in the inflammatory markers which are high in COVID-19 (C reactive protein d-dimer, lactate dehydrogenase, ferritin, and liver enzymes). A phase II clinical trial study (ClinicalTrials.gov: NCT04715932) evaluated the consequences of treatment with hesperidin on COVID-19-related infection. Participants will record their symptoms in a journal during the trial, and daily oral temperature readings will be taken to track treatment effects. In this study, 216 participants who met all exclusion and inclusion criteria were randomly assigned to receive either hesperidin 1000 mg once daily (q.d.) or a placebo (1:1 allocation ratio) for 14 days. Another clinical trial study (ClinicalTrials.gov: NCT00921934) demonstrated that shingles (herpes zoster infection) has long been successfully treated with antioxidants including higher doses of vitamin C. High doses of vitamin C can reduce symptoms in addition to the acute ones, even long-term effects, such as uncomfortable post-herpetic neuropathy, may be reduced or avoided completely. In multiple clinical trials, zinc and green tea supplementation both showed encouraging but variable outcomes for promoting immune function during cold and flu-like illness in non-hospitalized patients. Combination therapy in this clinical trial (ClinicalTrials.gov: NCT04898023) may have an even greater impact because antioxidants in green tea have been demonstrated to raise cellular zinc contents, which prevent virus reproduction. In this investigation, adult community patients participating in a randomized placebo-controlled trial will be assessed for their symptom duration and severity from cold and flu-like illnesses, including COVID-19, in relation to a combination supplementation utilizing recommended doses of zinc and green tea extract. According to a phase I clinical trial (ClinicalTrials.gov: NCT03870529) study, researchers assessed how effectively a high-dose vitamin A molecule treatedaffected ones with non–small cell lung cancer that can be removed by surgery. Vitamin A molecule may boost the quantity of germinal centres (immune centres that develop antibodies) in lymph and tumour tissues, which may be helpful for cancer patients. Themain goal of this study is to evaluate the proportion of tumours that were surgically removed and had germinal centres (GCs) in patients who received neoadjuvant vitamin A compound (vitamin A) versus controls. Additionally, the secondary goals include comparing the prevalence of GCs in nearby lymph nodes in neoadjuvant vitamin A lung cancer patients to controls, comparing the histopathological responses based on tumour necrosis in neoadjuvant vitamin A lung cancer patients to controls, and lastly comparing the overall survival of neoadjuvant vitamin A lung cancer patients to controls. A prior clinical trial study (ClinicalTrials.gov: NCT01910948) was conducted to analyse the impact of perioperative therapy of omega-3 polyunsaturated fatty acids on the immune system during surgery for gastric cancer. The results demonstrated that, in trialgroup patients, omega-3 polyunsaturated fatty acids can have the ability to lower initial postoperative inflammatory cytokine proclamation, postoperative fever period, decrease the prevalence of systemic inflammatory response syndrome, endorse the recovery of gastrointestinal function, enhance patients' nutritional status, lower infection and related complications after surgery, and lower the number of days spent in the hospital after surgery compared to the control groups. Another clinical trial study (ClinicalTrials.gov: NCT00197561) was conducted to determine if daily oral selenium supplementation, when compared to a placebo, improves immune response, lowers HIV-1 infection viral load at 6months postnatally, lowers the risk of genital flaking of HIV-1 infected cells at 36 weeks of gestation, and lowers the risk of mastitis at 6weeks postnatally in HIV-1-positive pregnant women. In this trial, researchers enrolled HIV-positive pregnant women and randomly assigned them to either selenium or a placebo. Standard prenatal care will be provided to all mothers, including prenatal multivitamin supplements and nevirapine in the anticipation of mother-to-child transmission. Researchers investigated how selenium supplements affect a range of intermediate outcomes that can be used to forecast the risks of HIV transmission and the disease

TABLE 3.1
Recent updates on clinical trials conducted using different bioactive compounds in disease management

Sl. No.	NCT Number https://clinicaltrials.gov/	Disease	Study Description	Phase	Study Start Date	Recruitment Status	FastPosted	Last Update Posted	Country
1	NCT04117022	Diabetic Retinopathy	The capability of the chromatic electroretinogram (chERG) to identify alterations in overall retinal function following supplementation with carotenoid vitamins in diabetic retinopathy patients	NA	2019-09-01	Unknown status	2019-10-01	2021-04-23	US
2	NCT04715932	COVID-19	This study's primary goal is to ascertain how COVID-19 symptoms respond to short-term hesperidin treatment	II	2021-02-18	Completed	2021-01-20	2022-04-08	Canada
3	NCT03870529	Non–Small Cell Lung Cancer	In this trial investigation, individuals with non–small cell lung cancer that can be surgically removed were treated with a high-dose vitamin A molecule, and the results can be promising	I	2019-08-19	Completed	2019-03-12	2023-07-17	USA
4	NCT01820299	Inflammation in Patients with Advanced Cancer	This study's main objective is to find phytochemicals from plants (oligomeric procyanidin complex and vitamin D3) that can safely lower systemic inflammation (inflammation throughout the body) in people with advanced cancer.	I	2013-03	Completed	2013-03-28	2018-05-18	USA

5	NCT01910948	Gastric Cancer	This study's objective is to assess the immunomodulatory consequences of perioperative omega-3 polyunsaturated fatty acid administration in surgery of gastric cancer	IIII	2013-06	Unknown status	2013-07-30	2013-07-30	NA
6	NCT02278965	Early-Stage Breast Cancer	In this trial, women with a history of early-stage breast cancer will be given a 1-year course of omega--3 fatty acids and metformin to see if it is feasible.	I	2015-01-22	Completed	2014-10-30	2022-08-17	USA
7	NCT00197561	HIV Infection	The goal of this study is to ascertain whether getting regular selenium supplements orally to HIV-1-positive pregnant women improves their immunological health and lowers their HIV-1 viral load	III	2003-09	Completed	2005-09-20	2010-11-11	Tanzania
8	NCT04751604	COVID-19	This goal of this study is to determine whether nicotinamide, a kind of vitamin B3, can promote the treatment of SARS-CoV-2 infection (COVID-19) through dietary intervention	NA	2021-02-01	Completed	2021-02-12	2022-11-08	Germany
9	NCT01477112	Oxidative Stress in Type 2 Diabetics	The goal was to examine the effects of carotene supplementation on iron metabolism, oxidative balance, and antioxidant plasma capacity in type 2 diabetics and healthy individuals at doses comparable to the daily nutritional requirement	NA	2010-01	Completed	2011-11-22	2011-11-22	Venezuela

(continued)

TABLE 3.1 (Continued)
Recent updates on clinical trials conducted using different bioactive compounds in disease management

Sl. No.	NCT Number https://clinicaltrials.gov/	Disease	Study Description	Phase	Study Start Date	Recruitment Status	FastPosted	Last Update Posted	Country
10	NCT02195960	Side Effects of Radiotherapy for Breast Cancer	The major goals of this clinical trial are to evaluate the relationship between the toxicity of breast cancer radiotherapy and the effect of consuming supplements high in polyphenols (anthocyanins)	NA	2014-06-09	Completed	2014-07-21	2021-01-27	Italy
11	NCT04622865	Moderate and Severe COVID-19	The purpose of the study is to determine if the combination of masitinib and isoquercetin is effective for treating adult hospital patients with moderate and severe COVID-19	II	2020-06-01	Recruiting	2020-11-10	2023-02-06	France
12	NCT05387252	Side Effects of COVID-19 Vaccines	The aim of the study is to determine if the combination of masitinib and isoquercetin is effective for treating adult hospital patients with moderate and severe COVID-19.	NA	2022-05-31	Not yet recruiting	2022-05-24	2022-05-24	NA
13	NCT04534075	Radiotherapy	The researchers postulate that increasing dietary fibre consumption during radiation therapy may result in improved gut health over the long term	III	2020-09-02	Recruiting	2020-09-01	2023-05-01	Sweden
14	NCT04357782	Novel Coronavirus Infection (COVID-19)	According to the researchers' hypotheses, high-dose intravenous vitamin C can besafe and tolerable in coronavirus disease 2019 (COVID-19) subjects when administered early or late	I and II	2020-04-16	Completed	2020-04-22	2022-02-25	US

			in the disease course and may lower the risk of respiratory failure necessitating mechanical ventilation and the emergence of acute respiratory distress syndrome in addition to lowering the requirement for supplemental oxygen						
15	NCT04171947	Imbalance in the Vaginal Environment	The purpose of this study is to evaluate the effects of vaginal ovules containing 2 mg of the tea extract Matuzalem on the subjective symptoms of bacterial vaginosis, such as pain, itching, discharge, redness, and odour	NA	2017-12-04	Completed	2019-11-21	2019-11-21	Czechia
16	NCT01111864	Microbiota and Iron Status Changes Following Iron Fortification of Supplemental Foods	The aim of the study is to determine whether home iron fortification of neonates in Kenya with a micronutrient powder will improve their iron status and change the composition and metabolic activity of their gut bacteria. Weekly active observation will be conducted to check on the babies' well-being	NA	2010-02	Completed	2010-04-28	2013-06-07	Kenya
17	NCT04494451	Multidrug-Resistant BacterialInfections	The objective is to assess the impacts of vitamin C on systemic haemodynamics and AKI outcomes in patients of cirrhosis and MDR infection	NA	2020-07-25	Unknown status	2020-07-31	2020-07-31	India

(continued)

TABLE 3.1 (Continued)
Recent updates on clinical trials conducted using different bioactive compounds in disease management

Sl. No.	NCT Number https://clinicaltrials.gov/	Disease	Study Description	Phase	Study Start Date	Recruitment Status	FastPosted	Last Update Posted	Country
18	NCT00877422	Elderly Pneumonia	The researchers of the current study propose that insufficient levels of serum vitamin D are linked to the development of pneumonia, and that vitamin D treatment may reduce pneumonia incidence and increase survival in institutionalized elderly people	NA	2008-08	Completed	2009-04-07	2009-04-08	Japan
19	NCT02122627	COPD (pre COVID)	The current study's objective is to assess the impact of vitamin D supplementation on the rate of COPD exacerbations in patients with a vitamin D deficiency	NA	2015-04-10	Completed	2014-04-24	2019-09-10	Netherlands
20	NCT04682574	COVID-19	Water-soluble vitamin C (ascorbic acid) has anti-inflammatory, immune-modulating, anti-oxidative, antithrombotic, and antiviral effects. Acute respiratory distress syndrome and sepsis patients should benefit from vitamin C treatment if these effects are taken into account. The goal of the current research is to assess how well vitamin C treats COVID-19 infection in patients.	NA	2020-11-02	Completed	2020-12-23	2022-05-06	Pakistan
21	NCT02119143	Common Cold	This study examines the effect of mineral and vitamin supplement on the prevalence of the common cold in a group of middle management	NA	2014-04	Completed	2014-04-21	2017-08-01	Austria

			workers.The number of duty days missed owing to the common cold or flu is the major consequence. Additionally, immunological parameters and redox biology/ oxidative stress markers will be identified. Utilizing questionnaires, the cohort's well-being will be assessed						
22	NCT04648917	Squamous Oesophageal Cell Cancer	This clinical trial seeks to examine the effectiveness and safety of caffeic acid in treating advanced oesophageal squamous cell carcinoma in Chinese patients	III	2019-05-01	Unknown status	2020-12-02	2020-12-03	China
23	NCT00656929	Viral Upper Respiratory Tract Infections	Supplementing with vitamin D3 to prevent viral upper respiratory tract infections: a randomized controlled trial	NA	2006-12	Completed	2008-04-11	2008-04-11	NA
24	NCT04428359	Treatment of Warts (MMR)	A study comparing the effectiveness of intralesional vitamin D3 with intralesional MMR (measles, mumps, rubella) vaccine in the treatment of warts	NA	2020-06-01	Completed	2020-06-11	2021-04-09	Nepal
25	NCT00921934	ViralInfection, Especially inthe Treatment of Shingles	Shingles and other viral infections are treated with intravenous vitamin C	NA	2009-04	Completed	2009-06-17	2017-12-21	Germany

(continued)

TABLE 3.1 (Continued)
Recent updates on clinical trials conducted using different bioactive compounds in disease management

Sl. No.	NCT Number https://clinicaltrials.gov/	Disease	Study Description	Phase	Study Start Date	Recruitment Status	FastPosted	Last Update Posted	Country
26	NCT05212480	Viral Infections (COVID-19 patients)	The clinical study's primary goal is to assess the effectiveness of zinc supplementation in non-critical patients. COVID-19 sufferers	NA	2022-02-15	Completed	2022-01-28	2023-03-16	Tunisia
27	NCT04898023	Viral Infections	Assessment of the impacts of supplementing with green tea extract and zinc on community respiratory viral infections'symptoms'length and severity	II	2022-09-21	Recruiting	2021-05-24	2023-06-02	US

progression. These clinical trials are essential for the development of new medical treatments with natural bioactive compounds against severe disease progression (Table 3.1).

3.7 CURRENT CHALLENGES AND FUTURE DIRECTIONS

This book chapter highlights recent progress in nanomedicine, including advanced technologies for the delivery of bioactive substances and novel diagnostic approaches. It is crucial for everyone to have access for medicine to achieve better health, but nowadays medical supplies are highly expensive. Therefore, the bioactive substances (flavonoid, phenolic compounds, dietary fibres, vitamins, and minerals), which are plentiful and found in a variety of plants, may be a promising option of molecules for the development of drugs and medical products having low cost. Different cultivars of the same species of medicinal plants may offer varying amounts of phenolic and flavonoid chemicals with biological activity. Nanotechnology was primarily used to improve the absorption, solubility, bioavailability, and controlled release of medications or drugs. Future research must focus on various known flavonoids, phenolic compounds, carotenoids, essential fatty acids, vitamins, and other bioactive compound molecular mechanisms and signalling pathways with the aimof applying this understanding in developing new drugs, by using specially designed nanocarriers like polymeric missiles, liposomal delivery, nano-emulsion dendrimers, quantum dots, nanocellulose, etc. The efficiency of these natural components can be considerably increased by those nano-delivery systems. Although regulatory frameworks and mechanism for nanomedicines require future advancements in safety or toxicity assessment, nanomedicine has already altered how we identify and administer medications in the biological system. Additional scientific experiments are required to clarify the molecular mechanism of drug distribution and close any remaining gaps in the use of plant bioactive chemicals in medical therapies.

REFERENCES

Abdol Wahab, N. R., Meor Mohd Affandi, M. M. R., Fakurazi, S., Alias, E., & Hassan, H. (2022). Nanocarrier system: state-of-the-art in oral delivery of astaxanthin. *Antioxidants (Basel), 11*(9), 1676. https://doi.org/10.3390/antiox11091676

Abdulhakeem, Z. R., Odda, A. H., & Abdulsattar, S. A. (2023). Pectin-based nanomaterials as a universal polymer for type 2 diabetes management. *Medical Journal of Babylon, 20*(1), 7–12.

Adefegha, S. A. (2018). Functional foods and nutraceuticals as dietary intervention in chronic diseases; novel perspectives for health promotion and disease prevention. *Journal of Dietary Supplements, 15*(6), 977–1009. https://doi.org/10.1080/19390211.2017.1401573

Ahmad Khorairi, A. N. S., Sofian-Seng, N.-S., Othaman, R., Abdul Rahman, H., Mohd Razali, N. S., Lim, S. J., & Wan Mustapha, W. A. (2023). A review on agro-industrial waste as cellulose and nanocellulose source and their potentials in food applications. *Food Reviews International, 39*(2), 663–688. https://doi.org/10.1080/87559129.2021.1926478

Ameta, S. K., Rai, A. K., Hiran, D., Ameta, R., & Ameta, S. C. (2020). Use of Nanomaterials in Food Science. In: Ghorbanpour M., Bhargava P., Varma A., Choudhary D.K. (eds) *Biogenic Nano-Particles and Their Use in Agro-Ecosystems*. Springer Singapore, p 457–488. https://doi.org/10.1007/978-981-15-2985-6_24

Amigo, L., & Hernández-Ledesma, B. (2020). Current evidence on the bioavailability of food bioactive peptides. *Molecules, 25*(19), 4479. https://doi.org/10.3390/molecules25194479

Amjad, E., Sokouti, B., & Asnaashari, S. (2022). A systematic review of anti-cancer roles and mechanisms of kaempferol as a natural compound. *Cancer Cell International, 22*(1), 1–22. https://doi.org/10.1186/s12935-022-02673-0

Arshad, M. S., Khalid, W., Ahmad, R., Khan, M., Ahmad, M. H., Safdar, S., Kousar, S., Munir, H., Shabbir, U., Zafarullah, M., Nadeem, M., Asghar, Z., & Suleria, H. (2021). Functional Foods and Human Health: An Overview. In: Ahmad M.S.A.a.M.H. (ed) *Functional Foods—Phytochemicals and Health Promoting Potential*. IntechOpen. http://dx.doi.org/10.5772/intechopen.99000

Askar, M. A., El Shawi, O. E., Mansour, N. A., & Hanafy, A. M. (2021). Breast cancer suppression by curcumin–naringenin-magnetic-nano-particles: in vitro and in vivo studies. *Tumor Biology, 43*(1), 225–247. https://doi.org/10.3233/tub-211506

Assadpour, E., & Jafari, S. M. (2019). Chapter 3—Nanoencapsulation: Techniques and Developments for Food Applications. In: López Rubio A., Fabra Rovira M. J., Martínez Sanz M., Gómez-Mascaraque L. G. (eds) *Nanomaterials for Food Applications*. Elsevier, p 35–61. https://doi.org/10.1016/B978-0-12-814130-4.00003-8

Aswathanarayan, J. B., & Vittal, R. R. (2019). Nanoemulsions and their potential applications in food industry. *Frontiers in Sustainable Food Systems, 3*, 95. https://doi.org/10.3389/fsufs.2019.00095

Bacha, K., Chemotti, C., Mbakidi, J.-P., Deleu, M., & Bouquillon, S. (2023). Dendrimers: synthesis, encapsulation applications and specific interaction with the stratum corneum—a review. *Macromol, 3*(2), 343–370. https://doi.org/10.3390/macromol3020022

Baek, J., Ramasamy, M., Willis, N. C., Kim, D. S., Anderson, W. A., & Tam, K. C. (2021). Encapsulation and controlled release of vitamin C in modified cellulose nanocrystal/chitosan nanocapsules. *Current Research in Food Science, 4*, 215–223. https://doi.org/10.1016/j.crfs.2021.03.010

Baksi, R., Singh, D. P., Borse, S. P., Rana, R., Sharma, V., & Nivsarkar, M. (2018). In vitro and in vivo anticancer efficacy potential of Quercetin loaded polymeric nanoparticles. *Biomedicine & Pharmacotherapy, 106*, 1513–1526. https://doi.org/10.1016/j.biopha.2018.07.106

Bharathiraja, S., Manivasagan, P., Quang Bui, N., Oh, Y.-O., Lim, I. G., Park, S., & Oh, J. (2016). Cytotoxic induction and photoacoustic imaging of breast cancer cells using astaxanthin-reduced gold nanoparticles. *Nanomaterials, 6*(4), 78. https://doi.org/10.3390/nano6040078

Bhat, E. A., Sajjad, N., Manzoor, I., & Rasool, A. (2019). Bioactive compounds in peanuts and banana. *Biochemistry & Analytical Biochemistry, 8*, 382. http://dx.doi.org/10.35248/2161-1009.19.8.382

Biswas, S., & Sengupta, S. (2022).Current use of nanoprotein and application in the development of food products for functional and nutritional benefits. *Journal of Microbiology, Biotechnology and Food Sciences, 11*(5), e1737–e1737. http://dx.doi.org/10.55251/jmbfs.1737

Bittleman, K. R., Dong, S., Roman, M., & Lee, Y. W. (2018). Folic acid-conjugated cellulose nanocrystals show high folate-receptor binding affinity and uptake by KB and breast cancer cells. *ACS Omega, 3*(10), 13952–13959. https://doi.org/10.1021/acsomega.8b01619

Bliss, D. Z., Savik, K., Jung, H. J., Whitebird, R., & Lowry, A. (2011). Symptoms associated with dietary fiber supplementation over time in individuals with fecal incontinence. *Nursing Research, 60*(3 Suppl), S58–S67. https://doi.org/10.1097%2FNNR.0b013e3182186d8c

Boonlao, N., Ruktanonchai, U. R., & Anal, A. K. (2022). Enhancing bioaccessibility and bioavailability of carotenoids using emulsion-based delivery systems. *Colloids and Surfaces B: Biointerfaces, 209*, 112211. https://doi.org/10.1016/j.colsurfb.2021.112211

Bourbon, A. I., Pereira, R. N., Pastrana, L. M., Vicente, A. A., & Cerqueira, M. A. (2019). Protein-based nanostructures for food applications. *Gels, 5*(1), 9. https://doi.org/10.3390%2Fgels5010009

Câmara, J. S., Albuquerque, B. R., Aguiar, J., Corrêa, R. C. G., Gonçalves, J. L., Granato, D., Pereira, J. A. M., Barros, L., & Ferreira, I. C. F. R. (2020). Food bioactive compounds and emerging techniques for their extraction: Polyphenols as a case study. *Foods, 10*(1), 37. https://doi.org/10.3390/foods10010037

Carotenuto, P., Pecoraro, A., Brignola, C., Barbato, A., Franco, B., Longobardi, G., Conte, C., Quaglia, F., Russo, G., & Russo, A. (2023). Combining β-Carotene with 5-FU via polymeric nanoparticles as a novel therapeutic strategy to overcome uL3-mediated chemoresistance in p53-deleted colorectal cancer cells. *Molecular Pharmaceutics, 20*(5), 2326–2340. https://doi.org/10.1021/acs.molpharmaceut.2c00876

Chavda, V. P., Patel, A. B., Mistry, K. J., Suthar, S. F., Wu, Z. X., Chen, Z. S., & Hou, K. (2022). Nano-drug delivery systems entrapping natural bioactive compounds for cancer: recent progress and future challenges. *Frontiers in Oncology, 12*, 867655. https://doi.org/10.3389/fonc.2022.867655

Chen, G., Yang, F., Fan, S., Jin, H., Liao, K., Li, X., Liu, G.-B., Liang, J., Zhang, J., & Xu, J.-F. (2022). Immunomodulatory roles of selenium nanoparticles: Novel arts for potential immunotherapy strategy development. *Frontiers in Immunology, 13*, 956181. https://doi.org/10.3389/fimmu.2022.956181

Chen, H., Weiss, J., & Shahidi, F. (2006). Nanotechnology in nutraceuticals and functional foods. *Food Technology (Chicago), 60*(3), 30–36.

Chis, A. A., Dobrea, C., Morgovan, C., Arseniu, A. M., Rus, L. L., Butuca, A., Juncan, A. M., Totan, M., Vonica-Tincu, A. L., & Cormos, G. (2020). Applications and limitations of dendrimers in biomedicine. *Molecules*, *25*(17), 3982. https://doi.org/10.3390%2Fmolecules25173982

Choi, S. J., & McClements, D. J. (2020). Nanoemulsions as delivery systems for lipophilic nutraceuticals: strategies for improving their formulation, stability, functionality and bioavailability. *Food Science and Biotechnology, 29*(2), 149–168. https://doi.org/10.1007/s10068-019-00731-4

Chopde, S., Datir, R., Deshmukh, G., Dhotre, A., & Patil, M. (2020). Nanoparticle formation by nanospray drying & its application in nanoencapsulation of food bioactive ingredients. *Journal of Agriculture and Food Research, 2*, 100085. http://dx.doi.org/10.1016/j.jafr.2020.100085

Crintea, A., Motofelea, A. C., Şovrea, A. S., Constantin, A.-M., Crivii, C.-B., Carpa, R., & Duţu, A. G. (2023). Dendrimers: advancements and potential applications in cancer diagnosis and treatment—an overview. *Pharmaceutics, 15*(5), 1406. https://doi.org/10.3390/pharmaceutics15051406

De Jong, W. H., & Borm, P. J. A. (2008). Drug delivery and nanoparticles: applications and hazards. *International Journal of Nanomedicine*, *3*(2), 133–149. https://doi.org/10.2147/ijn.s596

de la Rosa, L. A., Moreno-Escamilla, J. O., Rodrigo-García, J., & Alvarez-Parrilla, E. (2019). Chapter 12—Phenolic Compounds. In: Yahia E. M. (ed) *Postharvest Physiology and Biochemistry of Fruits and Vegetables*. Woodhead Publishing, pp. 253–271. https://doi.org/10.1016/B978-0-12-813278-4.00012-9

de Vos, P., Faas, M. M., Spasojevic, M., & Sikkema, J. (2010). Encapsulation for preservation of functionality and targeted delivery of bioactive food components. *International Dairy Journal, 20*(4), 292–302. https://doi.org/10.1016/j.idairyj.2009.11.008

Delgado-Vargas, F., Jiménez, A. R., & Paredes-López, O. (2000). Natural pigments: carotenoids, anthocyanins, and betalains—characteristics, biosynthesis, processing, and stability. *Critical Reviews in Food Science and Nutrition, 40*(3), 173–289. https://doi.org/10.1080/10408690091189257

Dewanjee, S., Chakraborty, P., Mukherjee, B., & DeFeo, V. (2020). Plant-based antidiabetic nanoformulations: the emerging paradigm for effective therapy. *International Journal of Molecular Sciences, 21*(6), 2217. https://doi.org/10.3390%2Fijms21062217

Dhingra, D., Michael, M., Rajput, H., & Patil, R. T. (2012). Dietary fibre in foods: a review. *Journal of Food Science and Technology, 49*(3), 255–266. https://doi.org/10.1007%2Fs13197-011-0365-5

Diez-Echave, P., Ruiz-Malagón, A. J., Molina-Tijeras, J. A., Hidalgo-García, L., Vezza, T., Cenis-Cifuentes, L., Rodríguez-Sojo, M. J., Cenis, J. L., Rodríguez-Cabezas, M. E., & Rodríguez-Nogales, A. (2021). Silk fibroin nanoparticles enhance quercetin immunomodulatory properties in DSS-induced mouse colitis. *International Journal of Pharmaceutics*, *606*, 120935. https://doi.org/10.1016/j.ijpharm.2021.120935

Ding, Q., Liu, W., Liu, X., Ding, C., Zhao, Y., Dong, L., Chen, H., Sun, S., Zhang, Y., & Zhang, J. (2022). Polyvinylpyrrolidone-modified taxifolin liposomes promote liver repair by modulating autophagy to inhibit activation of the TLR4/NF-κB signaling pathway. *Frontiers in Bioengineering and Biotechnology*, *10*, 860515. https://doi.org/10.3389/fbioe.2022.860515

Dini, S., Zakeri, M., Ebrahimpour, S., Dehghanian, F., & Esmaeili, A. (2021). Quercetin-conjugated superparamagnetic iron oxide nanoparticles modulate glucose metabolism-related genes and miR-29 family in the hippocampus of diabetic rats. *Scientific Reports, 11*(1), 8618. https://doi.org/10.1038/s41598-021-87687-w

El-Shahawy, A. A. G., Abdel-Moneim, A., Ebeid, A. S. M., Eldin, Z. E., & Zanaty, M. I. (2021). A novel layered double hydroxide-hesperidin nanoparticles exert antidiabetic, antioxidant and anti-inflammatory effects in rats with diabetes. *Molecular Biology Reports, 48*(6), 5217–5232. https://doi.org/10.1007/s11033-021-06527-2

Esfandyari-Manesh, M., Ghaedi, Z., Asemi, M., Khanavi, M., Manayi, A., Jamalifar, H., Atyabi, F., & Dinarvand, R. (2013). Study of antimicrobial activity of anethole and carvone loaded PLGA nanoparticles. *Journal of Pharmacy Research, 7*(4), 290–295. https://doi.org/10.3390%2Fjfb9010004

Fan, X., Ngo, H., & Wu, C. (2018). Natural and Bio-Based Antimicrobials: A Review Natural and Bio-Based Antimicrobials for Food Applications. ACS Symposium Series, vol 1287. American Chemical Society, p 1–24. doi:10.1021/bk-2018-1287.ch001

Gangwar, V., Garg, A., Lomore, K., Korla, K., Bhat, S. S., Rao, R. P., Rafiq, M., Kumawath, R., Uddagiri, B. V., & Kareenhalli, V. V. (2021). Immunomodulatory effects of a concoction of natural bioactive compounds—Mechanistic insights. *Biomedicines*, *9*(11), 1522. https://doi.org/10.3390/biomedicines9111522

Gharibzahedi, S. M. T., & Jafari, S. M. (2017). The importance of minerals in human nutrition: bioavailability, food fortification, processing effects and nanoencapsulation. *Trends in Food Science & Technology, 62*, 119–132. https://doi.org/10.1016/j.tifs.2017.02.017

Ginwala, R., Bhavsar, R., Chigbu, D. G. I., Jain, P., & Khan, Z. K. (2019). Potential role of flavonoids in treating chronic inflammatory diseases with a special focus on the anti-inflammatory activity of apigenin. *Antioxidants (Basel), 8*(2), 35. https://doi.org/10.3390%2Fantiox8020035

Goyal, R., Bala, R., Sindhu, R. K., Zehravi, M., Madaan, R., Ramproshad, S., Mondal, B., Dey, A., Rahman, M. H., & Cavalu, S. (2022). Bioactive based nanocarriers for the treatment of viral infections and SARS-CoV-2.*Nanomaterials*. 12, 1530.

Granja, A., Pinheiro, M., & Reis, S. (2016). Epigallocatechin gallate nanodelivery systems for cancer therapy, *8*(5), 307. https://doi.org/10.3390/nu8050307

Gu, C., Liu, Z., Yuan, X., Li, W., Zu, Y., & Fu, Y. (2017). Preparation of vitexin nanoparticles by combining the antisolvent precipitation and high pressure homogenization approaches followed by lyophilization for dissolution rate enhancement. *Molecules, 22*(11), 2038. https://doi.org/10.3390/molecules22112038

Gutiérrez, R. M. P., Gómez, J. T., Jerónimo, F. F. M., Paredes-Carrera, S. P., & Sánchez-Ochoa, J. C. (2023). Effects of selenium nanoparticles using potential natural compounds naringenin and baicalin for diabetes. *Biointerface Research in Applied Chemistry, 13*(6), 597. https://doi.org/10.33263/BRIAC136.597

Ha, H.-K., Rankin, S. A., Lee, M.-R., & Lee, W.-J. (2019). Development and characterization of whey protein-based nano-delivery systems: a review. *Molecules, 24*(18), 3254. https://doi.org/10.3390/molecules22112038

Hoshyar, R., Khayati, G. R., Poorgholami, M., & Kaykhaii, M. (2016). A novel green one-step synthesis of gold nanoparticles using crocin and their anti-cancer activities. *Journal of Photochemistry and Photobiology B: Biology, 159*, 237–242. https://doi.org/10.1016/j.jphotobiol.2016.03.056

Hussein, Y. H. A., & Youssry, M. (2018). Polymeric micelles of biodegradable diblock copolymers: enhanced encapsulation of hydrophobic drugs. *Materials, 11*(5), 688. https://doi.org/10.3390/ma11050688

Ignatova, M., Manolova, N., & Rashkov, I. (2007). Novel antibacterial fibers of quaternized chitosan and poly (vinyl pyrrolidone) prepared by electrospinning. *European Polymer Journal, 43*(4), 1112–1122. https://doi.org/10.1016/j.eurpolymj.2007.01.012

Imran, M., Revol-Junelles, A.-M., René, N., Jamshidian, M., Akhtar, M. J., Arab-Tehrany, E., Jacquot, M., & Desobry, S. (2012). Microstructure and physico-chemical evaluation of nano-emulsion-based antimicrobial peptides embedded in bioactive packaging films. *Food Hydrocolloids, 29*(2), 407–419. http://dx.doi.org/10.1016/j.foodhyd.2012.04.010

Islam, F., Saeed, F., Afzaal, M., Hussain, M., Ikram, A., & Khalid, M. A. (2023). Food grade nanoemulsions: promising delivery systems for functional ingredients. *Journal of Food Science and Technology, 60*(5), 1461–1471. https://doi.org/10.1007/s13197-022-05387-3

Jafari, Z., Bigham, A., Sadeghi, S., Dehdashti, S. M., Rabiee, N., Abedivash, A., Bagherzadeh, M., Nasseri, B., Karimi-Maleh, H., & Sharifi, E. (2021). Nanotechnology-abetted astaxanthin formulations in multimodel therapeutic and biomedical applications. *Journal of Medicinal Chemistry, 65*(1), 2–36. https://doi.org/10.1021/acs.jmedchem.1c01144

Jain, E., Tripathi, A. D., Agarwal, A., Mauraya, K. K., Rai, D. C., Mishra, R., & Singh, R. B. (2022). Anticancerous compounds in fruits, their extraction, and relevance to food functional foods and nutraceuticals in metabolic and non-communicable diseases. Elsevier, p 517–532. http://dx.doi.org/10.1016/B978-0-12-819815-5.00022-7

Jin, H., Zhao, Z., Lan, Q., Zhou, H., Mai, Z., Wang, Y., Ding, X., Zhang, W., Pi, J., & Evans, C. E. (2021). Nasal delivery of hesperidin/chitosan nanoparticles suppresses cytokine storm syndrome in a mouse model of acute lung injury. *Frontiers in Pharmacology, 11*, 592238. https://doi.org/10.3389%2Ffphar.2020.592238

Jin, Y., & Arroo, R. (2023). The protective effects of flavonoids and carotenoids against diabetic complications—a review of in vivo evidence. *Frontiers in Nutrition, 10*, 1020950. https://doi.org/10.3389/fnut.2023.1020950

Kaur, N., Chugh, V., & Gupta, A. K. (2014). Essential fatty acids as functional components of foods—a review. *Journal of Food Science and Technology, 51*(10), 2289–303. https://doi.org/10.1007%2Fs13197-012-0677-0

Kazmi, I., Al-Abbasi, F. A., Afzal, M., Altayb, H. N., Nadeem, M. S., & Gupta, G. (2021). Formulation and evaluation of kaempferol loaded nanoparticles against experimentally induced hepatocellular carcinoma: in vitro and in vivo studies. *Pharmaceutics, 13*(12), 2086. https://doi.org/10.3390/pharmaceutics13122086

Khater, S. I., Lotfy, M. M., Alandiyjany, M. N., Alqahtani, L. S., Zaglool, A. W., Althobaiti, F., Ismail, T. A., Soliman, M. M., Saad, S., & Ibrahim, D. (2022). Therapeutic potential of quercetin loaded nanoparticles: novel insights in alleviating colitis in an experimental DSS induced colitis model. *Biomedicines, 10*(7), 1654. https://doi.org/10.3390%2Fbiomedicines10071654

Kiss, É. (2020). Nanotechnology in food systems: a review. *Acta Alimentaria, 49*(4), 460–474. https://doi.org/10.1556/066.2020.49.4.12

Knijnenburg, J. T. N., Posavec, L., & Teleki, A. (2019). Chapter 4—Nanostructured Minerals and Vitamins for Food Fortification and Food Supplementation. In: López Rubio A., Fabra Rovira M. J., Martínez Sanz M., Gómez-Mascaraque L. G. (eds) *Nanomaterials for Food Applications*. Elsevier, p 63–98. https://doi.org/10.1016/B978-0-12-814130-4.00004-X

Konstantinidi, M., & Koutelidakis, A. E. (2019). Functional foods and bioactive compounds: a review of its possible role on weight management and obesity's metabolic consequences. *Medicines, 6*(3), 94. https://doi.org/10.3390/medicines6030094

Kumar, N., & Goel, N. (2019). Phenolic acids: Natural versatile molecules with promising therapeutic applications. *Biotechnology Reports, 24*, e00370. https://doi.org/10.1016/j.btre.2019.e00370

Li, C., Li, X., Jiang, Z., Wang, D., Sun, L., Li, J., & Han, Y. (2022). Flavonoids inhibit cancer by regulating the competing endogenous RNA network. *Frontiers in Oncology, 12*, 842790. https://doi.org/10.3389/fonc.2022.842790

Li, G., Zhang, Z., Liu, H., & Hu, L. (2021). Nanoemulsion-based delivery approaches for nutraceuticals: fabrication, application, characterization, biological fate, potential toxicity and future trends. *Food & Function, 12*(5), 1933–1953. https://doi.org/10.1039/D0FO02686G

Li, H., Zhang, J., Yang, L., Cao, H., Yang, Z., Yang, P., Zhang, W., Li, Y., Chen, X., & Gu, Z. (2023). Synergistic antimicrobial and antibiofilm nanoparticles assembled from naturally occurring building blocks. *Advanced Functional Materials, 33*(21), 2212193. https://doi.org/10.1002/adfm.202212193

Li, P., Bukhari, S. N. A., Khan, T., Chitti, R., Bevoor, D. B., Hiremath, A. R., SreeHarsha, N., Singh, Y., & Gubbiyappa, K. S. (2020). Apigenin-loaded solid lipid nanoparticle attenuates diabetic nephropathy induced by streptozotocin nicotinamide through Nrf2/HO-1/NF-kB signalling pathway. *International Journal of Nanomedicine, 15*, 9115–9124. https://doi.org/10.2147/ijn.s256494

Liu, C., Zhang, S., McClements, D. J., Wang, D., & Xu, Y. (2019). Design of astaxanthin-loaded core–shell nanoparticles consisting of chitosan oligosaccharides and poly (lactic-co-glycolic acid): enhancement of water solubility, stability, and bioavailability. *Journal of Agricultural and Food Chemistry, 67*(18), 5113–5121. https://doi.org/10.1021/acs.jafc.8b06963

Livney, Y. D. (2015). Nanostructured delivery systems in food: latest developments and potential future directions. *Current Opinion in Food Science, 3*, 125–135. https://doi.org/10.1016/j.cofs.2015.06.010

Loutfy, S. A., Elberry, M. H., Farroh, K. Y., Mohamed, H. T., Mohamed, A. A., Mohamed, E. B., Faraag, A. H. I., & Mousa, S. A. (2020). Antiviral activity of chitosan nanoparticles encapsulating curcumin against hepatitis C virus genotype 4a in human hepatoma cell lines. *International Journal of Nanomedicine, 15*, 2699–2715. https://doi.org/10.2147/ijn.s241702

Lushchak, O., Strilbytska, O., Koliada, A., Zayachkivska, A., Burdyliuk, N., Yurkevych, I., Storey, K. B., & Vaiserman, A. (2020). Nanodelivery of phytobioactive compounds for treating aging-associated disorders. *Geroscience, 42*(1), 117–139. https://doi.org/10.1007/s11357-019-00116-9

Luykx, D. M. A. M., Peters, R. J. B., van Ruth, S. M., & Bouwmeester, H. (2008). A review of analytical methods for the identification and characterization of nano delivery systems in food. *Journal of Agricultural and Food Chemistry, 56*(18), 8231–8247. https://doi.org/10.1021/jf8013926

Maity, S., & Chakraborti, A. S. (2020). Formulation, physico-chemical characterization and antidiabetic potential of naringenin-loaded poly D, L lactide-co-glycolide (N-PLGA) nanoparticles. *European Polymer Journal, 134*, 109818. https://doi.org/10.1016/j.eurpolymj.2020.109818

Mandpe, P., Prabhakar, B., & Shende, P. (2020). Role of liposomes-based stem cell for multimodal cancer therapy. *Stem Cell Reviews and Reports, 16*(1), 103–117. https://doi.org/10.1007/s12015-019-09933-z

Martins, C. C., Rodrigues, R. C., Mercali, G. D., & Rodrigues, E. (2022). New insights into non-extractable phenolic compounds analysis. *Food Research International, 157*, 111487. https://doi.org/10.1016/j.foodres.2022.111487

McClements, D. J. (2011). Edible nanoemulsions: fabrication, properties, and functional performance. *Soft Matter, 7*(6), 2297–2316. https://doi.org/10.1039/C0SM00549E

McClements, D. J. (2020). Recent advances in the production and application of nano-enabled bioactive food ingredients. *Current Opinion in Food Science, 33*, 85–90. https://doi.org/10.1016/j.cofs.2020.02.004

McClements, D. J., & Öztürk, B. (2021). Utilization of nanotechnology to improve the handling, storage and biocompatibility of bioactive lipids in food applications. *Foods, 10*(2), 365. https://doi.org/10.3390/foods10020365

Milani, J. (2019). Some New Aspects of Colloidal Systems in Foods. IntechOpen. http://dx.doi.org/10.5772/intechopen.75145

Milinčić, D. D., Popović, D. A., Lević, S. M., Kostić, A., Tešićž, L., Nedović, V. A., & Pešić, M. B. (2019). Application of polyphenol-loaded nanoparticles in food industry. *Nanomaterials (Basel), 9*(11), 1629. https://doi.org/10.3390%2Fnano9111629

Min, J. B., Kim, E. S., Lee, J.-S., & Lee, H. G. (2018). Preparation, characterization, and cellular uptake of resveratrol-loaded trimethyl chitosan nanoparticles. *Food Science and Biotechnology, 27*(2), 441–450. https://doi.org/10.1007%2Fs10068-017-0272-2

Mishra, S. B., & Kumari, N. (2021). Engineering of crystalline nano-suspension of lycopene for potential management of oxidative stress-linked diabetes in experimental animals. *BioNano Science, 11*(2), 345–354. https://link.springer.com/article/10.1007/s12668-021-00843-4

Mishra, S. B., Malaviya, J., & Mukerjee, A. (2015). Attenuation of oxidative stress and glucose toxicity by lutein loaded nanoparticles from *Spinacia oleracea* leaves. *Journal of Pharmaceutical Sciences and Pharmacology, 2*(3), 242–249. http://dx.doi.org/10.1166/jpsp.2015.1067

Mitchell, M. J., Billingsley, M. M., Haley, R. M., Wechsler, M. E., Peppas, N. A., & Langer, R. (2021). Engineering precision nanoparticles for drug delivery. *Nature Reviews Drug Discovery, 20*(2), 101–124. https://doi.org/10.1038/s41573-020-0090-8

Moghimi, R., Aliahmadi, A., Rafati, H., Abtahi, H. R., Amini, S., & Feizabadi, M. M. (2018). Antibacterial and anti-biofilm activity of nanoemulsion of *Thymus daenensis* oil against multi-drug resistant *Acinetobacter baumannii. Journal of Molecular Liquids, 265*, 765–770. https://doi.org/10.1016/j.molliq.2018.07.023

Mohanty, S., Konkimalla, V. B., Pal, A., Sharma, T., & Si, S. C. (2021). Naringin as sustained delivery nanoparticles ameliorates the anti-inflammatory activity in a Freund's complete adjuvant-induced arthritis model. *ACS Omega, 6*(43), 28630–28641. https://doi.org/10.1021/acsomega.1c03066

Mondal, S., Soumya, N. P. P., Mini, S., & Sivan, S. K. (2021). Bioactive compounds in functional food and their role as therapeutics. *Bioactive Compounds in Health and Disease, 4*(3), 24–39. https://doi.org/10.31989/bchd.v4i3.786

Montané, X., Kowalczyk, O., Reig-Vano, B., Bajek, A., Roszkowski, K., Tomczyk, R., Pawliszak, W., Giamberini, M., Mocek-Płóciniak, A., & Tylkowski, B. (2020). Current perspectives of the applications of polyphenols and flavonoids in cancer therapy. *Molecules, 25*(15), 3342. https://doi.org/10.3390%2Fmolecules25153342

Moratin, H., Ickrath, P., Scherzad, A., Meyer, T. J., Naczenski, S., Hagen, R., & Hackenberg, S. (2021). Investigation of the immune modulatory potential of zinc oxide nanoparticles in human lymphocytes. *Nanomaterials, 11*(3), 629. https://doi.org/10.3390/nano11030629

Muddineti, O. S., Ghosh, B., & Biswas, S. (2017). Current trends in the use of vitamin E-based micellar nanocarriers for anticancer drug delivery. *Expert Opinion on Drug Delivery, 14*(6), 715–726. https://doi.org/10.1080/17425247.2016.1229300

Müller, R. H., Radtke, M., & Wissing, S. A. (2002). Solid lipid nanoparticles (SLN) and nanostructured lipid carriers (NLC) in cosmetic and dermatological preparations. *Advanced Drug Delivery Reviews, 54*, S131–S155. https://doi.org/10.1016/S0169-409X(02)00118-7

Neethu, R. S., Reddy, M. V. N. J., Batra, S., Srivastava, S. K., & Syal, K. (2022). Vitamin C and its therapeutic potential in the management of COVID19. *Clinical Nutrition ESPEN, 50*, 8–14. https://doi.org/10.1016%2Fj.clnesp.2022.05.026

Niazvand, F., Orazizadeh, M., Khorsandi, L., Abbaspour, M., Mansouri, E., & Khodadadi, A. (2019). Effects of quercetin-loaded nanoparticles on MCF-7 human breast cancer cells. *Medicina, 55*(4), 114. https://doi.org/10.3390/medicina55040114

Nile, S. H., Baskar, V., Selvaraj, D., Nile, A., Xiao, J., & Kai, G. (2020). Nanotechnologies in food science: applications, recent trends, and future perspectives. *Nano-Micro Letters, 12*(45), 1–34. https://doi.org/10.1007/s40820-020-0383-9

Ninfali, P., Antonelli, A., Magnani, M., & Scarpa, E. S. (2020). Antiviral properties of flavonoids and delivery strategies. *Nutrients, 12*(9), 2534. https://doi.org/10.3390/nu12092534

Núñez-Gómez, V., González-Barrio, R., & Periago, M. J. (2023). Interaction between dietary fibre and bioactive compounds in plant by-products: impact on bioaccessibility and bioavailability. *Antioxidants, 12*(4), 976. https://doi.org/10.3390%2Fantiox12040976

Omer, A. K., Khorshidi, S., Mortazavi, N., & Rahman, H. S. (2022). A review on the antiviral activity of functional foods against COVID-19 and viral respiratory tract infections. *International Journal of General Medicine, 15*, 4817–4835. https://doi.org/10.2147%2FIJGM.S361001

Otchere, E., McKay, B. M., English, M. M., & Aryee, A. N. A. (2023). Current trends in nano-delivery systems for functional foods: a systematic review. *Peer J, 11*, e14980. https://doi.org/10.7717/peerj.14980

Othayoth, R., Khatri, K., Gadicherla, R., Kodandapani, S., & Botlagunta, M. (2023). Multivitamin–cisplatin encapsulated chitosan nanoparticles modulate DDX3X expression in cancer cell lines. *Nano Biomedicine and Engineering, 15*(1), 74–85. https://doi.org/10.26599/NBE.2023.9290008

Öztürk, B. (2017). Nanoemulsions for food fortification with lipophilic vitamins: production challenges, stability, and bioavailability. *119*(7), 1500539. https://doi.org/10.1002/ejlt.201500539

Pandey, K. B., & Rizvi, S. I. (2009). Plant polyphenols as dietary antioxidants in human health and disease. *Oxidative Medicine and Cellular Longevity, 2*(5), 270–278. https://doi.org/10.4161%2Foxim.2.5.9498

Pateiro, M., Gómez, B., Munekata, P. E. S., Barba, F. J., Putnik, P., Kovačević, D. B., & Lorenzo, J. M. (2021). Nanoencapsulation of promising bioactive compounds to improve their absorption, stability, functionality and the appearance of the final food products. *Molecules, 26*(6), 1547. https://doi.org/10.3390%2Fmolecules26061547

Pereira, M. C., Oliveira, D. A., Hill, L. E., Zambiazi, R. C., Borges, C. D., Vizzotto, M., Mertens-Talcott, S., Talcott, S., & Gomes, C. L. (2018). Effect of nanoencapsulation using PLGA on antioxidant and antimicrobial activities of guabiroba fruit phenolic extract. *Food Chemistry, 240*, 396–404. https://doi.org/10.1016/j.foodchem.2017.07.144

Plati, F., & Paraskevopoulou, A. (2022). Micro- and nano-encapsulation as tools for essential oils advantages'exploitation in food applications: the case of oregano essential oil. *Food and Bioprocess Technology, 15*(5), 1–29. https://doi.org/10.1007/s11947-021-02746-4

Rehman, A., Ahmad, T., Aadil, R. M., Spotti, M. J., Bakry, A. M., Khan, I. M., Zhao, L., Riaz, T., & Tong, Q. (2019). Pectin polymers as wall materials for the nano-encapsulation of bioactive compounds. *Trends in Food Science & Technology, 90*, 35–46. https://doi.org/10.1016/j.tifs.2019.05.015

Reque, P. M., & Brandelli, A. (2021). Encapsulation of probiotics and nutraceuticals: applications in functional food industry. *Trends in Food Science & Technology, 114*, 1–10. https://doi.org/10.1016/j.tifs.2021.05.022

Reza Mozafari, M., Johnson, C., Hatziantoniou, S., & Demetzos, C. (2008). Nanoliposomes and their applications in food nanotechnology. *Journal of Liposome Research, 18*(4), 309–327. https://doi.org/10.1080/08982100802465941

Rezende, E. S. V., Lima, G. C., & Naves, M. M. V. (2021). Dietary fibers as beneficial microbiota modulators: a proposed classification by prebiotic categories. *Nutrition, 89*, 111217. https://doi.org/10.1016/j.nut.2021.111217

Rivera-Madrid, R., Carballo-Uicab, V. M., Cárdenas-Conejo, Y., Aguilar-Espinosa, M., & Siva, R. (2020). 1—Overview of Carotenoids and Beneficial Effects on Human Health. In: Galanakis, C.M. (ed) *Carotenoids: Properties, Processing and Applications*. Academic Press, p 1–40. https://doi.org/10.1016/B978-0-12-817067-0.00001-4

Ruiz Canizales, J., Velderrain Rodríguez, G. R., Domínguez Avila, J. A., Preciado Saldaña, A. M., Alvarez Parrilla, E., Villegas Ochoa, M. A., & González Aguilar, G. A. (2018). Encapsulation to protect different bioactives to be used as nutraceuticals and food ingredients. In: Mérillon, J.-M., Ramawat, K.G. (eds) *Bioactive Molecules in Food*. Springer International Publishing, p 1–20. https://doi.org/10.1007/978-3-319-54528-8_84-1

Samanta, S., Banerjee, J., Das, B., Mandal, J., Chatterjee, S., Ali, K. M., Sinha, S., Giri, B., Ghosh, T., & Dash, S. K. (2022). Antibacterial potency of cytocompatible chitosan-decorated biogenic silver nanoparticles and molecular insights towards cell–particle interaction. *International Journal of Biological Macromolecules, 219*, 919–939. https://doi.org/10.1016/j.ijbiomac.2022.08.050

Sampathkumar, K., Tan, K. X., & Loo, S. C. J. (2020). Developing nano-delivery systems for agriculture and food applications with nature-derived polymers. *iScience, 23*(5), 101055. https://doi.org/10.1016/j.isci.2020.101055

Sandoval-Yañez, C., & Castro Rodriguez, C. (2020). Dendrimers: amazing platforms for bioactive molecule delivery systems. *Materials, 13*(3), 570. https://doi.org/10.3390%2Fma13030570

Santos, A., Veiga, F., & Figueiras, A. (2019). Dendrimers as pharmaceutical excipients: synthesis, properties, toxicity and biomedical applications. *Materials, 13*(1), 65. https://doi.org/10.3390/ma13010065

Sawpari, R., Samanta, S., Banerjee, J., Das, S., Dash, S. S., Ahmed, R., Giri, B., & Dash, S. K. (2023). Recent advances and futuristic potentials of nano-tailored doxorubicin for prostate cancer therapy. *Journal of Drug Delivery Science and Technology, 81*, 104212. https://doi.org/10.1016/j.jddst.2023.104212

Schoener, A. L., Zhang, R., Lv, S., Weiss, J., & McClements, D. J. (2019). Fabrication of plant-based vitamin D 3-fortified nanoemulsions: influence of carrier oil type on vitamin bioaccessibility. *Food & Function, 10*(4), 1826–1835. https://doi.org/10.1039/C9FO00116F

Shafique, B., Ranjha, M., Murtaza, M. A., Walayat, N., Nawaz, A., Khalid, W., Mahmood, S., Nadeem, M., Manzoor, M. F., Ameer, K., Aadil, R. M., & Ibrahim, S. A. (2022). Recent trends and applications of nanoencapsulated bacteriocins against microbes in food quality and safety. *Microorganisms, 11*(1). https://doi.org/10.3390/microorganisms11010085

Shah, A., Ul Ashraf, Z., Gani, A., Masoodi, F. A., & Gani, A. (2022). β-Glucan from mushrooms and dates as a wall material for targeted delivery of model bioactive compound: nutraceutical profiling and bioavailability. *Ultrasonics Sonochemistry, 82*, 105884. https://doi.org/10.1016/j.ultsonch.2021.105884

Shaikh, S. (2022). Sources and Health Benefits of Functional Food Components Current Topics in Functional Food. IntechOpen. https://doi.org/10.5772/intechopen.104091

Sharma, C., Dhiman, R., Rokana, N., & Panwar, H. (2017). Nanotechnology: an untapped resource for food packaging. *Frontiers in Microbiology, 8*, 1735. https://doi.org/10.3389/fmicb.2017.01735

Shishir, M. R. I., Xie, L., Sun, C., Zheng, X., & Chen, W. (2018). Advances in micro and nano-encapsulation of bioactive compounds using biopolymer and lipid-based transporters. *Trends in Food Science & Technology, 78*, 34–60. https://doi.org/10.1016/j.tifs.2018.05.018

Siddiqui, I. A., & Sanna, V. (2016). Impact of nanotechnology on the delivery of natural products for cancer prevention and therapy. *Molecular Nutrition Food Research, 60*(6), 1330–1341. https://doi.org/10.1002/mnfr.201600035

Sim, S., & Wong, N. K. (2021). Nanotechnology and its use in imaging and drug delivery. *Biomedical Reports, 14*(5), 1–9. https://doi.org/10.3892/br.2021.1418

Singh, H. (2016). Nanotechnology applications in functional foods; opportunities and challenges. *Preventive Nutrition and Food Science, 21*(1), 1. https://doi.org/10.3746%2Fpnf.2016.21.1.1

Singh, H., Thompson, A., Liu, W., & Corredig, M. (2012). Liposomes as Food Ingredients and Nutraceutical Delivery Systems. In: *Encapsulation Technologies and Delivery Systems for Food Ingredients and Nutraceuticals*. Elsevier, p 287–318. https://doi.org/10.1533/9780857095909.3.287

Singh, T., Shukla, S., Kumar, P., Wahla, V., Bajpai, V. K., & Rather, I. A. (2017). Application of nanotechnology in food science: perception and overview. *Frontiers in Microbiology, 8*, 1501. https://doi.org/10.3389/fmicb.2017.01501

Su, Q., Zhao, X., Zhang, X., Wang, Y., Zeng, Z., Cui, H., & Wang, C. (2022). Nano functional food: opportunities, development, and future perspectives. *International Journal of Molecular Sciences, 24*(1), 234. https://doi.org/10.3390/ijms24010234

Subramaniam, S., Selvaduray, K. R., & Radhakrishnan, A. K. (2019). Bioactive compounds: natural defense against cancer? *Biomolecules, 9*(12), 758. https://doi.org/10.3390/biom9120758

Subramanian, P. (2021). Lipid-based nanocarrier system for the effective delivery of nutraceuticals. *Molecules, 26*(18), 5510. https://doi.org/10.3390/molecules26185510

Sulaiman, G. M., Waheeb, H. M., Jabir, M. S., Khazaal, S. H., Dewir, Y. H., & Naidoo, Y. (2020). Hesperidin loaded on gold nanoparticles as a drug delivery system for a successful biocompatible, anti-cancer, anti-inflammatory and phagocytosis inducer model. *Scientific Reports, 10*(1), 9362. https://doi.org/10.1038/s41598-020-66419-6

Sun, D., Zhang, W., Mou, Z., Chen, Y., Guo, F., Yang, E., & Wang, W. (2017). Transcriptome analysis reveals silver nanoparticle-decorated quercetin antibacterial molecular mechanism. *ACS Applied Materials & Interfaces, 9*(11), 10047–10060. https://doi.org/10.1021/acsami.7b02380

Surendran, V., & Palei, N. N. (2022). Formulation and characterization of rutin loaded chitosan–alginate nanoparticles: antidiabetic and cytotoxicity studies. *Current Drug Delivery, 19*(3), 379–394. https://doi.org/10.2174/1567201818666211005090656

Tang, C.-H. (2021). Assembly of food proteins for nano-encapsulation and delivery of nutraceuticals (a mini-review). *Food Hydrocolloids, 117*, 106710. https://doi.org/10.1016/j.foodhyd.2021.106710

Taylor, T. M., Weiss, J., Davidson, P. M., & Bruce, B. D. (2005). Liposomal nanocapsules in food science and agriculture. *Critical Reviews in Food Science and Nutrition, 45*(7–8), 587–605. https://doi.org/10.1080/10408390591001135

Teodoro, A. J. (2019). Bioactive compounds of food: their role in the prevention and treatment of diseases. *Oxidative Medicine and Cell Longevity, 2019*, 3765986. https://doi.org/10.1155/2019/3765986

Thalhauser, S., & Breunig, M. (2020). Considerations for efficient surface functionalization of nanoparticles with a high molecular weight protein as targeting ligand. *European Journal of Pharmaceutical Sciences, 155*, 105520. https://doi.org/10.1016/j.ejps.2020.105520

Thies, C. (2012). 8—Nanocapsules as Delivery Systems in the Food, Beverage and Nutraceutical Industries. In: Huang, Q. (ed) Nanotechnology in the Food, Beverage and Nutraceutical Industries. Woodhead Publishing, p 208–256. https://doi.org/10.1533/9780857095657.2.208

Tiwari, R., & Takhistov, P. (2012). Nanotechnology-enabled delivery systems for food functionalization and fortification. In: Padua, G. W. & Wang, Q. (eds.) Nanotechnology Research Methods for Foods and Bioproducts, John Wiley & Sons, p 55–101. https://doi.org/10.1002/9781118229347.ch5

Toragall, V., & Baskaran, V. (2021). Chitosan–sodium alginate–fatty acid nanocarrier system: lutein bioavailability, absorption pharmacokinetics in diabetic rat and protection of retinal cells against H_2O_2 induced oxidative stress in vitro. *Carbohydrate Polymers, 254*, 117409. https://doi.org/10.1016/j.carbpol.2020.117409

Tutunchi, H., Naeini, F., Ostadrahimi, A., & Hosseinzadeh-Attar, M. J. (2020). Naringenin, a flavanone with antiviral and anti-inflammatory effects: a promising treatment strategy against COVID-19. *Phytotherapy Research, 34*(12), 3137–3147. https://doi.org/10.1002%2Fptr.6781

Ullah, A., Munir, S., Badshah, S. L., Khan, N., Ghani, L., Poulson, B. G., Emwas, A. H., & Jaremko, M. (2020). Important flavonoids and their role as a therapeutic agent. *Molecules, 25*(22), 5234. https://doi.org/10.3390/molecules25225243

Venkatanagaraju, E., Bharathi, N., Sindhuja, R. H., Chowdhury, R. R., & Sreelekha, Y. (2020). Extraction and purification of pectin from agro-industrial wastes. In: Masuelli, M. (ed) *Pectins—Extraction, Purification, Characterization and Applications*, p 1–15. http://dx.doi.org/10.5772/intechopen.85585

Voets, I. K., de Keizer, A., Cohen Stuart, M. A., Justynska, J., & Schlaad, H. (2007). Irreversible structural transitions in mixed micelles of oppositely charged diblock copolymers in aqueous solution. *Macromolecules, 40*(6), 2158–2164. https://doi.org/10.1021/ma0614444

Wadhwa, R., Paudel, K. R., Chin, L. H., Hon, C. M., Madheswaran, T., Gupta, G., Panneerselvam, J., Lakshmi, T., Singh, S. K., & Gulati, M. (2021). Anti-inflammatory and anticancer activities of naringenin-loaded liquid crystalline nanoparticles in vitro. *Journal of Food Biochemistry, 45*(1), e13572. https://doi.org/10.1111/jfbc.13572

Waheed Janabi, A. H., Kamboh, A. A., Saeed, M., Xiaoyu, L., BiBi, J., Majeed, F., Naveed, M., Mughal, M. J., Korejo, N. A., Kamboh, R., Alagawany, M., & Lv, H. (2020). Flavonoid-rich foods (FRF): a promising nutraceutical approach against lifespan-shortening diseases. *Iranian Journal of Basic Medical Sciences, 23*(2), 140–153. https://doi.org/10.22038%2FIJBMS.2019.35125.8353

Walia, A., Kumar, N., Singh, R., Kumar, H., Kumar, V., Kaushik, R., & Kumar, A. P. (2022). Bioactive compounds in *Ficus* fruits, their bioactivities, and associated health benefits: a review. *Journal of Food Quality, 2022*, 19. https://doi.org/10.1155/2022/6597092

Wang, Y., Ai, C., Wang, H., Chen, C., Teng, H., Xiao, J., & Chen, L. (2023). Emulsion and its application in the food field: an update review. *eFood, 4*(4), e102. https://doi.org/10.1002/efd2.102

Xia, I. F., Kong, H.-K., Wu, M. M. H., Lu, Y., Wong, K.-H., & Kwok, K. W. H. (2022). Selenium nanoparticles (SeNPs) immunomodulation is more than redox improvement: serum proteomics and transcriptomic analyses. *Antioxidants, 11*(5), 964. https://doi.org/10.3390/antiox11050964

Xiao, S., & Li, J. (2020). Study on functional components of functional food based on food vitamins. *Journal of Physics: Conference Series, 1549*(3), 032002. https://doi.org/10.1088/1742-6596/1549/3/032002

Xie, W., Tan, S., Ren, X., Yu, J., Yang, C., Xie, H., Ma, Z., Liu, Y., & Yang, S. (2023). Tumor-targeted astaxanthin nanoparticles for therapeutic application in vitro. *Colloid and Interface Science Communications, 55*, 100721. https://doi.org/10.1016/j.colcom.2023.100721

Xu, W., Ling, P., & Zhang, T. (2013). Polymeric micelles, a promising drug delivery system to enhance bioavailability of poorly water-soluble drugs. *Journal of Drug Delivery, 2013*. https://doi.org/10.1155/2013/340315

Yadav, N., Tripathi, A. K., & Parveen, A. (2022). PLGA-quercetin nano-formulation inhibits cancer progression via mitochondrial dependent Caspase-3, 7 and independent FoxO1 activation with concomitant PI3K/AKT suppression. *Pharmaceutics, 14*(7), 1326. https://doi.org/10.3390/pharmaceutics14071326

Yousefi, M., Jafari, S. M., Ahangari, H., & Ehsani, A. (2023). Application of nanoliposomes containing nisin and crocin in milk. *Advanced Pharmaceutical Bulletin, 13*(1), 134–142. https://doi.org/10.34172/apb.2023.014

Yuan, M., Zhang, G., Bai, W., Han, X., Li, C., & Bian, S. (2022). The role of bioactive compounds in natural products extracted from plants in cancer treatment and their mechanisms related to anticancer effects. *Oxidative Medicine and Cellular Longevity, 2022*, 1429869. https://doi.org/10.1155%2F2022%2F1429869

Zare, M., Norouzi Roshan, Z., Assadpour, E., & Jafari, S. M. (2021). Improving the cancer prevention/treatment role of carotenoids through various nano-delivery systems. *Critical Reviews in Food Science and Nutrition, 61*(3), 522–534. https://doi.org/10.1080/10408398.2020.1738999

Zhang, Z., Qiu, C., Li, X., McClements, D. J., Jiao, A., Wang, J., & Jin, Z. (2021). Advances in research on interactions between polyphenols and biology-based nano-delivery systems and their applications in improving the bioavailability of polyphenols. *Trends in Food Science & Technology, 116*, 492–500. https://doi.org/10.1016/j.tifs.2021.08.009

4 Role of Nanotechnology in Food Industries

Zahra Emam-Djomeh, Nima Mobahi, Mohammad Ekrami, and Elaheh Pourmohammad

4.1 INTRODUCTION

Nanotechnology is the control of matter on a near-atomic scale to create modern structures, materials, and gadgets. The innovation guarantees logical headway in numerous segments such as pharmaceutical, buyer items, vitality, materials, and fabricating (Primožič *et al.*, 2021). Nanotechnology alludes to building structures, gadgets, and frameworks. Nanomaterials (NMs) have a length scale between 1 and 100 nm. At this measure, materials start to show one-of-a-kind properties that influence physical, chemical, and natural behavior (Yu *et al.*, 2018). Investigating, creating, and utilizing these properties is at the heart of modern innovation. The nanostructured fabric finds its middle-of-the-road estimate between 1 and 1000 nm that can be prepared into other shapes (Nile *et al.*, 2020). The European Commission (EC) definition applies to all particulate NMs independent of their root, i.e., common, coincidental, or made. A fabric may be an NM if 50% or more of its constituent particles, in any case, whether they are unbound or a portion of agglomerates or totals, within the number-based molecule measure dispersion have one or more outside measurements between 1 and 100 nm (Thiruvengadam *et al.*, 2018, Ekrami *et al.*, 2022b). Nanotechnology offers appealing openings within the food industry such as for food security and quality control as well as the generation of unused food additives/supplements and other flavors. Within the food industry, nanotechnology can also be utilized for the generation of bundles with upgraded warm and/or mechanical properties and security (Hamad *et al.*, 2018). The assortment of nanostructures with differing properties makes them appropriate for expansion to foods as well as in bundling items that improve the dietary quality of foods. Accessible NMs are abused in agribusiness and food sciences (Jagtiani, 2022).

Nanotechnology within the food industry has created two wide applications of nanostructures, i.e., food fixings and sensors (Table 4.1).

Within the food industry, nano-food fixings cover a wide utility zone beginning from foodpreparation to food bundling (Singh *et al.*, 2017). In food preparation, nanostructures act as anticaking operators, antimicrobial operators, nano additives, nanocarriers, and nanocomposites, whereas, in food bundling, they possibly serve as nanosensors to screen the food quality (Nile *et al.*, 2020). The NMs are demonstrated to be as awesome enzymesupports by their expansive surface-to-volume proportions in comparison to the ordinary macro-sized back frameworks (Mei and Wang, 2020). It makes the proteins hyperactive, tough, and cost-effective. Later nanocarriers created by nanotechniques have the potential to act as particular and elite conveyance frameworks to transport the food-added substances into food fixings with no unsettling influence on their essential physicochemical properties and morphology (Sahani and Sharma, 2021). Here, molecule estimate is the key calculation that influences the conveyance rate of a bioactive compound to the target locales in vivo

DOI: 10.1201/9781003438168-4

TABLE 4.1
Various trends used in nanotechnology

Nanotechnology method	Type of nanotechnology	Application
Pathogen detection	NM nanoelectronics (carbon-based materials)	Detect a wide range of poisons, infections, chemical pollutants, and adulteration at an early stage
Inhibit biofilms	(Metal)NMs	Natural antibacterial potential provided by nanotechnological techniques offers an option for disinfecting food contact surfaces and preventing microbial biofilm formation
Improved packaging	(Carbon)nanotubes	Reduce foodborne illness and manage the gas exchange. Multilayer coatings and films improve their functionality. Nanoscale reinforcements may improve bioplastic packaging and extend perishable food shelf life.
Smart packaging	NMs	Smart packaging can detect biochemical or microbiological changes in food, preventing food fraud and adulteration.
Active packaging	NMs	Active packaging uses an eco-friendly bio-nano composite containing biological components to extend product life.
Nano additives	NMs (composites, coating)	Reduced calorie density, delayed digestion, increased satiety, and accessible vitamins, minerals, and nutrients.
Nanoencapsulation	NPs (chemical, physical)	Improve food flavor, texture, consistency, and durability.
Nanoemulsions	NPs	Encapsulation, protection, improving bioavailability, and target release of sensitive functional compounds
		Delivery of nutraceuticals, probiotics, flavors, and colors
Downstreamprocesses		Extraction of a relatively large molecular weight protein, bovine serum albumin

as it were submicron NPs can be acclimatized in a few cell lines effectively, and bigger estimate micro-particles are not permitted to pass through (Kumar *et al.*, 2020).

4.2 NANOENCAPSULATION

The food industry has confronted expanding challenges over a long time, particularly concerning food security and conservation amid capacity and conveyance. Since old times, human creatures have looked for proficient ways to bundle and protect food to guarantee the accessibility and quality of items for longer periods (Ashfaq *et al.*, 2022, Ekrami *et al.*, 2023b). Nanotechnology can have a greater and faster impact on all aspects of agriculture compared to other eco-friendly technologies and agricultural biotechnology (Nejat *et al.*, 2022). This can lead to synchronized benefits for the public, as well as legal, moral, and environmental effects. There are many uses for nanotechnology. In the agriculture sector, nano fertilizers, nano pesticides, nanosensors, and nanoformulations can be used (Ashraf *et al.*, 2021). These can also be used in the food industry, especially nanosensors and nanoformulations (Figure 4.1). Among the benefits of using nanotechnology in the food industry, we can mention the reduction of environmental pollution, reduction of diseases, reduction of process risks for factory workers, reduction of waste, and, most importantly, the reduction of water consumption (Biswas *et al.*, 2022).

In the food industry, nanotechnology is about using small particles that can detect harmful substances in food. These particles can be made from natural molecules like sugars or proteins (Rizvi *et al.*, 2022, Emam-Djomeh *et al.*, 2023b). They can be used as sensors to find germs or other things that could make people sick. They can also be used to keep track of where food comes from. Nanotechnology can help protect against things in the environment by using encapsulation

FIGURE 4.1 A brief overview of different types of nanotechnology in food processing.

systems. Moreover, it can be utilized in making food ingredients like flavors and antioxidants (Ranjha *et al.*, 2022). The aim is to enhance the performance of these ingredients while reducing the amount needed. As more new ingredients are added to food, scientists will continue to study and develop more ways to release and control the effects of beneficial nutrients in our bodies. Because NPs have a larger surface area compared to bigger particles of the same substance, they are believed to have more biological activity. This provides different ways to use food (Chellaram *et al.*, 2014). NPs can be used in functional foods as special ingredients. These ingredients, which can be found in some natural foods, have health benefits and could lower the risk of diseases like cancer. Nanotechnology can make bioactive compounds better by making their particles smaller. This helps them to be taken up by the body more easily and stay in the digestive system for longer (Ndlovu *et al.*, 2020). It also improves their ability to dissolve in liquids. Nanotechnology has the potential to be helpful in all parts of making and processing food. However, a lot of the ways how it is done are either too expensive or hard to do on a largescale for businesses. This text talks about how using really small techniques can save money in the food industry (Primožic *et al.*, 2021). It mentions some specific areas where these techniques can be helpful, like making new materials, making different kinds of food, and storing food. The article mainly talks about the ways that using these techniques can be profitable in the food industry both now and in the future (Mahmud *et al.*, 2022).

In simple terms, nanotechnology can be used in two ways for food. The first way involves physically processing the food, like grinding or milling it. For instance, using small-scale technologies to make things better, like making products feel nicer, keeping moisture in, making them crunchy or fragile, and enhancing their taste, smell, and appearance:

i. Interactive foods and beverages can give you the flavors and colors you want whenever you want them. This is possible because of tiny capsules called nanocapsules that burst when exposed to different microwave frequencies.
ii. The Nanotechnology Initiative has studied how nanotechnology can be used to clean and treat water. They have focused on areas like making better filters, preventing bacteria growth, and getting rid of harmful substances.
iii. We are making tiny versions of herbal plants by grinding them up into a fine powder or turning them into a liquid mixture.

iv. Making ganoderma spore into a very fine powder by breaking its cell walls and releasing useful substances.
v. The frying oil refining catalytic device, made of a very small ceramic material, stops the frying oil from becoming too hot and prevents bad smells.
vi. Tiny tubes made of milk protein that are rigid, hollow, and measured in micrometers have the potential to be used in creating new ingredients for thickening, forming gels, enclosing nanoscale materials, and controlling their release.
vii. Protein-coated nanocantilever is a small sensor made of silicon that can quickly detect viruses, bacteria, and other harmful things. It vibrates on its own at a specific frequency. It is a new and advanced type of sensor. When dirty things land on the devices, the tiny mass changes can make the nanocantilever shake differently and be found easily.
viii. Scientists have created a man-made DNA structure that looks like a tree. It is labeled with different colors to help identify harmful bacteria in food.
ix. A small and easy-to-carry device was created to detect harmful substances in food. This device uses tiny wires, special antibodies that can identify harmful bacteria or chemicals, and glowing antibodies that can highlight the presence of these substances. It can detect toxins, harmful bacteria, and chemicals all at the same time.
x. Silver NPs have been used in various products like bandages and refrigerators to stop the growth of bacteria and other small organisms.

It also helps make products last longer, by making connections between different things and doing other similar actions. Certain factories can make very small wheat flour particles known as nanosized wheat flour. These tiny particles can hold a lot of water because they have been developed to have more connections between the particles (He and Hwang, 2016).

Typically, the production techniques for NPs may be categorized into three groups: (1) chemical approaches, (2) physical methods, and (3) bio-assisted methods. For example, Figure 4.2 shows the schematics of different zinc oxide (ZnO) production methods.

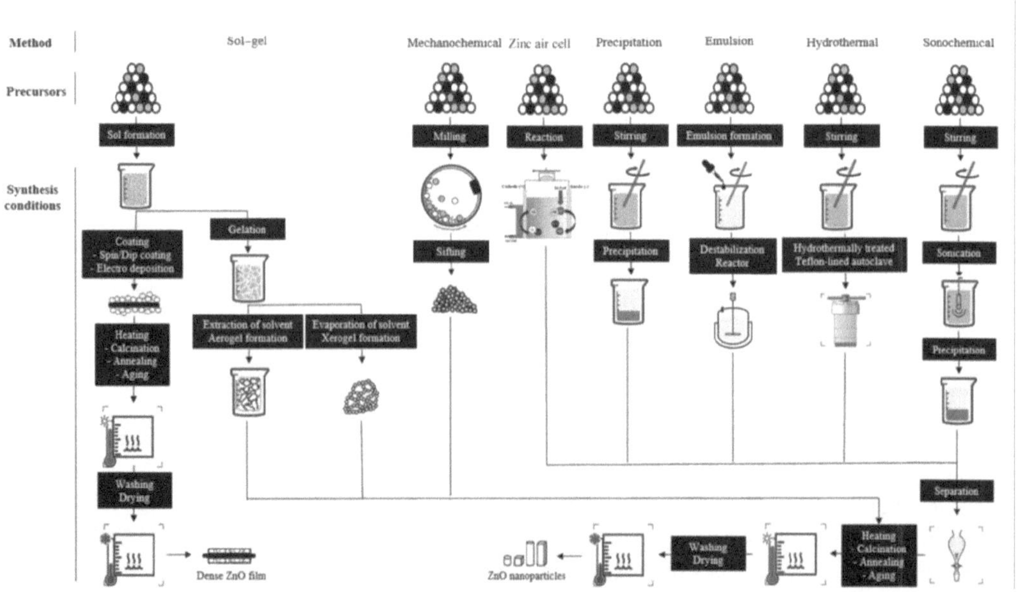

FIGURE 4.2 Schematics of different ZnO production methods.

Functional ingredients are important parts of food. They can be things like vitamins, antibacterials, antioxidants, things that give flavor, or things that keep food fresh. They can look and act differently (Subramani and Ganapathyswamy, 2020). Functional ingredients are often not used alone but instead are combined with other substances to make them easier to use. A delivery system has many jobs, and one of them is to take a working part to where it needs to go (Nile *et al.*, 2020). Delivery systems have two main roles when it comes to food: they make sure the taste, texture, and shelf life of the food stay good, and they also protect the food from getting ruined by chemicals or bacteria. Additionally, delivery systems control how the food's ingredients are released when they are in certain conditions (Tahir *et al.*, 2021). Nanocapsules are great tools for delivering useful substances because they can do all these tasks effectively. These small structures include mixtures, tiny oil drops, and particles made from natural substances (Pateiro *et al.*, 2021).

Nanoencapsulation is when a barrier is created to stop bad chemical reactions and to release vitamins or other healthy ingredients in a controlled way. The nanoencapsulation process is important for vitamins because it helps protect them and has certain key features (Mahato *et al.*, 2021). Based on the research of the last two decades, there are various methods for nanoencapsulation of bioactive compounds, which can be mentioned as follows:

i. Using natural components of food waste, especially for vitamins.
ii. Using nanocomposites and nanocarriers.
iii. Using the technology of applying new low-energy methods such as spontaneous emulsification instead of high-energy preparation methods to preserve bioactive substances against harsh processing conditions, reducing surfactant and removing cosurfactant.

4.2.1 Vitamins

Vitamins are used in nanoencapsulation. Vitamins are delicate substances, so they need to be protected from things that can harm them, such as heat and oxidants. Encapsulation is a new and hopeful way to protect the natural traits of vitamins for a longer period (Resende *et al.*, 2020). This process involves covering or capturing a natural material or a mix with another substance. Usually, the trapped thing is in a liquid form, but it can also be a gas or solid. The coating substance has different names like capsule, wall material, membrane, or carrier (Mohammadi *et al.*, 2023). One of the features of encapsulating vitamins is to protect them from the external environment such as the negative impact of light, thermal processes such as spray drying, and ultraviolet radiation during food processing, which creates a platform for enriching food products with a set of vitamins (Mujica-Álvarez *et al.*, 2020). Controlled release of vitamins in the human body according to their absorption site, improving flow properties, helping to reduce the rate of weight gain in the human body, measuring the exact level of vitamin delivery, and also removing the unpleasant taste of some vitamins pointed out (Maleki *et al.*, 2022).

Different methods have been proposed for different vitamins such as electrospinning (Couto *et al.*, 2023, Coelho *et al.*, 2021, Farahmand *et al.*, 2023), nanoprecipitation (Sokołowska *et al.*, 2022, Zhang *et al.*, 2020), nanoemulsification (Karbalaei-Saleh *et al.*, 2023, Gahruie *et al.*, 2020). A schematic diagram of the electrospinning process is shown in Figure 4.3.

4.2.2 Antibacterials

A "natural antimicrobial agent" is a natural substance that can fight against germs and bacteria (Ekrami *et al.*, 2022c). It comes from plants, animals, or tiny organisms like bacteria or fungi. Some examples of these substances are polyphenols, flavonoids, tannins, alkaloids, terpenoids, isothiocyanates, polypeptides, and their altered forms (Zhu *et al.*, 2021). Essential oils (EOs) are made by plants to help protect themselves from things outside of them. They work well against bacteria and are

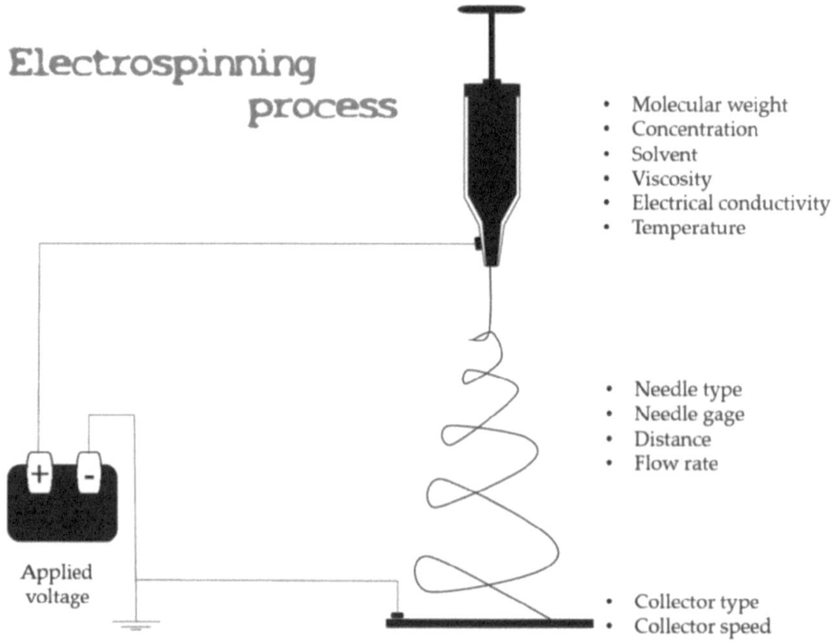

FIGURE 4.3 Schematic diagram of electrospinning process.

commonly used in things like cleaning products and beauty items (Negi and Kesari, 2022). EOs come from plants and can be made by using steam to extract them from different parts of the plant like flowers, leaves, stems, roots, and fruits. EOs are very strong-smelling liquids that easily evaporate (Ekrami *et al.*, 2019). They are typically attracted to fats and oils, are lightweight, and emit a strong scent. Furthermore, many EOs are not harmful to mammals and do not last for a long time. This means that they are quite safe for both our health and the environment (Ekrami *et al.*, 2022a). Among the nanometric epitome frameworks right now being utilized for the conveyance of bioactive compounds, nanoemulsions are especially reasonable for food applications (Beikzadeh *et al.*, 2020). EOs are natural substances from plants that have strong scents and can easily evaporate. They are mainly used for adding flavor and fragrance to products, as well as for their natural properties. In the past few years, scientists have been studying EOs and the natural compounds in them that can fight bacteria and fungi that cause foodborne illnesses (Amani *et al.*, 2021, Kour *et al.*, 2022).

When it comes to antimicrobials, encapsulation can make the amount of the helpful compounds higher in places where microorganisms are commonly found in food, such as areas with a lot of water or where liquid and solid meet (Ekrami-Djomeh et *al.*, 2023a). There are two main ways that plant EOs and their main components affect microorganisms (Mirzakhani *et al.*, 2018). The first way is by changing the structure and makeup of the microbial cells and mycelia, such as the cell membrane, cell wall, and organelles (Reis *et al.*, 2022). The second way is by reducing or stopping the production and growth of spores, which helps prevent harmful pathogens from harming future generations. Thyme, clove, cinnamon, oregano, nettle, and mallow EOs are the ones that have been researched the most. The greatest effect of these antibacterials is on all types of Gram-positive bacteria (Mukurumbira *et al.*, 2022).

4.2.3 Probiotics

Probiotics are small living things that can help us stay healthy when we consume enough of them. For instance, lactic acid bacteria are very well-known because they are found in many different things like food, plants, soil, and even humans (Koh *et al.*, 2022). They are good options for making food products, dietary supplements, and medicine. These probiotic mixtures are supposed to help the body by changing the bacteria in the gut, creating certain substances, getting rid of harmful substances from food, creating certain proteins, and controlling harmful bacteria (Rodrigues *et al.*, 2020). However, for probiotics to have these benefits, they have to resist things like stomach acid, being broken down by enzymes, being affected by bile salts, competing with other bacteria, and also being able to stick to the lining of the gut (Xu *et al.*, 2022). Encapsulation strategies are needed to both protect and promote the delivery of something to its intended destination. Several types of probiotic encapsulation are described below (Rajam and Subramanian, 2022, Gu *et al.*, 2022):

i. Bulk encapsulation systems are machines that are used to package a large quantity of materials or products together in a protective casing. The process of nano-encapsulating probiotics using a hydrocolloid system is commonly used to protect probiotics from things like low acidity, so they can survive better. This method uses special capsules to hold probiotics. These capsules are like tiny balls and have small holes that can let some things pass through. The capsules can release the probiotics when they come into contact with certain conditions. The usual covering materials for food are safe materials made from natural substances such as carbohydrates, proteins, and fats. The choice of material depends on different factors such as how well it works with the food and what properties it should have.
ii. Single-cell encapsulation systems: Microorganisms in nature have developed ways to survive in tough conditions over many years. Scientists have used this as inspiration to create a protective covering for living cells, called single-cell encapsulation, to help the cells withstand stress and provide extra functions. Different methods using materials like silica, graphene, and metal-organic frameworks have been created for single-cell encapsulation. No matter what materials are used, these methods of sealing have been created using gentle chemicals that do not harm the cells much. Using these methods, we can create shells that protect cells and break down easily. These shells can help cells better resist outside pressures and get the nutrients they need. They also allow for further customization of the cells.
iii. Layer-by-layer embodiment: Nanoscale thin films of charged polymers (polyelectrolytes) can be arranged by the substituting statement of polyanions and polycations on substrate surfaces. This concept of consecutive testimony of oppositely charged polymers ordinarily driven by electrostatic responses is superior to those known as the layer-by-layer get-together strategy and can be connected to planar and colloidal substrates.

4.3 FOOD PACKAGING

Since the prehistoric age, man has been trying to improve and devise better food preservation techniques. From the caveman trying to preserve the food by storing the fresh kill in caves that provided a dampened environment to keepingit from being spoiled to the refrigeration techniques of the 21st century, man has come a long way (Primožič *et al.*, 2021). Cellars and cold streams would also find their use in the preservation of food. Drying and fermentation processes existed since almost 10,000 BCE, and today we use the modified versions of these processes. Increasing climate change arrival imperatives, expanding populace, industrialization, efficiency, and tall postharvest misfortunes have put human life uncovered to multitudinous and uncommon challenges. This has led to endangering the food supply of more than two billion people in different countries, especially in poor are as (Mahmud *et al.*, 2022). The lack of possibility to send ready and semi-ready foods as well as raw materials due to the lack of necessary infrastructure such as cooling systems, refrigerators, and proper packaging is one of the causes of this food poverty. Agrarian nanotechnology has centered on an inquiry about and application to unravel rural and natural issues of maintainability, trim change, and efficiency upgrade (Alfei *et al.*, 2020). It appears that rural nanotechnology is exceptionally curious for creating nations due to the lessening of starvation, undernutrition, and mortality rates in children (Mei and Wang, 2020). Some advanced countries in North America, Europe, and East Asia appear to have evinced increasing interest in utilizing NMs for horticulture employment as uncovered using greater number of distributions and licenses (Emam-Djomeh *et al.*, 2023a). Nanotechnology has many benefits and features:

i. NPs of composites to a substance can improve its properties. These particles are very small, typically less than 100 nm in size. When added to a material, it can change its color, strength, conductivity, or other characteristics. Nanocomposites are used in various industries, such as electronics, medicine, and construction, to enhance the performance and functionality of different products. Silver, titanium dioxide, silicon dioxide, and nano-clay are being used in packaging materials to make sure that food stays safe. These substances help to change how air and smells can pass through the packaging, make it stronger and more resistant to heat, and protect against harmful ultraviolet (UV) rays. They also help to prevent the growth of bacteria and fungi on the surface of the packaging.
ii. Nylon nanocomposites are used in food packaging to block oxygen and carbon dioxide from entering. "Multi-layer polyethylene terephthalate (PET)" bottles are used for beer and other alcoholic drinks to maintain freshness and prevent any unwanted smells from getting in.

The first thing about the environmental benefits is this issue, which is very important due to the increasing pollution of the environment (Mei and Wang, 2020). Nanostructured packaging can be lighter, reduce the use of natural materials, and these packages can be reused more efficiently, which helps reduce energy waste and support the packaging framework. Its application in science, security, and quality of foods may be a concern of great magnitude and ought to continuously be recognized because it is straight forwardly related to shopper well-being (Fahmy *et al.*, 2020). However, it is difficult to ensure that nanotechnology packaging is safe because tiny particles from the packaging can get into the food. People are worried that nanostructures may harm people and the environment. The rules about how nanotechnology packaging is controlled are different in each country (Chausali *et al.*, 2022).

In the US, the Food and Drug Administration (FDA) oversees the use of nanotechnology packaging that comes into contact with food. The European Union (EU) has prescribed rules for controlling nanotechnology packaging. These rules are more detailed and cover more aspects than the rules in the US. So, the safety of packaging with nanotechnology is still being studied and learned about (Peidaei *et al.*, 2021). We need to do more research to as certain if nanostructures are safe or not, and if they have any advantages. Food packaging has been around for a very long time. In the beginning, packaging was made from things found in nature like leaves, shells, and animal skins. These materials were used to keep food safe and prevent it from going bad and touching things it should not (Mustafa and Andreescu, 2020). The packaging changed and got better as people found new materials and ways to make things (Figure 4.4).

In the past, people used ceramic and glass containers to keep their food safe and prevent any dirt or germs from getting into it. Smart packaging is a type of packaging that includes active and intelligent technologies (Peidaei *et al.*, 2021). Even though the ideas of intelligent and smart packaging are different, people often use these two terms interchangeably. For instance, smart sensors are substances that can keep food fresh and healthy. They can also check if the food is still good to eat (Ekrami *et al.*, 2022e). Polyphenols that change color in the presence of salt, like the pigments found in plants called anthocyanins and battalions, are a type of smart sensor (Ekrami *et al.*, 2018). These substances found in living organisms can prevent or slow down damage from free radicals and fight against harmful microorganisms. They can also change their color when the surrounding environment becomes more acidic or basic, which can indicate when food has gone bad (Mustafa and Andreescu, 2020).

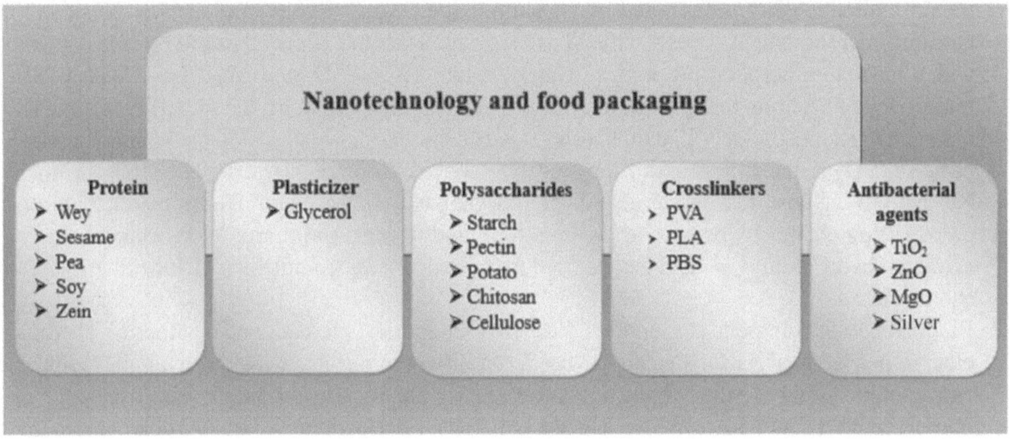

FIGURE 4.4 Overview of different types of nanotechnology methods in producing a variety of food packaging.

Packaging materials in nanotechnology are very different from conventional packaging (Shakouri *et al.*, 2023). Among the types of nanotechnology cases in food packaging industries, we can mention edible coatings and films, bio-nanocomposites, mats, fibers and nanofibers, and nanosprays (Babu, 2022). Bio-nanocomposites are packaging coatings that consist of aqueous solutions of polymeric materials such as carbohydrates, proteins, or other items, along with plasticizers such as glycerol, as well as cross-linkers such as polyvinyl alcohol (PVA), nanocrystalline cellulose, and other items along with nano elements, which are generally items such as antibacterial elements such as Some metals (silver and titanium), EOs, and plant extracts such as nettles and Malloware formed (Khezerlou *et al.*, 2021). Edible films are synthesized from primary materials such as starch, proteins, or pectin, supplemented with crosslinking agents and nanomaterials to improve mechanical properties and confer antibacterial functionalities (Biswas *et al.*, 2022). Fibers and nanofibers are produced from protein or carbohydrate water solutions along with some crosslinkers such as PVA using electrospinning or three-dimensional (3D) printing methods to create nano packaging.

Some NMs used in nanotechnology packaging are as follows (Ashfaq *et al.*, 2022, Singh *et al.*, 2023):

i. Main agents: Proteins, carbohydrates, and lipids are the most important components of nanocomposites and bio-nanocomposites. Normally, proteins are macromolecules that do not have nano dimensions, but with the help of increasing the purity percentage and applying processes such as ultrasound waves, they can be converted into micro and nano molecules. Fats also achieve this thickness with the help of processes such as ultrasound or mechanical–thermal processes.

ii. Cellulose nanocrystals (CNCs) and cellulose nanofibers (CNFs) are tiny structures with a shape like needles. We can make nanocellulose from various things like plants (wood, nonwoody, and agriculture wastes), bacteria, fungi, and tunicates. Different types of bacteria can make cellulose when the environment is right. They have a diameter between 2 and 25 nm and a length between 100 and 750 nm. They can be found in different natural materials like plants, agricultural waste, trees, and leftovers from those.

iiii. Nanosized antimicrobial agents are very effective because they have a lot of surface area compared to their size, which makes them better at killing microorganisms than larger ones. Metals that have antimicrobial properties are divided into two groups: Some commonly used or tested things are metal ions (like silver, copper, gold, and platinum) and metal oxide (like titanium dioxide (TiO_2), ZnO, and magnesium oxide (MgO). For example, silver NPs can kill harmful microorganisms because they have a larger area that allows them to release Ag^+ ions, which have antimicrobial properties. Materials like TiO_2, ZnO, and MgO have been used to make packaging films that kill germs (Ekrami *et al.*, 2021). Among the metallic oxides, TiO_2 has been used most widely in paints, foods, and cosmetics as well as food packaging materials, because TiO_2 is inert, non-toxic, inexpensive, and environmentally friendly with antimicrobial activity against a wide variety of microorganisms. These materials are better than other germ-killing chemicals because they are strong and stable. Antimicrobial substances that are extracted in different ways and can be used in biological nanocomposites include nisin, thymol and carvacrol.

iv. Biodegradable polymers are divided into two categories: natural and synthetic. There are other types of materials that can be used for different purposes. Some examples of these casesinclude starch or thermoplastic starch (TPS), chemically modified cellulose (e.g., cellulose acetate (CA) and cellulose acetate butyrate), polylactic acid (PLA), polycaprolactone (PCL), polyhydroxyalkanoate (PHA), polyhydroxybutyrate (PHB), and poly (butylene succinate) (PBS).

The improvement in the strength and flexibility of packaging material mixed with tiny particles can be credited to the very sturdy and long shape of NPs, as well as the strong connection they have with these particles (Wahab *et al.*, 2021). Polymer nanocomposites are good at stopping gases like oxygen and carbon dioxide, as well as water vapor. Research has shown that the ability of gases to pass through nanocomposites decreases when certain factors are considered (Ekrami *et al.*, 2022d). These factors include the type of NPs used, how well theymix with the polymer material, the shape of the clay particles, and the overall structure of the nanocomposites. In simple words, the polymer nanocomposites that have clay minerals spread out well and have a large shape would have the best ability to prevent gas from passing through (Cheng *et al.*, 2022). There is a direct relationship between the nano dimensions and the transverse bonds that form in the structure of nano-coatings. In this regard, some crosslinkers, such as PVA, can create high transverse links and strength and strongly prevent the passage of moisture and gases (Khezerlou *et al.*, 2021). Also, the creation of cross bonds in covalent, hydrogen, van der Waals, and ionic forms increases mechanical strength and resistance against physical damage to the coating and product. One of the characteristics of the presence of NPs can be mentioned to improve the surface characteristics of food coatings. By creating transverse bonds, these particles create a regular structure on the surface and other parts of the coatings and reduce the surface roughness and irregularity of the internal texture (Singh *et al.*, 2023).

The nano elements used in nano-coatings are diverse and have different properties, but most of their features are in increasing the storage time while keeping the product features in their original form (Jagtiani, 2022). Using different types of particles made from metals, oxides, and organic compounds is important to improve the ability of a material to prevent the growth of microbes. Increasing the strength and improving the texture of food packaging is one of the other desired goals for creating nanostructures in food coatings (Sahoo *et al.*, 2021).

4.4 FOOD SAFETY

In the recent years of nanotechnology, the speed and accuracy of some factors such as detecting the contamination of a product with the help of nanosensors or detection kits, preventing the growth of pathogenic agents with the help of antibacterial NMs and nano-coatings that prevent the reduction of shelf life and physical and microbial degradation of the producthave brought about huge benefits (Singh *et al.*, 2023). When food will not be expended promptly after generation, it must be contained in a bundle that serves various capacities. To ensure the protection of food from secondary contamination, oxygen, light and moisture, packaging must be able to withstand environmental and physical conditions (Sahoo *et al.*, 2021). This can be a tall arrangement for any fabric to fill. A critical issue in food bundling is relocation and porousness, which imperils the well-being of the item and in this way the well-being of the customer. So distant, no fabric has been found within the food bundling industry that is impenetrable to barometrical gases, water vapor, or common substances in bundled food or indeed the bundling fabric itself (Biswas *et al.*, 2022). In a few applications, tall boundaries to movement or gas dissemination are undesirable, such as in bundles for new natural products and vegetables whose shelflife is dependenton gettinga nonstop supply of oxygen for maintaining cellular breath. Plastics utilized for carbonated refreshment holders, on the other hand, must have tall oxygen and carbon dioxide obstructions to avoid oxidation and decarbonization of the refreshment substance. In other applications, the permeability of carbon dioxide is significantly less critical compared to that of oxygen or water vapor (Alfei *et al.*, 2020).

Due to the complexity of food storage conditions, modern packaging must provide special conditions and additional features for the duration of consumption (Thangadurai *et al.*, 2020). Foodborne diseases have become a global concern that has greatly affected food safety regulations (Mohammad *et al.*, 2022). Early and exact discovery of foodborne sickness is a prerequisite for avoiding, controlling, and moderating the effects of potential flare-ups. Conventional research facility strategies for the location of foodborne ailment depends basically on culture and biochemical recognizable proof of microbial specialists, nucleic corrosive location by polymerase chain response (PCR), immunological discovery such as enzyme-linkedimmunosorbent assay (ELISA) and horizontal stream immunoassay, or chromatography strategies for the location of poisons and chemical contaminants (Mustafa and Andreescu, 2020). Even though those conventional strategies are generally touchy and particular for the location of microorganisms and tried analytes, ostensibly, most of these tests are difficult, time-consuming, and require gifted research facility, which makes them inconsistent for point-of-care testing (Jagtiani, 2022, Primožič *et al.*, 2021).

Nanotechnology could be a quickly expanding field including NMs. Nanotechnology has applications in several regions of food security from foodpreparation to measures for the location of contaminants (Neme *et al.*, 2021). NMs have been utilized to expel contaminants, as antimicrobials, and as cancer prevention agents. In expansion to sensors for the discovery of pesticides, chemicals, overwhelming metals, foodborne pathogens, and poisons, nanotechnology is additionally utilized in extraction methods to adsorb the following sums of pesticides (Mustafa and Andreescu, 2020).

4.5 REGULATION OF NANOTECHNOLOGIES IN FOOD INDUSTRIES

With the rapid improvement of science, the entryways of different businesses such as agrarian generation, creature bolster and squander treatment, food preparation, and food bundling materials have been opened for the nearness of nanotechnology (Tarhan, 2020). A few of these nano processes have been commercialized for a long time, whereas numerous others are right now under investigation and improvement. Anticipated benefits of nanotechnology-based items in these divisions incorporate expanded proficiency in agrochemicals, expanded bioavailability of supplements, or more secure bundling materials (Nile *et al.*, 2020). In any case, such unused items or compounds may pose a chance to human well-being and the environment due to their particular properties and far-reaching utilization and presentation. For this reason, governments, state governments, national governments as well as worldwide and mainland unions such as the EU are attempting to set up laws to avoid conceivable or clear harm from the presence of nanotechnology within the food industry (Jagtiani, 2022).

There are endeavors around the world to direct the generation and secure utilization of NMs and nanotechnology by laws or by suggestions and rules (most of which are tragically non-binding).

As of now, there are no laws completely controlling the negative characteristics of NMs, not one or the other within the EU nor in any other nation (Ayala-Fuentes and Chavez-Santoscoy, 2021). Numerous nations consider the current laws to be adequate to control NMs and nanotechnology, which is due to different reasons, including the capacity to anticipate the transmission and defilement of COVID-19, the need for comprehensive information on the conceivable perils of this issue, and the search for building up laws in a few nations (Fajardo *et al.*, 2022). However, the European Parliament, non-governmental organizations (NGOs), the US FDA, the Organization for Economic Cooperation and Development. (OECD), and the International Organization for Standardization (ISO) are undertaking several endeavors that incorporate the definition of the term "NMs," enlistment or permitting methods, particular data prerequisites for chance evaluation, risk management, and controls to extend straightforwardness and traceability in utilization. Commerciallabeling or giving notice to the registry for items containing NMs is one of the foremost vital ways to educate customers (Tarhan, 2020).

The applications of nanotechnology in the food industry are increasing the bioavailability of bioactive compounds, improving solubility and changing the texture of products (Fajardo *et al.*, 2022). The specified rules are set or are being planned based on two categories, first, for the current food products where nanotechnology has been introduced and no change has been made in the nature of the food and has positively affected its excellent properties. The second category is related to modern foods based on added materials obtained from nanotechnology (Dera *et al.*, 2020). If we use NMs in food or animal feed, it will be regulated by specific laws depending on the type of NM used and the reason for using it (Sahoo *et al.*, 2021). The meanings of the words and where they come from are given in the extra information. Although it is not directly stated, the use of nanotechnology in making food is currently regulated by EC Regulation No 258/97. This law deals with new types of foods and ingredients used in food production. "Novel food" refers to food that was not commonly eaten in the EU before May 1997. This includes newly created or innovative food, as well as food made using new methods or technologies. This law is being changed and the new proposal should give stronger rules for controlling foods that have been modified using new technologies like nanotechnology. It will also cover foods, vitamins, minerals, and other substances that have been made using NMs (Singh *et al.*, 2023).

Based on the rules, scientists have been able to create genetically modified crops by using tiny carriers called nanocarriers. These modified crops have been grown successfully in plants like rice, tobacco, rapeseed, maize, wheat, onion, cotton, cowpea, spinach, and arugula. These small-sized agricultural products made with nanotechnology need to be checked using different methods and plans to make sure they do not have any harmful effects on people, animals, and the environment (Mohammad *et al.*, 2022). Different countries like the US, Europe, India, China, Canada, Australia, and others have made rules and guidelines to control genetically modified farming products that are carried by tiny particles called nanocarriers. These rules are about watching over tiny things made from plants all around the world. So, in the current discussion, we talked about different rules and regulations that countries have for products made from NC-based materials used in agriculture. This includes their advice and laws on making sure these products are safe for people to use all over the world (Lugani *et al.*, 2021).

We need to study and understand new foods, supplements, food additives, enzymes, and flavorings to determine if these products are safe to use. This means looking at things like how they change in size, how they interact with chemicals if they are toxic to cells, how they affect the physical and mechanical properties, their structure, and the level of bacteria they carry. In the EU, there are rules about how food additives are made (Ekrami and Emam-Djomeh, 2014). However, some cases have not been talked about yet, like when the size of particles in the ingredients used in the food industry is changed during the production of an additive that is being studied (Chaturvedi and Dave, 2020). Some small particles like SiO_2, TiO_2, and $CaCO_3$ are found in vegetables. Carbon (E153), silver (E174), gold (E175), and iron oxide (E172) were allowed in the 2010 laws. However, after more

TABLE 4.2
Advantages, disadvantages, and challenges of using nanotechnology in the food industry

Application	Advantages	Disadvantages	Challenges
Bio nanosensors	1. Food was protected from any pollutant and toxic effects of the sensing dye and enzymes by the matrix of nanosensors. 2. Nanosensors can operate cooperative interactions between analytes, enzymes, and reporter dyes.	Nanoparticle toxicity and behavior in humans are concerns. Nanoparticle size, content, concentration, and ambient parameters like humidity, wind speed, temperature, and light affect their environmental effects.	The food business values improving sensors, nanosensors, and indications. Nanosensors need a communication link to send receptor signals.
Improved packaging	TiO_2, clay, and silicate nanofillers reinforce biopolymers and increase oxygen absorption, antimicrobial activity, and biosensing. Food packaging uses nanofillers and biopolymers.	There exists a potential for the transfer of NPs from food packaging materials containing NMs to the food contained within said packaging.	Nanotechnology has improved food color, taste, texture, flavor, bioavailability of nutrients, and consistency. This technique must meet regulatory standards before being widely adopted in the food business.
Nano additives	Due to their large surface area, nano additions improve sensory perception, nutritional value, and bioavailability.	NMs in food contact human organs. Food concentration and quantity can enhance exposure.	The growing utilization of NMs in the context of food products, specifically as additives for flavor enhancement or coloration, has garnered significant interest from both the general public and governmental entities.

than 10 years, there is a need to make new laws in this area again. Even though the regulation 1935/2004 covers everything that touches food, only two laws mention NMs without defining what they are. The "Active and Intelligent Materials and Articles" regulation is not very specific about NPs and requires each case to be evaluated for risk. However, the "Plastic Food Contact Materials" regulation is more detailed and specific (Tarhan, 2020).

The current ways of assessing risks and testing NPs are mostly thought to be suitable. However, some parts, like how the samples are prepared, how they are described, how the amount of exposure is measured, and what effects they have, need better methods that have been checked and proven to be reliable. The group of scientists studying new health risks has suggested that each risk of NMs should be evaluated separately (Singh *et al.*, 2023).

4.6 ADVANTAGES, DISADVANTAGES, AND CHALLENGES

The expanding requestsfrom completely different parts of the world for adequate and secure food and the ceaseless harm of conclusion businesses in its different parts, especially the contamination of bundling materials cleared out within the environment by makers and buyers, are the major challenges we are confronting nowadays. The need for keen choices and more economical strategies forfood generation is exceptionally critical considering the persistent increment of the human populace and the quick exhaustion of worldwide assets. The implementation of nanotechnology in the food and packaging industries increases food safety, along with adding valuable features to the final product. In expansion, nano-based sensors are a perfect approach to examine all components that emphatically and straightforwardly influence the efficiency of the food industry. In this area, the positive and negative applications of nanotechnology within the food industry are inspected. The applications of nanotechnology in food handling and production lines incorporate nano-purification of squandered water to anticipate natural contamination to nonstop enhancement of quality through food additives, making food more valuable due to the expansion of advantageous substances and examining the impact of the nearness of NPs in food bundling. Finally, the quality, shortcomings, opportunities and risks of nanotechnology should be evaluated and discussed in each application to clearly define the potential of nanotechnology (Table 4.2).

4.7 CONCLUSION AND FUTURE SCOPE

Nanotechnology has colossal potential within the field of food businesses both in food bundling and processing applications. The application of nanotechnology in the food industry has an awesome advantage for food security, where there isa requirement for fast, genuine time, basic, and cost-effective strategy to distinguish defilement of foodborne pathogens, poisons, and chemicals. Nanotechnology can give high throughput and convenient demonstrative stages for screening pathogenic microorganisms and poisons specifically from food tests in the field and point-of-care clinical settings. Several of these innovations provide multiplexing capabilities for simultaneous testing of different high-affinity analytes, which can substantially reduce testing time and cost. In later a long time, nanotechnology-based tests have made strides persistently with the integration of mechanization and microfluidic frameworks, which combine all vital tests dealing with and investigation steps in a straightforward handheld biosensing gadget. It should not be overlooked that nanotechnology has not had a long history in the food industry and has played a prominent role. However, due to the ever-increasing requirements of the world, endeavors proceed to extend the quality and amount of the food industry. Finally, the qualities, shortcomings, openings, and dangers of nanotechnology are assessed and talked about to supply wide application and clearly see the potential of nanotechnology as well as future headings for nano-based agro-food applications toward maintainability.

REFERENCES

Alfei, S., Marengo, B. & Zuccari, G. (2020). Nanotechnology application in food packaging: a plethora of opportunities versus pending risks assessment and public concerns. *Food Research International*, **137**, 109664.

Amani, F., Sami, M. & Rezaei, A. (2021). Characterization and antibacterial activity of encapsulated rosemary essential oil within amylose nanostructures as a natural antimicrobial in food applications. *Starch-Stärke*, **73**, 2100021.

Ashfaq, A., Khursheed, N., Fatima, S., Anjum, Z. & Younis, K. (2022). Application of nanotechnology in food packaging: pros and cons. *Journal of Agriculture and Food Research*, **7**, 100270.

Ashraf, S. A., Siddiqui, A. J., Abd Elmoneim, O. E., Khan, M. I., Patel, M., Alreshidi, M., Moin, A., Singh, R., Snoussi, M. & Adnan, M. (2021). Innovations in nanoscience for the sustainable development of food and agriculture with implications on health and environment. *Science of the Total Environment*, **768**, 144990.

Ayala-Fuentes, J. C. & Chavez-Santoscoy, R. A. (2021). Nanotechnology as a key to enhance the benefits and improve the bioavailability of flavonoids in the food industry. *Foods*, **10**, 2701.

Babu, P. J. (2022). Nanotechnology mediated intelligent and improved food packaging. *International Nano Letters*, **12**, 1–14.

Beikzadeh, S., Akbarinejad, A., Swift, S., Perera, J., Kilmartin, P. A. & Travas-Sejdic, J. (2020). Cellulose acetate electrospun nanofibers encapsulating Lemon Myrtle essential oil as active agent with potent and sustainable antimicrobial activity. *Reactive and Functional Polymers*, **157**, 104769.

Biswas, R., Alam, M., Sarkar, A., Haque, M. I., Hasan, M. M. & Hoque, M. (2022). Application of nanotechnology in food: processing, preservation, packaging and safety assessment. *Heliyon*, **8**.

Chaturvedi, S. & Dave, P. N. (2020). Application of nanotechnology in foods and beverages. In: Nanoengineering in the Beverage Industry. Pp. 137–162. Elsevier.

Chausali, N., Saxena, J. & Prasad, R. (2022). Recent trends in nanotechnology applications of bio-based packaging. *Journal of Agriculture and Food Research*, **7**, 100257.

Chellaram, C., Murugaboopathi, G., John, A., Sivakumar, R., Ganesan, S., Krithika, S. & Priya, G. (2014). Significance of nanotechnology in food industry. *APCBEE Procedia*, **8**, 109–113.

Cheng, H., Chen, L., McClements, D. J., Xu, H., Long, J., Zhao, J., Xu, Z., Meng, M. & Jin, Z. (2022). Recent advances in the application of nanotechnology to create antioxidant active food packaging materials. *Critical Reviews in Food Science and Nutrition*, 1–16.

Coelho, S. C., Laget, S., Benaut, P., Rocha, F. & Estevinho, B. N. (2021). A new approach to the production of zein microstructures with vitamin B12, by electrospinning and spray drying techniques. *Powder Technology*, **392**, 47–57.

Couto, A. F., Favretto, M., Paquis, R. & Estevinho, B. N. (2023). Co-encapsulation of epigallocatechin-3-gallate and vitamin B12 in zein microstructures by electrospinning/electrospraying technique. *Molecules*, **28**, 2544.

Dera, M. W., Teseme, W. B. & Mulugeta Wegari Dera, W. B. T. (2020). Review on the application of food nanotechnology in food processing. *Am J Eng Technol Manag*, **5**, 41–47.

Ekrami, A., Ghadermazi, M., Ekrami, M., Hosseini, M. A., Emam-Djomeh, Z. & Hamidi-Moghadam, R. (2022a). Development and evaluation of *Zhumeria majdae* essential oil-loaded nanoliposome against multidrug-resistant clinical pathogens causing nosocomial infection. *Journal of Drug Delivery Science and Technology*, **69**, 103148.

Ekrami, M., Babaei, M., Fathi, M., Abbaszadeh, S. & Nobakht, M. (2023a). Ginger essential oil (*Zingiber officinale*) encapsulated in nanoliposome as innovative antioxidant and antipathogenic smart sustained-release system. *Food Bioscience*, **53**, 102796.

Ekrami, M., Ekrami, A., Esmaeily, R. & Emam-Djomeh, Z. (2022b). Nanotechnology-based formulation for alternative medicines and natural products: an introduction with clinical studies. In: Gopi, S., Balakrishnan, P. & Bračič, M. (eds) Biopolymers in Nutraceuticals and Functional Foods. Polymer Chemistry Series. Royal Society of Chemistry.

Ekrami, M., Ekrami, A., Hosseini, M. A. & Emam-Djomeh, Z. (2022c). Characterization and optimization of Salep mucilage bionanocomposite films containing *Allium jesdianum* Boiss. Nanoliposomes for antibacterial food packaging utilization. *Molecules*, **27**, 7032.

Ekrami, M., Ekrami, A., Moghadam, R. H., Joolaei-Ahranjani, P. & Emam-Djomeh, Z. (2022d). Food-based polymers for encapsulation and delivery of bioactive compounds. In: Gopi, S., Balakrishnan, P. &

Bračič, M. (eds) Biopolymers in Nutraceuticals and Functional Foods. Polymer Chemistry Series. Royal Society of Chemistry.

Ekrami, M., Emam-Djomeh, Z., Ghoreishy, S. A., Najari, Z. & Shakoury, N. (2019). Characterization of a high-performance edible film based on Salep mucilage functionalized with pennyroyal (*Mentha pulegium*). *International Journal of Biological Macromolecules*, **133**, 529–537.

Ekrami, M., Emam-Djomeh, Z., Joolaei-Ahranjani, P., Mahmoodi, S. & Khaleghi, S. (2021). Eco-friendly UV protective bionanocomposite based on Salep-mucilage/flower-like ZnO nanostructures to control photo-oxidation of Kilka fish oil. *International Journal of Biological Macromolecules*, **168**, 591–600.

Ekrami, M. & Emam-Djomeh, Z. (2014). Water vapor permeability, optical and mechanical properties of salep-based edible film. *Journal of Food Processing and Preservation*, **38**, 1812–1820.

Ekrami, M., Magna, G., Emam-Djomeh, Z., Saeed Yarmand, M., Paolesse, R. & DiNatale, C. (2018). Porphyrin-functionalized zinc oxide nanostructures for sensor applications. *Sensors*, **18**, 2279.

Ekrami, M., Roshani-Dehlaghi, N., Ekrami, A., Shakouri, M. & Emam-Djomeh, Z. (2022e). pH-responsive color indicator of saffron (*Crocus sativus* L.) anthocyanin-activated salep mucilage ediblefilm for real-time monitoring of fish fillet freshness. *Chemistry*, **4**, 1360–1381.

Ekrami, M., Shakouri, M., Nikkhou, S. & Emam-Djomeh, Z. (2023b). Extraction and physicochemical characterization of gum. In: M.S Sreekala, Lakshmipriya Ravindran, ... Sabu Thoma Handbook of Natural Polymers, Volume 1. Pp. 597–630. Elsevier.

Emam-Djomeh, Z., Ekrami, M. & Ekrami, A. (2023a). Overview of types of materials used for food component encapsulation. In: Lai, W.-F. (ed) Materials Science and Engineering in Food Product Development. Pp. 73–92. Wiley.

Emam-Djomeh, Z., Ekrami, M. & Ekrami, A. (2023b). Design and use of hydrogels for food component encapsulation. In: Materials Science and Engineering in Food Product Development. Pp. 211–226. Wiley.

Fahmy, H. M., Eldin, R. E. S., Serea, E. S. A., Gomaa, N. M., Abo Elmagd, G. M., Salem, S. A., Elsayed, Z. A., Edrees, A., Shams-Eldin, E. & Shalan, A. E. (2020). Advances in nanotechnology and antibacterial properties of biodegradable food packaging materials. *RSC Advances*, **10**, 20467–20484.

Fajardo, C., Martinez-Rodriguez, G., Blasco, J., Mancera, J. M., Thomas, B. & DeDonato, M. (2022). Nanotechnology in aquaculture: applications, perspectives and regulatory challenges. *Aquaculture and Fisheries*, **7**, 185–200.

Farahmand, E., Emam-Djomeh, Z., Ekrami, M. & Razavi, S. H. (2023). Polymethacrylate coated electrospun chitosan/PEO nanofibers loaded with thyme essential oil: A newfound potential for antimicrobial food packaging. *Journal of Food and Bioprocess Engineering*, **6**, 8–16.

Gahruie, H. H., Niakousari, M., Parastouei, K., Mokhtarian, M., Eş, I. & Mousavi Khaneghah, A. (2020). Co-encapsulation of vitamin D3 and saffron petals' bioactive compounds in nanoemulsions: effects of emulsifier and homogenizer types. *Journal of Food Processing and Preservation*, **44**, e14629.

Gu, Q., Yin, Y., Yan, X., Liu, X., Liu, F. & McClements, D. J. (2022). Encapsulation of multiple probiotics, synbiotics, or nutrabiotics for improved health effects: a review. *Advances in Colloid and Interface Science*, **309**, 102781.

Hamad, A. F., Han, J.-H., Kim, B.-C. & Rather, I. A. (2018). The intertwine of nanotechnology with the food industry. *Saudi Journal of Biological Sciences*, **25**, 27–30.

He, X. & Hwang, H.-M. (2016). Nanotechnology in food science: functionality, applicability, and safety assessment. *Journal of Food and Drug Analysis*, **24**, 671–681.

Jagtiani, E. (2022). Advancements in nanotechnology for food science and industry. *Food Frontiers*, **3**, 56–82.

Karbalaei-Saleh, S., Yousefi, S. & Honarvar, M. (2023). Optimization of vitamin B12 nano-emulsification and encapsulation using spontaneous emulsification. *Food Science and Biotechnology*, **33**, 399–415.

Khezerlou, A., Tavassoli, M., Alizadeh Sani, M., Mohammadi, K., Ehsani, A. & McClements, D. J. (2021). Application of nanotechnology to improve the performance of biodegradable biopolymer-based packaging materials. *Polymers*, **13**, 4399.

Koh, W. Y., Lim, X. X., Tan, T.-C., Kobun, R. & Rasti, B. (2022). Encapsulated probiotics: potential techniques and coating materials for non-dairy food applications. *Applied Sciences*, **12**, 10005.

Kour, P., Afzal, S., Gani, A., Zargar, M. I., Tak, U. N., Rashid, S. & Dar, A. A. (2022). Effect of nanoemulsion-loaded hybrid biopolymeric hydrogel beads on the release kinetics, antioxidant potential and antibacterial activity of encapsulated curcumin. *Food Chemistry*, **376**, 131925.

Kumar, P., Mahajan, P., Kaur, R. & Gautam, S. (2020). Nanotechnology and its challenges in the food sector: a review. *Materials Today Chemistry*, **17**, 100332.

Lugani, Y., Sooch, B. S., Singh, P. & Kumar, S. (2021). Nanobiotechnology applications in food sector and future innovations. In: Microbial Biotechnology in Food and Health. Pp. 197–225. Elsevier.

Mahato, D. K., Mishra, A. K. & Kumar, P. (2021). Nanoencapsulation for agri-food applications and associated health and environmental concerns. *Frontiers in Nutrition*, **8**, 663229.

Mahmud, J., Sarmast, E., Shankar, S. & Lacroix, M. (2022). Advantages of nanotechnology developments in active food packaging. *Food Research International*, **154**, 111023.

Maleki, G., Woltering, E. J. & Mozafari, M. (2022). Applications of chitosan-based carrier as an encapsulating agent in food industry. *Trends in Food Science & Technology*, **120**, 88–99.

Mei, L. & Wang, Q. (2020). Advances in using nanotechnology structuring approaches for improving food packaging. *Annual Review of Food Science and Technology*, **11**, 339–364.

Mirzakhani, M., Ekrami, M. & Moini, S. (2018). Chemical composition, total phenolic content and antimicrobial activities of *Zhumeria majdae*. *Journal of Food and Bioprocess Engineering*, **1**, 47–52.

Mohammad, Z. H., Ahmad, F., Ibrahim, S. A. & Zaidi, S. (2022). Application of nanotechnology in different aspects of the food industry. *Discover Food*, **2**, 12.

Mohammadi, M. A., Farshi, P., Ahmadi, P., Ahmadi, A., Yousefi, M., Ghorbani, M. & Hosseini, S. M. (2023). Encapsulation of vitamins using nanoliposome: recent advances and perspectives. *Advanced Pharmaceutical Bulletin*, **13**, 48.

Mujica-Álvarez, J., Gil-Castell, O., Barra, P. A., Ribes-Greus, A., Bustos, R., Faccini, M. & Matiacevich, S. (2020). Encapsulation of vitamins A and E as spray-dried additives for the feed industry. *Molecules*, **25**, 1357.

Mukurumbira, A., Shellie, R., Keast, R., Palombo, E. & Jadhav, S. (2022). Encapsulation of essential oils and their application in antimicrobial active packaging. *Food Control*, **136**, 108883.

Mustafa, F. & Andreescu, S. (2020). Nanotechnology-based approaches for food sensing and packaging applications. *RSC Advances*, **10**, 19309–19336.

Ndlovu, N., Mayaya, T., Muitire, C. & Munyengwa, N. (2020). Nanotechnology applications in crop production and food systems. *Int J Plant Breed*, **7**, 624–634.

Negi, A. & Kesari, K. K. (2022). Chitosan nanoparticle encapsulation of antibacterial essential oils. *Micromachines*, **13**, 1265.

Nejat, M. S., Ekrami, M. & Emam-Djomeh, Z. (2022). Microencapsulation Liposomal Technologies in Bioactive Functional Foods and Nutraceuticals. Biopolymers in Nutraceuticals and Functional Foods. Edited by Sreerag Gopi; Preetha Balakrishnan; Matej Bračič, 232-263. ISBN: 978-1-83916-804-8 https://doi.org/10.1039/9781839168048-00232

Neme, K., Nafady, A., Uddin, S. & Tola, Y. B. (2021). Application of nanotechnology in agriculture, postharvest loss reduction and food processing: food security implication and challenges. *Heliyon*, **7**. e08539. https://doi.org/10.1016/j.heliyon.2021.e08539

Nile, S. H., Baskar, V., Selvaraj, D., Nile, A., Xiao, J. & Kai, G. (2020). Nanotechnologies in food science: applications, recent trends, and future perspectives. *Nano-Micro Letters*, **12**, 1–34.

Pateiro, M., Gómez, B., Munekata, P. E., Barba, F. J., Putnik, P., Kovačević, D. B. & Lorenzo, J. M. (2021). Nanoencapsulation of promising bioactive compounds to improve their absorption, stability, functionality and the appearance of the final food products. *Molecules*, **26**, 1547.

Peidaei, F., Ahari, H., Anvar, A. & Ataei, M. (2021). Nanotechnology in food packaging and storage: a review. *Iranian Journal of Veterinary Medicine*, **15**. 122–153. DOI: 10.22059/IJVM.2021.310466.1005130

Primožic, M., Knez, Ž. & Leitgeb, M. (2021). (Bio)Nanotechnology in food science—food packaging. *Nanomaterials*, **11**, 292. https://doi.org/10.3390/nano11020292

Primožič, M., Knez, Ž. & Leitgeb, M. (2021). (Bio) Nanotechnology in food science—food packaging. *Nanomaterials*, **11**, 292.

Rajam, R. & Subramanian, P. (2022). Encapsulation of probiotics: past, present and future. *Beni-Suef University Journal of Basic and Applied Sciences*, **11**, 46.

Ranjha, M. M. A. N., Shafique, B., Rehman, A., Mehmood, A., Ali, A., Zahra, S. M., Roobab, U., Singh, A., Ibrahim, S. A. & Siddiqui, S. A. (2022). Biocompatible nanomaterials in food science, technology, and nutrient drug delivery: recent developments and applications. *Frontiers in Nutrition*, **8**, 778155.

Reis, D. R., Ambrosi, A. & DiLuccio, M. (2022). Encapsulated essential oils: a perspective in food preservation. *Future Foods*, **5**, 100126.

Resende, D., Lima, S. A. C. & Reis, S. (2020). Nanoencapsulation approaches for oral delivery of vitamin A. *Colloids and Surfaces B: Biointerfaces*, **193**, 111121.

Rizvi, S. S., Moraru, C. I., Bouwmeester, H., Kampers, F. W. & Cheng, Y. (2022). Nanotechnology and food safety. In: Ensuring Global Food Safety. Pp. 325–340. Elsevier.

Rodrigues, F., Cedran, M., Bicas, J. & Sato, H. (2020). Encapsulated probiotic cells: relevant techniques, natural sources as encapsulating materials and food applications—a narrative review. *Food Research International*, **137**, 109682.

Sahani, S. & Sharma, Y. C. (2021). Advancements in applications of nanotechnology in global food industry. *FoodChemistry*, **342**, 128318.

Sahoo, M., Vishwakarma, S., Panigrahi, C. & Kumar, J. (2021). Nanotechnology: current applications and future scope in food. *Food Frontiers*, **2**, 3–22.

Shakouri, M., Salami, M., Lim, L.-T., Ekrami, M., Mohammadian, M., Askari, G., Emam-Djomeh, Z. & McClements, D. J. (2023). Development of active and intelligent colorimetric biopolymer indicator: anthocyanin-loaded gelatin-basil seed gum films. *Journal of Food Measurement and Characterization*, **17**, 472–484.

Singh, R., Dutt, S., Sharma, P., Sundramoorthy, A. K., Dubey, A., Singh, A. & Arya, S. (2023). Future of nanotechnology in food industry: challenges in processing, packaging, and food safety. *Global Challenges*, **7**, 2200209.

Singh, T., Shukla, S., Kumar, P., Wahla, V., Bajpai, V. K. & Rather, I. A. (2017). Application of nanotechnology in food science: perception and overview. *Frontiers in Microbiology*, **8**, 268461.

Sokołowska, M., Marchwiana, M. & El Fray, M. (2022). Vitamin E-loaded polymeric nanoparticles from biocompatible adipate-based copolymer obtained using the nanoprecipitation method. *Polimery*, **67**, 543–551.

Subramani, T. & Ganapathyswamy, H. (2020). An overview of liposomal nano-encapsulation techniques and its applications in food and nutraceutical. *Journal of Food Science and Technology*, **57**, 3545–3555.

Tahir, A., Shabir Ahmad, R., Imran, M., Ahmad, M. H., Kamran Khan, M., Muhammad, N., Nisa, M. U., Tahir Nadeem, M., Yasmin, A. & Tahir, H. S. (2021). Recent approaches for utilization of food components as nano-encapsulation: a review. *International Journal of Food Properties*, **24**, 1074–1096.

Tarhan, Ö. (2020). Chapter-16 Safety and regulatory issues of nanomaterials in foods. *Handbook of Food Nanotechnology*, 655–703. Handbook of Food Nanotechnology Applications and Approaches. Edited by: Seid Mahdi Jafari. https://doi.org/10.1016/B978-0-12-815866-1.00016-9

Thangadurai, D., Sangeetha, J. & Prasad, R. (2020). Nanotechnology for Food, Agriculture, and Environment. Springer.

Thiruvengadam, M., Rajakumar, G. & Chung, I.-M. (2018). Nanotechnology: current uses and future applications in the food industry. *3 Biotech*, **8**, 1–13.

Wahab, A., Rahim, A. A., Hassan, S., Egbuna, C., Manzoor, M. F., Okere, K. J. & Walag, A. M. P. (2021). Chapter-10 Application of nanotechnology in the packaging of edible materials. *Preparation of Phytopharmaceuticals for the Management of Disorders*, 215–225. https://doi.org/10.1016/B978-0-12-820284-5.00004-6

Xu, C., Ban, Q., Wang, W., Hou, J. & Jiang, Z. (2022). Novel nano-encapsulated probiotic agents: encapsulate materials, delivery, and encapsulation systems. *Journal of Controlled Release*, **349**, 184–205.

Yu, H., Park, J.-Y., Kwon, C. W., Hong, S.-C., Park, K.-M. & Chang, P.-S. (2018). An overview of nanotechnology in food science: preparative methods, practical applications, and safety. *Journal of Chemistry*, 2018, Volume 2018, Article ID 5427978. https://doi.org/10.1155/2018/5427978

Zhang, H., Chen, B., Zhu, Y., Sun, C., Adu-Frimpong, M., Deng, W., Yu, J. & Xu, X. (2020). Enhanced oral bioavailability of self-assembling curcumin–vitamin E prodrug-nanoparticles by co-nanoprecipitation with vitamin E TPGS. *Drug Development and Industrial Pharmacy*, **46**, 1800–1808.

Zhu, Y., Li, C., Cui, H. & Lin, L. (2021). Encapsulation strategies to enhance the antibacterial properties of essential oils in food system. *Food Control*, **123**, 107856.

5 Nanotechnology in Food Packaging and Preservation

Anna Rafaela Cavalcante Braga,
Monica Masako Nakamoto, Monize Bürck,
Letícia Guerreiro da Trindade, and Marcelo Assis

5.1 INTRODUCTION

Food is an indispensable requirement of all living things. Due to social, technical, and technological factors, food production and demand do not match well in many countries. As a result, food industries must expand beyond their current capacities (European Commission, 2020).

Several actions have been taken to optimize food production and safety, including emerging technologies. Nanotechnology continuously evolves into a ground-breaking field of research because it offers plentiful potential for advanced nanomaterials for research in various disciplines. Incorporating nanotechnology into the food industry can directly expand food processing and preservation capabilities (Giaconia et al., 2020). In addition, nanotechnology improves thermal stability, solubility, food security, preservation capabilities, and the development of novel, highly bioavailable foods. It is a technology that operates with materials ranging in size from 1 to 100 nm (Parameswaranpillai et al., 2022).

In addition, their applications in packaging industries require specific research into the advanced properties of packaging films (Nakamoto et al., 2023). In the food packaging industry, nanomaterials can modify and outperform traditional food processing technologies for enhancing food and providing packaging films with superior mechanical properties.

Consequently, improvements and transformations in the food packaging industry are denoted by strategies investigating their broader expectations regarding food quality, packaging technology, and biosensing capabilities. Innovations in nanomaterials for food packaging are solely attributable to their eccentric properties, such as excellent optical, barrier, and thermal properties, antimicrobial activity, and advanced sensing properties, which affect their chemical, physical, and biological potential (Vasile & Baican, 2021). For nanomaterials to be utilized in various industrial applications, inimitable properties such as an exceptionally high surface-to-volume ratio, ultrafine particle size, high porosity, electrical conductivity, and diffusivity attracted considerable interest.

Nonetheless, optimizing nanomaterials in the food industry can improve nutritional values and extend shelf lives (Hosseini & Mahdi, 2020), thereby ensuring both the safety and integrity of food products. Consequently, this work aims to investigate nanomaterials' potential in advancing the food packaging industry by analyzing their current state of the art and capitalizing on their fascinating properties.

Critical components of this chapter include the development and characterization of nanomaterials for food packaging applications, techniques of nanostructure synthesis, nanomaterial-based food packaging, safety concerns, and future trends.

DOI: 10.1201/9781003438168-5

5.2 NANOMATERIALS USED IN FOOD PACKAGING

Sustainable approaches enabled biopolymeric materials from natural sources to constitute films, reducing synthetic materials derived from limited resources. It interestingly turned (I) nanocellulose from agricultural waste, such as soybean, orange peel, and sugar cane bagasse, and even bacterial fermentation, and (II) chitosan from crustaceans, fish, and seafood processing waste, insects, and fungi into suitable raw material for the food packaging industry (Espinosa et al., 2022; Li et al., 2023). Further, (III) starch also performs as recent biopolymeric materials used in food packaging, followed by incorporating (IV) metallic nanoparticles. (I), (II), and (III) are interlaced with biodegradability, and (IV) deals with improving the antioxidant and antimicrobial activities of films. Research papers that applied I–IV nanomaterials for food packaging in the last 2 years were summarized (Table 5.1 and Figure 5.1). Most studies developed films that could replace conventional plastics.

Part of the studies investigated fresh food packaging and resourcefully; one study developed paper material for cups or containers for liquid foods using 0.3% nanocellulose from soybean, 2% precipitated calcium carbonate (PCC), and 0.1% alkyl ketene dimer (AKD) with high hydrophobicity (contact angle > 150°) due to the interaction between PCC and AKD, increasing the coated paper roughness and, thus, the water contact angle (Li et al., 2023).

Additionally, it was observed that fewer studies carried out cytotoxic effects and/or migration analysis (Matteis et al., 2023; Mouzahim et al., 2023; Ndwandwe et al., 2022). In the coming years, research work must evaluate metallic nanoparticles released on food in touch with nanofilms and their average daily intake to guarantee the product's safety, effectiveness, and feasibility because even with acceptable levels of migration, consumer repeated exposure might have significance.

In the matter of biodegradability, more progress was made. A nanocomposite with improved hydrophobicity and tensile strength made by starch, polyvinyl alcohol, TiO_2 and elderberry extract (Jayakumar et al., 2022) severely disrupted the 60th day of soil biodegradability analysis but became folded and fragile at the 30th day. Apple pomace and sugar cane bagasse provided nanocellulose for films combined with PEG showed cracks and fragility on the 35th and 40th day of the biodegradation test (Garg et al., 2023; Madhavan et al., 2023).

Chitosan blended with Kaolin clay, Ag nanoparticles, and *Ficus carica* extract showed several fractures on the 7th day (Mouzahim et al., 2023). Although biodegradability is central to mitigating pollution, few studies in very recent years have carried out these tests, and none have simulated the largest disposal site, ocean conditions.

Finally, Ag, ZnO, TiO_2, and SiO_2 nanoparticle-loaded films are extensively reported as efficient antimicrobials. Amazingly, Ag and ZnO nanoparticles have also improved the ultraviolet (UV) light barrier, extending the shelf life of fresh products (Gasti et al., 2022; Zhai et al., 2022). Additionally, Ag nanoparticles loaded cellulose nanocrystals interacted with sulfuric compounds from chicken breast spoilage gases and changed their color, producing a good freshness indicator for consumers (Kwon & Ko, 2022). As characteristics influence applications, achieving food packaging films with different properties is essential.

5.3 TECHNIQUES OF NANOSTRUCTURE SYNTHESIS

Encapsulation of bioactive molecules in food packaging must not alter food products' sensory properties, color, or flavor. In this sense, there are many encapsulation methods; however, a single method cannot be considered universal because individual food components have their characteristic molecular structure. Electrospinning, reverse micelle, nanoemulsions, and sol–gel stand out among the existing methods and are illustrated in Figure 5.2.

TABLE 5.1
Nanomaterials used in food packaging in the last 2 years

Material	Source	Blend	Described potential application	Year	Reference
Nanocellulose	Soybean residue	Precipitate calcium carbonate and AKD	Containers or cups for liquid foods	2023	(Li et al., 2023)
	Sugar cane bagasse	Curry leaves ethanolic extract and PEG*	-	2023	(Madhavan et al., 2023)
	Apple pomace	Guar gum, PEG, and glutaraldehyde	Primary/secondary layer for multilayer packaging	2023	(Garg et al., 2023)
	Bacterial fermentation	Polyhydroxyalkanoate with glycerol or PEG		2023	(da Silva et al., 2023)
	Bacterial fermentation	Sunflower protein isolate	Fresh fruit packaging	2022	(Efthymiou et al., 2022)
	Orange peel waste	PVA**		2022	(Espinosa et al., 2022)
	White ginger	PVA and zinc oxide	Fresh fruit packaging	2022	(Rahmadiawan et al., 2022)
	Bamboo	Silver nanoparticles	Antimicrobial food packaging	2022	(Ren et al., 2022)
	Pullulan	Graphene		2022	(Shanmugam et al., 2022)
	Commercial cellulose nanocrystals	Silver nanoparticles	Intelligent food packaging for colorimetric freshness indicator	2022	(Kwon & Ko, 2022)
Starch	Potato starch	Lignin from moso bamboo and glycerol		2023	(Sun et al., 2023)
		PVA, titanium dioxide, elderberry extract	Intelligent food packaging wraps	2022	(Jayakumar, Radoor, Kim, Rhim, Parameswaranpillai, et al., 2022)
	Potato starch	Selenium	Active food packaging for fatty, acidic, and hydrophilic foods.	2022	(Ndwandwe et al., 2022)
		Polybutylene adipate-co-terephthalate, silver, and zinc oxide nanoparticles		2022	(Zhai et al., 2022)
		PVA, glycerol, copper oxide, and zinc oxide nanoparticles	Antimicrobial food packaging	2022	(Francis et al., 2022)
Chitosan		Cerium oxide nanoparticles	Food packaging for fatty foods such as nuts and meats	2023	(Purohit et al., 2023)
		Eugenol nanoemulsion, zinc oxide, and *Aloe vera* gel	Active food packaging	2023	(Basumatary et al., 2023)
		Kaolin clay, silver nanoparticles, and *Ficus carica* extract	Fresh apple slice packaging	2023	(Mouzahim et al., 2023)
		Aloe vera extract-mediated silver nanoparticles		2023	(Matteis et al., 2023)
		PVA and silver nanoparticles		2023	(Yang et al., 2023)

(continued)

TABLE 5.1 (Continued)
Nanomaterials used in food packaging in the last 2 years

Material	Source	Blend	Described potential application	Year	Reference
		Zinc oxide nanoparticles and stearic acid		2023	(Suyatma et al., 2023)
		PVA and iron titanium dioxide nanoparticles		2022	(Alghamdi, Abutalib, Rajeh, et al., 2022)
		Grape seed extract-mediated silver nanoparticles	Active food packaging	2022	(Zhao et al., 2022)
		Silica dioxide nanoparticles and gallic acid	Active food packaging	2022	(Dong et al., 2022)
		Pullulan, clove essential oil, zinc oxide.	Active food packaging	2022	(Gasti et al., 2022)
Silver	*Azadirachta indica* extract-mediated silver nanoparticles	Microcrystalline cellulose from sugarcane bagasse, starch, whey protein from buffalo milk, and PEG	Food packaging for vegetables	2023	(Pandian et al., 2023)
	In situ generated silver nanoparticles	Soluble soybean polysaccharide	Food packaging for light-sensitive products	2022	(Liu et al., 2022)

Note: * Polyethylene glycol; ** Polyvinyl alcohol.

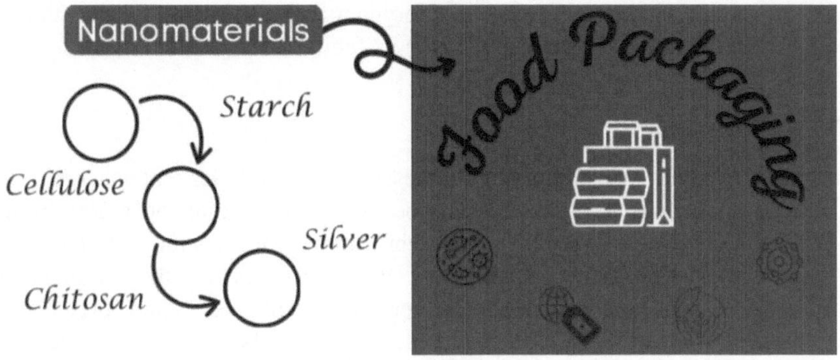

FIGURE 5.1 Schematic summary of nanomaterials used in food packaging in the last 2 years.

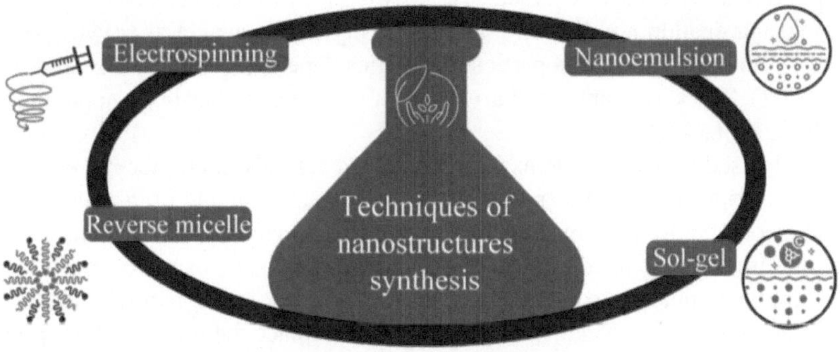

FIGURE 5.2 Schematic summary of techniques of nanostructure synthesis.

5.3.1 Electrospinning

Electrospinning is an efficient, simple, and low-cost technique for manufacturing continuous nonwoven fibers on a scale ranging from submicrons to micrometers (Mehta et al., 2017). This physical process forms these ultrafine fibers by subjecting a polymer solution or melts to a high-voltage supplier (Bhullar et al., 2015).

These barriers make the solution electrospinning technique more used to develop food packages, as it is more viable and versatile. In a classic horizontal laboratory solution electrospinning setup, a syringe containing the polymer solution is supported horizontally in a digitally controlled pump, which allows control and adjustment of the solution flow rate. Then, this syringe is connected to a cannula, and the solution is pumped from the syringe through the cannula to a metal needle, which is connected to a high-voltage power supply (5±25 kV). When the droplet at the tip of the needle has its surface tension overcome by the electrical force, a jet of charged polymer will be ejected from the Taylor cone. The electric field will accelerate this elongated jet toward a grounded and oppositely charged collector; during the process, the complete evaporation of the solvent occurs (Wang et al., 2021).

Furthermore, different additives with antioxidant and antimicrobial properties can be incorporated into the nanofibers and indicator dyes, drugs, and enzymes (Aman Mohammadi et al., 2020). These advantages, together with the diameter and microstructure of the obtained fibers, the non-requirement of the thermal process, and the accelerated solvent evaporation, make the electrospinning technique one of the most promising to producesmart food packaging. However, some barriers must be overcome for its wide use, such as frequently using toxic solvents, mechanical resistance of fibers, and rapid loss of volatile bioactive compounds (Forghani et al., 2021). The application of different electrospun fibers in smart food packaging is given in Table 5.2.

5.3.2 Reverse Micelle

Reverse micelles (RMs) are a technique for synthesizing nanoparticles that are part of the bottom-up method. In the bottom-up method, various processes synthesize nanoparticles from atomic or molecular species. The general method for RM fabrication implicates the complete dissolution of surfactants in nonpolar solvents (Sun & Bandara, 2019). Specifically, the surfactant reverses its orientation in a water-in-oil microemulsion and generates aggregates called RMs with nanometer size (Eastoe et al., 2006; Liu et al., 2014). At the core of RMs are trapped water droplets and microwater pools formed by water solubilization in their polar cores.

The microwater pools can be modified by altering the proportion of water and surfactants; inside the microwater pools, proteins, enzymes, and amide acids can be solubilized (Ashfaq et al., 2022). The parameters commonly used to characterize an RM system are size, shape, internal structure of aqueous cores, aggregation number, and microviscosity (Hasegawa et al., 1996). The choice of surfactant implies the variation of nanoparticle properties depending on their size and morphology (Arsene et al., 2021). The surfactants used are anionic, cationic, zwitterionic, nonionic, and mixed surfactants (Lépori et al., 2016).

The most widely used for these systems is sulfosuccinic acid bis (2-ethylhexyl) ester sodium salt (AOT), an anionic surfactant, which is not approved for use in food and drug materials (Fuglestad et al., 2016). The potential toxicity of the RM components, especially the nonpolar solvent, limits their applications, aiming to expand the use of this method in food, cosmetic, and pharmaceutical industries. Several biocompatible solvents such as isopropyl myristate, methyl laurate, ethyl myristate, ethyl palmitate, and ethyl oleate have been tested to replace the traditional nonpolar solvents (Dib et al., 2019).

On the other hand, AOT can be replaced by surfactants such as lecithin and lecithin derivatives, lactylate esters, glycerol fatty acid esters, partial glycerides and derivatives, sucrose esters, sorbitan fatty acid esters, or polyoxyethylenesorbitan fatty acid esters that are allowed to be used in food by The United States Code of Federal Regulations number and the European Union number (Flanagan & Singh, 2006).

5.3.3 Nanoemulsions

Nanoemulsions are appealing systems for utilization in food packaging because they have characteristics such as small droplet size (20 to 200 nm), transparent/translucent oil in water emulsions, tunable rheology, and exceptional stability (Harwansh et al., 2019). This system hasheterogeneous dispersions composed of at least two immiscible phases: a polar (commonly water) and a non-polar (commonly oil) phase. The polar phase may contain water-soluble or water-dispersed ingredients (sugars, salts, proteins, and polysaccharides), and in the non-polar phase, there are one or more water-insoluble substances (triglyceride oils, essential oils, lipid-soluble vitamins, lipid substitutes, flavors, colors, among others) (Naseema et al., 2021; Ozogul et al., 2022).

The most used techniques for the preparation of nanoemulsions are high-pressure homogenization, ultrasonication, phase inversion, and micro fluidization and involve leading components of oil, surfactant, water, and co-surfactant (Naseema et al., 2021).

The search for active polymeric films that can extend the shelf life of foods has intensified research into incorporating different types of nanoemulsions into biopolymers for food packaging. Because of that, there are three main advantages: (i) the release of bioactive substances that will prolong the shelf life of the foods; (ii) the use of biodegradable polymers, thus minimizing the use of plastics that cause serious environmental problems; and (iii) the fact that the addition of nanoemulsions in

TABLE 5.2
Application of different electrospun fibers in smart packaging of foodstuff

Electrospun fibers	Smart material	Food sample	Typeofsmart device	Feature	Reference
Hydrolyzed PVA	TiO_2, methylene blue, and glycerol	Meatball	Indicator	Oxygenindicators	(Yılmaz & Altan, 2021)
Zein	Cocktail of nature-derived antimicrobials	–	Bio-degradable packaging material	Antimicrobial	(Aytac et al., 2020)
Pullulan/ethylcellulose	Cinnamaldehyde	–	Active food packaging	Antimicrobial	(Yang et al., 2020)
Zein	Anthocyanins	–	Indicator	pH	(Prietto et al., 2018)
Polyvinylalcohol	Redcabbage (*Brassicaoleracea* L.) extract	Fruit (Rutab)	Biosensor	pH	(Maftoonazad & Ramaswamy, 2019)
Poly-L-lacticacid	Blueberryanthocyanins	Mutton	Sensor	Monitor meat foods freshness	(Sun et al., 2021)

biopolymers can lead to changes in properties that limit their use for food packaging such as poor thermo-mechanical and barrier properties (Mahmud et al., 2022). Table 5.3 presents some different nanoemulsions and their manufacturing method for food packaging purposes.

5.3.4 Sol–Gel

The use of nanomaterials in food packaging design is due to their mechanical, thermal, and barrier properties and the possibility that some nanoparticles have to prevent the growth of microbes that can harm the products. These nanomaterials can be synthesized by different strategies, such as hydrothermal, sol–gel, wet chemical, microemulsion, and sonochemical and biological (Chadha et al., 2022). The sol–gel synthesis (Table 5.4) has been gaining prominence for its versatility of processing and the facility with which its properties may be tailored to adjust a particular use.

This is a bottom-up method in which most nanoparticles can be synthesized. In a sol–gel method, the sol is formed by a mixture of solids in a colloid suspended in a liquid, while the gel is a solid macromolecule dissolved in a solvent (Chaudhary et al., 2020). After the synthesis, the nanoparticles can be recovered by sedimentation, ultra-filtration, or centrifugation, followed by drying to reduce the humidity (Adeyeye & Ashaolu, 2021). Other techniques have advantages and disadvantages. Among the advantages are low synthesis temperatures, nanoparticles with high purity, different sizes of particles that can be easily synthesized, and low energy consumption. The disadvantages are longer reaction duration and high cost (Rane et al., 2018).

5.4 NANOMATERIAL-BASED FOOD PACKAGING

As already mentioned, to provide protection to food from possible environmental contamination, dust, moisture, physical damage, and gases during transportation, storage, and sale, food packaging emerges as the primary mechanism for this purpose (Perera et al., 2022).

Annually, about 40% of the world's petroleum production is destined to manufacture plastics, with 50% of this production designated for use as packaging (Jayakumar et al., 2022). Despite having benefits related to their use, as most conventional plastics are synthesized from fossil fuel-based materials, they can cause negative impacts on the environment, especially concerning pollution and waste generation (Nakamoto et al., 2023), affecting more than 6000 living organisms by ingesting plastic waste and, consequently, affecting the environmental balance, possibly generating irreversible effects if they alter and cause changes in the habitat and aquatic ecosystems (Jayakumar et al., 2022).

To minimize these adverse effects, the use of biopolymers has been employed as a promising alternative in the elaboration of food packaging, using mainly polysaccharides, proteins, and lipids, alone or together (Agudelo-Cuartas et al., 2020; Hosseini et al., 2023; Lorevice et al., 2016; Xie et al., 2023). Unfortunately, these packages have inferior mechanical and physicochemical properties to conventional plastics, requiring strategies to overcome these obstacles. Nanotechnology is an approach that can improve the thermal stability and the mechanical and barrier properties of films (Mahmud et al., 2022). In addition, it presents promising results in applying active and intelligent packaging (Ghosh & Katiyar, 2021).

5.4.1 Enhancing Food Packaging

Films prepared with different biopolymers present different results regarding their mechanical and physicochemical properties, exposing certain advantages and disadvantages. Polysaccharides exhibit good film-forming characteristics due to their gel-forming capacity and viscosity (Montero Garcia et al., 2016). Films synthesized from polysaccharides can inhibit the migration of aromas and gases such as carbon dioxide and oxygen (Zhang et al., 2014), providing a delay in the ripening of fruits and, consequently, increasing their shelf life. However, due to their hydrophilic character,

TABLE 5.3
Diverse nanoemulsions and ways of production for food packaging purposes

Emulsification method	Essential oil	Model of food packaging	Polymer	Methods of film manufacture	Reference
Ultrasound probe sonicator	*Thymus daenensis*	Film	Hydroxylpropyl methylcellulose	Solvent casting	(Moghimi et al., 2017)
Ultrasound	Rosemary essential oil	Film	Gelatin	Solvent casting	(Xia et al., 2023)
Ultrasound probe sonicator	Citrus sinensis	Coating	Sodium alginate	–	(Das et al., 2020)
Magnetic stirrer	Cinnamaldehyde	Film	Polyvinyl alcohol	Electrospinning	(Shao et al., 2018)
Magnetic stirrer/ultrasound probe sonicator	Copaiba oil	Film	Pectin	Solvent casting	(Norcino et al., 2020)
Magnetic stirrer	Essential oils	Film	Arrowroot starch	Solvent casting	(de Oliveira Filho et al., 2021)
Ultrasound	*Cuminum cyminum* essential oil	Film	Gelatin	Electrospinning	(Mohajeri et al., 2023)

TABLE 5.4
Diverse nanomaterials synthesized by a sol–gel method for food packaging purposes

Nanomaterials	Polymer	Application	Model of food packaging	Methods of film manufacture	Reference
TiO_2-NPs	Low-density polyethylene	To enhance mechanical properties and permeability to air and water and achieve anti-bacterial properties	Film	Meltblending	(Youssef et al., 2023)
Silver and titanium dioxide nanoparticles	Polyethylene	To enhance antimicrobial property	Film	Meltmixing	(Lotfi et al., 2019)
ZnO-NPs	Poly(vinyl alcohol)	To enhance antimicrobial activity	Film	Casting	(Factori et al., 2021)
Ni/ZnONPs	Sodium alginate/polyethylene oxide	To enhance the antibacterial activity	Film	Casting	(Alghamdi, Abutalib, Mannaa, et al., 2022)
ZnO-NPs	Low-density polyethylene	To enhance antimicrobial activity	Film	Mixer/hydraulic press	(Rojas et al., 2019)
ZnO-nanorods (NRs)	Poly(ethylene oxide)/carboxymethyl cellulose	To enhance the antibacterial activity	Film	casting	(Tarabiah et al., 2022)
Ag/TiO_2-NPs	Chitosan/polyethylene oxide	To enhance structural, electrical, mechanical properties and antibacterial activity	Film	Casting	(Abutalib & Rajeh, 2021)

they present sensitivity to humidity and lower barrier resistance to water permeability (Trajkovska Petkoska et al., 2021).

Similarly to polysaccharides, proteins are also hydrophilic and exhibit fragilities toward moisture. However, they have good optical and mechanical properties, showing lower tensile strength and greater flexibility than polysaccharide-based films (Montero Garcia et al., 2016). Films made from lipids can provide brightness to food and protection against moisture transfer precisely because they are hydrophobic. However, they are fragile and have a strong flavor and potential occurrence of lipid oxidation (Rios et al., 2022).

As explained earlier, the main weaknesses found can be observed, and it can be inferred that using only one biopolymer provides delicate mechanical properties to the film and may be susceptible to different environmental conditions such as oxygen, temperature, and luminosity (Nakamoto et al., 2023). The combination of different materials can complement and fill these gaps and amplify the potential of the prepared film, and the incorporation of compounds, including additives, plasticizers, antimicrobials, and antioxidants, can also improve the functional performance of the packaging. However, directly incorporating active compounds in these films implies difficulties as most are chemically unstable, volatile, and incompatible with the polymer matrix. To overcome these challenges, nanoencapsulation of active agents can be successfully used (Khezerlou et al., 2021).

As already mentioned, nanotechnology is a mechanism that aims to improve the properties of films, offering several advantages, including increased shelf life and food quality and safety. The nanoparticles used can be employed in the form of nanotubes, nanocomposites, nanocapsules, nanofibers, nanoemulsions, and nanospheres (Perera et al., 2022) that have a dimension of 1 to 100 nm (Duncan, 2011).

Studies employing cellulose nanofibers (Bangar & Whiteside, 2021; Guo et al., 2023), α-tocopherol nanoemulsion (Agudelo-Cuartas et al., 2020), chitosan nanoparticles (Lorevice et al., 2016), eugenol nanoparticles (Lin et al., 2023), and silver nanoparticles (AgNPs; Mumtaz et al., 2023) were developed and obtained satisfactory and promising results in improving the properties and functionality of the films.

Although nanotechnology demonstrates several advantages, attention should be paid to the possible toxic effects of nanoparticles on human health. Even though nanoparticles on the surface of films are not harmful, their association and migration in food can have a negative impact on health and can be exposed through inhalation, dermal contact, and oral ingestion (Mahmud et al., 2022). This migration depends on numerous factors, such as particle size, solubility, environmental conditions, the interaction between the material and the nanoparticle, as well as its exposure time (Perera et al., 2022). Particles smaller than 300 nm in diameter pose greater risks to human health because they can precisely cross cellular barriers (Ahmad et al., 2023), can cause damage to cell membranes and organelles due to their ability to generate reactive oxygen species (ROS) (Mahmud et al., 2022), and affect the liver, kidneys, gastrointestinal tract, and immune system as they cannot be degraded and are soluble (Perera et al., 2022).

The European Commission has established Specific Migration Limits and General Migration Limits to incorporate nanoparticles in food materials (Ahmad et al., 2023). To ensure the safety of food and human health, it is crucial that further research is carried out in the development of packaging using nanotechnology and that more studies are conducted in the area to reach and find means and materials that do not present toxicity and that improve the characteristics of the films. A study conducted by das Neves et al. (2023) synthesized an amido-poly (butyl adipate co-terephthalate) film with AgNPs with *Fusarium oxysporum* components and tested it against different *Salmonella* serotypes, such as multidrug-resistant *Salmonella Saintpaul* and *Salmonella enteritidis*. In addition to this evaluation, they also investigated whether the film would enable the migration of AgNPs in chicken. As a result, they verified the inhibition of the growth of these bacteria and diagnosed migration values above the limits allowed by the European Food Safety Authority (EFSA), demonstrating the need to improve the techniques and concentrations of silver so that they are not harmful to health.

5.4.2 Active Packaging

Active packaging can be defined as a packaging system that aims to incorporate components that release or absorb substances inside or outside the packaged food (Jayakumar et al., 2022); the main characteristic of active packaging is the presence of two types of systems: active scavenging systems and active release systems (Flórez et al., 2022). The first system contains compounds involved in the release of substances with preservative properties, such as antioxidants and antimicrobials, and the second is related to the absorption of undesirable substances like oxygen, ethylene, and moisture (Monção et al., 2022), both helping to promote the quality and extend the shelf life of foods such as meat, vegetables, fruits, and seafood (Mahmud et al., 2022).

The addition of antimicrobials and antioxidants represents, for the most part, a significant part of active food packaging, inhibiting microbial growth and slowing down oxidative reactions, promoting food safety and extending shelf life (Jayakumar et al., 2022; Trajkovska Petkoska et al., 2021). The direct incorporation of bioactive compounds still presents certain limitations due to their chemical instability, low solubility, and easy oxidation under particular conditions, requiring the implementation of alternatives that guarantee their stability and enable their controlled release (Pires et al., 2023).

Using nanoparticles with biopolymers is a promising alternative to address these adversities. Metallic nanoparticles, for example, exhibit efficient antimicrobial activity against gram-positive and gram-negative bacteria, as in the study by Kowsalya et al. (2019), who synthesized AgNPs employing *Vitis vinifera* fruit peel and poly (vinyl alcohol) hybrid nanofibers together with AgNPs and evaluated their antimicrobial activity against *Bacillus cereus, Staphylococcus aureus, Escherichia coli,* and *Pseudomonas aeruginosa.* The results were satisfactory, as they demonstrated antimicrobial action against these bacteria. Another research using AgNPs also demonstrated antimicrobial activity against *E. coli* and *S. aureus*, in addition to presenting an efficient reinforcing effect on the films' mechanical, thermal, and barrier properties (Yu et al., 2016).

Organic nanoparticles also exhibit promising effects for active food packaging. Chitosan nanoparticles were incorporated into Chia gelatin–mucilage-based films in a study by Khodaman et al. (2022) to evaluate the prepared films' mechanical, physicochemical, and antimicrobial properties. Four different concentrations of chitosan nanoparticles were used (0%, 2%, 4%, and 6%) as the active compound, and the results showed that the addition of these nanoparticles caused significant changes in opacity, water vapor permeability, and antimicrobial properties against *S. aureus*.

Another research by Pereira et al. (2016) used nanocomposites based on whey protein concentrate and montmorillonite (MMT) incorporated with lycopene and observed that the addition of MMT in its highest concentration (2%) improved the barrier properties and the incorporation of lycopene provided antioxidant activity to the films. As mentioned previously, it is imperative that testing of nanoparticle migration in food and toxicity of nanoparticles is performed to maintain food and consumer safety.

5.4.3 Smart Packaging

Smart packaging, unlike active packaging, allows monitoring and verification of food quality through smart devices, such as indicators, sensors, and data carriers (Zhang et al., 2022), providing detection of possible chemical, physical, and biological changes (Babu, 2022; Terra et al., 2021), informing and alerting the consumer about the quality in which the product is, in realtime (Kuntzler et al., 2020). Often, the spoilage of food products involves temperature change, production of gases and acids, and variation in oxygen levels (Babu, 2022). Smart packaging contains sensors that detect changes in pH, gases, presence of food spoilage pathogens, toxins, food additives, colorants, preservatives in food (Mohammadi & Jafari, 2020), alerting the consumer whether the product is fit or unfit for consumption. The application of nanotechnology in developing sensors, indicators, and detectors brings successful results for food safety monitoring (Ashfaq et al., 2022).

Recent studies show some colorimetric indicators using anthocyanin (Duan et al., 2021; Liu et al., 2022), red radish color extract (Jang et al., 2023), and curcumin (Duan et al., 2021) as the main bioactive compound for smart packaging. All this research presented satisfactory results related to the change of coloration according to the different pH levels, evidencing that they are potential compounds to be used in this regard.

Although nanotechnology is being widely disseminated in the food area and presents an excellent proposal for improvements, few studies still elucidate the use of nanobiosensors in food packaging. As with any other technology, some materials may pose risks to human health. As previously mentioned, research should be developed in this area so that the negative effects are avoided, and the potential of nanotechnology for society can be exploited.

5.5 FUTURE PERSPECTIVES OF NANOTECHNOLOGY

Nanotechnology has played a significant role in developing advanced food packaging and extending the shelf life of food products. It offers unique opportunities for innovative solutions to address food safety, preservation, and quality control challenges. Using nanostructured materials and advanced manufacturing techniques, specific functionalities can be incorporated into the packaging to enhance the quality and safety of food, addressing critical issues such as barrier properties and microbial proliferation control (de Oliveira et al., 2023; Sohail et al., 2018). These outcomes directly impact the shelf life of food products, helping to prevent their rapid deterioration (Assis et al., 2021).

In addition, with the advent of active and smart packaging, nanotechnology also plays a vital role in enhancing food safety by enabling rapid and sensitive detection of foodborne pathogens and contaminants (Fang et al., 2017; Yam et al., 2005). Furthermore, these same nanosensors can provide insights into the quality and safety of various types of food, always related to their deterioration (Kuswandi et al., 2011). These nanoscale devices offer high sensitivity, selectivity, and specificity, allowing real-time monitoring and ensuring the safety of food products (Kuswandi et al., 2022). Another point to be highlighted within active packaging is that new approaches can be achieved for the targeted and controlled release of bioactive compounds in food (Almasi et al., 2021). Nanocarriers such as liposomes, solid lipid nanoparticles, and nanoemulsions can encapsulate bioactive substances such as vitamins, antioxidants, and antimicrobial agents (Falsafi et al., 2022; Vasile & Baican, 2021). This encapsulation protects the bioactive compounds from degradation and enhances their stability, bioavailability, and controlled release in food matrices.

The use of biomaterials for food packaging is another niche that has been extensively studied, especially after the severe acute respiratory syndrome coronavirus 2 (SARS-CoV-2) pandemic, which has changed the dynamics of plastic waste utilization and disposal (Jafarzadeh et al., 2023; Yuan et al., 2021). In this regard, biopolymers such as starch, cellulose, chitosan, and protein-based materials are being explored as alternatives to petroleum-based plastics. These biopolymers have the advantage of being derived from renewable sources and possessing inherent biodegradability, making them environmentally friendly options for food packaging and other food-related applications (Torres-Giner et al., 2021).

However, along with the tremendous potential, there are concerns regarding nanotechnology's regulatory and safety aspects in the food industry. As nanotechnology continues to advance in food science, it is crucial to assess the potential toxicological implications associated with using nanomaterials in food-related applications (Kumar, 2015). Understanding the toxicological aspects is essential to ensure the safety of consumers and the environment. One of the main concerns in nanotoxicology is the potential for nanomaterials to interact with biological systems and induce adverse effects. Due to their small size and large surface area, nanoparticles can exhibit unique physicochemical properties that may lead to unintended biological interactions and toxicity (Abbasi et al., 2023). These interactions can affect cellular structures, membrane integrity, and cellular functions, resulting in cellular damage or dysfunction. In general, one of the barriers to the toxicological assessment of nanomaterials is the lack of standardized testing and the variability in synthesizing nanostructured systems. This can lead to different results due to entirely different methodologies (Amorim et al., 2023; Restuccia et al., 2010).

The potential toxicity of nanomaterials, their potential migration from packaging materials into food, and their impact on human health and the environment must be carefully assessed (Restuccia et al., 2010; Yan et al., 2022). In this regard, robust risk assessment protocols and stringent regulatory frameworks should be in place to ensure the safe and responsible use of nanotechnology in the food sector, thereby enhancing its application in this field.

REFERENCES

Abbasi, R., Shineh, G., Mobaraki, M., Doughty, S., & Tayebi, L. (2023). Structural parameters of nanoparticles affecting their toxicity for biomedical applications: a review. *Journal of Nanoparticle Research*, *25*(3), 43. https://doi.org/10.1007/s11051-023-05690-w

Abutalib, M. M., & Rajeh, A. (2021). Enhanced structural, electrical, mechanical properties and antibacterial activity of Cs/PEO doped mixed nanoparticles (Ag/TiO_2) for food packaging applications. *Polymer Testing*, *93*, 107013. https://doi.org/10.1016/j.polymertesting.2020.107013

Adeyeye, S. A. O., & Ashaolu, T. J. (2021). Applications of nano-materials in food packaging: a review. *Journal of Food Process Engineering*, *44*(7). https://doi.org/10.1111/jfpe.13708

Agudelo-Cuartas, C., Granda-Restrepo, D., Sobral, P. J. A., Hernandez, H., & Castro, W. (2020). Characterization of whey protein-based films incorporated with natamycin and nanoemulsion of α-tocopherol. *Heliyon*, *6*(4), e03809. https://doi.org/10.1016/j.heliyon.2020.e03809

Ahmad, A., Qurashi, A., & Sheehan, D. (2023). Nano packaging – progress and future perspectives for food safety, and sustainability. *Food Packaging and Shelf Life*, *35*, 100997. https://doi.org/10.1016/j.fpsl.2022.100997

Alghamdi, H. M., Abutalib, M. M., Mannaa, M. A., Nur, O., Abdelrazek, E. M., & Rajeh, A. (2022). Modification and development of high bioactivities and environmentally safe polymer nanocomposites doped by Ni/ZnO nanohybrid for food packaging applications. *Journal of Materials Research and Technology*, *19*, 3421–3432. https://doi.org/10.1016/j.jmrt.2022.06.077

Alghamdi, H. M., Abutalib, M. M., Rajeh, A., Mannaa, M. A., Nur, O., & Abdelrazek, E. M. (2022). Effect of the Fe_2O_3/TiO_2 nanoparticles on the structural, mechanical, electrical properties and antibacterial activity of the biodegradable chitosan/polyvinyl alcohol blend for food packaging. *Journal of Polymers and the Environment*, *30*(9), 3865–3874. https://doi.org/10.1007/s10924-022-02478-2

Almasi, H., Jahanbakhsh Oskouie, M., & Saleh, A. (2021). A review on techniques utilized for design of controlled release food active packaging. *Critical Reviews in Food Science and Nutrition*, *61*(15), 2601–2621. https://doi.org/10.1080/10408398.2020.1783199

Aman Mohammadi, M., Hosseini, S. M., & Yousefi, M. (2020). Application of electrospinning technique in development of intelligent food packaging: a short review of recent trends. *Food Science & Nutrition*, *8*(9), 4656–4665. https://doi.org/10.1002/fsn3.1781

Amorim, M. J. B., Peijnenburg, W., Greco, D., Saarimäki, L. A., Dumit, V. I., Bahl, A., Haase, A., Tran, L., Hackermüller, J., Canzler, S., & Scott-Fordsmand, J. J. (2023). Systems toxicology to advance human and environmental hazard assessment: a roadmap for advanced materials. *Nano Today*, *48*, 101735. https://doi.org/10.1016/j.nantod.2022.101735

Arsene, M.-L., Răut, I., Călin, M., Jecu, M.-L., Doni, M., & Gurban, A.-M. (2021). Versatility of reverse micelles: from biomimetic models to nano (bio)sensor design. *Processes*, *9*(2), 345. https://doi.org/10.3390/pr9020345

Ashfaq, A., Khursheed, N., Fatima, S., Anjum, Z., & Younis, K. (2022). Application of nanotechnology in food packaging: pros and cons. *Journal of Agriculture and Food Research*, *7*, 100270. https://doi.org/10.1016/j.jafr.2022.100270

Assis, M., Simoes, L. G. P., Tremiliosi, G. C., Ribeiro, L. K., Coelho, D., Minozzi, D. T., Santos, R. I., Vilela, D. C. B., Mascaro, L. H., Andrés, J., & Longo, E. (2021). PVC-SiO_2-Ag composite as a powerful biocide and anti-SARS-CoV-2 material. *Journal of Polymer Research*, *28*(9), 361. https://doi.org/10.1007/s10965-021-02729-1

Aytac, Z., Huang, R., Vaze, N., Xu, T., Eitzer, B. D., Krol, W., MacQueen, L. A., Chang, H., Bousfield, D. W., Chan-Park, M. B., Ng, K. W., Parker, K. K., White, J. C., & Demokritou, P. (2020). Development of biodegradable and antimicrobial electrospun zein fibers for food packaging. *ACS Sustainable Chemistry & Engineering*, *8*(40), 15354–15365. https://doi.org/10.1021/acssuschemeng.0c05917

Babu, P. J. (2022). Nanotechnology mediated intelligent and improved food packaging. *International Nano Letters*, *12*(1), 1–14. https://doi.org/10.1007/s40089-021-00348-8

Bangar, S. P., & Whiteside, W. S. (2021). Nano-cellulose reinforced starch bio composite films – a review on green composites. *International Journal of Biological Macromolecules*, *185*, 849–860. https://doi.org/10.1016/j.ijbiomac.2021.07.017

Basumatary, I. B., Mukherjee, A., & Kumar, S. (2023). Chitosan-based composite films containing eugenol nanoemulsion, ZnO nanoparticles and aloe vera gel for active food packaging. *International Journal of Biological Macromolecules*, *242*, 124826. https://doi.org/10.1016/J.IJBIOMAC.2023.124826

Bhullar, S. K., Kaya, B., & Jun, M. B.-G. (2015). Development of bioactive packaging structure using melt electrospinning. *Journal of Polymers and the Environment*, *23*(3), 416–423. https://doi.org/10.1007/s10924-015-0713-z

Chadha, U., Bhardwaj, P., Selvaraj, S. K., Arasu, K., Praveena, S., Pavan, A., Khanna, M., Singh, P., Singh, S., Chakravorty, A., Badoni, B., Banavoth, M., Sonar, P., & Paramasivam, V. (2022). Current trends and future perspectives of nanomaterials in food packaging application. *Journal of Nanomaterials*, *2022*, 1–32. https://doi.org/10.1155/2022/2745416

Chaudhary, P., Fatima, F., & Kumar, A. (2020). Relevance of nanomaterials in food packaging and its advanced future prospects. *Journal of Inorganic and Organometallic Polymers and Materials*, *30*(12), 5180–5192. https://doi.org/10.1007/s10904-020-01674-8

da Silva, F. A. G. S., Matos, M., Dourado, F., Reis, M. A. M., Branco, P. C., Poças, F., & Gama, M. (2023). Development of a layered bacterial nanocellulose-PHBV composite for food packaging. *Journal of the Science of Food and Agriculture*, *103*(3), 1077–1087. https://doi.org/10.1002/JSFA.11839

das Neves, M. da S., Scandorieiro, S., Pereira, G. N., Ribeiro, J. M., Seabra, A. B., Dias, A. P., Yamashita, F., Martinez, C. B. dos R., Kobayashi, R. K. T., & Nakazato, G. (2023). Antibacterial activity of biodegradable films incorporated with biologically-synthesized silver nanoparticles and the evaluation of their migration to chicken meat. *Antibiotics*, *12*(1), 178. https://doi.org/10.3390/antibiotics12010178

Das, S., Vishakha, K., Banerjee, S., Mondal, S., & Ganguli, A. (2020). Sodium alginate-based edible coating containing nanoemulsion of *Citrus sinensis* essential oil eradicates planktonic and sessile cells of food-borne pathogens and increased quality attributes of tomatoes. *International Journal of Biological Macromolecules*, *162*, 1770–1779. https://doi.org/10.1016/j.ijbiomac.2020.08.086

de Oliveira Filho, J. G., Albiero, B. R., Cipriano, L., de Oliveira Nobre Bezerra, C. C., Oldoni, F. C. A., Egea, M. B., de Azeredo, H. M. C., & Ferreira, M. D. (2021). Arrowroot starch-based films incorporated with a carnauba wax nanoemulsion, cellulose nanocrystals, and essential oils: a new functional material for food packaging applications. *Cellulose*, *28*(10), 6499–6511. https://doi.org/10.1007/s10570-021-03945-0

de Oliveira, M. C., Assis, M., Simões, L. G. P., Minozzi, D. T., Ribeiro, R. A. P., Andrés, J., & Longo, E. (2023). Unraveling the intrinsic biocidal activity of the SiO_2-Ag composite against SARS-CoV-2: a joint experimental and theoretical study. *ACS Applied Materials & Interfaces*, *15*(5), 6548–6560. https://doi.org/10.1021/acsami.2c21011

Dib, N., Silber, J. J., Correa, N. M., & Falcone, R. D. (2019). Combination of a protic ionic liquid-like surfactant and biocompatible solvents to generate environmentally friendly anionic reverse micelles. *New Journal of Chemistry*, *43*(26), 10398–10404. https://doi.org/10.1039/C9NJ02268F

Dong, W., Su, J., Chen, Y., Xu, D., Cheng, L., Mao, L., Gao, Y., & Yuan, F. (2022). Characterization and antioxidant properties of chitosan film incorporated with modified silica nanoparticles as an active food packaging. *Food Chemistry*, *373*, 131414. https://doi.org/10.1016/J.FOODCHEM.2021.131414

Duan, M., Yu, S., Sun, J., Jiang, H., Zhao, J., Tong, C., Hu, Y., Pang, J., & Wu, C. (2021). Development and characterization of electrospun nanofibers based on pullulan/chitin nanofibers containing curcumin and anthocyanins for active-intelligent food packaging. *International Journal of Biological Macromolecules*, *187*, 332–340. https://doi.org/10.1016/j.ijbiomac.2021.07.140

Duncan, T. V. (2011). Applications of nanotechnology in food packaging and food safety: barrier materials, antimicrobials and sensors. *Journal of Colloid and Interface Science*, *363*(1), 1–24. https://doi.org/10.1016/j.jcis.2011.07.017

Eastoe, J., Hollamby, M. J., & Hudson, L. (2006). Recent advances in nanoparticle synthesis with reversed micelles. *Advances in Colloid and Interface Science*, *128–130*, 5–15. https://doi.org/10.1016/j.cis.2006.11.009

Efthymiou, M. N., Tsouko, E., Papagiannopoulos, A., Athanasoulia, I. G., Georgiadou, M., Pispas, S., Briassoulis, D., Tsironi, T., & Koutinas, A. (2022). Development of biodegradable films using sunflower protein isolates and bacterial nanocellulose as innovative food packaging materials for fresh fruit preservation. *Scientific Reports*, *12*(1), 1–13. https://doi.org/10.1038/s41598-022-10913-6

Espinosa, E., Rincón, E., Morcillo-Martín, R., Rabasco-Vílchez, L., & Rodríguez, A. (2022). Orange peel waste biorefinery in multi-component cascade approach: polyphenolic compounds and nanocellulose for food packaging. *Industrial Crops and Products*, *187*, 115413. https://doi.org/10.1016/j.indcrop.2022.115413

European Comission. (2020). A farm to fork strategy for a fair, healthy and environmentally-friendly food system. https://eur-lex.europa.eu/legal-content/EN/TXT/?uri=CELEX:52020DC0381

Factori, I. M., Amaral, J. M., Camani, P. H., Rosa, D. S., Lima, B. A., Brocchi, M., da Silva, E. R., & Souza, J. S. (2021). ZnO nanoparticle/poly(vinyl alcohol) nanocomposites via microwave-assisted sol–gel synthesis for structural materials, UV shielding, and antimicrobial activity. *ACS Applied Nano Materials*, *4*(7), 7371–7383. https://doi.org/10.1021/acsanm.1c01334

Falsafi, S. R., Karaca, A. C., Deng, L., Wang, Y., Li, H., Askari, G., & Rostamabadi, H. (2022). Insights into whey protein-based carriers for targeted delivery and controlled release of bioactive components. *Food Hydrocolloids*, *133*, 108002. https://doi.org/10.1016/j.foodhyd.2022.108002

Fang, Z., Zhao, Y., Warner, R. D., & Johnson, S. K. (2017). Active and intelligent packaging in meat industry. *Trends in Food Science & Technology*, *61*, 60–71. https://doi.org/10.1016/j.tifs.2017.01.002

Flanagan, J., & Singh, H. (2006). Microemulsions: a potential delivery system for bioactives in food. *Critical Reviews in Food Science and Nutrition*, *46*(3), 221–237. https://doi.org/10.1080/10408690590956710

Flórez, M., Guerra-Rodríguez, E., Cazón, P., & Vázquez, M. (2022). Chitosan for food packaging: recent advances in active and intelligent films. *Food Hydrocolloids*, *124*, 107328. https://doi.org/10.1016/j.foodhyd.2021.107328

Forghani, S., Almasi, H., & Moradi, M. (2021). Electrospun nanofibers as food freshness and time-temperature indicators: a new approach in food intelligent packaging. *Innovative Food Science & Emerging Technologies*, *73*, 102804. https://doi.org/10.1016/j.ifset.2021.102804

Francis, D. V., Thaliyakattil, S., Cherian, L., Sood, N., & Gokhale, T. (2022). Metallic nanoparticle integrated ternary polymer blend of PVA/starch/glycerol: a promising antimicrobial food packaging material. *Polymers*, *14*(7), 1379. https://doi.org/10.3390/polym14071379

Fuglestad, B., Gupta, K., Wand, A. J., & Sharp, K. A. (2016). Characterization of cetyltrimethylammonium bromide/hexanol reverse micelles by experimentally benchmarked molecular dynamics simulations. *Langmuir*, *32*(7), 1674–1684. https://doi.org/10.1021/acs.langmuir.5b03981

Garg, R., Rana, H., Singh, N., & Goswami, S. (2023). Guargum/nanocellulose based novel crosslinked antimicrobial film with enhanced barrier and mechanical properties for food packaging. *Journal of Environmental Chemical Engineering*, *11*(2), 109254. https://doi.org/10.1016/J.JECE.2022.109254

Gasti, T., Dixit, S., Hiremani, V. D., Chougale, R. B., Masti, S. P., Vootla, S. K., & Mudigoudra, B. S. (2022). Chitosan/pullulan based films incorporated with clove essential oil loaded chitosan-ZnO hybrid nanoparticles for active food packaging. *Carbohydrate Polymers*, *277*, 118866. https://doi.org/10.1016/J.CARBPOL.2021.118866

Ghosh, T., & Katiyar, V. (2021). Edible Food Packaging: An Introduction. In Nanotechnology in Edible Food Packaging (pp. 1–23). Springer. https://doi.org/10.1007/978-981-33-6169-0_1

Giaconia, M. A., Ramos, S. dos P., Pereira, C. F., Lemes, A. C., De Rosso, V. V., Braga, A. R. C., Amendoeira, M., Ramos, P., Fratelli, C., Cesar, A., Vera, V., Rosso, D., Rafaela, A., & Braga, C. (2020). Overcoming restrictions of bioactive compounds biological effects in food using nanometer-sized structures. *Food Hydrocolloids*, *107*(February), 105939. https://doi.org/10.1016/j.foodhyd.2020.105939

Guo, L., Li, M., Xu, Q., Jin, L., & Wang, Y. (2023). Bio-based films with high antioxidant and improved water-resistant properties from cellulose nanofibres and lignin nanoparticles. *International Journal of Biological Macromolecules*, *227*, 365–372. https://doi.org/10.1016/j.ijbiomac.2022.12.128

Harwansh, R. K., Deshmukh, R., & Rahman, M. A. (2019). Nanoemulsion: promising nanocarrier system for delivery of herbal bioactives. *Journal of Drug Delivery Science and Technology*, *51*, 224–233. https://doi.org/10.1016/j.jddst.2019.03.006

Hasegawa, M., Sugimura, T., Shindo, Y., & Kitahara, A. (1996). Structure and properties of AOT reversed micelles as studied by the fluorescence probe technique. *Colloids and Surfaces A: Physicochemical and Engineering Aspects*, *109*, 305–318. https://doi.org/10.1016/0927-7757(96)03463-2

Hosseini, H., & Mahdi, S. (2020). Introducing nano/microencapsulated bioactive ingredients for extending the shelf-life of food products. *Advances in Colloid and Interface Science*, *282*, 102210. https://doi.org/10.1016/j.cis.2020.102210

Hosseini, S. F., Mousavi, Z., & McClements, D. J. (2023). Beeswax: a review on the recent progress in the development of superhydrophobic films/coatings and their applications in fruits preservation. *Food Chemistry*, *424*, 136404. https://doi.org/10.1016/j.foodchem.2023.136404

Jafarzadeh, S., Forough, M., Kouzegaran, V. J., Zargar, M., Garavand, F., Azizi-Lalabadi, M., Abdollahi, M., & Jafari, S. M. (2023). Improving the functionality of biodegradable food packaging materials via porous nanomaterials. *Comprehensive Reviews in Food Science and Food Safety*, 22(4), 2850–2886. https://doi.org/10.1111/1541-4337.13164

Jang, J. H., Kang, H. J., Adedeji, O. E., Kim, G. Y., Lee, J. K., Kim, D. H., & Jung, Y. H. (2023). Development of a pH indicator for monitoring the freshness of minced pork using a cellulose nanofiber. *Food Chemistry*, *403*, 134366. https://doi.org/10.1016/j.foodchem.2022.134366

Jayakumar, A., Radoor, S., Kim, J. T., Rhim, J. W., Nandi, D., Parameswaranpillai, J., & Siengchin, S. (2022). Recent innovations in bionanocomposites-based food packaging films – a comprehensive review. *Food Packaging and Shelf Life*, *33*, 100877. https://doi.org/10.1016/j.fpsl.2022.100877

Jayakumar, A., Radoor, S., Kim, J. T., Rhim, J. W., Parameswaranpillai, J., Nandi, D., Srisuk, R., & Siengchin, S. (2022). Titanium dioxide nanoparticles and elderberry extract incorporated starch based polyvinyl alcohol films as active and intelligent food packaging wraps. *Food Packaging and Shelf Life*, *34*, 100967. https://doi.org/10.1016/J.FPSL.2022.100967

Khezerlou, A., Tavassoli, M., Alizadeh Sani, M., Mohammadi, K., Ehsani, A., & McClements, D. J. (2021). Application of nanotechnology to improve the performance of biodegradable biopolymer-based packaging materials. *Polymers*, *13*(24), 4399. https://doi.org/10.3390/polym13244399

Khodaman, E., Barzegar, H., Jokar, A., & Jooyandeh, H. (2022). Production and evaluation of physicochemical, mechanical and antimicrobial properties of chia (*Salvia hispanica* L.) mucilage-gelatin based edible films incorporated with chitosan nanoparticles. *Journal of Food Measurement and Characterization*, *16*(5), 3547–3556. https://doi.org/10.1007/s11694-022-01470-7

Kowsalya, E., MosaChristas, K., Balashanmugam, P., Tamil Selvi, A., & Rani, J. C. I. (2019). Biocompatible silver nanoparticles/poly(vinyl alcohol) electrospun nanofibers for potential antimicrobial food packaging applications. *Food Packaging and Shelf Life*, *21*, 100379. https://doi.org/10.1016/j.fpsl.2019.100379

Kumar, L. Y. (2015). Role and adverse effects of nanomaterials in food technology. *Journal of Toxicology and Health*, *2*(1), 2. DOI: 10.7243/2056-3779-2-2

Kuntzler, S. G., Costa, J. A. V., Brizio, A. P. D. R., & Morais, M. G. de. (2020). Development of a colorimetric pH indicator using nanofibers containing *Spirulina* sp. LEB 18. *Food Chemistry*, *328*(May 2019), 126768. https://doi.org/10.1016/j.foodchem.2020.126768

Kuswandi, B., Moradi, M., & Ezati, P. (2022). Food sensors: off-package and on-package approaches. *Packaging Technology and Science*, *35*(12), 847–862. https://doi.org/10.1002/pts.2683

Kuswandi, B., Wicaksono, Y., Jayus, A., Heng, L.Y., & Ahmad, M. (2011). Smart packaging: sensors for monitoring of food quality and safety. *Sensing and Instrumentation for Food Quality and Safety*, *5*(3–4), 137–146. https://doi.org/10.1007/s11694-011-9120-x

Kwon, S., & Ko, S. (2022). Colorimetric freshness indicator based on cellulose nanocrystal–silver nanoparticle composite for intelligent food packaging. *Polymers*, *14*(17), 3695. https://doi.org/10.3390/polym14173695

Lépori, C. M. O., Correa, N. M., Silber, J. J., & Falcone, R. D. (2016). How the cation 1-butyl-3-methylimidazolium impacts the interaction between the entrapped water and the reverse micelle interface created with an ionic liquid-like surfactant. *Soft Matter*, *12*(3), 830–844. https://doi.org/10.1039/C5SM02421H

Li, P., Zhou, M., Jian, B., Lei, H., Liu, R., Zhou, X., Li, X., Wang, Y., & Zhou, B. (2023). Paper material coated with soybean residue nanocellulose waterproof agent and its application in food packaging. *Industrial Crops and Products*, *199*, 116749. https://doi.org/10.1016/J.INDCROP.2023.116749

Lin, L., Mei, C., Shi, C., Li, C., Abdel-Samie, M. A., & Cui, H. (2023). Preparation and characterization of gelatin active packaging film loaded with eugenol nanoparticles and its application in chicken preservation. *Food Bioscience*, *53*, 102778. https://doi.org/10.1016/j.fbio.2023.102778

Liu, F., Wang, X., Zhao, X., Hu, H., Chen, F., & Sun, Y. (2014). Surface properties of walnut protein from AOT reverse micelles. *International Journal of Food Science & Technology*, *49*(2), 626–633. https://doi.org/10.1111/ijfs.12345

Liu, J., Ma, Z., Liu, Y., Zheng, X., Pei, Y., & Tang, K. (2022). Soluble soybean polysaccharide films containing in-situ generated silver nanoparticles for antibacterial food packaging applications. *Food Packaging and Shelf Life*, *31*, 100800. https://doi.org/10.1016/j.fpsl.2021.100800

Liu, L., Wu, W., Zheng, L., Yu, J., Sun, P., & Shao, P. (2022). Intelligent packaging films incorporated with anthocyanins-loaded ovalbumin-carboxymethyl cellulose nanocomplexes for food freshness monitoring. *Food Chemistry*, *387*, 132908. https://doi.org/10.1016/j.foodchem.2022.132908

Lorevice, M. V., Otoni, C. G., Moura, M. R. de, & Mattoso, L. H. C. (2016). Chitosan nanoparticles on the improvement of thermal, barrier, and mechanical properties of high- and low-methyl pectin films. *Food Hydrocolloids*, *52*, 732–740. https://doi.org/10.1016/j.foodhyd.2015.08.003

Lotfi, S., Ahari, H., & Sahraeyan, R. (2019). The effect of silver nanocomposite packaging based on melt mixing and sol–gel methods on shelf life extension of fresh chicken stored at 4 °C. *Journal of Food Safety*, *39*(3), e12625. https://doi.org/10.1111/jfs.12625

Madhavan, A., Reshmy, R., Arun, K. B., Philip, E., Sindhu, R., Nair, B. G., Awasthi, M. K., Pandey, A., & Binod, P. (2023). Murrayakoenigii extract blended nanocellulose-polyethylene glycol thin films for the sustainable synthesis of antibacterial food packaging. *Sustainable Chemistry and Pharmacy*, *32*, 101021. https://doi.org/10.1016/J.SCP.2023.101021

Maftoonazad, N., & Ramaswamy, H. (2019). Design and testing of an electrospun nanofiber mat as a pH biosensor and monitor the pH associated quality in fresh date fruit (Rutab). *Polymer Testing*, *75*, 76–84. https://doi.org/10.1016/j.polymertesting.2019.01.011

Mahmud, J., Sarmast, E., Shankar, S., & Lacroix, M. (2022). Advantages of nanotechnology developments in active food packaging. *Food Research International*, *154*, 111023. https://doi.org/10.1016/j.foodres.2022.111023

Matteis, V. De, Cascione, M., Costa, D., Martano, S., Manno, D., Cannavale, A., Mazzotta, S., Paladini, F., Martino, M., & Rinaldi, R. (2023). Aloe vera silver nanoparticles addition in chitosan films: improvement of physicochemical properties for eco-friendly food packaging material. *Journal of Materials Research and Technology*, *24*, 1015–1033. https://doi.org/10.1016/J.JMRT.2023.03.025

Mehta, P., Haj-Ahmad, R., Rasekh, M., Arshad, M. S., Smith, A., van der Merwe, S. M., Li, X., Chang, M.-W., & Ahmad, Z. (2017). Pharmaceutical and biomaterial engineering via electrohydrodynamic atomization technologies. *Drug Discovery Today*, *22*(1), 157–165. https://doi.org/10.1016/j.drudis.2016.09.021

Moghimi, R., Aliahmadi, A., & Rafati, H. (2017). Antibacterial hydroxypropyl methyl cellulose edible films containing nanoemulsions of *Thymus daenensis* essential oil for food packaging. *Carbohydrate Polymers*, *175*, 241–248. https://doi.org/10.1016/j.carbpol.2017.07.086

Mohajeri, P., Hematian Sourki, A., Mehregan Nikoo, A., & Ertas, Y. N. (2023). Fabrication, characterisation and antimicrobial activity of electrospun *Plantago psyllium* L. seed gum/gelatinenanofibres incorporated with *Cuminum cyminum* essential oil nanoemulsion. *International Journal of Food Science & Technology*, *58*(4), 1832–1840. https://doi.org/10.1111/ijfs.16324

Mohammadi, Z., & Jafari, S. M. (2020). Detection of food spoilage and adulteration by novel nanomaterial-based sensors. *Advances in Colloid and Interface Science*, *286*, 102297. https://doi.org/10.1016/j.cis.2020.102297

Monção, É. da C., Grisi, C. V. B., de Moura Fernandes, J., Souza, P. S., & de Souza, A. L. (2022). Active packaging for lipid foods and development challenges for marketing. *Food Bioscience*, *45*, 101370. https://doi.org/10.1016/j.fbio.2021.101370

Montero Garcia, M. P., Gómez-Guillén, M. C., López-Caballero, M. E., & Barbosa-Cánovas, G. V. (Eds.). (2016). *Edible Films and Coatings*. CRC Press. https://doi.org/10.1201/9781315373713

Mouzahim, M. E., Eddarai, E. M., Eladaoui, S., Guenbour, A., Bellaouchou, A., Zarrouk, A., & Boussen, R. (2023). Effect of kaolin clay and *Ficus carica* mediated silver nanoparticles on chitosan food packaging film for fresh apple slice preservation. *Food Chemistry*, *410*, 135470. https://doi.org/10.1016/J.FOODCHEM.2023.135470

Mumtaz, S., Ali, S., Mumtaz, S., Mughal, T. A., Tahir, H. M., & Shakir, H. A. (2023). Chitosan conjugated silver nanoparticles: the versatile antibacterial agents. *Polymer Bulletin*, *80*(5), 4719–4736. https://doi.org/10.1007/s00289-022-04321-z

Nakamoto, M. M., Assis, M., de Oliveira Filho, J. G., & Braga, A. R. C. (2023). Spirulina application in food packaging: gaps of knowledge and future trends. *Trends in Food Science & Technology*, 133(October 2022), 138–147. https://doi.org/10.1016/j.tifs.2023.02.001

Naseema, A., Kovooru, L., Behera, A. K., Kumar, K. P. P., & Srivastava, P. (2021). A critical review of synthesis procedures, applications and future potential of nanoemulsions. *Advances in Colloid and Interface Science*, *287*, 102318. https://doi.org/10.1016/j.cis.2020.102318

Ndwandwe, B. K., Malinga, S. P., Kayitesi, E., & Dlamini, B. C. (2022). Selenium nanoparticles–enhanced potato starch film for active food packaging application. *International Journal of Food Science & Technology*, *57*(10), 6512–6521. https://doi.org/10.1111/IJFS.15990

Norcino, L. B., Mendes, J. F., Natarelli, C. V. L., Manrich, A., Oliveira, J. E., & Mattoso, L. H. C. (2020). Pectin films loaded with copaiba oil nanoemulsions for potential use as bio-based active packaging. *Food Hydrocolloids*, *106*, 105862. https://doi.org/10.1016/j.foodhyd.2020.105862

Ozogul, Y., Karsli, G. T., Durmuş, M., Yazgan, H., Oztop, H. M., McClements, D. J., & Ozogul, F. (2022). Recent developments in industrial applications of nanoemulsions. *Advances in Colloid and Interface Science*, *304*, 102685. https://doi.org/10.1016/j.cis.2022.102685

Pandian, H., Senthilkumar, M, V. R., M, N., & S, S. (2023). *Azadirachta indica* leaf extract mediated silver nanoparticles impregnated nano composite film (AgNP/MCC/starch/whey protein) for food packaging applications. *Environmental Research*, *216*, 114641. https://doi.org/10.1016/J.ENVRES.2022.114641

Parameswaranpillai, J., Krishnankutty, R. E., Jayakumar, A., Rangappa, S. M., & Siengchin, S. (Eds.). (2022). *Nanotechnology-Enhanced Food Packaging*. Wiley. https://doi.org/10.1002/9783527827718

Pereira, R. C., de Deus Souza Carneiro, J., Borges, S. V., Assis, O. B. G., & Alvarenga, G. L. (2016). Preparation and characterization of nanocomposites from whey protein concentrate activated with lycopene. *Journal of Food Science*, *81*. https://doi.org/10.1111/1750-3841.13234

Perera, K. Y., Jaiswal, S., & Jaiswal, A. K. (2022). A review on nanomaterials and nanohybrids based bionanocomposites for food packaging. *Food Chemistry*, *376*, 131912. https://doi.org/10.1016/j.foodchem.2021.131912

Pires, J. R. A., Rodrigues, C., Coelhoso, I., Fernando, A. L., & Souza, V. G. L. (2023). Current applications of bionanocomposites in food processing and packaging. *Polymers*, *15*(10), 2336. https://doi.org/10.3390/polym15102336

Prietto, L., Pinto, V. Z., El Halal, S. L. M., de Morais, M. G., Costa, J. A. V., Lim, L.-T., Dias, A. R. G., & Zavareze, E. daR. (2018). Ultrafine fibers of zein and anthocyanins as natural pH indicator. *Journal of the Science of Food and Agriculture*, *98*(7), 2735–2741. https://doi.org/10.1002/jsfa.8769

Purohit, S. D., Priyadarshi, R., Bhaskar, R., & Han, S. S. (2023). Chitosan-based multifunctional films reinforced with cerium oxide nanoparticles for food packaging applications. *Food Hydrocolloids*, *143*, 108910. https://doi.org/10.1016/J.FOODHYD.2023.108910

Rahmadiawan, D., Abral, H., Yesa, W. H., Handayani, D., Sandrawati, N., Sugiarti, E., Muslimin, A. N., Sapuan, S. M., & Ilyas, R. A. (2022). White ginger nanocellulose as effective reinforcement and antimicrobial polyvinyl alcohol/ZnO hybrid biocomposite films additive for food packaging applications. *Journal of Composites Science*, *6*(10), 316. https://doi.org/10.3390/jcs6100316

Rane, A. V., Kanny, K., Abitha, V. K., & Thomas, S. (2018). Methods for Synthesis of Nanoparticles and Fabrication of Nanocomposites. In *Synthesis of Inorganic Nanomaterials* (pp. 121–139). Elsevier. https://doi.org/10.1016/B978-0-08-101975-7.00005-1

Ren, D., Wang, Y., Wang, H., Xu, D., & Wu, X. (2022). Fabrication of nanocellulose fibril-based composite film from bamboo parenchyma cell for antimicrobial food packaging. *International Journal of Biological Macromolecules*, *210*, 152–160. https://doi.org/10.1016/J.IJBIOMAC.2022.04.171

Restuccia, D., Spizzirri, U. G., Parisi, O. I., Cirillo, G., Curcio, M., Iemma, F., Puoci, F., Vinci, G., & Picci, N. (2010). New EU regulation aspects and global market of active and intelligent packaging for food industry applications. *Food Control*, *21*(11), 1425–1435. https://doi.org/10.1016/j.foodcont.2010.04.028

Rios, D. A. da S., Nakamoto, M. M., Braga, A. R. C., & da Silva, E. M. C. (2022). Food coating using vegetable sources: importance and industrial potential, gaps of knowledge, current application, and future trends. *Applied Food Research*, *2*(1), 100073. https://doi.org/10.1016/j.afres.2022.100073

Rojas, K., Canales, D., Amigo, N., Montoille, L., Cament, A., Rivas, L. M., Gil-Castell, O., Reyes, P., Ulloa, M. T., Ribes-Greus, A., & Zapata, P. A. (2019). Effective antimicrobial materials based on low-density polyethylene (LDPE) with zinc oxide (ZnO) nanoparticles. *Composites Part B: Engineering*, *172*, 173–178. https://doi.org/10.1016/j.compositesb.2019.05.054

Shanmugam, R., Mayakrishnan, V., Kesavan, R., Shanmugam, K., Veeramani, S., & Ilangovan, R. (2022). Mechanical, barrier, adhesion and antibacterial properties of pullulan/graphene bio nanocomposite coating on spray coated nanocellulose film for food packaging applications. *Journal of Polymers and the Environment*, *30*(5), 1749–1757. https://doi.org/10.1007/s10924-021-02311-2

Shao, P., Yan, Z., Chen, H., & Xiao, J. (2018). Electrospun poly(vinyl alcohol)/permutite fibrous film loaded with cinnamaldehyde for active food packaging. *Journal of Applied Polymer Science*, *135*(16), 46117. https://doi.org/10.1002/app.46117

Sohail, M., Sun, D.-W., & Zhu, Z. (2018). Recent developments in intelligent packaging for enhancing food quality and safety. *Critical Reviews in Food Science and Nutrition*, *58*(15), 2650–2662. https://doi.org/10.1080/10408398.2018.1449731

Sun, W., Liu, Y., Jia, L., Saldaña, M. D. A., Dong, T., Jin, Y., & Sun, W. (2021). A smart nanofibre sensor based on anthocyanin/poly-l-lactic acid for mutton freshness monitoring. *International Journal of Food Science & Technology*, *56*(1), 342–351. https://doi.org/10.1111/ijfs.14648

Sun, X., & Bandara, N. (2019). Applications of reverse micelles technique in food science: a comprehensive review. *Trends in Food Science & Technology*, *91*, 106–115. https://doi.org/10.1016/j.tifs.2019.07.001

Sun, X., Li, Q., Wu, H., Zhou, Z., Feng, S., Deng, P., Zou, H., Tian, D., & Lu, C. (2023). Sustainable starch/lignin nanoparticle composites biofilms for food packaging applications. *Polymers*, *15*(8), 1959. https://doi.org/10.3390/POLYM15081959

Suyatma, N. E., Gunawan, S., Putri, R. Y., Tara, A., Abbès, F., Hastati, D. Y., & Abbès, B. (2023). Active biohybrid nanocomposite films made from chitosan, ZnO nanoparticles, and stearic acid: optimization study to develop antibacterial films for food packaging application. *Materials 2023*, 16(3), 926. https://doi.org/10.3390/MA16030926

Tarabiah, A. E., Alhadlaq, H. A., Alaizeri, Z. M., Ahmed, A. A. A., Asnag, G. M., & Ahamed, M. (2022). Enhanced structural, optical, electrical properties and antibacterial activity of PEO/CMC doped ZnO nanorods for energy storage and food packaging applications. *Journal of Polymer Research*, *29*(5), 167. https://doi.org/10.1007/s10965-022-03011-8

Terra, A. L. M., Moreira, J. B., Costa, J. A. V., & Morais, M. G. de. (2021). Development of time-pH indicator nanofibers from natural pigments: an emerging processing technology to monitor the quality of foods. *LWT*, *142*(September 2020), 111020. https://doi.org/10.1016/j.lwt.2021.111020

Torres-Giner, S., Figueroa-Lopez, K. J., Melendez-Rodriguez, B., Prieto, C., Pardo-Figuerez, M., & Lagaron, J. M. (2021). Emerging Trends in Biopolymers for Food Packaging. In *Sustainable Food Packaging Technology* (pp. 1–33). Wiley. https://doi.org/10.1002/9783527820078.ch1

Trajkovska Petkoska, A., Daniloski, D., D'Cunha, N. M., Naumovski, N., & Broach, A. T. (2021). Edible packaging: sustainable solutions and novel trends in food packaging. *Food Research International*, *140*, 109981. https://doi.org/10.1016/j.foodres.2020.109981

Vasile, C., & Baican, M. (2021). Progresses in Food packaging, food quality, and safety – controlled-release antioxidant and/or antimicrobial packaging. *Molecules*, *26*(5), 1263. https://doi.org/10.3390/molecules26051263

Wang, Y., Xu, H., Wu, M., & Yu, D.-G. (2021). Nanofibers-based food packaging. *ES Food & Agroforestry*, 7, 1–24. https://doi.org/10.30919/esfaf598

Xia, J., Sun, X., Jia, P., Li, L., Xu, K., Cao, Y., Lü, X., & Wang, L. (2023). Multifunctional sustainable films of bacterial cellulose nanocrystal-based, three-phase pickeringnanoemulsions: a promising active food packaging for cheese. *Chemical Engineering Journal*, *466*, 143295. https://doi.org/10.1016/j.cej.2023.143295

Xie, H., Wang, Y., Ouyang, K., Zhang, L., Hu, J., Huang, S., Sun, W., Zhang, P., Xiong, H., & Zhao, Q. (2023). Development of chitosan/rice protein hydrolysates/ZnO nanoparticles films reinforced with cellulose nanocrystals. *International Journal of Biological Macromolecules*, *236*, 123877. https://doi.org/10.1016/j.ijbiomac.2023.123877

Yam, K. L., Takhistov, P. T., & Miltz, J. (2005). Intelligent packaging: concepts and applications. *Journal of Food Science*, *70*(1), R1–R10. https://doi.org/10.1111/j.1365-2621.2005.tb09052.x

Yan, M. R., Hsieh, S., & Ricacho, N. (2022). Innovative food packaging, food quality and safety, and consumer perspectives. *Processes*, *10*(4), 747. https://doi.org/10.3390/pr10040747

Yang, D., Liu, Q., Gao, Y., Wan, S., Meng, F., Weng, W., & Zhang, Y. (2023). Characterization of silver nanoparticles loaded chitosan/polyvinyl alcohol antibacterial films for food packaging. *Food Hydrocolloids*, *136*, 108305. https://doi.org/10.1016/J.FOODHYD.2022.108305

Yang, Y., Zheng, S., Liu, Q., Kong, B., & Wang, H. (2020). Fabrication and characterization of cinnamaldehyde loaded polysaccharide composite nanofiber film as potential antimicrobial packaging material. *Food Packaging and Shelf Life*, *26*, 100600. https://doi.org/10.1016/j.fpsl.2020.100600

Yılmaz, M., & Altan, A. (2021). Optimization of functionalized electrospun fibers for the development of colorimetric oxygen indicator as an intelligent food packaging system. *Food Packaging and Shelf Life*, *28*, 100651. https://doi.org/10.1016/j.fpsl.2021.100651

Youssef, A. M., Abd El-Aziz, M. E., & Morsi, S. M. M. (2023). Development and evaluation of antimicrobial LDPE/TiO_2 nanocomposites for food packaging applications. *Polymer Bulletin*, *80*(5), 5417–5431. https://doi.org/10.1007/s00289-022-04346-4

Yu, H.-Y., Yang, X.-Y., Lu, F.-F., Chen, G.-Y., & Yao, J.-M., 2016. Fabrication of multifunctional cellulose nanocrystals/poly(lactic acid) nanocomposites with silver nanoparticles by spraying method. *Carbohydrate Polymers*, *140*, 209–219. https://doi.org/10.1016/j.carbpol.2015.12.030

Yuan, X., Wang, X., Sarkar, B., & Ok, Y. S. (2021). The COVID-19 pandemic necessitates a shift to a plastic circular economy. *Nature Reviews Earth & Environment*, *2*(10), 659–660. https://doi.org/10.1038/s43017-021-00223-2

Zhai, X., Zhou, S., Zhang, R., Wang, W., & Hou, H. (2022). Antimicrobial starch/poly(butylene adipate-co-terephthalate) nanocomposite films loaded with a combination of silver and zinc oxide nanoparticles for food packaging. *International Journal of Biological Macromolecules*, *206*, 298–305. https://doi.org/10.1016/J.IJBIOMAC.2022.02.158

Zhang, M., Biesold, G. M., Choi, W., Yu, J., Deng, Y., Silvestre, C., & Lin, Z. (2022). Recent advances in polymers and polymer composites for food packaging. *Materials Today*, *53*, 134–161. https://doi.org/10.1016/j.mattod.2022.01.022

Zhang, Y., Rempel, C., & Mclaren, D. (2014). Edible Coating and Film Materials. In *Innovations in Food Packaging* (pp. 305–323). Elsevier. https://doi.org/10.1016/B978-0-12-394601-0.00012-6

Zhao, X., Tian, R., Zhou, J., & Liu, Y. (2022). Multifunctional chitosan/grape seed extract/silver nanoparticle composite for food packaging application. *International Journal of Biological Macromolecules*, *207*, 152–160. https://doi.org/10.1016/J.IJBIOMAC.2022.02.180

6 Nanosensors to Detect Food Contaminants

Selenay Sadak, İpek Küçük, Didem Nur Ünal, Cem Erkmen, and Bengi Uslu

6.1 INTRODUCTION

Globalization, which developed with the increase in population, brought industrialization. The possibility of food contamination can be viewed as industrialization's primary concern. Every step of the food manufacturing process, including food processing, transportation, and storage, is susceptible to contamination. These contaminations can be categorized into microbiological, chemical, and physical pollutants. One of the primary triggers of foodborne disease is pathogenic microorganisms (such as fungi, viruses, bacteria, and parasites), as well as toxins from organic materials and microorganisms. Heavy metal ions, veterinary drugs, aromatic hydrocarbons, and pesticides are a few examples of chemical contaminants. Hair, plants and plant residues are among the most common physical pollutants (Elfadil et al. 2021).

Today, food analysis is of great importance as the presence of increasing contaminants can cause threatening effects on human health. Therefore, rapid detection of toxins is important to prevent possible risks. The complex food sample matrix makes it extremely difficult to identify contaminants. Although high-performance liquid chromatography and gas chromatography are quite common in food analysis today, the long analysis time of these analytical methods, the need for expensive instruments, and the complex sample pretreatment procedures present researchers with various challenges. New detection methods have been developed to provide fast and precise detection in complex matrix situations where traditional techniques do not meet the need. Sensors offer benefits in terms of affordability, simplicity, ease of miniaturization, and scalability in mass production. Moreover, they find practical applications as point-of-care (POC) tools in different areas. Consequently, significant efforts have been directed toward developing highly selective sensor systems for detecting several molecules. In recent times, nanomaterials have gained significant prominence within the field of sensor technology. Utilizing materials at the nanoscale is believed to enhance the performance of sensor systems, owing to their unique and appealing physical and chemical properties. Nanomaterials exhibit distinct characteristics compared to their bulk materials, opening up opportunities for novel applications and the potential for improved performance. Therefore, nanosensors that can provide rapid quantitative and qualitative analysis have been the solution to researchers' challenges (Kaur et al. 2023).

This chapter discusses the advantages and future development of current nanosensors for the analysis of food contaminants such as pesticides, metals, mycotoxins, and antibiotics. In the first part, information about electrochemical sensors is given and the advantages of using electrochemical sensors in food analysis are emphasized. In the second part, the detection of food contaminants is examined under the heading of optical sensors. Both electrochemical sensor and optical sensor studies carried out in the last five years have been tabulated with analyte, method, determination limit, linear range, and real sample application parameters.

 DOI: 10.1201/9781003438168-6

6.1.2 Food Safety in the Developing World

Contaminants are substances that can enter food at any point throughout the production, transportation, storage, processing, or distribution steps of food production or from the environment and are not included on purpose. In recent years, population growth and industrialization have led to the spread and increase of contaminants in the food sector. The presence of food contaminants poses a serious risk to human health and prevents socio-economic development in the world. Many potentially toxic substances in the environment can contaminate food consumed by humans. These risk factors affecting food safety are grouped under three main headings: physical, microbiological, and chemical (Ranjan et al. 2016).

6.1.2.1 Biological Threatening

Biological contaminations: The leading biological hazards that may develop in food are microorganisms and mycotoxins.

Escherichia Coli, Staphlococcus aureus, Clostridium botulinum, Clostridium perfirigens and Listeria monocytogenes are among the important bacteria that cause bacterial hazards. These microorganisms can be transmitted to foods through dust, soil, air, various animals, raw foods, garbage, tools used in food production, equipment, and people. Just as bacteria themselves can pose a threat to human health, their toxins also pose a great risk. *Salmonella* species, which are found in the intestines of many farms and poultry, can reproduce quickly in poorly cooked or raw meat, chicken, eggs, and milk. *E. coli* can be transmitted to humans through animal foods. This microorganism is raw and not well-cooked minced meat, unpasteurized milk, feces-contaminated waters, poorly washed vegetables, and it can easily reproduce in fruits. The most toxic pollutants, such as those produced by microorganisms, are mycotoxins, which remain in food even after the biological source has been destroyed (Kim et al. 2003, Zhou et al. 2021).

Mycotoxins are known as secondary metabolites. Species belonging to the genera *Aspergillus, Penicillium*, and *Fusarium* constitute mycotoxins that cause serious health problems in humans. The indicator of contamination of food with mycotoxins is the "Carry Over" properties of these toxins. Mycotoxins are known to have carcinogenic, teratogenic, tremorgenic, hemorrhagic, dermatitic, hepatotoxic, nephrotoxic, and neurotoxic effects on humans (Sabuncuoğlu et al. 2008).

Naturally produced mycotoxins aflatoxin B1 and aflatoxin M1 have been found to have carcinogenic effects on humans. Also, even low-dose aflatoxin exposure, vomiting can cause liver and kidney destruction. Aflatoxins have often been detected in milk and dairy products (Sameiyan et al. 2021), corn, and peanut creams (Q. Wang et al. 2021). Ochratoxin A, which is frequently contaminated in red wine (Pacheco et al. 2015) and bakery products (Li et al. 2022), is produced by *Aspergillus* and *Penicillium* species. Ochratoxin A can be transferred from contaminated animal feed to humans and cause urinary tract tumors and nephropathy.

6.1.2.2 Chemical Threatening

The most important sources of chemical contamination in food are pesticides, heavy metals, and pharmaceutical drug residues. As chemical substances have the property of accumulating in the body, their accumulation in various organs of the body over time can lead to many systemic diseases (Li et al. 2019).

Heavy metals are defined as metals in the third or higher period of the periodic table and having a density greater than 5 g/cm^3 in terms of physical property. Long-term consumption of foods contaminated with heavy metals such as Hg^{2+}, Cd^{2+}, and Pb^{2+} causes serious diseases such as nervous disorders and cognitive impairments. The effects of sectors such as industry, agriculture, and mining increase the amount of heavy metals in the marine ecosystem (Liaquat et al. 2022). Because heavy metals are non-biodegradable, marine organisms (fish, shellfish, crustaceans) can accumulate these metals in potentially toxic concentrations. At the same time, there is an accumulation of heavy metals in plants grown with contaminated water used for irrigation in agricultural areas. It is important for public health that both drinking water and plants irrigated with contaminated water are routinely analyzed (Liaquat et al. 2022).

Pesticides: They are biological or chemical substances used in agricultural production to protect fruits and vegetables against these pests by killing, preventing, or controlling insects, weeds, bacteria, fungi, and other pests. Exposure of humans to high amounts of pesticides; can cause acute poisoning, cancer, and negative effects such as reproduction, adverse effects on the nervous system and immune system, and birth defects. The most dangerous pesticides are chlorinated hydrocarbon insecticides and organic phosphorus insecticides. Chlorinated hydrocarbon pesticides are quite stable, and due to their fat solubility, they accumulate and get stored in fatty tissues. Organochlorinated pesticides, such as dichlorodiphenyltrichloroethane, have been used in agriculture and vector-borne disease control for many years. Chlorpyrifos is another type of organophosphate pesticide that affects vision and causes other neurological toxic effects in humans (Khan et al. 2023).

Active substances such as tetracyclines, phenicols, fluoroquinolones, nitrofurans, gentamicin, beta-lactams, sulfonamides, and methicillin are applied to prevent diseases in animals or to reduce the risk of disease, to accelerate development, to increase yield with injuries caused by feed. When antibiotics are not fully metabolized and excreted in the organism, they can form residues in the food obtained from these animals and threaten human health by causing risks such as antibiotic resistance, anaphylactic shock, drug allergy, tissue damage, gastrointestinal disorders, and neurological damage in humans (Okoye et al. 2022).

Examples of physical pollutants are the undesirable pollutants in the environment, such as dust, soil, stones, glass, paper, hair, feathers, cigarette ash, flies, insects, and plastic, which are undesirable in food, resulting from inadequate hygiene practices (Coccia and Bontempi 2023).

6.1.3 A Quick Approach to Food Analysis – Nanosensors

The names of substances known as nanomaterials or nanoparticles have been more prominent in recent years due to advancements made in the field of nanotechnology. Materials that range in size from 1 to 100 nm are known as nanomaterials. As a result of their small diameters, these materials exhibit a wide range of optical, magnetic, and electrical characteristics. Nanomaterials have a higher surface area because their structure is incredibly tiny. By doing so, the interaction distance between the nanometer-sized material and the sample is reduced while also increasing the sensitivity of sensors made using nanomaterials (Zhang et al. 2024). The structural properties of the sensing materials used in the production of nanosensors have a critical impact on device performance. The controllable size, shape, and surface area of nanomaterials give them unique properties. By taking advantage of the controllable sizes of nanomaterials, target-specific nanosensors can be developed. Their large surface area increases the adsorption power of their surfaces, allowing the determination of much lower amounts of substances. On the surface of the bare electrode, nanomaterials can be used to increase weak electron transfer. Nanomaterials are attached to electrodes; it gives them superior selectivity, low detection limit and high sensitivity. Thanks to their durable and easy-to-prepare structure, their reproducibility is high. The use of nanomaterials in nanosensors greatly reduces the analysis time. Given that time is so important in food microbiology, the fact that nanosensors reduce the time needed to identify contaminants from days to hours or even minutes is an important result. The reduced analysis time provides an advantage over traditional methods in the diagnosis and treatment of conditions that harm human health caused by food contaminants.

Nanomaterials come in a wide variety of forms, sizes, and shapes nowadays. The first of these is metallic nanomaterials, which garners a lot of interest because to the variety of applications they have. Due to their nanoscale structure, metallic nanoparticles exhibit great strength, hardness, and wear resistance. They may also remain active for extended periods of time, and due to their capacity for optical and heat-based treatment systems, they have received considerable interest in food analysis as well as in the disciplines of biomedicine, engineering, and nanomedicine. Since 2005, it has been well known that nanomaterials have been employed extensively in a variety of pharmaceutical studies and medical applications, with metallic nanoparticles being used particularly frequently in these fields due to their antibacterial qualities. Due to their size and surface coating qualities, which have a toxic effect, these compounds are assumed to have an antibacterial effect (Vimbela et al. 2017). These nanoparticles are quite robust; however, they are not very ductile. This could restrict the usage locations (Păduraru et al. 2022, Guo et al. 2020).

One of the most prevalent elements in the world, carbon, is also extensively employed as a nanomaterial. Many distinct forms of carbon nanomaterials are created by bonding carbon atoms in different ways to create various allotropes. Examples of these include fullerenes, graphene, carbon nanotubes, and nanodiamonds. The variety of carbon nanomaterials is attributed to their small size, exceptional mechanical strength, electrical and thermal conductivity, optical characteristics, and vast surface areas that are nearly equal to a number of fundamental biomolecules. Carbon nanoparticles offer a wide range of applications as a result of these benefits. Carbon nanotubes are used for molecular recognition, fullerene derivatives, which are bioactive compounds, are used in solar energy studies and drug development procedures, while graphene, which has inert properties, is used in electronic applications as well as in the treatment of cancer. Graphene quantum dots, which have photoluminescence as a property, are also useful in bioimaging.

Polymeric nanomaterials, which are composed of nanoparticles formed of synthetic or natural polymers, are a different kind of nanomaterials. The development of polymeric particles dates back to the 1970s, when microscale medication delivery devices were created using polymers. These polymers later become smaller, reaching the nanoscale. Nanomedicine, bioavailability research, and controlled drug release are all areas where polymeric nanoparticles are frequently used. Biodegradable polymers are used to create these nanoparticles. These chemicals are both safe and very biocompatible. With regards to procedures like gene transfers and the creation of vaccines, biocompatibility is crucial. Chitosan helps plants absorb protein and water more easily, and polystyrene is a common material used in floriculture as an example of these polymers. Polymers are employed often in agriculture and for nutritional content analysis as a result (Wang and Sun 2021, Vasile 2019).

Nanomaterials are employed in changing sensors as well as in many other fields due to their benefits. Studies on the detection of food contaminants using sensors modified with nanomaterials have also been extensively conducted. In this chapter, nanosensors developed for food analysis applications are transferred under two main headings, electrochemically and optically, according to the type of signal transmission (Süfer and Karakaya 2011).

6.2 APPLICATIONS OF ELECTROCHEMICAL NANOSENSORS IN FOOD SAMPLES

Devices called electrochemical sensors depend on redox processes on the electrode surface to identify the target analyte. The target analyte's electrochemical characteristics influence the measuring method chosen. There are many electrochemical methods used for the determination of various analytes. The most used amperometric technique among these is chronoamperometry, whereas the most popular voltammetric techniques are cyclic voltammetry (CV), square wave voltammetry (SWV), and differential pulse voltammetry (DPV). Electrochemical techniques have several benefits, including the capacity to test with a tiny amount of sample, quick measurement, and simple sample preparation. It is frequently used to gather qualitative data for the study using CV. Detailed information about the thermodynamics of the redox process can be obtained by moving the voltage applied to the working electrode forward and backward. A square wave's frequency and step height are changed to establish the effective scanning speed in SWV, an incredibly quick and sensitive technology. With this method, a square wave is overlaid on the working electrode's base ladder potential as the excitation signal. Compared to DPV, the approach has a higher sensitivity. The use of DPV is equally common, though. DPV employs the application of pulses to a linear potential ramp of a fixed size. The examined materials are identified using the method's peak potential. Chronoamperometry, an amperometric technique, is very frequently utilized in analyses in addition to the voltammetric techniques already discussed. In this method, a current big enough to start the reaction is applied before looking at the change in current over time (Hoyos-Arbeláez et al. 2017).

DPV is the most widely used technique for electroactive analytes due to its high sensitivity, simplicity, and rapid response. If the target analyte is not electroactive, film conductivity or porosity can be altered by electrode modifications such as antibodies, enzymes, peptides, aptamers, and molecularly imprinted polymers (MIPs). MIP modification is carried out using redox probes, such as ferrocyanide and hexaamine ruthenium chloride; it can be monitored indirectly using CV and electrochemical impedance spectroscopy (EIS) (Süfer and Karakaya 2011). In this context, Table 6.1 presents the applications of electrochemical nanosensors and biosensors designed for the detection of food contaminants, selected from studies conducted in the last five years.

Patulin is a toxic secondary metabolite produced by some types of fungi, such as *Aspergillus* and *Penicillium*, and it has been determined in many studies that it will cause DNA damage at high concentrations (de Melo et al. 2012). A reported mycotoxin called patulin is considered a major contaminant in apples and apple-derived products. According to the World Health Organization, the maximum daily concentration of patulin in fruits and juices should be below 0.4 $\mu g/kg^{-1}$ body weight. Hatamluyi et al. (2020) modified the glassy carbon electrode (GCE) surface with nitrogen-doped graphene quantum dots (N-GQDs) and AuNPs-functionalized Cu–metal organic framework (Au/Cu-MOF) to electrochemically determine patulin. Subsequently, the MIP film was formed on N-GQDs and Au/Cu-MOF by electropolymerization using aniline as the functional monomer. The ultra-high porous structure of the MOFs and their large internal surface area have increased the binding zones on the MIP layer. At the same time, the high conductivity and electrocatalytic activity of N-GQDs effectively increased the conductivity of the developed sensor. The developed sensor has been successfully applied to the apple juice sample. The detection limit and linear range were determined as 0.0007, 0.001–70.0 M, respectively. The results showed that the sensor is highly sensitive and stable (Hatamluyi et al. 2020) (Figure 6.1).

The discharge of pharmaceuticals and their metabolites into the environment causes significant effects due to their toxic nature. Dimetridazole (DMZ) is an antibiotic used in the treatment of bacterial and protozoal infection in poultry. Detection of DMZ contamination in poultry is important because it causes carcinogenic effects. Especially in the respiratory tract, emphysema can cause serious damage such as lung damage. Selvi et al. (2021) synthesized the flower-like Mn-SnO@rGO nanocomposite by hydrothermal and ultrasonication method. SnO_2 was chosen because it is a low cost and non-toxic material. Because Mn ions are polyvalent, they are easily embedded in the crystal lattice on the surface of SnO_2. At the same time, on the two-dimensional (2D) reduced graphene oxide surface, flower-like MnSO nanoparticles were uniformly dispersed due to their synergistic effects. DMZ electroreduction was performed on the flower-like Mn-SnO@rGO nanocomposite modified GCE surface. The heteroaromatic nitro group in the DMZ structure is reduced to a hydroxylamine derivative. The prepared flower-like Mn-SnO@rGO nanocomposite has been successfully applied to detect DMZ in contaminated water, milk and egg samples. The newly developed sensor offered a cost-effective approach to the detection of nitrous imidazole-based drugs, especially in food samples (Figure 6.2) (Selvi et al. 2021).

Salmonella is mainly transmitted through the consumption of contaminated raw or uncooked meat, especially meat and eggs, and it causes serious health problems. Detecting salmonella contamination and quantity in food products precisely and quickly is of great importance for human health. Typhoid fever associated with food poisoning can occur after undetected contamination. In

TABLE 6.1
Detection of food contaminants with electrochemical nanosensors

Target molecule	Surface layer content	Measurement principle	Dynamic range	LOD	Real samples	References
Tetracycline	MIOPPy-AuNP/SPCE	DPV	1–20 μM	0.65 μM	Shrimp samples	(Devkota et al. 2018)
Ciprofloxacin	PBE	DPV	9.90–220 μM	4.96 μM	Honey and milk samples	(de Souza et al. 2022)
Ciprofloxacin	Ch-AuMIP/GCE	DPV	1–100 μM	210 nM	Mineral and tap water, milk, and pharmaceuticals	(Surya et al. 2020)
Paraoxon	Ce/UiO-66@MWCNTs/GCE	DPV	0.01–150 nM	4 pM	Cabbage and spinach	(Mahmoudi et al. 2019)
Sulfathiazole	MIP/CuS/Au@COF/GCE	DPV	1.0×10^{-4}–1.0×10^{-11}M	4.3×10^{-12}M	Fodder	(Y. Sun et al. 2020)
Ampicillin	Fe_3N-Co_2N/CC/MIP/GCE	Amperometry	5.56×10^{-9}–1.9×10^{-3}M	3.65×10^{-10}M	Milk samples	(Z. Liu et al. 2020)
Erythromycin	SDBS/SPCEs	Amperometry	0.1–50 mM	0.14 mM	Drinking water	(Veseli et al. 2019)
Dimetridazole	Mn-doped SnO_2/rGO/GCE	DPV	0.009–1291 μM	2 nM	Milk and egg samples	(Selvi et al. 2021)
Cloxacillin	GO-Au MIP/SPCE	DPV	110–750 nM	36 nM	Milk samples	(Jafari et al. 2019)
Al^{+3}	Pd/C/Au-SPE	EIS	5–100 μM	5 μM	Tomato and orange juice	(Poudyal et al. 2022)
Pb^{+2}	SP/MCH/GE	DPV	–	0.034 μM	Fresh fruit and vegetables	(Wang et al. 2020)
Ni^{+2}	AuNPs/GCE	DPV	0.01–3.8 mM	2.4 mM	Drinking water	(H. Shi et al. 2022)
Hg^{+2}	Cu-MOF/GCE	DPV	0.1–50 nM	0.0633 nM	Canned tuna fish and tap water	(Singh et al. 2020)
Hg^{+2}	rGOS@SnO_2/GCE	DPV	0.25–705.3 μM	1.27 nM	Fish extract, drinking water	(Govindasamy et al. 2020)
Zearalenone	BOMC-IL-Au NPs/GCE	SWV	0.0005–1 nM	0.1 pM	Corn, rice, and beer	(Hu et al. 2020)
Zearalenone	PtNi@Co-MOF/AuE	DPV	10 fM to10 nM	1.37 fM	Maize samples	(He and Yan 2020)
Patulin	Au@Cu-MOF/GCE	DPV	1 pM to70 nM	0.7 pM	Apple juice	(Hatamluyi et al. 2020)
Ochratoxin A	cDNA-MCH-Apt/AuE	EIS	0.05–10 nM	0.05 nM	Malt sample	(Hou et al. 2022)
Ochratoxin A	Apt/CDs-BP/GCE	EIS	0.1×10^{-7}–10 nM	0.03×10^{-8}nM	Wheat and grape juice samples	(Li et al. 2021)
Ochratoxin A	POPD-GNSs-rDNA/h-DNA/MWCNTs–GNSs/GCE	DPV	2 pM to 1 nM	1 pM	Coffee samples	(Lv et al. 2022)
Ochratoxin A	4-CP/PGE	EIS	0.1–6.4 ppb	0.100 ppb	Spiked beer samples	(Gökçe et al. 2020)
Ochratoxin A	P1-labeled Au NPs@UiO-66/GCE	DPV	0.001–100 nM	330 fM	Corn kernel samples	(Li et al. 2022)

(continued)

TABLE 6.1 (Continued)
Detection of food contaminants with electrochemical nanosensors

Target molecule	Surface layer content	Measurement principle	Dynamic range	LOD	Real samples	References
Ochratoxin A	MIP/MWCNT/GCE	DPV	0.050–1.0 μM	0.041 μM	Beer, white wine, and red wine	(Pacheco et al. 2015)
T^{-2} toxin	AuNPs/cSWNTs/CS/GCE	DPV	0.01–100 μM	0.13 μM	Swine meat	(Y. Wang et al. 2018)
Patulin fungal toxin	rGO/SnO_2 /GCE	DPV	50–600 nM	0.6635 nM	Apple juice	(Shukla et al. 2020)
Aflatoxin M1	Poly-NR/GCE	EIS	5–120 nM	0.5 nM	Cow milk samples	(Smolko et al. 2018)
Aflatoxin B1	COOH–GO–COOH–MWNT–SPE	DPV	0.1 fM to 100 pM	15.14 ag ml	Milk samples	(Wang et al. 2022)
Aflatoxin B1	PtNP/MIL–101(Fe)/GCE	EIS	0.01–80 nM	2 pM	Powder and pasteurized milk samples	(Jahangiri-Dehaghani et al. 2020)
Aflatoxin B1	PANI@MIP/CNC-CNT/hypodermic needle sensor	CV	0–25 nM	3 nM	Milk sample	(Wood and Mugo 2022)
Aflatoxin B1	Au-PANI-ITO	EIS	0.1–100 nM	0.05 nM	Corn	(Yagati et al. 2018)
Aflatoxin B1	Fe_3O_4@Au-Apt/ SPCE	EIS	0.02–50 nM	0.015 nM	Spiked peanuts samples	(C. Wang et al. 2018)
S. typhimurium	rGO-CNT/GCE	DPV	10^{-1}–10^{-8} cfu/mL	10^{-1} cfu/mL	Chicken meat samples	(Appaturi et al. 2020)
***E. coli* O157:H7**	CDs/ZnO nanorod/PANI/SPCE	DPV	1.3×10^{-18}–10×10^{-12}M	1.3×10^{-18}M	Water samples	(Pangajam et al. 2020)
Chlorpyrifos	Au@ZnWO4-GCE	DPV	0.001–100 μM	0.000350 μM	Water sample	(Shad et al. 2023)
Chlorpyrifos	AChE/Z1200/EμPAD	EIS	10–1000 nM ´	3 nM	Tomato juice	(Nagabooshanam et al. 2021)
Profenofos	CNTs@SiO_2-MIP/GCE	Amp	0.01–200 μM	0.002 μM	Various vegetables	(Amatatongchai et al. 2019)
Malathion	MIP–AU/SPE	DPV	3×10^{-6}–3×10^{-12} M	2×10^{-16}M	Olive fruits and oils	(Aghoutane et al. 2020)
Acetamiprid	Cu-MOF/GCE	DPV	0.1 pM–10 nM	2.9 fM	Green tea and black tea	(Qiao et al. 2019)
Dichlorodiphenyltrichloroet hane	PDA@Fe_3O_4-MIP MNPs/GCE	EIS	1×10^{-11}–1×10^{-3} M	6×10^{-12}M	Radish juice	(Miao et al. 2020)
3-chloropropane-1,2-diol	Cys-AgNPs/AuE	DPV	2.5–200 nM	2.4 nM	Palm oil and smoked mackerel samples	(Martin et al. 2021)
3-Monochloropropane-1,2-diol	MIP/NPG/GCE	CV	10^{-16}–10^{-7} M	3.5×10^{-17}M	Soy sauce samples	(Cheng et al. 2022)
3-Monochloropropane-1,2-Diol	AuN/p-ATP/GCE	DPV	6.61×10^{-4}–2.30×10^{-3}μM	3.30×10^{-4}μM	Edible oil	(Arris et al. 2022)
Trichloroacetic acid	Ag/SPCE	SWV	100–4295 μM	23.8 μM	Bread waste	(Duan et al. 2022)
Ractopamine	AuNPs@COF/GCE	CV	1.2–1600 μM	0.12 μM	Meat samples	(Yang et al. 2023)

Paraquat	PtNPs@MIP/SPGrE/PVC/SPE	ASV	0.05–1000 μM	0.02 μM	Various vegetables	(Somnet et al. 2021)
Bisphenol A	NiRu-MOF/SPE	DPV	0.0–460.0 μM	8.0 nM	Tomato paste, ketchup sauce, chili powder	(Dourandish et al. 2023)
Bisphenol	MIP/PPy@LSG	DPV	0.05–20 μM	8.0 nM	Tap, mineral water	(Beduk et al. 2020)
H_2O_2	GL-VS_2/GCE	AMP	up to −260μM	0.029 μM	Milk sample	(Karthik et al. 2018)
Rhodamine B	Ag/graphene/SPCE	SWV	2–100 μM	1.94 μM	Crackers	(Kartika et al. 2021)
Amaranth	MIP/MWCNT/GCE	LSV	0.007–1.0 μM and 0.40–17 μM	0.0004 μM	Various fruit drinks	(Wu et al. 2021)

Abbreviations: AChE:Acetylcholinesterase, Al^{+3}:Aluminum, Apt: Aptamer, ASV: Aodic stripping voltammetry, Au NP: Gold nanoparticles, AuE: Gold electrode, Au-SPE: Gold screen-printed electrode, BOMC: Boron-doped ordered mesoporous carbon, C: Carbon, CC: Carbon cloth, CD: Carbon dot, cDNA: Complementary DNA, CDs-BP: Carbon dots–black phosphorus nanohybrid, Ce: Cerium, Ch: Chitosan, CNC: Cellulose nanocrystals, COD: Covalent organic frameworks, COFs: Covalent organic frameworks, Co-MOF: Cobalt metal–organic framework, COOH–GO: Carboxylated graphene oxide, cSWNTs: Conjugated on covalently functionalized, CTAB:Cetyltrimethylammonium bromide, Cu-MOF: Functionalized copper–metal organic framework, CuS: Copper sulfide, CV: Cyclic voltammetry, Cys-AgNPs: Cysteine-coated silver nanoparticles, DPV: Differential pulse voltammetry, EIS: Electrochemical impedance spectroscopy, EμPAD: Electrochemical micro Paper Analytical Device, GCE: Glassy carbon electrode, GL-VS_2: Grass-like vanadium disulfide, GNSs: Gold nanospheres, H_2O_2: Hydrogen peroxide, h-DNA: Hairpin-DNA, Hg^{+2}: Mercury, IL: Ionic liquid, ITO: İndium tin oxide, LSG: Laser scribed graphene, LSV: Linear sweep voltammetry, MCH: 6-mercapto-1-hexanol, MIL: Metal–organic frameworks, MIOPPy: Molecularly imprinted overoxidized polypyrrole, MIP: Molecularly imprinted polymer, MNPs: Magnetic nanoparticles, Mn–SnO_2: Manganese-doped tin oxide, MWCNT: Multi-walled carbon nanotube, N-GQDs: Nitrogen-doped graphene quantum dots, NPG: Nanoporous gold, PANI: Polymerizing aniline, p-ATP:p-aminothiophenol, Pb^{+2}: Lead, PBE: Paper-based electrode, Pd: Palladium, PDA: Polydopamine, PGE: Pencil graphite electrode, Poly-NR: Poly neutral red, POPDGNSs:Polymer–nanoparticle composites, PPy: Polypyrrole, PtNPs: Platinum nanoparticles, PVC: Polyvinyl chloride, rDNA:Report DNA, rGO:Reduced graphene oxide, rGO-CNT:Reduced graphene oxide-carbon nanotubes, SnO_2:Tin oxide, SDBS: Sodium dodecylbenzenesulfonate, SP: Signal probe, SPCE: Screen-printed carbon electrodes, SPGrE:Screen-printed graphene paste electrode, SWV: Square wave voltammetry, UiO-66: Zirconium metal–organic framework, ZnO: Zincoxide, $ZnWO_4$: Zinc tungstate

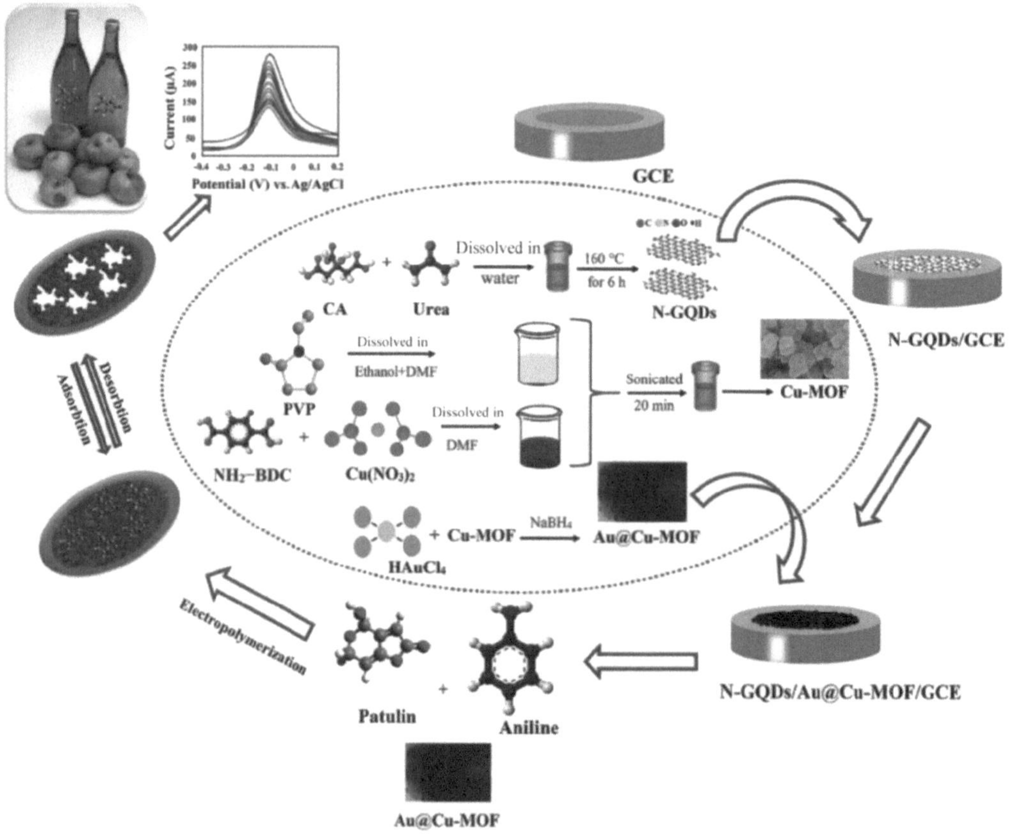

FIGURE 6.1 A schematic preparation of MIP/Au@Cu-MOF/N-GQs/GCE and electrochemical detection of Patulin. Reprinted with permissionfrom Hatamluyi et al. (2020).

another study, Appaturi et al. (2020) developed a biosensor for accurate and rapid electrochemical detection of *Salmonella typhimurium.* As seen in Figure 6.3, on the GCE the graphene oxide-carbon nanotube nanocomposite was then modified with amino modified DNA aptamers. Amino modified aptamer was used as a binder to provide binding between rGO-CNT-aptamer-bacteria in this study. A significantly more sensitive and stable aptasensor is produced by the increased surface area of the rGO-CNT hybrid nanocomposite, which enables covalent interaction between the carboxyl group of the rGO-CNT and amino-modified aptamers. The detection limit of *S. typhimurium* and the linear range detected by DPV using the developed aptasensor were found to be 10^1 cfu/mL,10^1–10^8cfu/mL, respectively. The toxic pathogen was also detected thanks to the aptasensor developed from the real chicken sample (Appaturi et al. 2020).

Numerous nanoparticles are used in the examination of various chemicals as a result of advancements in the field of nanobiotechnology. Among these, gold nanoparticles play a significant role. This is due to the distinctive characteristics and varied surface functionalities of these nanoparticles. Due to their narrow size distribution, efficient surface modifications, biocompatibility, and electrochemical properties, they are widely used in both biomedical investigations and drug analysis. As a result of their ability to connect to biomolecules without altering their characteristics, nanoparticles are used in DNA sensor investigations. Additionally, because they include a large number of atoms that speed up electron transfer in electrochemical investigations and have the capacity to be oxidized or reduced, they are frequently chosen in biorecognition applications (Rasheed and Sandhyarani 2017). Molecular imprinting is now used to electrodes that have been modified

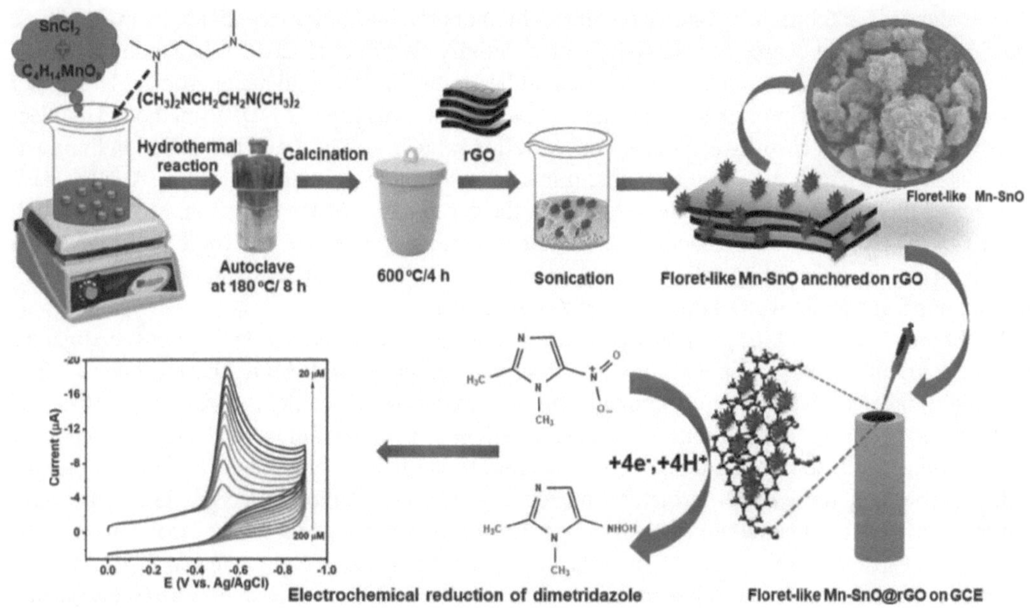

FIGURE 6.2 Schematic synthesis of floret-like assembled Mn-SnO@rGO nanocomposite and its utilization as an electrocatalyst for DMZ detection. Reprinted with permission from Selvi et al. (2021).

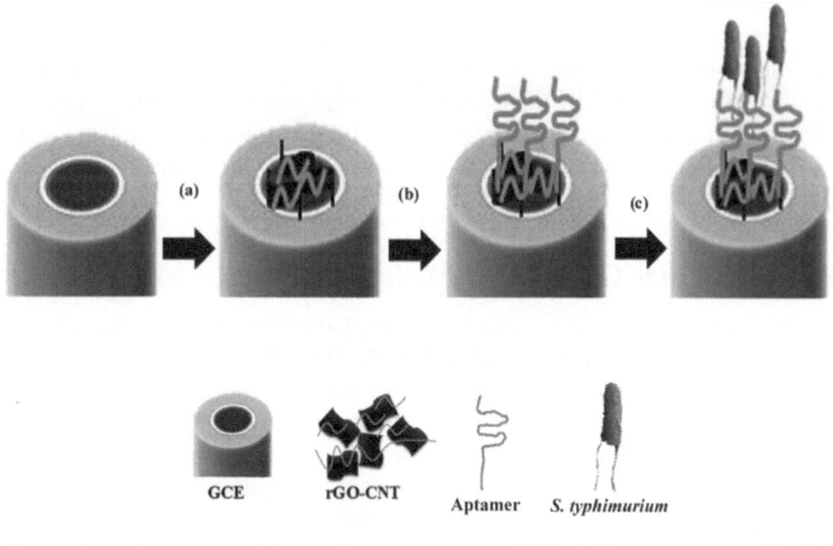

FIGURE 6.3 A schematicdiagram of the modification process of GCE. (a) Drop casting of rGO-CNT on GCE, (b) rGO-CNT/GCE modified with amino-aptamer, (c) ssDNA/rGO-CNT/GCE incubation with *S. typhimurium*. Reprinted with permission from Appaturi et al. (2020).

with nanomaterials and are employed in a variety of analyses. To generate cavities with a precise form and size where the analyte may be detected, an electrode surface must first be coated with a matrix and then removed. This process is known as molecular imprinting. Molecularly imprinted polymers, in essence, are synthetic polymer-structured receptors that function similarly to natural receptors. MIP-based electrochemical sensors have good detection capabilities for a wide range of compounds, including hormones, chemicals, biomarkers, and pharmaceuticals. It makes these decisions with a great degree of selectivity and sensitivity (Waffo et al. 2018). In a study by Surya et al. (2020), an MIP-based sensor which was modified with chitosan gold nanoparticles. The sensor was utilized to detect ciprofloxacin antibiotic. The analysis was made by DPV method. The electrode was studied successfully at a linear concentration range 1–100 μM and the detection limit was found as 210 nM. According to the results, it has been seen that gold nanoparticles (AuNPs) have accelerated the electron transfer, thus resulting in the acceleration of the detection. Moreover, 66% sensitivity for ciprofloxacin was achieved with the developed sensor. The electrode performed high selectivity and reproducibility. Also it was easy to prepare and apply successfully to determine CIP in milk, mineral and tap water samples (Surya et al. 2020).

MOFs are solid crystalline substances created by covalent coordination bonding between ligands, which are organic units, and metal ions, which are an inorganic unit. MOFs have a huge surface area, high porosity, and excellent thermal stability, making them excellent materials for electrode modification in various applications. They have a variety of uses, including catalysis, drug analysis, energy storage, and the identification of biomolecules, because they combine the qualities of organic and inorganic materials in their structural makeup, but the use of these materials in electrochemical research may be constrained by their poor mechanical stability and low electrical conductivity. However, because of their attributes, including size, morphological structure, and surface area, they are frequently combined with other materials (J. Zhang et al. 2021). Sing et al. (2020) designed a GCE with copper and metal–organic framework to detect mercury ions. The measurements were taken with DPV 0.1 M phosphate buffer. In light of the findings, the linear concentration range was found as 0.1–50 nM and the detection limit as 0.0633 nM. Moreover, EIS measurements were taken for examining the modified electrode more detailly. The results of these measurements have shown that Cu–MOF provided a better charge transfer kinetic through showing less amount of charge transfer resistance. Also, the geometric shape that results from the arrangement of bonds generated between metal ions and ligands gives rise to the porous structure of MOF nanocubes. More –OH and NH components are diffused by the mercury as a result of their geometric shape. As a result, the Cu-MOF draws in additional mercury molecules. The sensor was also applied for detection of mercury ions in canned tuna fish and tap water. The sensor has shown a wide linear range with a low detection limit and good selectivity, stability and repeatability (Singh et al. 2020).

6.3 APPLICATIONS OF OPTICAL NANOSENSORS IN FOOD SAMPLES

Among various sensing methods, optical detections by fluorescence or color changes stand out for their simplicity and low detection limits (Ahmad et al. 2019). Detectors known as optical sensors track changes in a material's optical characteristics and translate those alterations into an electronic signal, such as light. Recently, nanomaterials have often been used in colorimetric measurements, as well as being used in electrochemical measurements to enhance the detection of chemical and biochemical analytes (Sabui et al. 2022). Sensors require two basic characteristics: a recognition element and a transducer element. Due to their significant surface area, high electrical conductivity, magnetic properties, special physicochemical characteristics, and ease of modification of their size and structure, nanoparticles like gold nanoparticles, silver nanoparticles, magnetic nanoparticles, and quantum dots are quite remarkable for signal converters (Li et al. 2019). There are numerous studies that take advantage of the unique optical properties of nanostructures in the detection of food contaminations (Shenashen et al. 2022). Fluorescent sensors offer researchers parameters such as selectivity, and sensitivity. At the same time, they have surpassed conventional approaches due to their simplicity in production and ability to provide quick analysis (Li et al. 2019). Fluorescence-based sensors are commonly used in a variety of applications, including food and drug analysis.

In this context, Table 6.2 presents the applications of optical nanosensors and biosensors designed for the detection of food contaminants, selected from studies conducted in the last five years.

TABLE 6.2
Detection of food contaminants with optical nanosensors

Target molecule	Signal transducer	Measurement principle	Dynamic range	LOD	Real samples	References
Lincomycin	ZIS-HNCs@TiO_2	Photoluminescent	0.0001–0.1 nM	0.084 pM	Milk samples	(J. Peng et al. 2021)
Thiabendazole	Zr-MOF (Tb^{+3}@1)	Fluorescence	0–80 μM	0.271 μM	Orange samples	(X. Peng et al. 2021)
Chloramphenicol	PCN-222	Fluorescence	0.1 pM to10 nM	0.08 pM	Milk and shrimp samples	(S. Liu et al. 2020)
Doxycycline	Eu-In-BTEC	Fluorescence	0–6 μM	47 nM	Fish samples	(Yu et al. 2020)
Ceftazidime	GQDs@Fe_3O_4/MIP	Fluorescence	0.10–10.0 μM	0.05 μM	Milk samples	(Bunkoed et al. 2020)
Mafenide	MIL101-MMIP-GQDs	Spectrofluorimetry	0.10–25.0 μM	0.10 μM	Milk samples	(Orachorn and Bunkoed 2021)
Sulfisoxazole	MIL101-MMIP-CdTe QDs					
Marbofloxacin	OH-HNTs@N-GQDs@ Fe_3O_4@MIP	Fluorescence	0.10–25.0 μM	0.03 μM	Milk samples	(Pongprom et al. 2023)
Kanamycin	UCNPs-Apt/BHQ3-cDNA	Fluorescence	0.05–50 μM	18.9 nM	Milk samples	(Y. Zhang et al. 2021)
Tetracycline	UiO-66@SiO_2-Cit-Eu	Fluorescence	0.1–6 μM	17.9 μM	Honey and milk samples	(Jia et al. 2019)
Tetracycline	Zr-MOF/Cit-Eu	Fluorescence	3–60 μM	0.3 μM	Fish, pork, beef, and chicken samples	(Y. Zhao et al. 2023)
Tetracycline	Pd NPs/Apt/cDNA	Fluorescence	100–500 nM	45 nM	Raw milk samples	(S.R. Ahmed et al. 2021)
Chloramphenicol	AuNCs-FQICTS	Photoluminescence	0.5–50 nM	0.046 nM	Chicken samples	(Xiong et al. 2022)
Amantadine			0.05–1 nM	0.048 nM		
Ascorbic acid	Zn • (BA) • (BBI)	Fluorescence	5×10^{-5}M–5×10^{-5} M	1.6 ppb	Goat serum samples	(W. Liu et al. 2020)
Chloramphenicol				12 ppb		
Ceftriaxone				3.9 ppb		
Thiabendazole	Au@AgNRs/CMC/qPCR	Fluorescence	0.25–100 ppm	0.17 ppm	Nectarine and lemon	(Hu et al. 2023)
Nitrofurazone and nitrofurantoin	RhB@Tb-dcpcpt	Luminescence	–	–	Water samples	(Yu et al. 2019)
Amantadine	CDs	Fluorescence	0.03–1.2 nM	0.02 nM	Chicken	(Dong et al. 2019)
Cd^{+2}	NP-1	Fluorescence	0.10–60.0 μM	0.09μM	Black tea and rice samples	(Vildan et al. 2020)
CN^-	PVDF/BTPA	Fluorescence	1–10 μM	6.4 nM	Potatoes and cassava roots	(Enbanathan et al. 2023)
Hg^{+2}	Compd-2	Fluorescence	–	27×10^{-9} M	Human urine samples	(Kumar et al. 2020)
Hg^{+2}	Fe_3O_4/CDs	Fluorescence	0.003–0.01 μM	0.3 nM	Water samples	(Xie et al. 2021)
Fe^{+3}	MIL-101-NH_2	Fluorescence	0.01–0.2 mM	1.8 μM	Water samples	(Lv et al. 2019)

(continued)

TABLE 6.2 (Continued)
Detection of food contaminants with optical nanosensors

Target molecule	Signal transducer	Measurement principle	Dynamic range	LOD	Real samples	References
Cu^{+2}	PVDF-g-PAA-CDs	Fluorescence	–	0.001 μM	Water samples	(Zhang et al. 2019)
Hg^{+2}				0.001 μM		
Fe^{+3}				1 μM		
Cr (VI)	CDs@g-C_3N_4	Fluorescence	0.5–3 nM	0.54 nM	Water samples	(Radhakrishnan et al. 2020)
Cu (II)			0.1–1 nM	0.18 nM		
Pb (II)			0.3–1 nM	0.2 nM		
Escherichia coli O157:H7	DNA(FAM)+UiO66	Fluorescence	1.3×10^{2}–6.5×10^{4}cfu/mL	4.0×10^{1}cfu/ mL	Spring water, skim milk, and orange juice	(X. Sun et al. 2020)
Salmonella typhimurium	Fe_3O_4@chitosan	Fluorescence	10^{3}–10^{6}cfu/mL	100 cfu/mL	Lettuce samples	(Guo et al. 2020)
Escherichia coli O157:H7	FMNBs	Fluorescence	2.5×10^{2}–5.0×10^{5}cfu/mL	239 cfu/mL	Milk samples	(Huang et al. 2019)
Botulinum neurotoxin type A	MagQD NPs	Fluorescence	0–100 ng/mL	2.52 pg/mL	Milk and grape juice samples	(Wang et al. 2019)
Staphylococcal enterotoxin B				2.86 pg/mL		
Aflatoxin B1	CdSe/ZnS QDs	Fluorescence	5–150 pM	–	Dark Soy Sauce	(Guo et al. 2019)
Aflatoxin B1	AuNPs/GQDs/DTNs	Fluorescence	1–1000 pM	0.492 pM	Peanut oil samples	(Q. Wang et al. 2021)
Aflatoxin M1	BBA-cDNA	Fluorescence	0.7–10 nM	0.5 nM	Milk samples	(Sameiyan et al. 2021)
Ochratoxin A	DNA Apt	Fluorescence	2.5–1250 nM	2.5 nM	Red wine sample	(Li et al. 2020)
Ochratoxin A	–	Fluorescence	0.03–0.78 nM	0.02 nM	Rice samples	(Yu et al. 2019)
Zearalenone	MIP	Fluorescence	10–100 nM	10 nM	Cereal sample	(Sergeyeva et al.)
Zearalenone	DNA-templated AgNCs and MOF-derived magnetic porous Fe_3O_4/ carbon octahedra	Fluorescence	0.01–250 nM	2×10^{-3} nM	Maize and Wheat samples	(Sun et al. 2021)
Picric acid	CQDs/PANI	Fluorescence	0.37–1.42 μM	0.056 μM	Tap water samples	(H.M. Ahmed et al. 2021)
Carbofuran	B–Ser–His-GQDs	Fluorescence	1.0×10^{-13}–1.0×10^{-7} M	3.4×10^{-14} M	Cucumber andcabbage	(N. Wang et al. 2021)
Bisphenol	Fe_3O_4@SiO_2-MIP	Luminescence	1.0×10^{-9}–1.0×10^{-4} M	3.4×10^{-10} M	Water samples	(J. Shi et al. 2022)
4-Nitrophenol	MF-MIPs	Fluorescence	0.08–10 μM	23.45 nM	Fish samples	(Zhu et al. 2021)
Bisphenol A	BTSIXO/Fe^{+3}	Colorimetric	0.1–150 ppm	0.02 ppm	Fish and milk samples	(Sundaram et al. 2021)
Paraquat	PVA/Tb-MOF@ paper strip	Luminescence	0–50 μM	2.84 μM	Drinking water, cabbage, cowpea, vegetable-rinsed water	(Wiwasuku et al. 2022)

Omethoate	S-GQD-Apt	Fluorescence	4.7×10^{-6}–4.7×10^{-4}M	4.7×10^{-9} M	–	(Nair et al. 2021)
Pyrethroid	NRF/DBF	Fluorescence	5–150 µM	1.5 µM	Fruit juices	(Fang et al. 2022)
3-Nitropropionic acid	MOF	Fluorescence	–	1.0 µM	Sugar cane samples	(Yang et al. 2020)
Glyphosate	CQD@CdTe	Photoluminescence	0–1000 nM	2 pM	Cucumber, capsicum, and ginger samples	(Bera and Mohapatra 2020)
Borax	CDs	Fluorescence	100–500 µM	1.5 µM	Fish samples	(Prathumsuwan et al. 2019)
H_2O_2	ZV-Mn NPs	Colorimetric	10–280 µM	0.2 µM	Milk samples	(Rauf et al.)
Carbendazim	UCNPs-MnO_2	Luminescence	0.1–5000 nM	0.05 nM	Apple, cucumber, matcha powder, and water samples	(Ouyang et al. 2021)
Paraquat	PP6@graphene	Fluorescence	0.2–2.0 µM and 2.0–18.0 µM	0.06 µM	Water samples	(Qian et al. 2019)

Abbreviations: AgNCs: Aptamer-modified silver nanoclusters, Apt: Aptamer, Au@AgNRs: Au@Ag nanorods, AuNCs: Gold nanocluster, AuNPs: Gold nanoparticles, B–Ser–His-GQDs: Serine and histidine-functionalized graphene quantum dots, BBA: Bivalent binding aptamer, BHQ3: Black Hole Quencher-3, BTPA: 3-(3-(benzo[d]thiazol-2-yl)-2-hydroxyphenyl)-2-(4′-(diphenylamino)-[1,1′ -biphenyl]-4-yl)acrylonitrile, BTSIXO: 3′,6′-bis(diethylamino)-2-((3,4,5 trimethyl benzylidene) amino) spiro isoindoline-1,9′-xanthen]-3-one, Cd^{+2}: Cadmium ion, CDs: Carbon dots, CdTe QDs: Cadmium telluride quantum dots, CdTe: Cadmium telluride, CMC: Carboxymethylcellulose, CN^-: Cyanide ion, Compd-2: diphenyl ethertagged thiocarbohydrazide based receptor, CQDs: Carbon quantum dots, Cr: Chromium, Cu^-: Cupric cation, Cu: Copper, DBF: 4,5-bis-dihydroxyboron fluorescein, DTNs: DNA tetrahedron nanostructures, Eu-In-BTEC: In-codoped Europium based MOFs, FAM: DNA probe, Fe_3O_4: Iron oxide (II, III) magnetic nanoparticles, Fe^{+3}: Iron(III) ion, FMNBs: Fluorescent magnetic nanobeads, FQICTS: Fluorescence quenching immunochromatographic test strip, g-C_3N_4: Graphitic-carbon nitride, GQDs: Graphene oxide quantum dots, H_2O_2: Hydrogen peroxide, Hg^{+2}: Mercury ion, MagQD NPs: Magnetic quantum dot nanoparticles, MF-MIPs: Magnetic and fluorescent molecularly imprinted polymers, MIL-101: Materials of Institute Lavoisier, MIP: Molecularly imprinted polymer, MMIP: Magnetic molecularly imprinted polymer, MnO_2: Manganese dioxide, MOFs: Metal–organic frameworks, NP-1: Novel pyrene modified nanocrystalline cellulose, NRF: Non-covalent ratiometric fluorophore, OH-HNTs: Hydroxylated-halloysite, PAA: poly(acrylic acid), PANI: Polyaniline, Pb: Lead, PCN 222: Zirconium-porphyrin MOF, Pd NPs: Palladium nanoparticles, PP6: Phosphate pillar [6]arene, PVDF: poly(vinylidene fluoride), PVDF: Polyvinylidene flüoride, qPCR: Quantitative-polymerase-chain-reaction, RhB: Rhodamine B, S-GQD: Sulfur-doped graphene quantum dot, SiO_2: Silicon dioxide, Tb-MOF: $[Tb_2(H_2btec)(btec)(H_2O)]\cdot 4H_2O$, Tb^{+3}: Terbium (III), Tbdcpcpt: $[Me_2NH_2][Tb_3(dcpcpt)_3(HCOO)]$, TiO_2: Titanium oxide, UCNPs: Upconversion nanoparticles, UCNPs: Upconverting nanoparticles, ZIS-HNCs: Hollow $ZnIn_2S_4$ nanocages, [Zn • (BA) • (BBI)]: A two-dimensional zinc(II)-based metal–organic framework, Zr: Zirconium, ZV-Mn NPs: Zero valent manganese nanoparticles

A type of foodborne bacteria called *S. typhimurium* is frequently spread by eating raw vegetables (Pham et al. 2022). Researchers often utilize fluorescence detection for quick and accurate detections because the detection and quantification of *S. typhimurium* make substantial sense in reducing risk factors which threaten human health. In their study, Guo et al. (2020) first prepared magnetic nanoparticles with a Fe_3O_4/chitosan core shell structure. The carbon quantum dots were then covalently bonded with chitosan to form a fluorescence emitting fluorescence probe Fe_3O_4 chitosan to form magnetic fluorescence composite nanoparticles (FMNCs) (FMNCs). This material was bound to the aptamers of *S. typhimurium* and named a new fluorescence probe, FMNCs-Apt. This new fluorescence probe showed a linear relationship of 10^{3}–10^{6}cfu/mL under optimum conditions. The developed probe was also verified by applying it to lettuce samples, and the detection limit of the method was found to be 138 cfu/mL (Guo et al. 2020).

Micromechanical cleavage, epitaxial growth, and chemical processing are only a few of the ways to make graphene. The most researched method is chemical processing, which entails three steps: oxidizing graphite to graphitic oxide using H_2SO_4 and $KMnO_4$, obtaining graphene oxide through ultrasonic dispersion, and then chemically reducing it to produce graphene nanosheets using various explosive or toxic reducing agents. However, this method poses both an environmental threat and because graphene nanolayers have strong π–π fitting interactions, they are very difficult to disperse in aqueous solutions (Suresh et al. 2022). Qiana et al., with an easy and green synthesis in one step, have developed graphene colloids with a unique crystal structure and fairly good dispersibility in aqueous solution. In this method, without oxidation and reduction processes, the shedding and stabilization of the graphene nanolayer were achieved by phosphate batteries. Paraquat (PQ) (methyl cellogen) is a bipyridyllium quaternary ammonium compound widely used in agriculture for weed removal (Z. Zhao et al. 2023). PQ, a pesticide banned by many countries to date, is taken into the body through skin contact, respiration and sinidrime, causing serious health problems leading to death. A straightforward, accurate, and controlled competitive host–guest identification technique has been developed for quantitative detection by combining the fluorescence quenching feature of graphene with the molecular recognition capability of PP6. The chemical recognition capacity of PP6 and the fluorescence quenching feature of graphene are combined in the PP6@graphene nanocomposite. To determine the presence of PQ, a competitive fluorescence test based on the chemical recognition interaction between PP6@graphene and acridine orange (AO) was developed. The interaction between AO and PP6@graphene causes the fluorescence of AO to be completely quenched, and the quenched fluorescence gets better with the addition of PQ because PP6 has been shown to have a stronger ability to incorporate it against PQ than AO (Qian et al. 2019).

Mycotoxins are secondary metabolites produced by filamentous fungi, and Aflatoxin B1 (AFB)1 is the substance with the highest toxicity. The International Agency for Research on Cancer (IARC, 2002) has classified AFB1 as a group I carcinogen due to its mutagenic, teratogenic, immunosuppressive, and carcinogenic effects. Wang et al. (2021) designed a DNA walker, DNA tetrahedron nanostructures (DTNs), and a fluorescence aptasensor based on network hybridization chain reaction (HCR) for the detection of AFB1. The DNA walker structure was employed as a signal amplifier induced by AFB1 targets, where the DNA walker chain could move along a 3DAuNPtrace and assemble complementary sequences of DTNs, leading to the termination of the endonuclease and the release of a significant amount of DTNs. Four single-stranded DNAs with various fulcrums were used to make DTNs. DTNs increased the assembly speed of graphene oxide quantum dots (GQDs) based on nanotrees by offering alternative reaction orientations for HCR (Figure 6.4). The fluorescence at 515 nm gradually increased as AFB1 was introduced. With a detection limit as low as 0.492 pg/mL, this approach could selectively monitor AFB1 in a linear range from 1 to 1000 pg/mL. The produced aptasensor was successfully applied in peanut oil samples and the AFB1 recovery results were found to be in the range of 87.56%–105.28% (Q. Wang et al. 2021).

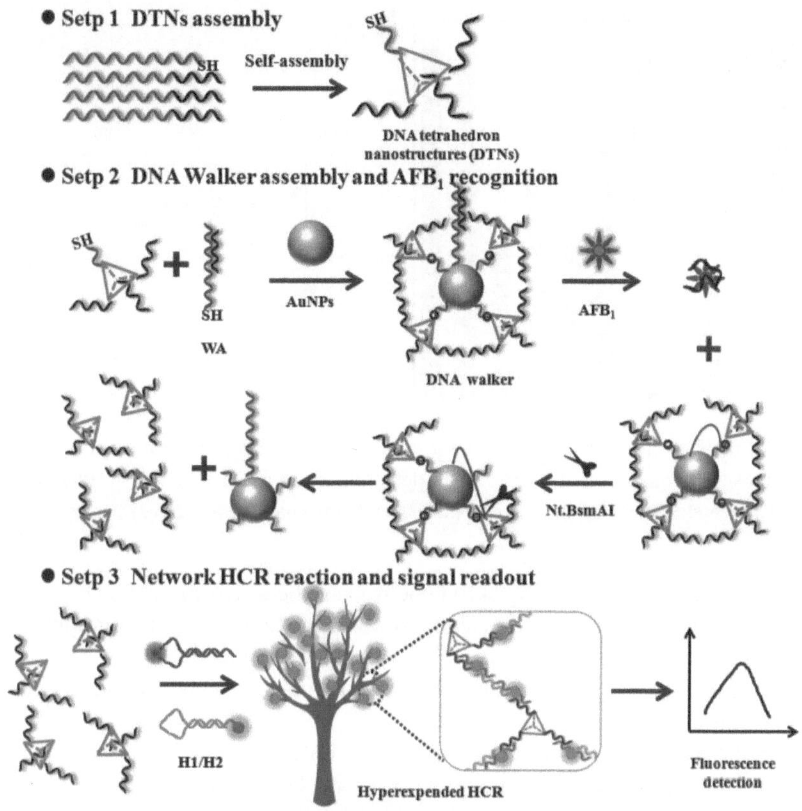

FIGURE 6.4 Schematic illustration of DNA nanomaterial-assisted dual amplified aptasensor for sensitive AFB1 detection based on DNA walker, DTNs and network HCR. Step 1: the assembly of DTNs. Step 2: the assembly of DNA walker and the first signal amplification based on targets-mediated DNA walker. Step 3: the second signal amplification based on network HCR. Reprinted with permission from Q. Wang et al. 2021.

Leaving industrial wastes to groundwater can cause heavy metal poisoning and pose a serious risk for both the environment and human health. In addition to their carcinogenic activities, the accumulation of these contaminates in the food chain has major health adverse effects, including kidney failure, brain damage, digestive system disruption, and impacts on the neuromuscular system. For the first time, carbon dots (CDs) were grafted onto the surface of polyvinylidene fluoride-polyacrylic (PVDF-g-PAA) membrane in this study to create a composite membrane of PVDF-g-PAA-CDs with stable fluorescence. The fluorescence of CDs on the surface of PAA-CD composite membrane is quenched as a result of the complexation of Cu^{+2}, Hg^{+2}, and Fe^{+3} with covalent bonds including $-NH_2$ and –COOH on the surface of CDs, which can effectively transfer electrons or energy and accelerate up the disappearance of electron/hole non-radiative recombination (Figure 6.5) (Zhang et al. 2019).

6.4 CONCLUSIONS AND FUTURE PERSPECTIVES

Due to the increase in population, access to safe food is getting harder day by day. Due to the rapid consumption of resources and the increasing need for food, the food industry and managers have started to seek various solutions to deliver safe food to people. Serious problems and diseases caused by spoiled and contaminated foods, leading to death, show that food analysis should be done seriously in routine applications. For this reason, it is an important need to carry out quality control analyzes in the entire process from the production process of the food to the delivery to

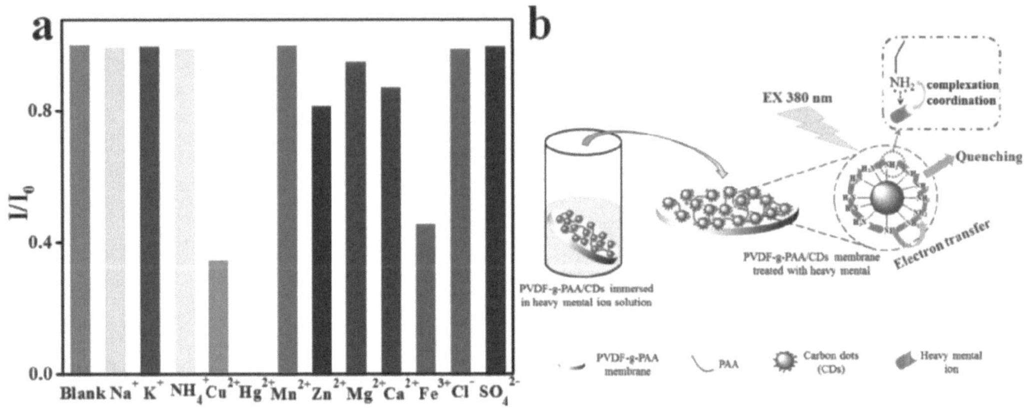

FIGURE 6.5 (a) I/I_0 upon the addition of different heavy irons to select the iron sensing assay for PVDF-g-PAA-CDs composite membrane; (b) fluorescence detection mechanism of PVDF-g-PAA-CDs composite membranes for heavy metal ions.Reprinted with permission from Zhang et al. (2019).

the consumer. To meet this need, it is vital to benefit from the developing technology and develop fast, simple and accessible analysis methods and devices. Although chromatographic methods are widely used in industry today in food analysis, these methods can be disadvantageous compared to today's developing technologies due to the long analysis time, the need for expensive equipment, and sample preparation procedures.

In addition to these methods, advances in nanotechnology and materials science have paved the way for the design of electrochemical and optical sensors. These sensors draw attention with their practical and fast measurement principles that enable portable, real-time and on-site food analysis. In recent years, many electrochemical and optical based nanosensors have been developed for the detection of biological, chemical and physical food contaminants. These sensors are more sensitive and selective for particular microbial contaminants, and some of them might be expanded to detect different residues, by altering different nanomaterials – for example, quantum dots, carbon nanotubes, metallic nanoparticles, graphene, nanomaterial based for the increasing sensitivity of this sensors. Future efforts should focus on some characteristics, such as cost, size, accuracy and reproducibility. Much more research is still required for the practical use of nanomaterial-based sensors in food analysis. In this sense, scientists continue to develop these nanosensors by taking advantage of the unique power of nanomaterials for purposes such as detecting a wide variety of food contaminants such as bacteria, fungi, drug residues, free pollutants, etc., reaching very low limit of detection LOD values, using smaller sample volumes, and developing miniaturized devices. In the light of these studies, it is expected that electrochemical and optical-based nanosensors will be effective analytical devices for food safety and detection of food pollutants, pathogens, etc. in the future.

ACKNOWLEDGEMENTS

Cem Erkmen thanks to the financial support from The Scientific and Technological Research Council of Turkey (TUBITAK) under the BIDEB/2218 postdoctoral scholarship program. Didem Nur Unal The Council of Higher Education (YOK) is greatly appreciated for providing scholarships under the special 100/2000 scholarship program to Didem Nur Unal. Didem Nur Unal and Ipek Kucuk also thanks to the financial support from TUBITAK under the BIDEB/2211-A doctoral scholarship program. Selenay Sadak thanks to the financial support from TUBITAK under the BIDEB/2210-A master scholarship program.

REFERENCES

Aghoutane Y, Diouf A, Österlund L, Bouchikhi B, El Bari N. 2020. Development of a molecularly imprinted polymer electrochemical sensor and its application for sensitive detection and determination of malathion in olive fruits and oils. Bioelectrochemistry [Internet]. 132:107404. https://doi.org/10.1016/j.bioelechem.2019.107404

Ahmad OS, Bedwell TS, Esen C, Garcia-Cruz A, Piletsky SA. 2019. Molecularly imprinted polymers in electrochemical and optical sensors. Trends Biotechnol [Internet]. 37(3):294–309. https://doi.org/10.1016/j.tibtech.2018.08.009

Ahmed HM, Ghali M, Zahra W, Ayad MM. 2021. Preparation of carbon quantum dots/polyaniline nanocomposite: towards highly sensitive detection of picric acid. Spectrochim Acta Part A Mol Biomol Spectrosc [Internet]. 260:119967. https://doi.org/10.1016/j.saa.2021.119967

Ahmed SR, Kumar S, Ortega GA, Srinivasan S, Rajabzadeh AR. 2021. Target specific aptamer-induced self-assembly of fluorescent graphene quantum dots on palladium nanoparticles for sensitive detection of tetracycline in raw milk. Food Chem [Internet]. 346(December 2020):128893. https://doi.org/10.1016/j.foodchem.2020.128893

Amatatongchai M, Sroysee W, Sodkrathok P. 2019. Novel three-dimensional molecularly imprinted polymer-coated carbon nanotubes (3D-CNTs@MIP) for selective detection of profenofos in food. Anal Chim Acta [Internet]. 1076:64–72. https://doi.org/10.1016/j.aca.2019.04.075

Appaturi JN, Pulingam T, Thong KL, Muniandy S, Ahmad N, Leo BF. 2020. Rapid and sensitive detection of *Salmonella* with reduced graphene oxide-carbon nanotube based electrochemical aptasensor. Anal Biochem [Internet]. 589(July 2019):113489. https://doi.org/10.1016/j.ab.2019.113489

Arris FA, Mohan D, Sajab MS. 2022. Facile synthesis of 3D printed tailored electrode for 3-monochloropropane-1, 2-diol (3-MCPD) sensing. Micromachines. 13(3). https://doi.org/10.3390/mi13030383

Beduk T, Lahcen AA, Tashkandi N, Salama KN. 2020. One-step electrosynthesized molecularly imprinted polymer on laser scribed graphene bisphenol a sensor. Sensors Actuators B Chem [Internet]. 314(March):128026. https://doi.org/10.1016/j.snb.2020.128026

Bera MK, Mohapatra S. 2020. Ultrasensitive detection of glyphosate through effective photoelectron transfer between CdTe and chitosan derived carbon dot. Colloids Surfaces A [Internet]. 596(January):124710. https://doi.org/10.1016/j.colsurfa.2020.124710

Bunkoed O, Raksawong P, Chaowana R, Nurerk P. 2020. A nanocomposite probe of graphene quantum dots and magnetite nanoparticles embedded in a selective polymer for the enrichment and detection of ceftazidime. Talanta [Internet]. 218(May):121168. https://doi.org/10.1016/j.talanta.2020.121168

Cheng W, Zhang Q, Wu D, Yang Y, Zhang Y, Tang X. 2022. A facile electrochemical method for rapid determination of 3-chloropropane-1, 2-diol in soy sauce based on nanoporous gold capped with molecularly imprinted polymer. Food Control. 134(September 2021):108750. https://doi.org/10.1016/j.foodcont.2021.108750

Coccia M, Bontempi E. 2023. New trajectories of technologies for the removal of pollutants and emerging contaminants in the environment. Environ Res [Internet]. 229:115938. https://doi.org/https://doi.org/10.1016/j.envres.2023.115938

Devkota L, Nguyen LT, Vu TT, Piro B. 2018. Electrochemical determination of tetracycline using AuNP-coated molecularly imprinted overoxidized polypyrrole sensing interface. Electrochim Acta [Internet]. 270:535–542. https://doi.org/10.1016/j.electacta.2018.03.104

Dong B, Li H, Mujtaba G, Yu X, Yu W, Wen K. 2019. Fluorescence immunoassay based on the inner-filter effect of carbon dots for highly sensitive amantadine detection in foodstuffs. Food Chem [Internet]. 294(April):347–354. https://doi.org/10.1016/j.foodchem.2019.05.082

Dourandish Z, Sheikhshoaie I, Maghsoudi S. 2023. Synthesis of NiRu-metal organic framework nanosheets: as active catalyst for the fabrication of rapid and simple electrochemical sensor for the determination of sudan I in presence of bisphenol A. J Food Meas Charact [Internet]. 17(2):1877–1888. https://doi.org/10.1007/s11694-022-01614-9

Duan W, Baez-Gaxiola MR, Gich M, Fernández-Sánchez C. 2022. Detection of chlorinated organic pollutants with an integrated screen-printed electrochemical sensor based on a carbon nanocomposite derived from bread waste. Electrochim Acta. 436(October):141459. https://doi.org/10.1016/j.electacta.2022.141459

Elfadil D, Lamaoui A, Della PF, Amine A, Compagnone D. 2021. Molecularly imprinted polymers combined with electrochemical sensors for food contaminants analysis. Molecules. 26(15):1–40. https://doi.org/10.3390/molecules26154607

Enbanathan S, Munusamy S, Ponnan S, Jothi D, Manoj S, Iyer K. 2023. Talanta AIE active luminous dye with a triphenylamine attached benzothiazole core as a portable polymer film for sensitively detecting CN– ions in food samples. Talanta [Internet]. 264(March):124726. https://doi.org/10.1016/j.talanta.2023.124726

Fang F, Zhao Q, Fan R, Wang H, Zhu J, Wang X. 2022. An efficiently ratiometric fluorescent probe based on bis-dihydroxyboron fluorescein complexes for detection of pyrethroid residues in fruit juices. Microchem J [Internet]. 172(PB):106954. https://doi.org/10.1016/j.microc.2021.106954

Gökçe G, Ben AS, Nemčeková K, Catanante G, Raouafi N, Marty JL. 2020. Aptamer-modified pencil graphite electrodes for the impedimetric determination of ochratoxin A. Food Control. 115(March):4–10. https://doi.org/10.1016/j.foodcont.2020.107271

Govindasamy M, Sriram B, Wang SF, Chang YJ, Rajabathar JR. 2020. Highly sensitive determination of cancer toxic mercury ions in biological and human sustenance samples based on green and robust synthesized stannic oxide nanoparticles decorated reduced graphene oxide sheets. Anal Chim Acta [Internet]. 1137:181–190. https://doi.org/10.1016/j.aca.2020.09.014

Guo L, Shao Y, Duan H, Leng Y, Huang X, Xiong Y. 2019. Magnetic quantum dot nanobead-based fluorescent immunochromatographic assay for the highly sensitive detection of Aflatoxin B_1 in dark soy sauce. Anal Chem[Internet]. 91(7):4727–4734. https://doi.org/10.1021/acs.analchem.9b00223

Guo Z, Huang X, Li Z, Shi J, Zhai X, Hu X, Liang N, Zou X. 2020. Rapid and highly sensitive detection of *Salmonella typhimurium* in lettuce by using magnetic fluorescent nanoparticles. Anal Meth[Internet]. 12:5861–5868. https://doi.org/10.1039/d0ay01744b

Hatamluyi B, Rezayi M, Beheshti HR, Boroushaki MT. 2020. Ultra-sensitive molecularly imprinted electrochemical sensor for patulin detection based on a novel assembling strategy using Au@Cu-MOF/N-GQDs. Sensors Actuators, B Chem [Internet]. 318(May):128219. https://doi.org/10.1016/j.snb.2020.128219

He B, Yan X. 2020. Ultrasensitive electrochemical aptasensor based on $CoSe_2$/AuNRs and 3D structured DNA-PtNi@Co-MOF networks for the detection of zearalenone. Sensors Actuators, B Chem [Internet]. 306(December 2019):127558. https://doi.org/10.1016/j.snb.2019.127558

Hou Y, Long N, Jia B, Liao X, Yang M, Fu L, Zhou L, Sheng P, Kong W. 2022. Development of a label-free electrochemical aptasensor for ultrasensitive detection of ochratoxin A. Food Control [Internet]. 135(January):108833. https://doi.org/10.1016/j.foodcont.2022.108833

Hoyos-Arbeláez J, Vázquez M, Contreras-Calderón J. 2017. Electrochemical methods as a tool for determining the antioxidant capacity of food and beverages: a review. Food Chem. 221:1371–1381. https://doi.org/10.1016/j.foodchem.2016.11.017

Hu B, Pu H, Sun D. 2023. Flexible Au@AgNRs/CMC/qPCR film with enhanced sensitivity, homogeneity and stability for in-situ extraction and SERS detection of thiabendazole on fruits. Food Chem [Internet]. 423(February):135840. https://doi.org/10.1016/j.foodchem.2023.135840

Hu X, Wang C, Zhang M, Zhao F, Zeng B. 2020. Ionic liquid assisted molecular self-assemble and molecular imprinting on gold nanoparticles decorated boron-doped ordered mesoporous carbon for the detection of zearalenone. Talanta [Internet]. 217(April):121032. https://doi.org/10.1016/j.talanta.2020.121032

Huang Z, Peng J, Han J, Zhang G, Huang Y, Duan M. 2019. A novel method based on fluorescent magnetic nanobeads for rapid detection of *Escherichia coli* O157:H7. Food Chem [Internet]. 276(235):333–341. https://doi.org/10.1016/j.foodchem.2018.09.164

Jafari S, Dehghani M, Nasirizadeh N, Baghersad MH, Azimzadeh M. 2019. Label-free electrochemical detection of *Cloxacillin* antibiotic in milk samples based on molecularly imprinted polymer and graphene oxide-gold nanocomposite. Meas J Int Meas Confed [Internet]. 145:22–29. https://doi.org/10.1016/j.measurement.2019.05.068

Jahangiri-Dehaghani F, Zare HR, Shekari Z. 2020. Measurement of aflatoxin M1 in powder and pasteurized milk samples by using a label-free electrochemical aptasensor based on platinum nanoparticles loaded on Fe-based metal–organic frameworks. Food Chem [Internet]. 310(July 2019):125820. https://doi.org/10.1016/j.foodchem.2019.125820

Jia L, Guo S, Xu J, Chen X, Zhu T, Zhao T. 2019. A ratiometric fluorescent nano-probe for rapid and specific detection of tetracycline residues based on a dye-doped functionalized nanoscaled metal–organic framework. Nanomaterials (Basel), 9, 9766. doi: 10.3390/nano9070976.

Karthik R, Vinoth Kumar J, Chen SM, Sundaresan P, Mutharani B, Chi Chen Y, Muthuraj V. 2018. Simple sonochemical synthesis of novel grass-like vanadium disulfide: a viable non-enzymatic electrochemical sensor for the detection of hydrogen peroxide. Ultrason Sonochem [Internet]. 48(July):473–481. https://doi.org/10.1016/j.ultsonch.2018.07.008

Kartika AE, Setiyanto H, Manurung RV, Jenie SNA, Saraswaty V. 2021. Silver nanoparticles coupled with graphene nanoplatelets modified screen-printed carbon electrodes for rhodamine B detection in food products. ACS Omega. 6(47):31477–31484. https://doi.org/10.1021/acsomega.1c03414

Kaur G, Bhari R, Kumar K. 2023. Nanobiosensors and their role in detection of adulterants and contaminants in food products. Crit Rev Biotechnol [Internet]. 44(4):547–561. https://doi.org/10.1080/07388551.2023.2175196

Khan MU, Javed AR, Ihsan M, Tariq U. 2023. A novel category detection of social media reviews in the restaurant industry. Multimed Syst. 29(3):1825–1838. https://doi.org/10.1007/s00530-020-00704-2

Kim S-B, Yavuz Corapcioglu M, Kim D-J. 2003. Effect of dissolved organic matter and bacteria on contaminant transport in riverbank filtration. J Contam Hydrol [Internet]. 66(1):1–23. https://doi.org/https://doi.org/10.1016/S0169-7722(03)00025-1

Kumar A, Kumar D, Chhibber M. 2020. Determination of mercury ions in aqueous medium and urine sample using thiocarbohydrazide based sensor. Chemistry Select. 5(43):13738–13747. https://doi.org/10.1002/slct.202002914

Li K, Qiao X, Zhao H, He Y, Sheng Q, Yue T. 2021. Ultrasensitive and label-free electrochemical aptasensor based on carbon dots-black phosphorus nanohybrid for the detection of ochratoxins A. Microchem J [Internet]. 168(May):106378. https://doi.org/10.1016/j.microc.2021.106378

Li Y, Wang Z, Sun L, Liu L, Xu C, Kuang H. 2019. Nanoparticle-based sensors for food contaminants. TrAC – Trends Anal Chem [Internet]. 113:74–83. https://doi.org/10.1016/j.trac.2019.01.012

Li Y, Zhang N, Wang H, Zhao Q. 2020. Fluorescence anisotropy-based signal-off and signal-on aptamer assays using lissamine rhodamine B as a label for ochratoxin A. J Agric Food Chem. 68(14):4277–4283. https://doi.org/10.1021/acs.jafc.0c00549

Li YL, Xie FT, Yao C, Zhang GQ, Guan Y, Yang YH, Yang JM, Hu R. 2022. A DNA tetrahedral nanomaterial-based dual-signal ratiometric electrochemical aptasensor for the detection of ochratoxin A in corn kernel samples. Analyst. 147:4578–4586. https://doi.org/10.1039/d2an00934j

Liaquat H, Imran M, Latif S, Hussain N, Bilal M. 2022. Multifunctional nanomaterials and nanocomposites for sensing and monitoring of environmentally hazardous heavy metal contaminants. Environ Res [Internet]. 214:113795. https://doi.org/https://doi.org/10.1016/j.envres.2022.113795

Liu S, Bai J, Huo Y, Ning B, Peng Y, Li S, Han D, Kang W, Gao Z. 2020. Biosensors and bioelectronics A zirconium-porphyrin MOF-based ratiometric fluorescent biosensor for rapid and ultrasensitive detection of chloramphenicol. Biosens Bioelectron [Internet]. 149(August 2019):111801. https://doi.org/10.1016/j.bios.2019.111801

Liu W, Qu X, Zhu C, Gao Y, Mao C, Song J, Niu H. 2020. A two-dimensional zinc (II)-based metal-organic framework for fluorometric determination of ascorbic acid, chloramphenicol and ceftriaxone. Microchim Acta 187, 136. https://doi.org/10.1007/s00604-019-3979-3

Liu Z, Fan T, Zhang Y, Ren X, Wang Y, Ma H, Wei Q. 2020. Electrochemical assay of ampicillin using Fe_3N-Co_2N nanoarray coated with molecularly imprinted polymer. Microchim Acta. 187(8). https://doi.org/10.1007/s00604-020-04432-2

Lv L, Hu J, Chen Q, Xu M, Jing C, Wang X. 2022. A switchable electrochemical hairpin-aptasensor for ochratoxin A detection based on the double signal amplification effect of gold nanospheres. New J Chem. 46(9):4126–4133. https://doi.org/10.1039/d1nj05729d

Lv S, Liu J, Li C, Zhao N, Wang Z. 2019. A novel and universal metal-organic frameworks sensing platform for selective detection and efficient removal of heavy metal ions. Chem Eng J [Internet]. 375(June):122111. https://doi.org/10.1016/j.cej.2019.122111

Mahmoudi E, Fakhri H, Hajian A, Afkhami A, Bagheri H. 2019. High-performance electrochemical enzyme sensor for organophosphate pesticide detection using modified metal-organic framework sensing platforms. Bioelectrochemistry [Internet]. 130:107348. https://doi.org/10.1016/j.bioelechem.2019.107348

Martin AA, Fodjo EK, Eric-Simon ZV, Gu Z, Yang G, Albert T, Kong C, Wang HF. 2021. Cys-AgNPs modified gold electrode as an ultrasensitive electrochemical sensor for the detection of 3-chloropropane-1, 2-diol. Arab J Chem [Internet]. 14(9):103319. https://doi.org/10.1016/j.arabjc.2021.103319

de Melo FT, de Oliveira IM, Greggio S, Dacosta JC, Guecheva TN, Saffi J, Henriques JAP, Rosa RM. 2012. DNA damage in organs of mice treated acutely with patulin, a known mycotoxin. Food Chem Toxicol. 50(10):3548–3555. https://doi.org/10.1016/j.fct.2011.12.022

Miao J, Liu A, Wu L, Yu M, Wei W, Liu S. 2020. Magnetic ferroferric oxide and polydopamine molecularly imprinted polymer nanocomposites based electrochemical impedance sensor for the selective separation and sensitive determination of dichlorodiphenyltrichloroethane (DDT). Anal Chim Acta [Internet]. 1095:82–92. https://doi.org/10.1016/j.aca.2019.10.027

Nagabooshanam S, Sharma S, Roy S, Mathur A, Krishnamurthy S, Bharadwaj LM. 2021. Development of field deployable sensor for detection of pesticide from food chain. IEEE Sens J. 21(4):4129–4134. https://doi.org/10.1109/JSEN.2020.3030034

Nair R V, Chandran PR, Mohamed AP, Pillai S. 2021. Sulphur-doped graphene quantum dot based fluorescent turn-on aptasensor for selective and ultrasensitive detection of omethoate. Anal Chim Acta [Internet]. 1181:338893. https://doi.org/10.1016/j.aca.2021.338893

Okoye CO, Nyaruaba R, Ita RE, Okon SU, Addey CI, Ebido CC, Opabunmi AO, Okeke ES, Chukwudozie KI. 2022. Antibiotic resistance in the aquatic environment: analytical techniques and interactive impact of emerging contaminants. Environ Toxicol Pharmacol [Internet]. 96:103995. https://doi.org/https://doi.org/10.1016/j.etap.2022.103995

Orachorn N, Bunkoed O. 2021. Nanohybrid magnetic composite optosensing probes for the enrichment and ultra-trace detection of mafenide and sulfisoxazole. Talanta [Internet]. 228(December 2020):122237. https://doi.org/10.1016/j.talanta.2021.122237

Ouyang Q, Wang L, Ahmad W, Rong Y, Li H, Hu Y, Chen Q. 2021. A highly sensitive detection of carbendazim pesticide in food based on the upconversion-MnO_2 luminescent resonance energy transfer biosensor. Food Chem [Internet]. 349(January):129157. https://doi.org/10.1016/j.foodchem.2021.129157

Pacheco JG, Castro M, Machado S, Barroso MF, Nouws HPA, Delerue-matos C. 2015. Molecularly imprinted electrochemical sensor for ochratoxin A detection in food samples. Sensors Actuators B Chem [Internet]. 215:107–112. https://doi.org/10.1016/j.snb.2015.03.046

Păduraru DN, Ion D, Niculescu AG, Muşat F, Andronic O, Grumezescu AM, Bolocan A. 2022. Recent developments in metallic nanomaterials for cancer therapy, diagnosing and imaging applications. Pharmaceutics. 14(2). https://doi.org/10.3390/pharmaceutics14020435

Pangajam A, Theyagarajan K, Dinakaran K. 2020. Highly sensitive electrochemical detection of *E. coli* O157:H7 using conductive carbon dot/ZnO nanorod/PANI composite electrode. Sens Bio-Sensing Res [Internet]. 29(November 2019):100317. https://doi.org/10.1016/j.sbsr.2019.100317

Peng J, Yang J, Chen B, Zeng S, Zheng D, Chen Y, Gao W. 2021. Design of ultrathin nanosheet subunits $ZnIn_2S_4$ hollow nanocages with enhanced photoelectric conversion for ultrasensitive photoelectrochemical sensing. Biosens Bioelectron [Internet]. 175(November 2020):112873. https://doi.org/10.1016/j.bios.2020.112873

Peng X, Bao G, Zhong Y, Zhang L, Zeng K, He J, Xiao W, Xia Y, Fan Q, Yuan H. 2021. Highly sensitive and rapid detection of thiabendazole residues in oranges based on a luminescent Tb^{3+}-functionalized MOF. Food Chem [Internet]. 343(October 2020):128504. https://doi.org/10.1016/j.foodchem.2020.128504

Pham T-TD, Phan LMT, Park J, Cho S. 2022. Review – electrochemical aptasensor for pathogenic bacteria detection. J Electrochem Soc [Internet]. 169(8):87501. https://doi.org/10.1149/1945-7111/ac82cd

Pongprom A, Chansud N, Sa-nguanprang S, Jullakan S, Bunkoed O. 2023. Magnetic fluorescent probe of hydroxylated-halloysite and nitrogen-doped graphene quantum dots in molecularly imprinted polymer to enrich and determine marbofloxacin. Microchem J [Internet]. 187(October 2022):108389. https://doi.org/10.1016/j.microc.2023.108389

Poudyal DC, Dhamu VN, Paul A, Samson M, Muthukumar S, Prasad S. 2022. A novel single step method to rapidly screen for metal contaminants in beverages, a case study with aluminum. Environ Technol Innov [Internet]. 28:102691. https://doi.org/10.1016/j.eti.2022.102691

Prathumsuwan T, Jaiyong P, In I, Paoprasert P. 2019. Label-free carbon dots from water hyacinth leaves as a highly fluorescent probe for selective and sensitive detection of borax. Sensors Actuators B Chem [Internet]. 299(April):126936. https://doi.org/10.1016/j.snb.2019.126936

Qian X, Zhou X, Gao W, Li J, Ran X, Du G, Yang L. 2019. One-step and green strategy for exfoliation and stabilization of graphene by phosphate pillar [6] arene and its application for fluorescence sensing of paraquat. Microchem J [Internet]. 150(June):104203. https://doi.org/10.1016/j.microc.2019.104203

Qiao X, Xia F, Tian D, Chen P, Liu J, Gu J, Zhou C. 2019. Ultrasensitive "signal-on" electrochemical aptasensor for assay of acetamiprid residues based on copper-centered metal-organic frameworks. Anal Chim Acta [Internet]. 1050:51–59. https://doi.org/10.1016/j.aca.2018.11.004

Radhakrishnan K, Sivanesan S, Panneerselvam P. 2020. Turn-on fluorescence sensor based detection of heavy metal ion using carbon dots@graphitic-carbon nitride nanocomposite probe. J Photochem Photobiol A Chem [Internet]. 389(November 2019):112204. https://doi.org/10.1016/j.jphotochem.2019.112204

Ranjan S, Dasgupta N, Lichtfouse E. 2016. Nanoscience in Food and Agriculture 3. Springer. https://doi.org/10.1007/978-3-319-48009-1

Rasheed PA, Sandhyarani N. 2017. Electrochemical DNA sensors based on the use of gold nanoparticles: a review on recent developments. Microchim Acta. 184(4):981–1000. https://doi.org/10.1007/s00604-017-2143-1

Rauf S, Ali N, Tayyab Z, Shah MAKY, Yang CP, Hu JF, 2020. Ionic liquid coated zerovalent manganese nanoparticles with stabilized and enhanced peroxidase-like catalytic activity for colorimetric detection of hydrogen peroxide. Mater Res Express. 7:035018. https://doi.org/10.1088/2053-1591/ab7f10

Sabui P, Mallick S, Singh KRB, Natarajan A, Verma R, Singh J, Singh RP. 2022. Potentialities of fluorescent carbon nanomaterials as sensor for food analysis. Luminescence. (October 2022):1047–1063. https://doi.org/10.1002/bio.4406

Sabuncuoğlu SA, Baydar T, Giray B, Şahin G. 2008. Mikotosinler: Toksik etkileri, degredasyonlari, oluşumlarinin önlenmesi ve zararli etkilerinin azaltilmasi. Hacettepe Univ J Fac Pharm. 28(1):63–92.

Sameiyan E, Khoshbin Z, Lavaee P, Ramezani M, Alibolandi M, Abnous K, Taghdisi SM. 2021. A bivalent binding aptamer-cDNA on MoS2 nanosheets based fluorescent aptasensor for detection of aflatoxin M1. Talanta [Internet]. 235(July):122779. https://doi.org/10.1016/j.talanta.2021.122779

Selvi SV, Rajaji U, Chen SM, Jebaranjitham JN. 2021. Floret-like manganese doped tin oxide anchored reduced graphene oxide for electrochemical detection of dimetridazole in milk and egg samples. Colloids Surfaces A Physicochem Eng Asp [Internet]. 631(June):127733. https://doi.org/10.1016/j.colsurfa.2021.127733

Sergeyeva T, Yarynka D, Dubey L, Dubey I, Piletska E, Linnik R, Antonyuk M, Ternovska T, Brovko O, Piletsky S, El A. Sensor based on molecularly imprinted polymer membranes and smartphone for detection of fusarium contamination in cereals. Sensors, 20(15), 4304; https://doi.org/10.3390/s20154304.

Shad NA, Munawar A, Javed Y, Rakha A, Riaz A, Din SU, Zareef I, Sajid MM, Khan MF, Akhtar S, Salman M. 2023. In-field deployable and facile nanosensor for the detection of pesticides residues. Anal Chim Acta [Internet]. 1259(October 2022):341204. https://doi.org/10.1016/j.aca.2023.341204

Shenashen MA, Emran MY, El Sabagh A, Selim MM, Elmarakbi A, El-Safty SA. 2022. Progress in sensory devices of pesticides, pathogens, coronavirus, and chemical additives and hazards in food assessment: food safety concerns. Prog Mater Sci [Internet]. 124(May 2019):100866. https://doi.org/10.1016/j.pmatsci.2021.100866

Shi H, Fu L, Chen F, Zhao S, Lai G. 2022. Preparation of highly sensitive electrochemical sensor for detection of nitrite in drinking water samples. Environ Res. 209(February): 3–8. https://doi.org/10.1016/j.envres.2022.112747

Shi J, Zhang X, Zhang Q, Yang P. 2022. Ultrasensitive and highly selective detection of bisphenol a using core–shell magnetic molecularly imprinted quantum dots electrochemiluminescent probe. Bull Environ Contam Toxicol [Internet]. 108(2):379–385. https://doi.org/10.1007/s00128-021-03351-z

Shukla S, Haldorai Y, Khan I, Kang SM, Kwak CH, Gandhi S, Bajpai VK, Huh YS, Han YK. 2020. Bioreceptor-free, sensitive and rapid electrochemical detection of patulin fungal toxin, using a reduced graphene oxide@SnO_2 nanocomposite. Mater Sci Eng C [Internet]. 113(April):110916. https://doi.org/10.1016/j.msec.2020.110916

Singh S, Numan A, Zhan Y, Singh V, Van Hung T, Nam ND. 2020. A novel highly efficient and ultrasensitive electrochemical detection of toxic mercury (II) ions in canned tuna fish and tap water based on a copper metal-organic framework. J Hazard Mater [Internet]. 399(May):123042. https://doi.org/10.1016/j.jhazmat.2020.123042

Smolko V, Shurpik D, Porfireva A, Evtugyn G, Stoikov I, Hianik T. 2018. Electrochemical aptasensor based on poly(neutral red) and carboxylated pillar[5]arene for sensitive determination of aflatoxin M1. Electroanalysis. 30(3):486–496. https://doi.org/10.1002/elan.201700735

Somnet K, Thimoonnee S, Karuwan C, Kamsong W, Tuantranont A, Amatatongchai M. 2021. Ready-to-use paraquat sensor using a graphene-screen printed electrode modified with a molecularly imprinted polymer coating on a platinum core. Analyst. 146(20):6270–6280. https://doi.org/10.1039/d1an01278a

de Souza CC, Alves GF, Lisboa TP, Matos MAC, Matos RC. 2022. Low-cost paper-based electrochemical sensor for the detection of ciprofloxacin in honey and milk samples. J Food Compos Anal [Internet]. 112(April):104700. https://doi.org/10.1016/j.jfca.2022.104700

Süfer Ö, Karakaya S. 2011. Gıda Endüstrisi ve Nanoteknoloji: Durum Tespiti ve Gelecek. Food Ind Nanotechnol Recent Situat Futur [Internet]. 9(6):81–88. http://search.ebscohost.com/login.aspx?direct=true&AuthType=ip,url,cookie,uid&db=aph&AN=80414164&site=ehost-live&scope=site

Sun X, Wang Y, Zhang L, Liu S, Zhang M, Wang J, Ning B, Peng Y, He J, Hu Y, Gao Z. 2020. CRISPR-Cas9 triggered two-step isothermal amplification method for *E. coli* O157:H7 detection based on a metal–organic framework platform. Anal Chem. 92(4):3032–3041. https://doi.org/10.1021/acs.analchem.9b04162

Sun Y, Gao H, Xu L, Waterhouse IN, Zhang H. 2020. Ultrasensitive determination of sulfathiazole using a molecularly imprinted electrochemical sensor with CuS micro flowers as an electron transfer probe and Au@COF for signal amplification. J Photochem Photobiol A Chem. 332(May):127376. https://doi.org/10.1016/j.foodchem.2020.127376

Sun Y, Zhang Y, Wang Z. 2021. A "turn-on" FRET aptasensor based on the metal-organic framework-derived porous carbon and silver nanoclusters for zearalenone determination. Sensors Actuators B Chem [Internet]. 347(1800):130661. https://doi.org/10.1016/j.snb.2021.130661

Sundaram E, Manna A, Lakshmi K, Sivasamy V. 2021. Colorimetric detection and bio-magnification of bisphenol A in fish organs. Food Chem [Internet]. 345(May 2020):128627. https://doi.org/10.1016/j.foodchem.2020.128627

Suresh R, Rajendran S, Kumar PS, Hoang TKA, Soto-Moscoso M, Jalil AA. 2022. Recent developments on graphene and its derivatives based electrochemical sensors for determinations of food contaminants. Food Chem Toxicol [Internet]. 165(March):113169. https://doi.org/10.1016/j.fct.2022.113169

Surya SG, Khatoon S, Lahcen AA, Nguyen ATH, Dzantiev BB, Tarannum N, Salama KN. 2020. A chitosan gold nanoparticles molecularly imprinted polymer based ciprofloxacin sensor. RSC Adv. 10(22):12823–12832. https://doi.org/10.1039/d0ra01838d

Tümay SO, Vildan S, Demirbas E, Ahmet S. 2020. Fluorescence determination of trace level of cadmium with pyrene modified nanocrystalline cellulose in food and soil samples.Food Chem Toxicol. 146(October):1–11. https://doi.org/10.1016/j.fct.2020.111847

Veseli A, Mullallari F, Balidemaj F, Berisha L, Švorc Ľ, Arbneshi T. 2019. Electrochemical determination of erythromycin in drinking water resources by surface modified screen-printed carbon electrodes. Microchem J [Internet]. 148(January):412–418. https://doi.org/10.1016/j.microc.2019.04.086

Vimbela GV., Ngo SM, Fraze C, Yang L, Stout DA. 2017. Antibacterial properties and toxicity from metallic nanomaterials. Int J Nanomedicine [Internet]. 12:3941–3965. https://doi.org/10.2147/IJN.S134526

Waffo AFT, Yesildag C, Caserta G, Katz S, Zebger I, Lensen MC, Wollenberger U, Scheller FW, Altintas Z. 2018. Fully electrochemical MIP sensor for artemisinin. Sensors Actuators, B Chem. 275(August):163–173. https://doi.org/10.1016/j.snb.2018.08.018

Wang C, Qian J, An K, Ren C, Lu X, Hao N, Liu Q, Li H, Huang X, Wang K. 2018. Fabrication of magnetically assembled aptasensing device for label-free determination of aflatoxin B1 based on EIS. Biosens Bioelectron [Internet]. 108(February):69–75. https://doi.org/10.1016/j.bios.2018.02.043

Wang C, Xiao R, Wang S, Yang X, Bai Z, Li X, Rong Z, Shen B, Wang S. 2019. Magnetic quantum dot based lateral flow assay biosensor for multiplex and sensitive detection of protein toxins in food samples. Biosens Bioelectron [Internet]. 146(September):111754. https://doi.org/10.1016/j.bios.2019.111754

Wang N, Li R, Wang Q, Yang Y, Li N, Li Z. 2021. Functionalized graphene quantum dots with strong yellow fluorescence emissions for highly sensitive detection of carbofuran in cucumber and. New J Chem45:17258–17265. https://doi.org/10.1039/d1nj02325j

Wang P, Luo B, Liu K, Wang C, Dong H, Wang X, Hou P, Li A. 2022. A novel COOH-GO-COOH-MWNT/pDA/AuNPs based electrochemical aptasensor for detection of AFB1. RSC Adv. 12(43):27940–27947. https://doi.org/10.1039/d2ra03883h

Wang Q, Zhao F, Yang Q, Wu W. 2021. Graphene oxide quantum dots based nanotree illuminates AFB1: dual signal amplified aptasensor detection AFB1. Sensors Actuators, B Chem [Internet]. 345(May):130387. https://doi.org/10.1016/j.snb.2021.130387

Wang W, Xu Y, Cheng N, Xie Y, Huang K, Xu W. 2020. Dual-recognition aptazyme-driven DNA nanomachine for two-in-one electrochemical detection of pesticides and heavy metal ions. Sensors Actuators, B Chem [Internet]. 321(June):128598. https://doi.org/10.1016/j.snb.2020.128598

Wang Y, Sun H. 2021. Polymeric nanomaterials for efficient delivery of antimicrobial agents. Pharmaceutics. 13(12):2108. https://doi.org/10.3390/pharmaceutics13122108

Wang Y, Zhang L, Peng D, Xie S, Chen D, Pan Y, Tao Y, Yuan Z. 2018. Construction of electrochemical immunosensor based on gold-nanoparticles/carbon nanotubes/chitosan for sensitive determination of T-2 toxin in feed and swine meat. Int J Mol Sci. 19(12):3895. https://doi.org/10.3390/ijms19123895

Wiwasuku T, Chuaephon A, Habarakada U, Boonmak J, Puangmali T, Kielar F, Harding DJ, Youngme S. 2022. A water-stable lanthanide-based MOF as a highly sensitive sensor for the selective detection of paraquat in agricultural products. ACS Sustainable Chem Eng.10(8):2761–2771. https://doi.org/10.1021/acssuschemeng.1c07966

Wood M, Mugo SM. 2022. A MIP-enabled stainless-steel hypodermic needle sensor for electrochemical detection of aflatoxin B1. Anal Methods. 14(21):2063–2071. https://doi.org/10.1039/d1ay02084f

Wu Y, Li G, Tian Y, Feng J, Xiao J, Liu J, Liu X, He Q. 2021. Electropolymerization of molecularly imprinted polypyrrole film on multiwalled carbon nanotube surface for highly selective and stable determination of carcinogenic amaranth. J Electroanal Chem [Internet]. 895(June):115494. https://doi.org/10.1016/j.jelechem.2021.115494

Xie R, Qu Y, Tang M, Zhao J, Chua S, Li T, Zhang F, Wheatley AEH, Chai F. 2021. Carbon dots-magnetic nanocomposites for the detection and removal of Hg^{2+}. Food Chem [Internet]. 364(December 2020):130366. https://doi.org/10.1016/j.foodchem.2021.130366

Xiong J, He S, Wang Z, Xu Y, Zhang L, Zhang H. 2022. Dual-readout fluorescence quenching immunochromatographic test strips for highly sensitive simultaneous detection of chloramphenicol and amantadine based on gold nanoparticle-triggered photoluminescent nanoswitch control. J Hazard Mater [Internet]. 429(November 2021):128316. https://doi.org/10.1016/j.jhazmat.2022.128316

Yagati AK, Chavan SG, Baek C, Lee MH, Min J. 2018. Label-free impedance sensing of aflatoxin B1 with polyaniline nanofibers/Au nanoparticle electrode array. Sensors (Switzerland). 18(5):1–14. https://doi.org/10.3390/s18051320

Yang S, Liu W, Li G, Bu R, Li P, Gao E. 2020. A pH-sensing fluorescent metal–organic framework: pH-triggered fluorescence transition and detection of mycotoxin. Inorg. Chem. 2020, 59(20):15421–15429. https://doi.org/10.1021/acs.inorgchem.0c02419

Yang S, Yang R, He J, Zhang Y, Yuan Y, Yue T, Sheng Q. 2023. Au nanoparticles functionalized covalent-organic-framework-based electrochemical sensor for sensitive detection of ractopamine. Foods. 12(4):842. https://doi.org/10.3390/foods12040842

Yu L, Chen H, Yue J, Chen X, Sun M, Hou J, Alamry KA, Marwani HM, Wang X, Wang S. 2020. Talanta Europium metal–organic framework for selective and sensitive detection of doxycycline based on fluorescence enhancement. Talanta [Internet]. 207(August 2019):120297. https://doi.org/10.1016/j.talanta.2019.120297

Yu M, Xie Y, Wang X, Li Y, Li G. 2019. Highly water-stable dye@Ln-MOFs for sensitive and selective detection toward antibiotics in water. ACS Appl Mater Interfaces [Internet]. 11(23):21201–21210. https://doi.org/10.1021/acsami.9b05815

Zhang D, Jiang W, Zhao Y, Dong Y, Feng X, Chen L. 2019. Carbon dots rooted PVDF membrane for fluorescence detection of heavy metal ions. Appl Surf Sci [Internet]. 494(March):635–643. https://doi.org/10.1016/j.apsusc.2019.07.141

Zhang J, Gao L, Zhang Y, Guo R, Hu T. 2021. A heterometallic sensor based on Ce@Zn-MOF for electrochemical recognition of uric acid. Microporous Mesoporous Mater. 322(May):111126. https://doi.org/10.1016/j.micromeso.2021.111126

Zhang M, Xu F, Cao J, Dou Q, Juan W, Jing W, Yang L, Chen W. 2024. Research advances of nanomaterials for the acceleration of fracture healing. Bioact Mater. 31(June 2023):368–394. https://doi.org/10.1016/j.bioactmat.2023.08.016

Zhang Y, Liu R, Hassan M, Li H, Ouyang Q, Chen Q. 2021. Fluorescence resonance energy transfer-based aptasensor for sensitive detection of kanamycin in food. Spectrochim Acta Part A Mol Biomol Spectrosc [Internet]. 262:120147. https://doi.org/10.1016/j.saa.2021.120147

Zhao Y, Liu M, Zhou S, Yan Z, Tian J, Zhang Q, Yao Z. 2023. Smartphone-assisted ratiometric sensing platform for on-site tetracycline determination based on europium functionalized luminescent Zr-MOF. Food Chem [Internet]. 425(March):136449. https://doi.org/10.1016/j.foodchem.2023.136449

Zhao Z, Kim E, Bentley WE, Payne GF. 2023. Spectroelectrochemical testing of a proposed mechanism for a redox-based therapeutic intervention: ascorbate treatment of severe paraquat poisoning. Adv Redox Res [Internet]. 8:100068. https://doi.org/https://doi.org/10.1016/j.arres.2023.100068

Zhou YH, Mujumdar AS, Vidyarthi SK, Zielinska M, Liu H, Deng LZ, Xiao HW. 2021. Nanotechnology for food safety and security: a comprehensive review. Food Rev Int [Internet]. 39(7):3858–3878. https://doi.org/10.1080/87559129.2021.2013872

Zhu W, Zhou Y, Liu S, Luo M, Du J, Fan J, Xiong H, Peng H. 2021. A novel magnetic fluorescent molecularly imprinted sensor for highly selective and sensitive detection of 4-nitrophenol in food samples through a dual-recognition mechanism. Food Chem [Internet]. 348(July 2020):129126. https://doi.org/10.1016/j.foodchem.2021.129126

7 Role of Nanotechnology in Detection of Heavy Metals in Food Products

Gurveer Kaur, Chirasmita Panigrahi, and Swati Agrawal

7.1 INTRODUCTION

Heavy metals are familiar environmental hazardous wastes because of their longevity in the atmosphere, high level of toxicity, absorption, and bioaccumulation in the human body. Over the past few decades, the development and affluence of industrial and urbanization activities have tremendously enhanced, thereby increasing the health concerns associated with heavy metal exposure (Yang et al., 2018). The pathways through which metals enter the human body can be many including skin exposure (Hoang et al., 2021), via drinking water (Mondal et al., 2021), and so on, among which food ingestion (Rai et al., 2019) is the most common direct pathway. Statistics from 2015 depicted that the consumption of heavy metals has brought about >2,000,000 illnesses, 10,000,000 disability-adjusted lives, and over 56,000 deaths globally (Gibb et al., 2019). Arsenic (As), cadmium (Cd), mercury (Hg), lead (Pb), and chromium (Cr) are five extensively studied harmful heavy metals (Kumar et al., 2022). Also, As, Cd, Pb, and Cr are non-biodegradable hazardous metals resulting from some industrial discharges or mineral sources (Qing et al., 2020). With their health risk contribution and high detection frequency, the heavy metals have been a research focal point to explore (Wei & Cen, 2020). While investigating the health hazards associated with consuming food, it was seen that cereals, vegetables, and aquatic foods were the major dietary intake sources of As with the total contribution being 86.46% (Wei et al., 2019). The cumulative disability-adjusted life years for coronary heart disease (CHD) and cancers accredited to arsenic consumption through food is nearly 1.7 and 52 million each years, respectively (Oberoi et al., 2019). Therefore, the health hazards of heavy metal contact through consumption of food demand special emphasis.

The environment and living things are seriously at risk due to the widespread use of very hazardous heavy metal ions in many different industries. The serious environmental pollution triggered by heavy metals happens to be long-term, concealed and persistent as metals are nonbiodegradable and have a lengthy half-life (Nabulo et al., 2011; Ali et al., 2019). Their inability to biodegrade causes them to accumulate in human tissues as they pass the food chain, severely harming health. The abundance of heavy metals in recently harvested fruits and vegetables poses several health disorders, such as kidney dysfunction, carcinogenesis, immune system malfunctioning, and even death as a result of bioaccumulation and biomagnification (Ahmed et al., 2019; Shamsudduha et al., 2019). Metal consumption can eventually cause oxidative stress leading to adverse harmful effects (Li et al., 2019) and diseases (Fu & Xi, 2020), particularly in prenatal maternal cases, where some undesirable birth outcomes like, low birth weight or any disorder (Laine et al., 2015). The detection of the heavy metals in food is of the utmost importance to ensure safety and purity of food, and protection of human health.

DOI: 10.1201/9781003438168-7

Numerous nanotechnology methods for sensing heavy/toxic metals have been advanced in recent years that are noteworthy for food safety assurance. Nanotechnological strategies are looking for attention owing to the advantages related to abolition of the antithetical significance of heavy metals. With the development in domain of nanotechnology, nanomaterials for detection and removal of heavy metal been meticulously conceptualized and manufactured, offering countless advantages (He et al., 2019). This chapter discusses the heavy metal contamination in various food categories and extensively elaborates nanotechnology-based metal detection techniques, applications of nanosensors in food processing, their limitations and safety regulations involved.

7.2 HEAVY METAL CONTAMINATION IN FOODS

Many environmental contaminants including toxic metals are regarded the prime sources of water pollution with metals during agricultural and industrial waste discharges of pesticides, oil and coal combustion, fertilizers, and plastics. Food crops which are one of the most crucial components of our nutrition, may embrace a variety of essential yet hazardous metals, based on the condition of soil and nature of the applied growth medium (Waqas et al., 2015). Edible vegetables accounting for around 90% of the overall intake in humans' diet mostly contribute to the metal ingestion. Cereals and marine foods are the chief dietary sources of Pb, Cr, Cd, Hg, and As. Metal pollution of animal feed and its constituents also denotes a critical problem for animal health and induces the build-up of toxic metals in the non-vegetarian food chain such as milk, egg, and meat. An overview of the contribution of various food groups and heavy metals to health risks is depicted in Figure 7.1. Cereals (49.5%) and aquatic foods (41.1%) presented the foremost health risks caused by heavy metals and arsenic was found to be the primary heavy metal (Jin et al., 2023).

7.2.1 Cereals

Collado-López et al. (2022) summarized the existing evidence about the abundance of heavy metal content in various food groups in different countries around the world. They reported that, in the cereals food group, rice was the most described food subgroup and cadmium was the most studied heavy metal. Salama and Radwan (2005) identified the concentrations of certain metals, i.e., Cd, Cu, Pb, and Zn in rice, wheat, barley and maize samples gathered from markets in Alexandria. The maize samples showed higher levels of Zn (15.45 mg kg^{-1}) and Cd (0.14 mg kg^{-1}). Meanwhile, the peak levels of Cu (1.96 mg kg^{-1}) and Pb (0.40 mg kg^{-1}) were detected in barley and wheat samples, respectively. El-Hassanin et al. (2020) analyzed the effect of source of irrigation water on Cd and Pb levels in maize grains. They noticed that levels of toxic metals like, Cd and Pb in case of freshwater irrigation were up to 0.01 and 0.04 mg kg^{-1}, respectively. However, low-quality water irrigation yielded higher concentrations of Pb (0.26–0.55 mg kg^{-1}) and Cd (0.038–0.112 mg kg^{-1}). The amount of mineral bio-accessibility also varies depending on the cleaning procedure and cooking method (made with or without draining water). Raw rice grains washed many times and using high volume of water for cooking decreased the As content in cooked rice (Kumarathilaka et al., 2019).

7.2.2 Fruits and Vegetables

According to the reports of Collado-López et al. (2022), in the vegetables group, all subgroups were investigated, and found fruiting and leafy vegetables to be the most reported, and Cd followed by Pb is the maximum studied metal in these subclasses. With regard to the roots and tubers category, potatoes trailed by carrots were the most examined subgroups, and Cd, Pb followed by As were the most reported heavy metals. In fruits grouping, tropical fruits were the most evaluated subgroup.

Fruits and vegetables grown at the wayside are supposed to absorb the airborne hazardous metals, specially from exhaust emissions (Feng et al., 2011). Also, Shahid et al. (2017)revealed that heavy

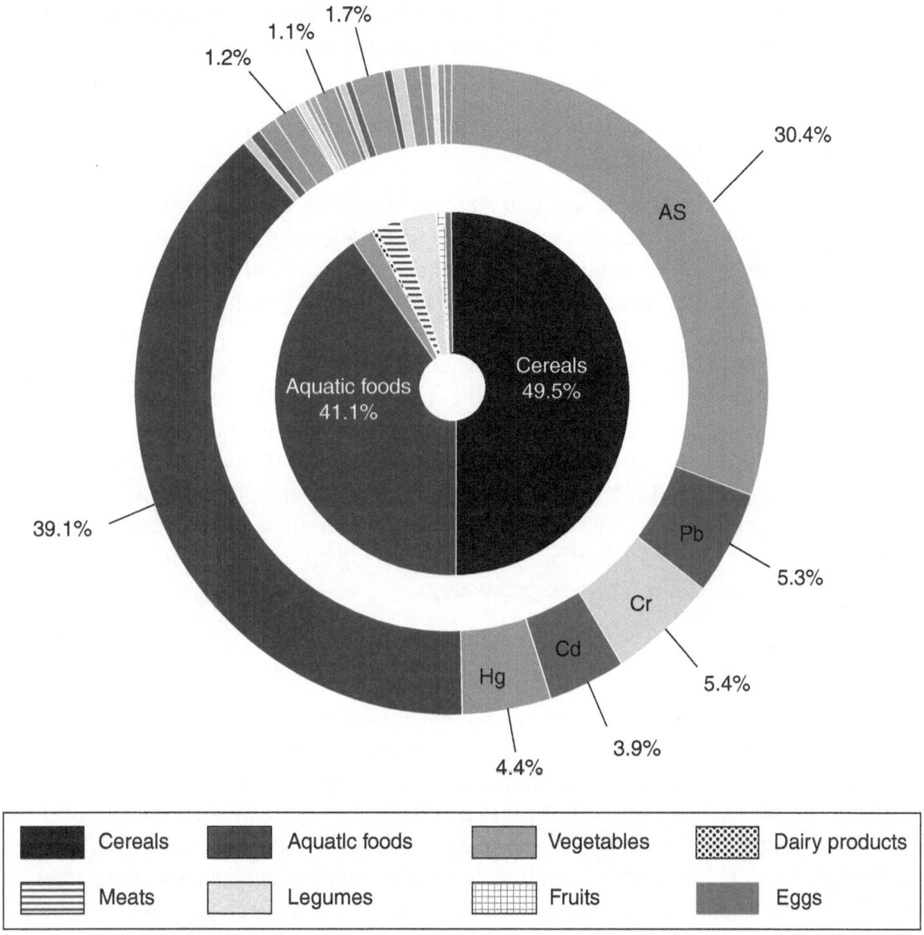

FIGURE 7.1 Contribution of different categories of foods and heavy metals to health risks. (The inner ring shows the contribution of each food ingredient, and the outer ring shows the contribution of each heavy metal.) (From Jin et al., 2023.)

metals may be piled up and collected on the leafy parts of plants. Abd El-Rahman et al. (2015) examined the heavy metal (Cd, Pb, Al, and As) contamination of vegetable samples (cucumber, tomatoes, watercress, lettuce, and potatoes) gathered from some regions in Egypt. The higher accumulation of metals in tuber vegetables (potato) followed by leafy vegetables (watercress and lettuce) in contrast to fruit vegetables (cucumber and tomatoes), was the most important examination of this study. In addition, Eissa and Negim (2018) disclosed the relocation of some heavy metals (Zn, Pb, Cu, Cd, and Ni) from a metal-contaminated soil to spinach and lettuce. They observed that the gathered heavy metals in roots of those vegetables were greater than those deposited in their shoots. Thus, it may be hypothesized that, metals are mostly transferred from soil to the plants and the closer the proximity of parts to the soil, higher the amount of metal.

Radwan and Salama (2006) detected Cu, Zn, Pb, and Cd in distinct fruits (apple, melon, banana, grapefruit, date, orange, peach, watermelon, and strawberries) collected from Alexandria city, Egypt. The content of metals varied from 1.2 to 18.3 mg kg^{-1} for Cu, 1.36 to 10.5 mg kg^{-1} for Zn, 0.05 to 0.87 mg kg^{-1} for Pb, < 0.002 to 0.05 mg kg^{-1} for Cd. Kandil et al. (2020) determined some metals in fruit samples received from Giza governorates. They revealed that the concentrations of metals differed based on the metal kind, fruit type, and metal–sample association. For example, the highest concentrations of Pb were 0.22, 0.12, and 0.05 mg kg^{-1} in strawberry, pomegranate, and orange, respectively. Meanwhile, in orange, strawberry, and pomegranate, the maximum levels of Cu were 5.5, 3.5, and 1.9 mg kg^{-1}, respectively.

7.2.3 Milk and Dairy Products

The existence of toxic metals in milk and dairy products poses a significant risk as milk is extensively consumed by children and infants. Cashman (2011) disclosed that the levels of metals in milk products depend on many components, such as, genetic characters of animal, stage of lactation, nutritional feed of the animal, metal contamination from the equipment during dairy processing, environmental factors, and manufacturing practices. Main sources of Cu accumulation in milk or milk products are Cu alloys used in different equipment, elevated Cu levels in water and animal feed. Also, existence of Pb in milk may be linked to the industrial air pollution in dairy farms regions. Enb et al. (2009) evaluated the heavy metal contents in milk of buffalo and cow in Egyptian animal farms. They observed that concentrations of 4.37, 0.98, 0.21, and 0.08 mg kg^{-1}, respectively of Zn, Fe, Cu, and Mn are found in buffalo's milk which were higher than their levels as 3.97, 0.79, 0.19, and 0.05 mg kg^{-1}, respectively in cow's milk. Also, the concentrations of hazardous metals such as Cd and Pb were higher in the milk of buffalo (0.12 and 0.08 mg kg^{-1}, respectively) as compared with the milk of cow (0.09 and 0.04 mg kg^{-1}, respectively).

7.2.4 Egg

Elliott et al. (2017) examined the levels of metal contamination in various animal feeds. They specified that approximately 24, 19, and 15% of samples from poultry feed were adulterated with Cd, As, and Pb, respectively. The contamination quantities of ruminant feed samples with Cd, As, and Pb were 21, 27, and 16%, respectively. Kabeer et al. (2021) determined the correlation of the heavy metals found in poultry feed with their accumulation in egg. The content of accumulated Cr, Pb, and Se in feed were 1.30, 2.59, and 0.94 ppm, respectively. Meanwhile, corresponding levels in egg white were 0.180, 0.658, and 0.216 ppm. Greater levels were noted in egg yolk as 0.701, 0.262, and 0.266 ppm, respectively.

Hussien and Nosir (2017) evaluated some metal concentrations (Cr, Cu, Pb, and Cd) in egg samples found from El-Monefia Governorate to be 2.12, 1.61, 0.70, and 0.31 mg kg^{-1}, respectively. The concentrations of As, Fe, Cu, Cd, and Pb in brown shell egg samples gathered from Mansoura city were evaluated by Al-Ashmawy (2013). They detected that the mean levels of Fe, Cu, and Pb in egg samples were 30.04, 0.78, and 0.19 mg kg^{-1}, respectively, while As and Cd were not noticed in any tested egg sample. Meanwhile, the mean levels of Pb, Cu, and Cd in egg sample of Assiut governorate were 0.59, 0.57, and 0.02 mg kg^{-1}, respectively (Ahmed, 2017).

Also, Hashish et al. (2012) determined the metal levels in egg samples in distinct geographic regions in Egypt as North Delta (Alexandria governorate), South Delta (Cairo governorate), Middle Delta (Menoufia governorate), and Upper Egypt (Sohag governorate). They recorded the greatest levels of As, Pb, Mn, Cu, and Zn in Menoufia governorate egg samples as 0.033, 0.303, 1.59, 6.7, and 64.3mg kg^{-1}, respectively. Meanwhile, the greatest levels of Cd and Fe were noticed in Alexandria governorate's egg samples at 86.1 and 0.015 mg kg^{-1}, respectively.

7.2.5 Meat

The major sources of heavy metals in animal systems are contaminated feed, water and soil, polluted air, filthy equipment, and improper manufacturing practices (Caggiano et al., 2005). Darwish et al. (2015) reported the dependency of the levels of metals in meat on the animal age. They observed that the water content and protein concentration of meat were decreased, while fat and ash contents were increased with an increase in the animal age, and this led to augmentation in metal levels. Darwish et al. (2010) examined the levels of Cd and Pb in the liver, kidneys, and muscles of cattle and sheep. The highest amount of both metals was observed in kidneys and liver, while the lowest values of both Cd and Pb were found in their muscles. Also, the determined metals in cattle organs were recorded higher compared to those detected in sheep organs. They observed that levels of Cd and Pb in the organs of aged animals were more than those of young animals. El-Wehedy et al. (2018) studied the toxic metal levels in different meats served at Egyptian hospitals, such as raw meat, raw chicken, cooked meat, and cooked chicken. The cooked chicken sample noted the highest mean levels of Pb, Cd, and Asas 0.421, 0.202, and 0.122 mg kg^{-1}, respectively. They further added that no significant difference in metal levels between beef and chicken samples was found. However, the cooked samples showed a significant rise in metal levels compared to raw samples and this may be ascribed to the evaporation and loss of water from the cooked tissue.

7.2.6 Fish and Sea Foods

Aquatic foods only make up to 6.4% of daily meals, but they can have 41% of heavy metal exposure. The different fishes accumulate metals from water either via direct consumption of water or by the uptake through the skin, gills, and digestive tract (Shoukry et al., 2016). Jin et al. (2023) disclosed a standard-exceeding ratio of 23% of As content in aquatic foods, especially seafoods, with the largest concentration attaining 1.23 mg kg^{-1}. Kim et al. (2019) revealed that high consumption of dietary mercury via shellfish and fish was allied with an amplified risk of colorectal cancer, having an odds ratio of 3.14. Hamada et al. (2018) determined the levels of Pb, Hg, and Cd in Nile tilapia fillet samples received from Menoufia, Egypt. They stated that the levels of Pb, Hg, and Cd were 0.34, 0.73, and 0.11 mg kg^{-1}, respectively in small-sized wild tilapia, while 0.54, 1.18, and 0.16 mg kg^{-1}, respectively in large-sized wild tilapia.

It is seen that the fish organs can amass metals several times greater than their levels in water (Abdel-Mohsien & Mahmoud, 2015). They described that the levels of Pb, Cr, and Cd in fish samples of the Nile River were significantly greater than their concentrations in water with the bioaccumulation factor ranging between 8 and 123. Moreover, as per reports of Helmy et al. (2018), the accumulation rate of metals varied with the kind of fish. It has been revealed that the *Claris gariepinus* fish species recorded higher levels of Pb, Hg, and Pb compared with *Oreochromis niloticus* and *Mugil cephalus*. Also, the accumulation rate of metals was based on the variety of metals (Abdelhamid et al., 2013) wherein the bioaccumulated Pb in fish muscles was noticed to be higher than Cu, Zn, and Fe. The levels of metals in fish were also found to be affected by growth conditions and water properties. In this aspect, Shokr et al. (2019) evaluated the levels of Pb and Hg in freshwater fish (*Oreochromis niloticus* and *Clarias gariepinus*) and marine water fish (*Pagruspagrus* and *Sardina pilchardus*). They observed that the levels of Hg (0.89–1.10 mg kg^{-1}) and Pb (0.49–0.65 mg kg^{-1}) in freshwater fish were higher than their respective levels in marine water fish 0.58–0.73 mg kg^{-1} and 0.28–0.34 mg kg^{-1}.

7.3 NANOTECHNOLOGY-BASED HEAVY METAL DETECTION TECHNIQUES

Emerging nanotechnology offers promising solutions for analyzing and removing heavy metals from complex matrices. Numerous techniques utilizing nanomaterials have been developed for heavy metal analysis, including colorimetric, electrochemical, fluorescent, and biosensing technologies. Nanotechnology has enabled these methods to have a broad linear range, excellent selectivity, high sensitivity, low detection, and quantification limits, making them highly effective in detecting even trace amounts of heavy metals in food and water resources. To tackle heavy metal removal, researchers have employed different categories of nanomaterials, such as magnetic nanoparticles

(NPs), metal oxide NPs, carbon nanotubes, graphene, and derivatives. These nanomaterials possess unique properties that facilitate the efficient adsorption and elimination of heavy metal contaminants (Gong et al., 2021).

7.3.1 Colorimetric

Colorimetric analysis is a popular approach for detecting the presence of numerous analytes, including heavy metals, by utilizing the color shift of a dye. This technique is frequently used in combination with nanotechnology in the food enterprise to identify and measure chemical contaminations. Nanotechnology-based colorimetry has been developed as a sensitive and effective approach for detecting heavy metals in food. This method makes use of nanomaterials like carbon nanotubes and metal NPs to enable colorimetric sensing of heavy metal ions. These NPs have distinctive optical characteristics, such as localized surface plasmon resonance (LSPR), which can be modified to alter their color in response to the presence of heavy metal ions (Zhang et al., 2012). In this approach to identify heavy metals, metal NPs, which are thought to be particularly efficient nanomaterials, are usually used. The basic idea behind this method is shown by the color alterations that take place as a result of the metal NPs' varying sizes. For instance, monodispersed gold nanoparticles (AuNPs) with a diameter under 30 nm show a unique red color when present in an aqueous solution (Hyder et al., 2022). This method involves the functionalization of metal NPs, such as AuNPs or silver nanoparticles (AgNPs), with particular ligands or receptors that bind to the target heavy metal ions only. A chemical reaction takes place when these functionalized NPs come into touch with the heavy metal-containing food sample, resulting in a noticeable color change. The qualitative and semi-quantitative signal of the presence and concentration of heavy metals in the food product is provided by this color change. Aqueous solutions containing heavy metals can aggregate metal NPs like AuNPs, AgNPs, and copper nanoparticles (CuNPs), which causes a shift in their localized surface plasmon resonance band and alters the color of the solution. Similar to AuNPs, AgNPs, and CuNPs have plasmonic features that make them popular choices for colorimetric detection. AgNPs smaller than 30 nm look yellow in their unaggregated condition, while CuNPs between 10 and 20 nm appear red (Zhong et al., 2021).

Chromogenic substances might be of three different types: organic, inorganic, or other substances. Each type has distinct benefits and drawbacks, including those related to sensitivity, selectivity, stability, and cost. Each chromogenic substance has its own distinct optical and chemical characteristics, which results in completely diverse colorimetric sensing methods. Various chromogenic materials include:

- Organic materials, such as DNA, peptides, polymers, and colors.
- Inorganic materials: metal NPs, quantum dots (QDs), metal-organic frameworks, etc.
- Other materials: carbon nanotubes, graphene, enzymes, etc.

There are two types of sensing techniques generally used for the colorimetric approach of heavy metal detection: direct and indirect. The chromogenic materials and metal ions interact directly in direct methods, changing color as a result. Utilizing other substances or methods that could indirectly impact the chromogenic materials' color is known as an indirect strategy.

Heavy metal ions bond to the surface of NPs, causing aggregation or dispersion and a noticeable color shift that may be seen with the naked eye. The presence and level of heavy metals in food samples are determined by the color change as a qualitative or quantitative indication. Nanotechnology-based colorimetric methods offer a practical and quick method for finding heavy metals (Tu et al., 2022). However, typical colorimetric techniques frequently have low detection efficiencies due to the chelation of metal ions to enzymes (Tu et al., 2022). AgNPs have LSPR characteristics that make it possible to detect heavy metal ions in a sensitive and targeted manner (Roto et al., 2019). The selectivity and sensitivity of lead (Pb^{2+}) analysis are further enhanced by the addition of dithizone to AgNPs (Roto et al., 2019). AgNPs that have been green-synthesized from plant extracts have also been utilized to selectively detect heavy metal ions in aqueous solutions using colorimetry (Karthiga & Anthony, 2013). According to Karthiga and Anthony (2013), these AgNPs demonstrated the selective detection of particular heavy metal ions, such as lead (Pb^{2+}) and mercury (Hg^{2+}) (Hong et al., 2017).

Carbon nanotubes (CNTs) have also been used in colorimetry based on nanotechnology to detect heavy metals. CNTs are appropriate for biosensing applications because of their high surface area, superior electrical conductivity, and biocompatibility (Deka et al., 2023). The immobilization of specialized receptors that bind to heavy metal ions, such as antibodies or DNA probes, can be accomplished using functionalized carbon nanotubes as platforms. Heavy metal ion binding to the CNT surface's receptors results in a change in the nanotubes' electrical properties, which may be detected and linked to the amount of heavy metals present in the food sample. Carbon quantum dots (CQDs) were investigated as fluorescence sensors in a recent study by Tian et al. (2023) for the precise detection of heavy metal ions, even those found in food additives and contaminants. Due to their unique optical characteristics and low toxicity, CQDs are especially well suited for use in applications in the inspection of food safety.

These colorimetric techniques based on nanotechnology have several advantages for identifying heavy metals in food products. Colorimetric techniques are suitable for on-site or testing at the point of care because they are simple to use, affordable, and simple to interpret. These techniques provide quick and visible detection without the need for specialized equipment because they rely on a particular indication or probe changing color when heavy metals are present. They enable the detection of hazardous metals at low concentrations because of their great sensitivity, selectivity, and speed of detection. Furthermore, the utilization of nanomaterials allows for the construction of portable and low-cost detecting instruments.

7.3.2 Electrochemical and Biochemical Techniques

Electrochemical detection is frequently carried out with a three-electrode setup that includes a working electrode (sensing electrode, WE), a counter electrode, and a reference electrode. The basic concept of electrochemical sensors is the conversion of chemical signals into electrical signals, which are subsequently recognized by a potentiostat as output transducer signals (Ding et al., 2021). There are oxidation or reduction takes place on the surface of the sensing electrode due to the presence of the heavy metal ions, further electron loss or gain takes place. The conversion of the generated chemical signals into electrical signals enables the identification of the link between the amount of heavy metal ions and current. The electrochemical workstation then examines the generated electrical signals to determine a correlation between the amount of heavy metals present and the current, allowing for efficient detection (Wu et al., 2021). The working electrode's quality has a significant impact on the measurement procedure as a whole. To improve the sensitivity and selectivity of the detection process, the electrode surface must be properly changed. To do this, a range of materials or materials in combination can be used to increase the electrode's sensitivity, decrease the detection limit, and improve overall detection efficiency. As a result, various nanomaterials have been established for amplifying electrochemical signals to achieve ultrasensitive detection (Nam et al., 2022). Voltammetry, amperometry, potentiometry, coulometry, and impedance are the most common electrochemical detection techniques. These methods permit the quantitative correlation between specific electrical characteristics (such as current, resistance, conductance, and electric potential) and the amount or concentration of the targeted heavy metal. There are two major strategies for improving electrochemical signals by means of nanomaterials: electrode modifiers and signal tags (Qian et al., 2021; Tajik et al., 2022).

7.3.2.1 Electrode Modifiers

The effectiveness of electrochemical detection relies heavily on the characteristics of the working electrode. To increase the selectivity and sensitivity of sensors for metal ion detection, chemical or biochemical modifications of the electrode with electron mediators are employed to enable site-specific identification of the target analyte (Raju et al., 2023). The detection limit of the sensing electrode can be increased by modification using various organic, inorganic, and biological materials (Dai et al., 2018). Inorganic nanomaterials effectively boost the electrode surface's specific surface area when used, which facilitates a greater fixation of heavy metal ions. The efficacy of electrochemical detection is increased by the catalytic properties of these NPs, which facilitate the deposition of heavy metal ions. However, inorganic nanomaterials face barriers to wider use due to their comparatively expensiveness and constrained scalability.

However, changes to organic materials on electrodes promote the heavy metal ions deposition after forming metal complexes via coordination bonds. However, organic materials can unfortunately only be used to a limited extent in the electrochemical detection of heavy metal ions due to their low stability and probable toxicity. The working electrode has been changed to enhance the detection efficiency. These created electrodes demonstrate outstanding electrochemical activity, according to research in electrochemical studies (Yao et al., 2019). Notably, screen-printing technology has been used to build disposable electrochemical sensors, taking into mind practicality and cost-effectiveness, in response to the growing demand for on-the-spot detection in food analysis.

Nanomaterials have emerged as interesting possibilities for electrode modification due to their enormous surface area, potential for various changes, and unique quantum mechanical features. As electrode modifiers, a variety of nanomaterials have been used, including silicon-based, carbon-based, and metallic NPs. The nanoscale design of the electrode surface, which promotes more catalytic activity, better conductivity, a bigger active surface area, and faster electrode kinetics, is responsible for the better performance of nanomaterials-based sensors (Waheed et al., 2018). Such developments, including the insertion of organic ligands onto the surface, and the creation of nanostructures, have made it feasible to miniaturize, which is the basis for lab-on-chip (LOC) technologies (Waheed et al., 2018).

Overall, the electrochemical method offers a number of benefits and has become a powerful sensing instrument for detecting heavy metal ions (HMI). These features include accurate outcomes, ease of use, low cost, exceptional sensitivity, and multiplexed and on-site detection ability (Zhang et al., 2016). Electrochemical sensing outperforms spectroscopic techniques due to its ease of use and minimum apparatus needs. Growing interest has been shown in the incorporation of electrochemical devices into automated fluidic structures for monitoring a variety of heavy metals.

7.3.2.2 Signal Tags

Electrochemical signals can be effectively enhanced by incorporating signal tags, which are labeled elements adhered to the electrode surface to facilitate the generation of electrochemical signals for analyte analysis. Nanomaterials with exceptional electrochemical activity hold promise as ideal candidates for producing catalytic signals by catalyzing biochemical reactions (Yu et al., 2019; Nam et al., 2022). For instance, Wei et al. (2015) developed thymine-functionalized AgNPs as signal tags for detecting mercury. In this system, the presence of Hg persuaded the accretion of Ag-T NPs, leading to an amplification of the electrochemical signals (Nam et al., 2022).

7.3.2.3 Electrochemiluminescence

Electrochemiluminescence (ECL) is a process where electrochemically produced species undertake vast energetic electron transfer reactions, leading to the emanation of light from excited states. It has become popular due to its simple setup, versatility, and high sensitivity. ECL-based identification methods for metal ions, particularly those utilizing nanostructured materials, have gained importance in recent years. Various metal nanostructures, like silica nanoparticles (SiNPs), AuNPs, and QDs, are employed for this purpose. SiNPs have proven to be a promising platform for ECL-based detection, mainly because of their surface chemistry, which allows easy functionalization and modification (Aragay & Merkoci, 2012). Recently, a research group developed an electrochemiluminescence (ECL) sensor for lead ions (Pb^{2+}) using Au nanoclusters (Fang et al., 2011). They observed ECL emission from gold nanoclusters at 1.45V vs. Ag/AgCl when triethylamine was used as the co-reactant for Pb^{2+} ions. The system showed selectivity for Pb^{2+} ions due to the greater affinity of Pb^{2+} to sulfur atoms present in BSA-protected nanoclusters. Additionally, cysteine was put as a masking agent to prevent intrusions from other metals like Cd^{2+}. While this work demonstrates the feasibility of the ECL sensor for Pb^{2+} detection, further studies are needed to explore its analytical performance and potential interference in real sample applications (Aragay & Merkoci, 2012).

7.3.2.4 Other Strategies

Conductimetry and biofuel cells have also been explored as electrical and electrochemical strategies for heavy metal detection. A bioinspired crystal growth concept was developed using a peptide nanotube recognition platform for lead ions (Pb^{2+}) detection (De La Rica et al., 2010). The peptide nanotubes selectively bind Pb^{2+} ions and act as templates for the growth of Pb crystals through molecular recognition. When an electrochemical reduction step is finished, the nanotube metallization induces alteration in electrical properties, enhancing the conductivity between electrodes, which can be utilized as an electrochemical signal for the detection of Pb^{2+} ions (Aragay & Merkoci, 2012).

Preconcentration and stripping processes are combined in the electrochemical technique known as anodic stripping voltammetry (ASV) (Ye et al., 2020). Researchers have developed a nanocomposite named fluorinated graphene/gold nanocage (FGP/AuNC) for use in ASV to enable simultaneous identification of heavy metals (Tan et al., 2020). The electrochemical sensor displays distinct and separate stripping peaks for lead and cadmium lead by integrating a bismuth layer (Yao et al., 2019). A group of functional materials called metal–organic frameworks (MOFs), which are known for their exceptional structural tunability has also recently attracted a lot of attention in electrochemical sensing for a variety of analytes (Zhang et al., 2019; Wang et al., 2019). MOFs have gained a lot of attention because of their flexibility in applications involving electrochemical sensing (Zhang et al., 2020; Li et al., 2019).

The reliability, usability, and cost-effectiveness of electrochemical technologies are improved by a number of factors, particularly for on-site applications. These techniques enable simple procedures to be carried out and are especially well suited for the construction of tiny, portable devices that are ideal for in-situ contamination monitoring. The ability to monitor water samples in real-time is also made feasible by them, which is difficult to perform with some spectroscopic techniques (Pujol et al., 2014). They also offer rapid analytical results. However, it is critical to note that electrochemical techniques have limitations. They are less sensitive and have more limited detection capabilities when compared to other spectroscopic and optical approaches. To address this problem and boost their effectiveness in detecting heavy metal ions, additional advancements in electrode design are needed.

7.3.3 Fluorescence Technique

A cutting-edge method for finding heavy metals in food is the fluorescence technique based on nanotechnology. This method makes use of specialized nanomaterials, such as CQDs or other fluorescent NPs, that may particularly attach to heavy metal ions (Maghsoudi et al., 2021; Yao et al., 2019).

In addition to identifying other pollutants such as food additives, bacteria, pesticide residues, antibiotics, and nutritional components, fluorescence sensors have been widely used in the study of heavy metals in food samples. However, due to limitations such as optical decolorization and the presence of highly hazardous heavy metals in QDs, the application of organic fluorescent dyes and QDs in the food business is limited (Ngafwan et al., 2022). Researchers have created biosensors based on nanotechnology to detect heavy metals in food to get around these difficulties. To improve sensing performance and create integrated detection systems, these biosensors make use of hybridized nanomaterials and nanocomposites (Maghsoudi et al., 2021). The biosensor surface platform may considerably increase the linear dynamic range and detection limit for heavy metals by integrating NPs and nanocomposites.

MOFs, in addition to biosensors based on nanomaterials, have shown potential in the chemical detection approach of heavy metals in food and water. Especially luminescent metal–organic frameworks (LMOFs) provide quantitative fluorescence detection and enhanced selectivity for toxic metal ions (Yao et al., 2019). The potential of MOFs goes beyond detection as they can also be used to eradicate heavy metals from aqueous solutions, improving food safety, and protecting the environment.

These innovative approaches make use of nanomaterials like graphene-based nanotechnology (Sharma et al., 2021), CQDs (Tian et al., 2023), and metal-based QDs (Han et al., 2020). These NPs experience certain interactions that lead to the emission of fluorescent signals when they come into contact with a food sample containing heavy metals. The concentration and number of heavy metals present in the food have a direct correlation with the fluorescence intensity. This approach makes it possible to precisely and sensitively detect heavy metal pollutants in food products by measuring the emitted fluorescence, which improves food safety and quality control procedures.

These NPs are ideally suited for sensing ions of heavy metals in food products due to their optical characteristics and minimal toxicity. Forster resonance energy transfer (FRET) is a method that is frequently used for fluorescence-based detection (Han, 2023). The sensor development for detecting ions of heavy metals in food matrices has extensively taken advantage of FRET, which depends on the energy transfer between a fluorescent donor and acceptor. The overlap, i.e., spectral between the acceptor's absorption and the donor's emission as well as the ideal separation between the donor and acceptor are important considerations for FRET-based sensors (Han, 2023). These advances offer sensitive and effective ways for heavy metal detection in food products, which has the potential to improve the safety and quality of food to the dependence on mixing CQDs in aqueous solutions or water, the practical use of CQDs-based fluorescence sensors for on-site detection has been constrained. However, by immobilizing CQDs within a hydrogel matrix and integrating photonic functions, significant attempts have been made to overcome this constraint. Direct detection of Hg^{2+} in water samples has been shown using photonic devices made of hydrogel waveguides doped with CQDs (Guo et al., 2017). Furthermore, Pb^{2+} and Cr^{3+} have been effectively recognized in aqueous environments using graphene QDs, which have demonstrated fluorescence stability for a prolonged length of time (about 4–5 months) (Raj et al., 2021). Notably, the presence of ions of heavy metal can reduce QD fluorescence. Cadmium sulfide (CdS) QDs, for instance, can preferentially interact with Cu^{2+} and, as a result, diminish the fluorescence intensity, a feature used in the investigation of environmental materials. The synthesis of a double-emission ratio fluorescence nanoprobe (RFN) involved three steps: fluorescein isothiocyanate (FITC) doping of SiO_2, coating of FITC-SiO_2 microspheres with amino groups, and surface-based assembly of AuNCs (Han et al., 2018). The RFN showed potential for monitoring heavy metal ions, contaminants, and environmental protection while successfully detecting trace quantities of Hg^{2+} in environmental water samples (Luo et al., 2019). Additionally, under appropriate circumstances, graphene and its derivatives could act as fluorescence donors to produce fluorescence signals as well as energy receptors to satisfy a variety of fluorophores (Fu et al., 2012). The highly effective quenching ability of AuNPs allows for the use of AuNPs-functionalized graphene as a highly efficient fluorescent detecting technique for Pb^{2+} in natural water samples.

The advantages of nanotechnology-based fluorescence approaches include excellent sensitivity, selectivity, and the potential for real-time or quick detection of heavy metals in food samples. Fluorescent technology based on nanotechnology also offers the multiplexing capacity, and improved detection capabilities for heavy metal analysis. Fluorescent-based analyses provide more quantitative precision compared to colorimetric techniques, making them ideal for applications requiring precise concentration measurements. These characteristics make it a very valuable and promising method for finding heavy metals in food products. Nanotechnology-based fluorescence approaches are very beneficial for maintaining food safety and quality control in a variety of industries because of their potent characteristics. By making use of these benefits, this technology significantly contributes to protecting public health and reducing the contamination of heavy metals in the food supply.

7.3.4 Other Nanomaterial-Based Methodologies

Upconversion nanoparticles (UCNPs) have garnered significant attention as fluorescent biomarkers because of their advantageous optical and chemical characteristics, such as low toxicity, large stokes shift, photostability, and scintillation, among others (Lim et al., 2006). Lanthanide-based UCNPs can radiate short-wavelength light in distinct colors when excited by near-infrared light, leading to improved sensitivity by avoiding autofluorescence. A "turn-on" nanosensor was formulated using FRET between long-strand aptamer-functionalized UCNPs and short-strand aptamer-functionalized AuNPs. During the non-existence of Hg^{2+} ions, FRET between the UCNPs and AuNPs occurs due to the particular matching between the two aptamers, following to fluorescence quenching of the UCNPs. This technique has been successfully applied to detect Hg^{2+} ions in tap water and milk samples with high accuracy and selectivity (Liu et al., 2018; Gong et al., 2021). Metal NPs have been employed both as electrochemical platforms' modifiers as well as amplifying tags for innovative biosensing systems, particularly those based on DNA, for the detection of heavy metals such as Hg^{2+} and Pb^{2+}(Miao et al., 2009; Shen et al., 2008). DNA probes are utilized for heavy metal detection based on two renowned phenomena: DNA hybridization and the affinity of heavy metal cations for thymine–thymine mismatches in complementary sequences (Miyake et al., 2006). Additionally, DNAzymes, which can be specifically catalyzed by metal ions, have also been extensively used in this context (Willner et al., 2008; Aragay & Merkoci, 2012). Researchers have increasingly shown interest in using natural adsorbents, including inorganic materials and minerals, as electrochemical surfaces for the detection of heavy metal ions. One such material is nanosized hydroxyapatite (NHAP), $Ca_{10}(PO_4)_6(OH)_2$, which is a bio-ceramic similar to the mineral component of bones. NHAP offers excellent biocompatibility and possesses unique multi-adsorbing sites, making it a suitable candidate for constructing three-dimensional network structures in electrochemical sensors for heavy metals (Aragay & Merkoci, 2012).

7.4 NANOMATERIALS IN SENSOR DEVELOPMENT

Trace heavy metal ions have been detected using a variety of techniques over the years, including atomic emission spectroscopy (AES), inductively coupled plasma mass spectrometry (ICP-MS), chromatography, ICP-AES, and atomic absorption spectroscopy (AAS) (Shrivastava et al., 2023). These conventional techniques are successful, but because of their time-consuming nature, high cost, and intricate processes, they cannot be used widely. Detecting several metal ions simultaneously provides additional challenges. Therefore, there is a need for the development of innovative sensors that can provide quick, reliable, sensitive, cost-effective, and practical analysis of different heavy metal atoms. While some innovative methods, viz. colorimetry, mass spectrometry (MS), electrochemistry, surface-enhanced Raman scattering (SERS), and fluorescence, offer low detection limits and precise measurements of the number of heavy metals in food, they have constraints like expensive equipment, time-consuming sample preparation and pre-treatment, high operational rates, and more detection time (Gong et al., 2021). As a result, these techniques do not fully satisfy the requirements for inexpensive, quick, and on-site measurements and are best suited for laboratory analysis. In view of these challenges, recent technological breakthroughs offer great prospects for developing new detection technologies for detecting trace metal ions more efficiently.

7.4.1 Optical Sensors

Optical chemical sensors (OCSs) offer immense potential for on-site identification of heavy metals, falling under the category of chemical sensors that use electromagnetic radiation to produce an analytical signal in a transduction element. These sensors function by detecting variations in specific optical properties (e.g., absorption, transmission, emission, lifetime) caused by the involvement of the sample with radiation when an immobilized indicator (organic dye) binds to the analyte (Gruber et al., 2017). This interaction can result in variations in the optical properties of nanomaterial-based chemical sensors through processes like target-induced aggregation or anti-aggregation and surface modifications of nanomaterials. Designing effective OCSs depends on the careful selection of suitable solid support (matrix), appropriate functionalization, material morphology, and immobilization techniques (Ding et al., 2016). The review primarily focuses on three classes of optical sensors: fluorescent sensors, SERS sensors, and surface plasmon resonance (SPR) sensors (Ullah et al., 2018).

7.4.1.1 Fluorescent Sensors

Fluorescent sensors were designed to detect analytes based on the changes induced in the physicochemical characteristics of a fluorophore. These changes typically involve alterations in the fluorescence intensity, fluorescence lifetime, and fluorescence anisotropy, which are closely linked to charge transfer and energy transfer procedures. The core components of a fluorescence nanosensor consist of a fluorescent element responsible for signaling the binding event and a receptor element designed for specific binding to the target analyte. By combining these elements, fluorescence nanosensors can provide highly sensitive and selective identification of different analytes (Ullah et al., 2018). FRET is an optical phenomenon where energy is transferred from a donor fluorophore to an acceptor fluorophore through a non-radiative pathway. In recent years, significant attempts have been made to utilize nanomaterials as FRET sensors. Compared to organic fluorophores, QDs exhibit superior fluorescence intensity, broader excitation and emission ranges, and greater photostability (Gong et al., 2021). QDs have gained popularity as an alternative to dye-based fluorophores for their exceptional optical properties, viz. a broad array of absorption spectra, narrow tunable emissions, high quantum yield, and resistance to photobleaching. CQDs possess stronger fluorescence properties and decreased toxicity compared to traditional QDs (Gong et al., 2021). QDs like CdSe, CdS, and PbS contain toxic elements, making them less desirable for certain applications. In contrast, graphene oxide (GO) is non-toxic and exhibits carboxylic acid groups on its surface, which allows for easy covalent attachment of amine-functionalized DNA and other biomolecules. GO also serves as an efficient and cost-effective fluorescence quencher, making it an attractive alternative to QDs in metal ion sensing. GO's water solubility and surface functionalization capabilities, facilitated by oxygen-containing groups, have allowed for fine-tuning of its fluorescence properties, leading to its increased use in heavy metal ion sensing applications (Ullah et al., 2018).

Graphene and its derivatives possess dual capabilities in fluorescence sensing. They can act as energy receptors to suppress distinct fluorophores and also as fluorescence donors to generate fluorescence signals under specific conditions. AuNPs-functionalized graphene serves as an efficient fluorescent sensing platform for detecting Pb^{2+} in environmental water samples (Gong et al., 2021). Graphene oxide (GO) contains carboxyl groups, hydroxyl groups, and epoxy groups that generally result in non-emission and non-radiative recombination of local electron–hole pairs. By covalently grafting allylamine on the surface of GO, a fluorescence nanosensor can be designed, which shows highly sensitive and selective distinction of Fe^{3+} from Fe^{2+} and other metal ions via electron transfer-induced fluorescence quenching (Gong et al., 2021). Similarly, GO functionalized with 8-hydroxyquinoline (8-HQ) has been formulated for fluorescence identification of Zn^{2+}. In the current research, researchers covalently grafted the surface of mesoporous silica nanoparticles (MSNPs) with an ethylenediamine derivative of a rhodamine 6G silica particle (RSSP). This hybrid material was utilized as a fluorescent sensor for the selective identification of Fe^{3+} in water (Ullah et al., 2018).

7.4.1.2 SERS Sensors

Plasmonic nanostructures like AuNPs and AgNPs exhibit coherent oscillations of their conduction electrons when illuminated with specific wavelengths of light, resulting in a greatly enhanced electromagnetic field called localized surface plasmon resonance (LSPR). This LSPR effect, which depends on the accumulation state, type, size, and shape of the plasmonic NPs, as well as the refractive index of the surrounding medium, causes changes in the color of the colloid suspension. LSPR, acting as a highly sensitive optical transducer, can be utilized in combination with SERS to amplify the Raman scattering signal up to 10,000 times. SERS allows for the ultrasensitive identification of analytes by significantly increasing the Raman scattering cross-section of molecules closely associated with noble metal nanostructures (Ullah et al., 2018). The creation of "hot spots" for SERS enhancement occurs when plasmonic fields of neighboring NPs couple, facilitated by the aggregation of metal NPs. As an ultrasensitive sensing method, SERS offers advantages over fluorescent methods, including narrower peak widths in the Raman spectra, the capability to identify single molecules, differentiation of distinct molecular orientations, and distinguishing highly similar molecules. SERS has proven to be a powerful tool for biosensing, overcoming limitations related to fluorescence-based methods, like photobleaching of organic dyes and quenching instigated by complex sample matrices. However, a drawback of SERS sensing using nanostructures is the potential detachment of these molecules from metallic NPs, leading to poor reproducibility. To address this issue, researchers are actively exploring functionalization approaches for nanostructures, which has garnered significant interest (Ullah et al., 2018).

7.4.1.3 Plasmonic Sensors

When noble metal NPs are exposed to incident electromagnetic radiation, their free conduction electrons undergo oscillations, resulting in SPR. For instance, 20 nm-sized monodispersed AuNPs in an aqueous solution possess a "red wine" color due to a strong SPR peak in their absorption spectrum. However, if these NPs aggregate strongly, they may lose their characteristic nano-regime properties, causing a shift in the SPR peak and resulting in a color change of the solution. This principle forms the basis for the development of colorimetric sensors used for detecting different analytes. Yang et al., (2017) developed a colorimetric sensor based on 15-nm-diameter AuNPs for highly specific and sensitive detection of Cu^{2+}, Hg^{2+}, and Ag^{+} ions in water. When metal ions are not present, the addition of o-phenylenediamine (OPDA) to AuNPs caused their accumulation through NH–gold chemistry, leading to a color variation from red to blue. However, when Hg^{2+}, Cu^{2+}, and Ag^{+} ions were introduced, they specifically reduced the AuNPs, converting them into metal NPs. These metal NPs initiated catalytic redox reactions between the metal ions and OPDA, resulting in the oxidation of OPDA and the generation of 2,3-diaminophenazine (OPDAox) and polymer (poly-OPDA). Consequently, the original red color of AuNPs was retained due to the insufficient amount of OPDA remaining in the solution (Ullah et al., 2018).

7.4.2 Colorimetric Sensors

Colorimetric analysis is a widely utilized technique to detect the existence of chemical contaminants, involving heavy metals, based on the color variation of a dye. This method is often combined with nanotechnology to improve its sensitivity and accuracy, particularly in the food industry. Metal NPs are commonly utilized for detecting heavy metals in this context. The color variation in the solution depends on the size of the metal NPs. For example, monodispersed AuNPs with a diameter smaller than 30 nm exhibit a red color in an aqueous solution (Hyder et al., 2022). When heavy metals are present, they accelerate the accumulation of AuNPs, causing a red shift in the localized surface plasmon resonance band, resulting in a noticeable color change in the solution (Gong et al., 2021; Tharmaraj & Yang, 2014). Similarly, AgNPs and CuNPs are also utilized for colorimetric detection due to their plasmonic properties similar to AuNPs (Shrivastava et al., 2022; Zhu et al., 2012). Unaggregated AgNPs with sizes smaller than 30 nm appear yellow in solution, while CuNPs with sizes varying from 10 to 20 nm display a red color (González et al., 2014; Markin & Markina, 2019; Nam et al., 2022).

In a research conducted by Yang et al. (2017), a new approach for the triple detection of Hg^{2+}, Cu^{2+}, and Ag^{+} was introduced. The concept involved the use of AuNPs coupled with OPDA. This combination of AuNPs and OPDA demonstrated the capability to simultaneously detect and differentiate Hg^{2+}, Cu^{2+}, and Ag^{+} ions in a sensitive and selective manner, providing a promising platform for multi-metal ion detection (Nam et al., 2022). AuNPs are equipped with localized surface plasmon resonances (LSPR), which enable the analysis of certain heavy metal ions based on the dispersion of AuNPs. A sensor for detecting Hg^{2+} was constructed using 4-mercaptophenylboronic acid (MPBA) and AuNPs. Initially, the AuNPs are secured by a coating of citrate ions (-negatively charged), which cause electrostatic repulsion to prevent particle accumulation (Gong et al., 2021). MPBA acts as an aggregation agent and possesses a greater attraction for AuNPs through Au–S crosslinking. As a result, the solution of AuNPs changes color from red to blue. However, in the presence of increasing concentrations of Hg^{2+}, the thiol group of MPBA prefers to bind with Hg^{2+} rather than polymerize with AuNPs. This preference results in the solution's color shifting back from blue to red (Gong et al., 2021). These Hg^{2+} nano-sensors demonstrate high detection specificity and have been successfully applied to test water samples (Gong et al., 2021). The use of AuNPs in conjunction with different materials, such as CQDs and graphene oxide, demonstrates their versatility in detecting and analyzing heavy metal ions in various environmental samples. For detecting Cu^{2+}, AgNPs functionalized with thiomalic acid (TMA) have been employed. In the existence of Cu^{2+}, carboxylic acid groups in TMA favorably network with Cu^{2+} to form stable complexes, leading to the aggregation of AgNPs and causing a red shift in the localized surface plasmon resonance (LSPR) band. These sensors have been successfully applied for real-time monitoring of Cu^{2+} in different environmental water samples (Gong et al., 2021). These modified AgNPs demonstrate great potential for detecting heavy metal ions with high sensitivity and selectivity in various environmental samples.

7.4.3 Biosensors

Biosensing approaches based on nanotechnology offer a quick and simple monitoring procedure that enables quick action to prevent eating contaminated food and reduce health risks (Maghsoudi et al., 2021). Additionally, many heavy metals can be detected simultaneously using this approach. It is feasible to conduct a thorough analysis of the existence of heavy metals in food. The design of the nanotechnology-based biosensing technique makes it feasible to simultaneously detect and quantify a number of heavy metals in a single test, allowing for a thorough assessment of heavy metal contamination in food (Maghsoudi et al., 2021).

Nanobiosensors, which are analytical devices used to detect biological products, are at the forefront of revolutionizing the healthcare industry, forensic medicine, homeland security, and environmental protection. They have also found applications in the food and drink industry for remote sensing of water quality and the determination of drug residue in food. For environmental protection, nanobiosensors are utilized in detecting pesticides, and heavy metal ions in river water, and conducting genome analysis of organisms. The use of nanomaterials, like gold nanoparticles, magnetic nanoparticles, carbon nanotubes, and QDs, has played a critical role in enhancing the performance and sensitivity of biosensors. These nanomaterials have been extensively researched for their applications in biosensors, bridging the gap between biological detection and material science (Pandit et al., 2016). Biosensors, as defined by IUPAC (International Union of Pure and Applied Chemistry), are self-contained integrated devices designed to give particular semi-quantitative or quantitative analytical information utilizing a biological recognition component in direct spatial connection with a transducer element. Nanomaterials-based biosensors represent a cutting-edge technology that integrates material science, molecular engineering, chemistry, and biotechnology. These biosensors have the potential to significantly enhance the sensitivity and specificity of biomolecule identification due to the exceptional properties of nanomaterials. The application of nanobiosensors has revolutionized chemical and biological analysis, enabling speedy detection of multiple substances in vivo. Magnetic nanomaterials, carbon materials, oxide nanoparticles, and metallophthalocyanines are among the nanomaterials utilized to improve the electrochemical signals in biosensors. Functional nanoparticles can bind to biological molecules like proteins, peptides, and nucleic acids, allowing for specific biomolecular detection in biosensors (Pandit et al., 2016).

Nanostructured thin films offer exciting opportunities to fabricate electrochemical sensors and biosensors with enhanced detection capabilities. One of the key advantages of nanostructured thin films is their ability to significantly improve the detection limit in biosensing devices. This can be achieved by incorporating compatible materials, including natural polymers, which contribute to both high detection sensitivity and the preservation of the structural integrity and biocatalytic activity of biomolecules (Pandit et al., 2016). The main objective behind utilizing these materials is to combine the enhanced detection capabilities of nanoscale films while ensuring that the biomolecules remain intact and active, enabling accurate and reliable biosensing.

Heavy metals, particularly bisphenols, may be easily found in food samples because of the development of biosensors based on nanotechnology. MOFs in conjunction with the tyrosinase enzyme are one potential strategy (Zhou et al., 2022). MOFs are highly porous materials with customizable characteristics that are perfect for selectively trapping particular heavy metal ions. However, the tyrosinase enzyme has the ability to catalyze the oxidation of phenolic substances, including bisphenols, producing a signal that can be measured (Zhou et al., 2022). Functionalizing MOFs to enable targeted ions of heavy metal binding is the first step in the construction of bionanosensors. The ions of heavy metal bond to the surface of the MOFs in a precise way when a food sample containing heavy metals is added to the biosensor. This change in certain characteristics is first assessed and quantified to determine the levels of heavy metal in the food sample (Xiaoyu et al., 2022). Heavy metals in food may be quickly and accurately detected using biologically based nanotechnology techniques.

The cost of biologically based nanotechnology for heavy metal detection in food could hinder its broad application, especially in resource-constrained settings. Furthermore, the complexity of bio-nanotechnology-based detection methods could bring challenges. Designing and constructing biological sensors that properly integrate nanomaterials with biological recognition components demands specialized knowledge in both nanotechnology and biotechnology (Yang et al., 2019). These issues might hinder the widespread application of such technologies in specific locations and industries due to their resource and experience limits.

7.5 APPLICATIONS OF NANOSENSORSIN FOOD PROCESSING AND PACKAGING INDUSTRIES

Nanotechnology has shown promise in the food industry, with a range of uses including the detection of heavy metals in food processing and packaging (He & Hwang, 2016). Its application has the potential to improve nutritional delivery and flavor, promote food security, prolong shelf life, and detect illnesses, pollutants, and pesticides (He & Hwang, 2016). Inorganic metal nanoparticles, metal oxides, and nano-organic materials with bioactive substances are a few examples of nanostructured materials (NSMs) that can be used in food systems (Bajpai et al., 2018). These NSMs can be included in the materials used for food packaging, creating functional and intelligent systems. The possibility of detecting heavy metals has been made possible by the use of nanoparticles in food packaging. One application of carbon nanotubes is the development of packaging materials that can detect microorganisms, dangerous proteins, and food spoilage. Similarly, metal nanoparticles, carbon nanotubes, and QDs can serve as the building blocks for biosensors that assess bacteria and conduct tests for the safety of food (Bajpai et al., 2018). To ensure the safety and quality of food items, heavy metals in food must be detected using these nanosensors. The use of nanotechnology in food production and packaging can assist the industry by improving monitoring capabilities and enhancing customer safety.

Vilela et al. (2016) used graphene oxide-based microbots (GOx-microbots) for the efficient arrest, transfer, and elimination of heavy metals, especially lead, from water. These GOx-microbots are active self-propelled devices that remove heavy metal contaminants in a highly effective manner, allowing for their collection for recycling in the process. As promising nanomaterials with exceptional adsorption characteristics for heavy metals removal from aqueous solutions, graphene and its composites have also come to light (Vilela et al., 2016). These nanoparticles have demonstrated strong adsorption capacities and can be used in sensing systems for the proper detection of heavy metals in liquids. Due to their adaptability and effectiveness, these nanomaterials hold enormous promise for reducing heavy metals in polluted water bodies and fostering environmental sustainability. Laser-induced breakdown spectroscopy (LIBS) has been demonstrated to be effective for the analysis of heavy metals in natural water. However, problems, such as droplet splashing and laser energy degradation are encountered during the detection of water samples utilizing LIBS (You et al., 2023). A unique method based on agarose films has been suggested for heavy metals detection sensitively to overcome difficulties (You et al., 2023). A promising approach to improve the precision and dependability of heavy metal detection using LIBS in water samples is provided by this innovative liquid-solid conversion technique.

Fluorinated graphene/gold nanocage (FGP/AuNC) is an entirely novel nanocomposite that was created for the simultaneous measurement of numerous heavy metals utilizing square wave anodic stripping voltammetry (Tan et al., 2020). The Zn^{2+}, Cd^{2+}, Pb^{2+}, Cu^{2+}, and Hg^{2+} in real food samples, including peanut, rape bolt, and tea, were found using the FGP/AuNC electrode. This technique was effective in precisely estimating the amount of heavy metal ions present in agricultural food products. As an electrode, FGP/AuNC nanocomposite offers a dependable and effective method for the sensitive detection of heavy metals, making it an important instrument for food safety analysis and quality control (Tan et al., 2020).

To accurately detect lead in leafy vegetables (spinach) at extremely low levels (parts-per-trillion, ppt), an innovative and highly sensitive electrochemical sensor has been created (Zhang et al., 2020). A porphyrin-functionalized metal-organic framework (porph@MOF) and a DNAzyme that is Pb^{2+}-dependent are both essential components of this novel sensor. The porph@MOFhas shown outstanding peroxidase-mimicking properties as well as impressive stability over a broad temperature and pH range. The detection limit for Pb^{2+} was established to be a remarkable 5 pM (equal to 1 ppt by weight) using porph@MOF as the mimic catalyst in the sensing system. With this level of sensitivity, the sensor might be used to regularly check for lead contamination in leafy vegetables (Chinese cabbage, spinach, etc.), providing an important tool for protecting the environment and food safety.

The simultaneous detection of Ag^{+} and Hg^{2+} in a variety of liquids, i.e., drinking water, red wine, lake water, and orange juice was achieved by developing a novel electrochemical biosensor using DNA-modified Fe_3O_4@Au magnetic nanoparticles (Miao et al., 2017). Across a range of liquid matrices, this adaptable biosensor demonstrated good performance, making it a trustworthy instrument for detecting heavy metal contamination in a variety of drinks and water sources. For the identification and quantification of Hg^{2+} and iodide in various water samples (mineral, tap, and river water), as well as fish samples (canned), fluorescence-based carbon dots were used (Tabaraki & Sadeghinejad, 2018). Iodide and mercury ions each have a limit of detection of 74 and 63 nM, respectively, by this approach of detection. In particular, for chromium, copper, and cobalt ions in very low concentrations from polluted water, metal oxide, and magnetic nanoparticles have emerged as advantageous methods for finding and removing harmful ions of heavy metal (Gong et al., 2021).

The levels of cadmium and lead in milk samples were simultaneously measured using a new bismuth oxide (in nanoparticles) modified carbon composite electrode and the ionic liquid n-octyl pyridinium hexafluorophosphate (Ping et al., 2012). With excellent detection limits of 0.15 and 0.21 g L^{-1} for cadmium and lead, respectively, the composite electrode demonstrated 3.0 to 30.0 g L^{-1} for both cadmium and lead ions, a wider linear response range. Further testing of the electrode's effectiveness in identifying trace metal ions in milk samples and other liquid samples produced acceptable findings. The mercuric ions in milk samples were detected by graphitic carbon nitrides nanosheets up to 0.3 nM concentration limit by fluorescence detection technique (Zhuang et al., 2017). The detection of lead and cadmium in newborn milk powder was carried out electrochemically using a carbon (nano porous) based screen-printed electrode (Chen et al., 2023). Due to its substantial adsorption capacity and superior mass transport, the lower limit of detection was achieved for ions of cadmium and lead.

The identification of heavy metals in various water and food samples has been mostly accomplished using optical sensors, with DNA-functionalized gold nanoparticles (DNA-AuNPs) serving as a common platform. These DNA-AuNPs have been used to find Hg^{2+} in aqueous and meat samples, among other heavy metals. These nano-based optical sensors have a limit of detection (LOD) of 100 nM and are frequently used to detect Hg^{2+} in agricultural and food items (Mustafa & Andreescu, 2020). Detecting a variety of metal ions in food, such as Hg^{2+}, Cu^{2+}, and Fe^{3+}, is another application for optical sensors. The fluorescent sensor responds selectively and differently to various metal ions, enabling simultaneous detection of all ions of heavy metals. According to Maghsoudi et al. (2021), optical biosensors have successfully detected arsenic with a remarkable detection limit of 6 ppb at concentrations ranging from 50 to 700 ppb in aqueous meals.

Food packaging labels made with nanotechnology have a promising future in detecting heavy metals. Carbon nanotubes (CNTs), for instance, have been used in ultrasensitive electrical biosensing, presenting a method to amplify enzyme-linked electrical detection of proteins and DNA (Han et al., 2020). By adding certain receptors or ligands for heavy metal ions to the CNTs, this method has been modified to detect heavy metals in food packaging. Another strategy is to use labels made of nanoparticles on food packaging. For example, it was possible to add ligands or receptors that preferentially bind to heavy metal ions to AuNPs. Colorimetric or fluorescence-based assays are used to identify this binding event. These nanoparticles have been used successfully as labels in food packaging to identify heavy metals. In addition, some nanomaterials, such as those based on graphene, have promising electrical and optical characteristics that make them adequate for use in sensing applications in the food packaging sector (Sood & Sharma, 2019). To selectively attach to heavy metal ions and provide a quantifiable signal, graphene can be functionalized with certain receptors or ligands. By identifying and preventing heavy metal contamination in packaged food goods, these nanotechnology-based labels provide creative ways to ensure quality and safety of food. These labels can be specifically targeted to heavy metal ions and offer a visual or electronic indicator of their presence in food packaging.

7.6 LIMITATIONS OF NANOMATERIAL-BASED SENSORS IN DETECTION

The heavy metal detection in food has shown potential for sensors based on nanomaterials. There are a few constraints, though, that must be taken into account.

a. *Sensitivity of nanomaterial-based sensors*: The sensitivity of the sensors may not be sufficient to detect trace quantities of heavy metals, despite the fact that nanomaterials like carbon nanotubes (CNTs) and graphene offer greater surface area and signal augmentation capabilities. In comparison to nanomaterial-based sensors, conventional methods like liquid chromatography and atomic absorption spectrometry have higher sensitivity (Chrouda, 2023). To ensure accurate heavy metal detection in food, it is crucial to address this constraint and strive towards improving the sensitivity of nanomaterial-based sensors.
b. *Specificity and selectivity of sensors*: Contrary to conventional approaches, nanomaterial-based sensors may have difficulty differentiating between various heavy metal complexes. This distinction is crucial as the molecule affects how dangerous certain heavy metals are. The overall accuracy and reliability of heavy metal detection could be compromised by false-positive or false-negative results caused by this interference. Consequently, the lack of compound specificity in nanomaterial-based sensors may restrict their capacity to precisely assess the possible health concerns related to heavy metal contamination in food (Yoshii et al., 2023).
c. *Stability and durability of sensors*: As nanomaterials are susceptible to deterioration and fouling, the functioning and reliability of the sensors may be affected over time. The practical application of sensors based on nanomaterial for heavy metal detection in food depends on ensuring the long-term stability and resilience of these devices (Han et al., 2020).
d. *Detection range*: To detect heavy metals at very low or high concentrations, nanomaterial-based sensors may have a limited detection range. This drawback can make it difficult to use them in circumstances when exact measurements of heavy metals are necessary (Gong et al., 2021).
e. *Complexity in sensor fabrication and sample preparation*: Nanomaterial-based sensors may be constrained by their complexity and manufacturing procedure. Some of these sensors might need complex fabrication processes, which limits their scalability and usefulness for point-of-care diagnostics (Rao et al., 2022). Sample preparation processes for some nanomaterial-based sensors can be complex and time-consuming, raising the possibility of sample contamination. The complexity of sample preparation may limit the usefulness and effectiveness of the sensors.
f. *Toxicity concerns*: Some nanomaterials are harmful, especially in large doses or under certain situations. The health of the customer may be in danger if these nanoparticles penetrate or release heavy metals or toxins into the food product (Waheed et al., 2018).
g. *Cost*: Nanomaterial-based sensors may not be widely used because of their high production and implementation costs, which is especially true in areas with limited resources.

7.7 SAFETY AND REGULATIONS

Strict safety rules apply to the use of nano-based sensors for heavy metal's detection in food. Prior to going on the market, these sensors go through extensive risk analyses to determine any potential dangers (He & Hawang, 2016; Ko, 2015). To ensure that the nanomaterials employed in the sensors do not contaminate food or pose health problems, they must adhere to standards on food contact materials. For customers to be aware of the use of nanomaterials in food packaging or other items, accurate labeling and transparency are necessary (Gong et al., 2021). For testing, approval, and evaluation, standardization and guidelines have been established to guarantee uniform safety assessments. Environmental impact evaluations may be carried out to analyze any potential environmental release, whereas toxicology and biocompatibility studies evaluate the safety of nanomaterials (Deka et al., 2023). International safety standard standardization makes it easier to adopt a single strategy, and post-market surveillance tracks safety performance after products enter the market.

Safety rules are essential for the use of nano-based sensors for identifying heavy metals in food to protect consumers and ensure food safety. The Ministry of Food and Drug Safety (MFDS) actively participates in the development of safety management and assessment frameworks for the use of nanomaterials in medicines, food, drinks, and packaging (Ko, 2015). The MFDS is currently investigating nano-safety in food and food packaging and has plans to quickly develop new safety standards and guidelines pertaining to nano-food packaging (Ko, 2015).

The possible dangers related to the use of nanoparticles in food applications must be taken into account in addition to regulatory initiatives. The usage of some nanomaterials in manufacturing, packaging, and as food additives may have different effects on people and animals. Regarding the production, processing, packaging, and consumption of nanotechnology in the food industry, safety issues must be addressed and regulatory laws must be put in place (He & Hwang, 2016). The effective use of nanotechnology for the detection of heavy metals in food while preserving the health and well-being of the general public can be achieved by carefully addressing safety and regulatory issues.

7.8 CONCLUSIONS

Metal ions are prominent entities that are found almost everywhere, in water, soil, air, and food. Due to their non-biodegradability nature, prolonged half-life, and potential buildup in the body, they can be reasons for numerous hazardous health risks. The abundance of heavy metals thus, becomes a serious threat to humans and a global alarm for the environment. Heavy metals tend to be accumulated at any stage of the food chain and their presence in daily consumed food even at very low concentrations can be toxic which is a big concern for human health. Raw, untreated, or minimally processed foods are known to be good sources of vitamins and nutrients and are the preferred forms for a healthy diet, so ensuring their quality composition and safety becomes the highest priority. In regard to heavy metal detection, nanotechnology-based approaches provide adaptable and efficient solutions to enhance food safety, quality assurance, and consumer protection in the food processing and packaging industries. The successful application of nanomaterial-based sensors in food safety applications will depend on overcoming the constraints, including enhancing specificity, extending the detection range, ensuring stability, streamlining sample preparation, cutting costs, minimizing interference, establishing standard protocols, and addressing toxicity issues.

REFERENCES

Abd El-Rahman, H.S.M., El-Dakak, A.M.N.H., & Zein, H. (2015). Investigation and evaluation on heavy metal contaminations of green salads and potato fried in different restaurants and fresh vegetables in some Egyptian governorates. *International Journal of Environmental Monitoring and Analysis*, *3*(2), 28–37. https://doi.org/10.11648/j.ijema.20150302.11

Abdelhamid, A. M., Gomaa, A. H., & El-Sayed, H. G. M. (2013). Studies on some heavy metals in the river Nile water and fish at Helwan area, Egypt. *Egyptian Journal of Aquatic Biology and Fisheries*, *17*(2), 105–126.

Abdel-Mohsien, H. S., & Mahmoud, M. A. M. (2015). Accumulation of some heavy metals in *Oreochromis niloticus*from the Nile in Egypt: potential hazards to fish and consumers. *Journal of Environmental Protection*, *6*(9), 1003–1013. https://doi.org/10.4236/jep.2015.69089

Ahmed, A. A. H., Mohammed, E. E. P., Yahia, D., & Faried, A. S. M. (2017). Lead, cadmium and copper levels in table eggs. *Journal of Advanced Veterinary Research*, *7*(3), 66–70.

Ahmed, A. S., Sultana, S., Habib, A., Ullah, H., Musa, N., Hossain, M. B., Rahman, M., & Sarker, M. S. I. (2019). Bioaccumulation of heavy metals in some commercially important fishes from a tropical river estuary suggests higher potential health risk in children than adults. *Public Library of Science (PLoS) One, 14*(10), e0219336. https://doi.org/10.1371/journal.pone.0219336

Al-Ashmawy, M. A. M. (2013). Trace elements residues in the table eggs rolling in the Mansoura City markets Egypt. *International Food Research Journal*, *20*(4), 1783–1787.

Ali, H., & Khan, E. (2019). Trophic transfer, bioaccumulation, and biomagnification of non-essential hazardous heavy metals and metalloids in food chains/webs – concepts and implications for wildlife and human health. *Human and Ecological Risk Assessment: An International Journal, 25*(6), 1353–1376. https://doi.org/10.1080/10807039.2018.1469398

Aragay, G., & Merkoci, A. (2012). Nanomaterials application in electrochemical detection of heavy metals. *Electrochimica Acta,* 84, 49–61. https://doi.org/10.1016/j.electacta.2012.04.044

Bajpai, V., Kamle, M., Shukla, S., Mahato, D., Chandra, P., Hwang, S., & Yu, J. (2018). Prospects of using nanotechnology for food preservation, safety, and security. *Journal of Food and Drug Analysis, 4*(26), 1201–1214. https://doi.org/10.1016/j.jfda.2018.06.011

Caggiano, R., Sabia, S., D'Emilio, M., Macchiato, M., Anastasio, A., Ragosta, M., & Paino, S. (2005). Metal levels in fodder, milk, dairy products, and tissues sampled in ovine farms of Southern Italy. *Environmental Research, 99*(1), 48–57. https://doi.org/10.1016/j.envres.2004.11.002

Cashman, K.D. (2011). Milk salts, trace elements, nutritional significance. In *Encyclopedia of Dairy Science,* 2nd edition, Vol. 3, ed. by Fuquay, J. W., Fox, P. F. & McSweeney, P. L. H. Academic Press, pp. 933–940.

Chen, H., Yao, Y., Zhang, C., & Ping, J. (2023). Determination of heavy metal ions in infant milk powder using a nanoporous carbon modified disposable sensor. *Foods, 12*(4), 730. https://doi.org/10.3390/foods12040730

Chrouda, A. (2023). A novel electrochemical sensor based on sodium alginate-decorated single-walled carbon nanotubes for the direct electrocatalysis of heavy metals ions. *Polymers for Advanced Technologies, 6*(34), 1807–1816. https://doi.org/10.1002/pat.6002

Collado-López, S., Betanzos-Robledo, L., Téllez-Rojo, M. M., Lamadrid-Figueroa, H., Reyes, M., Ríos, C., & Cantoral, A. (2022). Heavy metals in unprocessed or minimally processed foods consumed by humans worldwide: a scoping review. *International Journal of Environmental Research and Public Health, 19*(14), 8651. https://doi.org/10.3390/ijerph19148651

Dai, X., Wu, S., & Li, S. (2018). Progress on electrochemical sensors for the determination of heavy metal ions from contaminated water. *Journal of the Chinese Advanced Materials Society, 6,* 91–111. https://doi.org/10.1080/22243682.2018.1425904

Darwish, W. S., Hussein, M. A., El-Desoky, K. I., Ikenaka, Y., Nakayama, S., Mizukawa, H., & Ishizuka, M. (2015). Incidence and public health risk assessment of toxic metal residues (cadmium and lead) in Egyptian cattle and sheep meats. *International Food Research Journal, 22*(4).

Darwish, W. S., Morshdy, A. E., Ikenaka, Y., Ibrahim, Z. S., Fujita, S., & Ishizuka, M. (2010). Expression and sequence of CYP1A1 in the camel. *Journal of Veterinary Medical Science, 72*(2), 221–224. https://doi.org/10.1292/jvms.09-0319

Deka, A., Kumar, K., & Basumatary, S. (2023). Monitoring strategies for heavy metals in foods and beverages: limitations for human health risks. In Heavy Metals – Recent Advances, ed by Almayyahi, B. A. https://doi.org/10.5772/intechopen.110542

De La Rica, R., Mendoza, E., & Matsui, H. (2010). Bioinspired target-specific crystallization on peptide nanotubes for ultrasensitive Pb ion detection. *Small, 6*(16), 1753–1756.

Ding, Q., Li, C., Wang, H. J., Xu, C. L., & Kuang, H. (2021). Electrochemical detection of heavy metal ions in water. *Chemical Communications, 57*(59), 7215. http://dx.doi.org/10.1039/D1CC00983D

Ding, Y., Wang, S., Li, J., & Chen, L. (2016). Nanomaterial-based optical sensors for mercury ions. *TrAC Trends in Analytical Chemistry, 82,* 175–190. https://doi.org/10.1002/smll.201000489

Eissa, M. A., & Negim, O. E. (2018). Heavy metals uptake and translocation by lettuce and spinach grown on a metal-contaminated soil. *Journal of Soil Science and Plant Nutrition, 18*(4), 1097–1107. http://dx.doi.org/10.4067/S0718-95162018005003101

El-Hassanin, A. S., Samak, M. R., Abdel-Rahman, G. N., Abu-Sree, Y. H., & Saleh, E. M. (2020). Risk assessment of human exposure to lead and cadmium in maize grains cultivated in soils irrigated either with low-quality water or freshwater. *Toxicology Reports, 7,* 10–15. https://doi.org/10.1016/j.toxrep.2019.11.018

Elliott, S., Frio, A., & Jarman, T. (2017). Heavy metal contamination of animal feedstuffs – a new survey. *Journal of Applied Animal Nutrition, 5,* e8. https://doi.org/10.1017/jan.2017.7

El-Wehedy, S. E., Darwish, W. S., Tharwat, A. E., & Hafez, A. E. (2018). Estimation and health risk assessment of toxic metals and antibiotic residues in meats served at hospitals in Egypt. *Journal of Veterinary Science and Technology, 9*(2). DOI: 10.4172/2157-7579.1000524

Enb, A., Abou Donia, M. A., Abd-Rabou, N. S., Abou-Arab, A. A. K., & El-Senaity, M. H. (2009). Chemical composition of raw milk and heavy metals behaviour during processing of milk products. *Global Veterinaria, 3*(3), 268–275. https://doi.org/10.1039/C0CC04180G

Fang, Y.M., Song, J., Li, J., Wang, Y.W., Yang, H.H., Sun, J.J., & Chen, G.N. (2011). Electrogenerated chemiluminescence from Au nanoclusters. *Chemical Communications, 47*(8), 2369–2371.

Feng, J., Wang, Y., Zhao, J., Zhu, L., Bian, X., & Zhang, W. (2011). Source attributions of heavy metals in rice plant along highway in Eastern China. *Journal of Environmental Sciences, 23*(7), 1158–1164. https://doi.org/10.1016/S1001-0742(10)60529-3

Fu, X., Lou, T., Chen, Z., Lin, M., Feng, W., & Chen, L. (2012). "Turn-on" fluorescence detection of lead ions based on accelerated leaching of gold nanoparticles on the surface of graphene. *ACS Applied Materials & Interfaces, 4*(2), 1080–1086. https://doi.org/10.1021/am201711j

Fu, Z., & Xi, S. (2020). The effects of heavy metals on human metabolism. *Toxicology Mechanisms and Methods, 30*(3), 167–176. https://doi.org/10.1080/15376516.2019.1701594

Gibb, H.J., Barchowsky, A., Bellinger, D., Bolger, P.M., Carrington, C., Havelaar, A.H., Oberoi, S., Zang, Y., O'Leary, K., & Devleesschauwer, B. (2019). Estimates of the 2015 global and regional disease burden from four foodborne metals – arsenic, cadmium, lead and methylmercury. *Environmental Research, 174*, 188–194. https://doi.org/10.1016/j.envres.2018.12.062

Gong, Z., Chan, H. T., Chen, Q., & Chen, H. (2021). Application of nanotechnology in analysis and removal of heavy metals in food and water resources. *Nanomaterials (Basel, Switzerland), 11*(7), 1792. https://doi.org/10.3390/nano11071792

González, A., Noguez, C., Beránek, J., & Barnard, A. (2014). Size, shape, stability, and color of plasmonic silver nanoparticles. *The Journal of Physical Chemistry C, 118*(17), 9128–9136. https://doi.org/10.1021/jp5018168

Gruber, P., Marques, M. P., Szita, N., & Mayr, T. (2017). Integration and application of optical chemical sensors in microbioreactors. *Lab on a Chip, 17*(16), 2693–2712. https://doi.org/10.1039/C7LC00538E

Guo, J., Zhou, M., & Yang, C. (2017). Fluorescent hydrogel waveguide for on-site detection of heavy metal ions. *Scientific Reports, 7*(1), 7902. https://doi.org/10.1038/s41598-017-08353-8

Hamada, M. G., Elbayoumi, Z. H., Khader, R. A., & M Elbagory, A. R. (2018). Assessment of heavy metal concentration in fish meat of wild and farmed *Nile tilapia* (*Oreochromis niloticus*), Egypt. *Alexandria Journal for Veterinary Sciences, 57*(1). https://doi.org/10.5455/ajvs.295019

Han, A., Hao, S., Yang, Y., Li, X., Luo, X., Fang, G., & Wang, S. (2020). Perspective on recent developments of nanomaterial based fluorescent sensors: applications in safety and quality control of food and beverages. *Journal of Food and Drug Analysis, 4*(28), 487–508. https://doi.org/10.38212/2224-6614.1270

Han, B., Li, Y., Hu, X., Yan, Q., Jiang, J., Yu, M., & He, G. (2018). Paper based visual detection of silver ions and l-cysteine with a dual-emissive nano system of carbon quantum dots and gold nanoclusters. *Analytical Methods, 10*(32), 3945–3950. https://doi.org/10.1039/C8AY01044G

Han, Z. (2023). Time-resolved fret-based immunosensor for the ultrasensitive and rapid detection of Cd^{2+}. *Journal of Agricultural and Food Chemistry, 29*(71), 11195–11203. https://doi.org/10.1021/acs.jafc.3c01535

Hashish, S. M., Abdel-Samee, L. D., & Abdel-Wahhab, M. A. (2012). Mineral and heavy metals content in eggs of local hens at different geographic areas in Egypt. *Global Veterinaria, 8*(3), 298–304.

Helmy, N. A., Maarouf, A. A., Hassan, M. A., & Hassanien, F. S. (2018). Detection of heavy metals residues in fish and shellfish. *Benha Veterinary Medical Journal, 34*(2), 255–264.

He, X., Deng, H., & Hwang, H. M. (2019). The current application of nanotechnology in food and agriculture. *Journal of Food and Drug Analysis, 27*(1), 1–21. https://doi.org/10.1016/j.jfda.2018.12.002

He, X. & Hwang, H. (2016). Nanotechnology in food science: functionality, applicability, and safety assessment. *Journal of Food and Drug Analysis, 4*(24), 671–681. https://doi.org/10.1016/j.jfda.2016.06.001

Hoang, H.G., Chiang, C.F., Lin, C., Wu, C.Y., Lee, C.W., Cheruiyot, N.K., Tran, H.T., & Bui, X.T. (2021). Human health risk simulation and assessment of heavy metal contamination in a river affected by industrial activities. *Environmental Pollution, 285*, 117414. https://doi.org/10.1016/j.envpol.2021.117414

Hong, M., Wang, M., Wang, J., Xu, X., & Lin, Z. (2017). Ultrasensitive and selective electrochemical biosensor for detection of mercury (II) ions by nicking endonuclease-assisted target recycling and hybridization chain reaction signal amplification. *Biosensors & Bioelectronics, 94*, 19–23. https://doi.org/10.1016/j.bios.2017.02.031

Hussien, H., & Nosir, S. (2017). Assessment of heavy metals residues in some food stuffs and its biocontrol by probiotic strain *Enterococcus facium* (E980) in an experimental model. *Alexandria Journal for Veterinary Sciences, 52*(1), 87–96.

Hyder, A, Buledi, J. A., Nawaz, M., Rajpar, D. B., Shah, Z.U., Orooji, Y., Yola, M. L., Karimi-Maleh, H., Lin, H., & Solangi, A. R. (2022). Identification of heavy metal ions from aqueous environment through gold, silver and copper nanoparticles: an excellent colorimetric approach. *Environmental Research, 205*, 112475. https://doi.org/10.1016/j.envres.2021.112475

Jin, J., Zhao, X., Zhang, L., Hu, Y., Zhao, J., Tian, J., Ren, J., Lin, K., & Cui, C. (2023). Heavy metals in daily meals and food ingredients in the Yangtze River Delta and their probabilistic health risk assessment. *Science of the Total Environment, 854*, 158713. https://doi.org/10.1016/j.scitotenv.2022.158713

Kabeer, M. S., Hameed, I., Kashif, S. U. R., Khan, M., Tahir, A., Anum, F., Khan, S., & Raza, S. (2021). Contamination of heavy metals in poultry eggs: a study presenting relation between heavy metals in feed intake and eggs. *Archives of Environmental & Occupational Health, 76*(4), 220–232. https://doi.org/10.1080/19338244.2020.1799182

Kandil, M. A., Khorshed, M. A., Saleh, I. A., & Eshmawy, M. R. (2020). Investigation of heavy metals in fruits and vegetables and their potential risk for Egyptian consumer health. *Plant Archives, 20*(1), 1453–1463.

Karthiga, D. & Anthony, S. (2013). Selective colorimetric sensing of toxic metal cations by green synthesized silver nanoparticles over a wide pH range. *RSC Advances, 37*(3), 16765. https://doi.org/10.1039/c3ra42308e

Kim, E., Wickramasuriya, S. S., Shin, T. K., Cho, H. M., Macelline, S. P., Lee, S. D., ... & Heo, J. M. (2019). Bioaccumulation and toxicity studies of lead and mercury in laying hens: effects on laying performance, blood metabolites, egg quality and organ parameters. *The Journal of Poultry Science, 56*(4), 277–284. https://doi.org/10.2141/jpsa.0180118

Ko, S. (2015). Nano-food packaging: an overview of market, migration research, and safety regulations. *Journal of Food Science, 5*(80), R910–R923. https://doi.org/10.1111/1750-3841.12861

Kumar, R., Ivy, N., Bhattacharya, S., Dey, A., & Sharma, P., (2022). Coupled effects of microplastics and heavy metals on plants: uptake, bioaccumulation, and environmental health perspectives. *Science of the Total Environment, 836*, 155619. https://doi.org/10.1016/j.scitotenv.2022.155619

Kumarathilaka, P., Seneweera, S., Ok, Y. S., Meharg, A., & Bundschuh, J. (2019). Arsenic in cooked rice foods: assessing health risks and mitigation options. *Environment International, 127*, 584–591. https://doi.org/10.1016/j.envint.2019.04.004

Laine, J.E., Bailey, K.A., Rubio-Andrade, M., Olshan, A.F., Smeester, L., Drobná, Z., Herring, A.H., Stýblo, M., García-Vargas, G.G., & Fry, R.C. (2015). Maternal arsenic exposure, arsenic methylation efficiency, and birth outcomes in the Biomarkers of Exposure to ARsenic (BEAR) pregnancy cohort in Mexico. *Environmental Health Perspectives, 123*, 186–192. https://doi.org/10.1289/ehp.1307476

Li, C., Zhou, K., Qin, W., Tian, C., Qi, M., Yan, X., & Han, W. (2019). A review on heavy metals contamination in soil: effects, sources, and remediation techniques. *Soil and Sediment Contamination: An International Journal. 28*, 380–394. https://doi.org/10.1080/15320383.2019.1592108

Lim, S. F., Riehn, R., Ryu, W. S., Khanarian, N., Tung, C.-K., Tank, D., & Austin, R. H. (2006). In vivo and scanning electron microscopy imaging of upconverting nanophosphors in *Caenorhabditis elegans. Nano Letters, 6*(2), 169–174. https://doi.org/10.1021/nl0519175

Luo, Z., Xu, H., Ning, B., Guo, Z., Li, N., Chen, L., & Zheng, B. (2019). Ratiometric fluorescent nanoprobe for highly sensitive determination of mercury ions. *Molecules, 24*(12), 2278. https://doi.org/10.3390/molecules24122278

Liu, Y., Ouyang, Q., Li, H., Chen, M., Zhang, Z., & Chen, Q. (2018). Turn-on fluoresence sensor for Hg^{2+} in food based on FRET between aptamers-functionalized upconversion nanoparticles and gold nanoparticles. *Journal of Agricultural and Food Chemistry, 66*(24), 6188–6195. https://doi.org/10.1021/acs.jafc.8b00546

Maghsoudi, A., Hassani, S., Mirnia, K., & Abdollahi, M. (2021). Recent advances in nanotechnology-based biosensors development for detection of arsenic, lead, mercury, and cadmium. *International Journal of Nanomedicine, 16*, 803–832. https://doi.org/10.2147/ijn.s294417

Markin, A. V., & Markina, N. E. (2019). Experimenting with plasmonic copper nanoparticles to demonstrate color changes and reactivity at the nanoscale. *Journal of Chemical Education, 96*(7), 1438–1442. https://doi.org/10.1021/acs.jchemed.8b01050

Miao, P., Liu, L., Li, Y., & Li, G. (2009). A novel electrochemical method to detect mercury (II) ions. *Electrochemistry Communications, 11*(10), 1904–1907. https://doi.org/10.1016/j.elecom.2009.08.013

Miao, P., Tang, Y., & Wang, L. (2017). DNA modified Fe_3O_4@au magnetic nanoparticles as selective probes for simultaneous detection of heavy metal ions. *ACS Applied Materials & Interfaces, 9(*4), 3940–3947. http://doi.org/10.1021/acsami.6b14247

Miyake, Y., Togashi, H., Tashiro, M., Yamaguchi, H., Oda, S., Kudo, M., Tanaka, Y., Kondo, Y., Sawa, R., Fujimoto, T., Machinami, T., & Ono, A. (2006). $Mercury^{II}$-mediated formation of thymine–Hg^{II}–thymine base pairs in DNA duplexes. *Journal of the American Chemical Society, 128*(7), 2172–2173. https://doi.org/10.1021/ja056354d

Mondal, D., Rahman, M.M., Suman, S., Sharma, P., Siddique, A.B., Rahman, M.A., Bari, A., Kumar, R., Bose, N., Singh, S.K., Ghosh, A., & Polya, D.A., 2021. Arsenic exposure from food exceeds that from drinking water in endemic area of Bihar, India. *Science of the Total Environment, 754*, 142082. https://doi.org/10.1016/j.scitotenv.2020.142082

Mustafa, F., & Andreescu, S. (2020). Nanotechnology-based approaches for food sensing and packaging applications. *RSC Advances, 10*(33), 19309–19336. doi: 10.1039/d0ra01084g

Nabulo, G., Black, C. R., & Young, S. D. (2011). Trace metal uptake by tropical vegetables grown on soil amended with urban sewage sludge. *Environmental Pollution, 159*, 368–376. https://doi.org/10.1016/j.envpol.2010.11.007

Nam, N. N., Do, H. D. K., Trinh, K. T. L., & Lee, N. Y. (2022). Recent progress in nanotechnology-based approaches for food monitoring. *Nanomaterials, 12*(23), 4116. https://doi.org/10.3390/nano12234116

Ngafwan, N., Rasyid, H., Abood, E., Abdelbasset, W., Al-Shawi, S., Bokov, D., & Jalil, A. (2022). Study on novel fluorescent carbon nanomaterials in food analysis. *Food Science and Technology, 42*. https://doi.org/10.1590/fst.37821

Oberoi, S., Devleesschauwer, B., Gibb, H.J., & Barchowsky, A. (2019). Global burden of cancer and coronary heart disease resulting from dietary exposure to arsenic, 2015. *Environmental Research, 171*, 185–192. https://doi.org/10.1016/j.envres.2019.01.025

Pandit, S., Dasgupta, D., Dewan, N., & Prince, A. (2016). Nanotechnology based biosensors and its application. *The Pharma Innovation, 5*(6), 18–25.

Ping, J., Wu, J., & Ying, Y. (2012). Determination of trace heavy metals in milk using an ionic liquid and bismuth oxide nanoparticles modified carbon paste electrode. *Chinese Science Bulletin, 57*(15), 1781–1787. http://doi.org/10.1007/s11434-012-5115-1

Pujol, L., Evrard, D., Serrano, K. G., Freyssinier, M., Cizsak, A. R., & Gros, P., (2014). Electrochemical sensors and devices for electrochemical assay in water: the French groups contribution. *Frontiers in Chemistry. 2*(19), 1–24. https://doi.org/10.3389/fchem.2014.00019

Qian, L., Durairaj, S., Prins, S., & Chen, A. (2021). Nanomaterial-based electrochemical sensors and biosensors for the detection of pharmaceutical compounds. *Biosensors and Bioelectronics, 175*, 112836. https://doi.org/10.1016/j.bios.2020.112836

Qing, Y., Yang, J., Zhu, Y., Li, Y., Ma, W., Zhang, C., Li, X., Wu, M., Wang, H., Kauffman, A.E., Xiao, S., Zheng, W., & He, G. (2020). Cancer risk and disease burden of dietary cadmium exposure changes in Shanghai residents from 1988 to 2018. *Science of the Total Environment, 734*, 139411. https://doi.org/10.1016/j.scitotenv.2020.139411

Radwan, M. A. & Salama, A. K. (2006). Market basket survey for some heavy metals in Egyptian fruits and vegetables. *Food and Chemical Toxicology, 44*(8), 1273–1278. https://doi.org/10.1016/j.fct.2006.02.004

Rai, P.K., Lee, S.S., Zhang, M., Tsang, Y.F., & Kim, K.H. (2019). Heavy metals in food crops: health risks, fate, mechanisms, and management. *Environmental International, 125*, 365–385. https://doi.org/10.1016/j.envint.2019.01.067

Raj, S. K., Yadav, V., Bhadu, G. R., Patidar, R., Kumar, M., & Kulshrestha, V. (2021). Synthesis of highly fluorescent and water-soluble graphene quantum dots for detection of heavy metal ions in aqueous media. *Environmental Science and Pollution Research, 28*, 46336–46342. https://doi.org/10.1007/s11356-020-07891-5

Raju, C. V., Cho, C. H., Rani, G. M., Manju, V., Umapathi, R., Huh, Y. S., & Park, J. P. (2023). Emerging insights into the use of carbon-based nanomaterials for the electrochemical detection of heavy metal ions. *Coordination Chemistry Reviews, 476*, 214920. https://doi.org/10.1016/j.ccr.2022.214920

Rao, A., Singh, R., Sreejith, S., Jose, V., & Pethe, S. (2022). Ultrasensitive detection of heavy metal ions with scalable singular phase thin film optical coatings. *Advanced Optical Materials, 8*(10), 2102623. https://doi.org/10.1002/adom.202102623

Roto, R., Mellisani, B., Kuncaka, A., Mudasir, M., & Suratman, A. (2019). Colorimetric sensing of Pb^{2+} ion by using Ag nanoparticles in the presence of dithizone. *Chemosensors, 3*(7), 28. https://doi.org/10.3390/chemosensors7030028

Salama, A. K. & Radwan, M. A. (2005). Heavy metals (Cd, Pb) and trace elements (Cu, Zn) contents in some foodstuffs from the Egyptian market. *Emirates Journal of Food and Agriculture*, 34–42. https://doi.org/10.9755/ejfa.v12i1.5046

Sharma, P., Pandey, V., Sharma, M., Patra, A., Singh, B., Mehta, S., & Husen, A. (2021). A review on biosensors and nanosensors application in agroecosystems. *Nanoscale Research Letters, 1*(16). https://doi.org/10.1186/s11671-021-03593-0

Shen, L., Chen, Z., Li, Y., He, S., Xie, S., Xu, X., Liang, Z., Meng, X., Li, Q., Zhu, Z., Li, M., Chris Le, X., & Shao, Y. (2008). Electrochemical DNAzyme sensor for lead based on amplification of DNA–Au bio-bar codes. *Analytical Chemistry, 80*(16), 6323–6328. https://doi.org/10.1021/ac800601y

Shahid, M., Dumat, C., Khalid, S., Schreck, E., Xiong, T., & Niazi, N. K. (2017). Foliar heavy metal uptake, toxicity and detoxification in plants: a comparison of foliar and root metal uptake. *Journal of Hazardous Materials, 325*, 36–58. https://doi.org/10.1016/j.jhazmat.2016.11.063

Shamsudduha, M., Joseph, G., Haque, S. S., Khan, M. R., Zahid, A., & Ahmed, K. M. (2019). Multi-hazard groundwater risks to water supply from shallow depths: challenges to achieving the sustainable development goals in Bangladesh. *Exposure and Health, 12*, 657–670. https://doi.org/10.1007/s12403-019-00325-9

Shokr, L. A., Hassan, M. A., & Elbahy, E. F. (2019). Heavy metals residues (mercury and lead) contaminating Nile and marine fishes. *Benha Veterinary Medical Journal, 36*(2), 40–48. https://doi.org/10.21608/BVMJ.2019.12543.1007

Shoukry, H. M., Ali, H., & Fayed, A. M. S. (2016). Detection of harmful residues in some fish species. *Egyptian Journal of Chemistry and Environmental Health, 2*(2), 363–381. https://doi.org/10.21608/EJCEH.2016.254338

Shrivastava, P., Jain, V. K., & Nagpal, S. (2023). Nanoparticle intervention for heavy metal detection: a review. *Environmental Nanotechnology, Monitoring & Management, 17,* 100667. https://doi.org/10.1016/j.enmm.2022.100667

Sood, S., & Sharma, C. (2019). Levels of selected heavy metals in food packaging papers and paperboards used in India. *Journal of Environmental Protection, 3*(10), 360–368. https://doi.org/10.4236/jep.2019.103021

Srivastava, M., Srivastava, S. K., Ojha, R. P., & Prakash, R. (2022). Smartphone-assisted colorimetric sensor based on nanozyme for on-site glucose monitoring. *Microchemical Journal, 182*, 107850. https://doi.org/10.1016/j.microc.2022.107850

Tabaraki, R., & Sadeghinejad, N. (2018). Microwave assisted synthesis of doped carbon dots and their application as green and simple turn off–on fluorescent sensor for mercury (II) and iodide in environmental samples.*Ecotoxicology and Environmental Safety, 153*, 101–106. https://doi.org/10.1016/j.ecoenv.2018.01.059

Tajik, S., Dourandish, Z., Nejad, F. G., Beitollahi, H., Jahani, P. M., & Di Bartolomeo, A. (2022). Transition metal dichalcogenides: synthesis and use in the development of electrochemical sensors and biosensors. *Biosensors and Bioelectronics*, 114674. https://doi.org/10.1016/j.bios.2022.114674

Tan, Z., Wu, W., & Feng, C. (2020). Simultaneous determination of heavy metals by an electrochemical method based on a nanocomposite consisting of fluorinated graphene and gold nanocage. *Microchimica Acta, 187*, 414. https://doi.org/10.1007/s00604-020-04393-6

Tharmaraj, V., & Yang, J. (2014). Sensitive and selective colorimetric detection of Cu^{2+} in aqueous medium via aggregation of thiomalic acid functionalized Ag nanoparticles. *Analyst, 139*(23), 6304–6309. https://doi.org/10.1039/C4AN01449A

Tian, J., Zhao, X., Wang, Y., & Hasan, M. (2023). Advances in fluorescent sensing carbon dots: an account of food analysis. *ACS Omega, 10*(8), 9031–9039. https://doi.org/10.1021/acsomega.2c07986

Tu, X., Ge, L., Deng, L., & Zhang, L. (2022). Morphology adjustment and optimization of CuS as enzyme mimics for the high efficient colorimetric determination of Cr(IV) in water. *Nanomaterials, 12*(12), 2087. https://doi.org/10.3390/nano12122087

Ullah, N., Mansha, M., Khan, I., & Qurashi, A. (2018). Nanomaterial-based optical chemical sensors for the detection of heavy metals in water: recent advances and challenges. *TrAC Trends in Analytical Chemistry, 100*, 155–166. https://doi.org/10.1016/j.trac.2018.01.002

Vilela, D., Parmar, J., Zeng, Y., Zhao, Y., & Sanchez, S. (2016). Graphene-based microbots for toxic heavy metal removal and recovery from water. *Nano Letters*, *4*(16), 2860–2866. https://doi.org/10.1021/acs.nanolett.6b00768

Waheed, A., Mansha, M., & Ullah, N. (2018). Nanomaterials-based electrochemical detection of heavy metals in water: current status, challenges and future direction. *TrAC Trends in Analytical Chemistry*, *105*, 37–51. https://doi.org/10.1016/j.trac.2018.04.012

Wang, L., Peng, X., Fu, H., Huang, C., Li, Y., & Liu, Z. (2019). Recent advances in the development of electrochemical aptasensors for detection of heavy metals in food. *Biosensors and Bioelectronics*, 111777. https://doi.org/10.1016/j.bios.2019.111777

Waqas, M., Li, G., Khan, S., Shamshad, I., Reid, B. J., Qamar, Z., & Chao, C. (2015). Application of sewage sludge and sewage sludge biochar to reduce polycyclic aromatic hydrocarbons (PAH) and potentially toxic elements (PTE) accumulation in tomato. *Environmental Science and Pollution Research*, *22*, 12114–12123. https://doi.org/10.1007/s11356-015-4432-8

Wei, J. & Cen, K., (2020). Assessment of human health risk based on characteristics of potential toxic elements (PTEs) contents in foods sold in Beijing, China. *Science of the Total Environment*, *703*, 134747. https://doi.org/10.1016/j.scitotenv.2019.134747

Wei, T., Dong, T., Wang, Z., Bao, J., Tu, W., & Dai, Z. (2015). Aggregation of individual sensing units for signal accumulation: conversion of liquid-phase colorimetric assay into enhanced surface-tethered electrochemical analysis. *Journal of the American Chemical Society*, *137*(28), 8880–8883. https://doi.org/10.1021/jacs.5b04348

Wei, J., Gao, J., & Cen, K. (2019). Levels of eight heavy metals and health risk assessment considering food consumption by China's residents based on the 5th China total diet study. *Science of the Total Environment*, *689*, 1141–1148. https://doi.org/10.1016/j.scitotenv.2019.06.502

Willner, I., Shlyahovsky, B., Zayats, M., & Willner, B. (2008). DNAzymes for sensing, nanobiotechnology and logic gate applications. *Chemical Society Reviews*, *37*(6), 1153–1165. https://doi.org/10.1039/B718428J

Wu, Q., Bi, H. M., & Han, X. J. (2021). Research progress of electrochemical detection of heavy metal ions. *Chinese Journal of Analytical Chemistry*, *49*, 330–340. https://doi.org/10.1016/S1872-2040(21)60083-X

Xiaoyu, X., Yang, Y., Yuning, W., & Qian, K. (2022). Nanomaterial-based sensors and strategies for heavy metal ion detection. *Green Analytical Chemistry*, *2*, 100020. https://doi.org/10.1016/j.greeac.2022.100020

Yang, J., Hou, B., Wang, J., Tian, B., Bi, J., Wang, N., & Huang, X. (2019). Nanomaterials for the removal of heavy metals from wastewater. *Nanomaterials*, 3(9), 424. https://doi.org/10.3390/nano9030424

Yang, J., Zhang, Y., Zhang, L., Wang, H., Nie, J., Qin, Z., & Xiao, W. (2017). Analyte-triggered autocatalytic amplification combined with gold nanoparticle probes for colorimetric detection of heavy-metal ions. *Chemical Communications*, *53*(54), 7477–7480. https://doi.org/10.1039/C7CC02198D

Yang, Q., Li, Z., Lu, X., Duan, Q., Huang, L., & Bi, J. (2018). A review of soil heavy metal pollution from industrial and agricultural regions in China: pollution and risk assessment. *Science of the Total Environment*, *642*, 690–700. https://doi.org/10.1016/j.scitotenv.2018.06.068

Yao, Y., Wu, H., & Ping, J. (2019). Simultaneous determination of Cd(II) and Pb(II) ions in honey and milk samples using a single-walled carbon nanohorns modified screen-printed electrochemical sensor. *Food Chemistry, 274*, 8–15. https://doi.org/10.1016/j.foodchem.2018.08.110

Ye, W., Li, Y., Wang, J., Li, B., Cui, Y., Yang, Y., & Qian, G. (2020). Electrochemical detection of trace heavy metal ions using a Ln-MOF modified glass carbon electrode. *Journal of Solid State Chemistry*, *281*, 121032. https://doi.org/10.1016/j.jssc.2019.121032

Yoshii, T., Nishitsugu, F., Kikawada, K., Maehashi, K., & Ikuta, T. (2023). Identification of cadmium compounds in a solution using graphene-based sensor array. *Sensors, 3*(23), 1519. https://doi.org/10.3390/s23031519

You, Z., Li, X., Huang, J., Chen, R., Jiyu, P., Kong, W., & Liu, F. (2023). Agarose film-based liquid–solid conversion for heavy metal detection of water samples by laser-induced breakdown spectroscopy. *Molecules*, *6*(28), 2777. https://doi.org/10.3390/molecules28062777

Yu, L., Cui, X., Li, H., Lu, J., Kang, Q., & Shen, D. (2019). A ratiometric electrochemical sensor for multiplex detection of cancer biomarkers using bismuth as an internal reference and metal sulfide nanoparticles as signal tags. *Analyst*, *144*(13), 4073–4080. https://doi.org/10.1039/C9AN00775J

Zhang, B., Chen, J., Zhu, H., Yang, T., Zou, M., Zhang, M., & Du, M. (2016). Facile and green fabrication of size-controlled AuNPs/CNFs hybrids for the highly sensitive simultaneous detection of heavy metal ions.*Electrochimica Acta, 196*, 422–430. https://doi.org/10.1016/j.electacta.2016.02.163

Zhang, H., Jin, M., & Xia, Y. (2012). Noble-metal nanocrystals with concave surfaces: synthesis and applications. *AngewandteChemie, 31*(51), 7656–7673. https://doi.org/10.1002/anie.201201557

Zhang, X., Huang, X., Xu, Y., Wang, X., Guo, Z., Huang, X., Li, Z., Shi, J., & Zou, X. (2020). Single-step electrochemical sensing of ppt-level lead in leaf vegetables based on peroxidase-mimicking metal-organic framework. *Biosensors & Bioelectronics, 168*, 112544. https://doi.org/10.1016/j.bios.2020.112544

Zhang, Y., Zhang, L., Wang, L., Wang, G., Komiyama, M., & Liang, X. (2019). Colorimetric determination of mercury(II) ion based on DNA-assisted amalgamation: a comparison study on gold, silver and Ag@Au nanoplates. *Microchimica Acta, 186*(11), 713. https://doi.org/10.1007/s00604-019-3873-z

Zhong, Y. Q., Ning, T. J., Cheng, L., Xiong, W., Wei, G. B., Liao, F. S., Ma, G. Q., Hong, N., Cui, H. F., & Fan, H. (2021). An electrochemical Hg^{2+} sensor based on signal amplification strategy of target recycling. *Talanta, 223*(1), 121709. https://doi.org/10.1016/j.talanta.2020.121709

Zhou, J., Lv, X., Jia, J., Din, Z., Cai, S., He, J., & Cai, J. (2022). Nanomaterials-based electrochemiluminescence biosensors for food analysis: recent developments and future directions. *Biosensors, 11*(12), 1046. https://doi.org/10.3390/bios12111046

Zhu, D., Li, X., Liu, X., Wang, J., & Wang, Z. (2012). Designing bifunctionalized gold nanoparticle for colorimetric detection of Pb^{2+} under physiological condition. *Biosensors and Bioelectronics, 31*(1), 505–509. https://doi.org/10.1016/j.bios.2011.11.026

Zhuang, Q., Sun, L., & Ni, Y. (2017). One-step synthesis of graphitic carbon nitride nanosheets with the help of melamine and its application for fluorescence detection of mercuric ions. *Talanta, 164*, 458–462. https://doi.org/10.1016/j.talanta.2016.12.004

8 Nanosensors for Pathogen Detection in Food Products

Ruslan Mehadi Galib and Mahabub Alam

8.1 INTRODUCTION: THE USE OF NANOSENSORS

Ensuring the safety of food remains a critical global concern, exerting significant influence on the overall health and well-being of individuals worldwide. The consumption of tainted food possesses the potential to give rise to severe health implications and outbreaks of foodborne ailments. Consequently, there exists a pressing need to formulate and execute proficient strategies aimed at vigilantly monitoring, identifying, and mitigating the presence of pathogens and contaminants within food commodities. Conventional techniques employed in the monitoring of food safety have traditionally relied upon a fusion of sensory assessments, microbiological cultures, and chemical assays. Regrettably, these methodologies frequently grapple with constraints pertaining to speed, sensitivity, specificity, and cost, thereby necessitating an exploration of inventive avenues. The guarantee of food product safety hinges upon the imperative creation of prompt, perceptive, and efficacious detection methodologies (Fung et al., 2018; Javaid et al., 2021).

Numerous disciplines are involved in the study of nanoparticles, and modern nanotechnologies are increasingly being used in agricultural and food research (Figure 8.1). These nanoscale devices take advantage of the special qualities that nanomaterials possess. For example, improved surface-to-volume ratios, particular surface chemistry, quick or delayed particle aggregation, electrical/heat conductivity, and distinctive optical capabilities (Adam & Gopinath, 2022; Mohammad et al., 2022; Shawon et al., 2020).

Nanosensors can be created to interact specifically with target pathogens, providing selective and incredibly accurate detection even at low concentrations. Incorporating nanomaterials into sensing platforms can also improve the signal transduction processes, resulting in higher detection limits and fewer false positives (Bruce-Tagoe & Danquah, 2023; Yang & Duncan, 2021).

8.1.1 Importance of Pathogen Detection

Microbial agents, encompassing bacteria, viruses, and parasites, pose a noteworthy peril to the integrity of our food reservoir. Adulterated food has the capacity to incite episodes of alimentary canal-based maladies, impacting individuals as well as broader communities. The aftermath of such incidents encompasses financial setbacks due to product withdrawals and waning consumer trust, while also extending to grave afflictions, hospital admittances, and potentially fatal outcomes. In an era of heightened global interdependence facilitated by international commerce and mobility, the prospect of swift transnational diffusion of foodborne pathogens has emerged as a considerable apprehension (Agnihotri et al., 2022; Mishra et al., 2023; Nguyen & Kim, 2020).

DOI: 10.1201/9781003438168-8

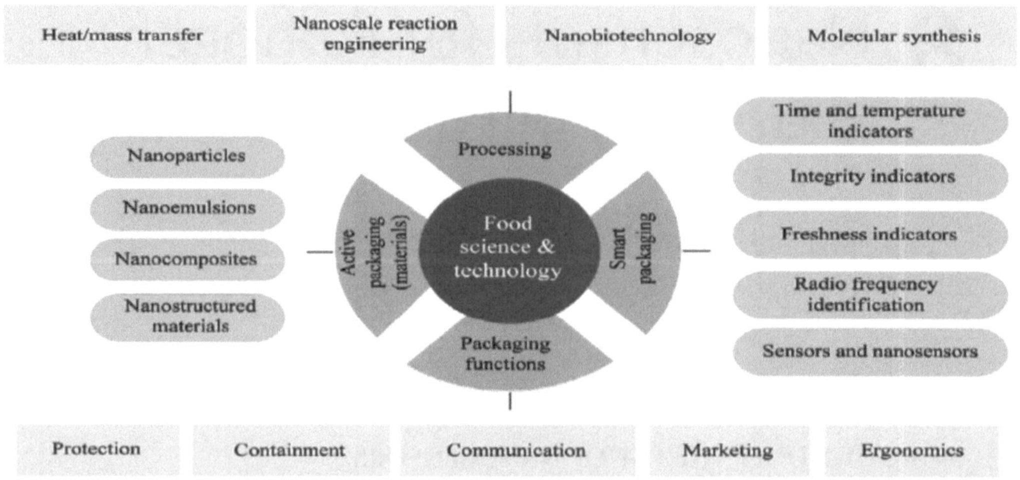

FIGURE 8.1 Various uses of nanotechnology in the field of food science and technology. Reprinted with written permission from Shawon et al. (2020).

Ensuring the safety of our food supply and safeguarding public health hinges upon the adept identification of pathogenic microorganisms present in food items. At every step of food production, processing, handling, and distribution, there is a continuous risk of contamination from harmful microorganisms like bacteria, viruses, and parasites. Quickly and reliably detecting these harmful agents is crucial to avoid foodborne illness outbreaks and minimize harm to public health. In the pursuit of enhancing precision in identifying pathogens within food samples, researchers leverage the distinctive attributes and dynamics offered by nanomaterials (Cesewski & Johnson, 2020; Malhotra et al., 2014). By enabling early pathogen detection, it has possible potential to halt the spread of foodborne infections (Mishra et al., 2023; Patel et al., 2023).

8.1.2 Major Difficulties with Pathogen Detection Methods

Microbiological culture techniques, while reliable, can take several days to yield results, during which time contaminated products may have already entered the market or reached consumers. Furthermore, these methods can be labor-intensive and require specialized training. For identifying specific infections and contaminants, conventional chemical tests like high-performance liquid chromatography (HPLC) and enzyme-linked immunosorbent assay (ELISA) have been used (Leva-Bueno et al., 2020; Nguyen & Kim, 2020; Saravanan et al., 2021).

When the microbial load is low, the inadequate sensitivity of culture-based approaches can result in an underestimation of the presence of foodborne pathogens (Solís et al., 2022). On the other hand, pathogen detection is difficult due to the complications and variety of food mediums involved. Food ingredients like lipids, proteins, and carbs can affect the assay's sensitivity and interfere with the detection process. Precise pathogen identification can be made more difficult by sample composition fluctuation and the probable presence of inhibitory compounds (Foddai & Grant, 2020; Sharafeldin & Davis, 2021).

The use of nanosensors in monitoring agricultural ecosystems is thoroughly explored in this chapter, with a focus on their usage in identifying pathogens and guaranteeing food safety. It explores the crucial role that nanotechnology plays in ensuring food safety, the special properties of nanomaterials for sensor development, and their use in several areas of food inspection. Potential issues, such as nanoparticle toxicity, will be discussed, as well as suggested tactics to encourage the use of nanosensors in the food trade. It will also consider the challenges of applying nanosensors in actual food safety contexts and suggest potential future research and development strategies.

8.2 FUNCTION OF NANOTECHNOLOGY IN FOOD

Numerous industries have been transformed by nanotechnology, and there is no exception in the industry of food (Table 8.1). The use of nanotechnology in food has sparked the creation of cutting-edge methods that enhance food quality, safety, and shelf life (Adam & Gopinath, 2022; Jafarizadeh-Malmiri et al., 2019). Using nanocomposites, nanocoatings, and nanoemulsions, packaging materials are being developed that offer better moisture and gas barriers, extending shelf life and minimizing food deterioration (Biswas et al., 2022; Emamhadi et al., 2020). Nanoencapsulation techniques can improve the stability and effectiveness of fragile bioactive substances like vitamins, antioxidants, and probiotics. Enzymatic processes including lipid oxidation, taste creation, and enzymatic hydrolysis can all be sped up by the use of nanocatalysts (Saini et al., 2022; Tavakoli et al., 2022). Proteins, carbohydrates, and colors are separated and purified from other food ingredients using nanofiltration and nanosieving membranes. Food safety for people with dietary restrictions or allergies is ensured through the use of nano-biosensors, which can identify specific biomarkers and allergens (Keçili et al., 2019; Shawon et al., 2020).

8.3 ADVANCES IN FOOD SAFETY DRIVEN BY NANOTECHNOLOGY

Nanomaterials' distinctive qualities and capacities have facilitated the creation of pioneering methods for identifying and reducing allergies, pollutants, and foodborne pathogens. To detect pathogens accurately, nano-biosensors use a variety of detection methods, including photosensitive, electrochemical, and biosensing techniques (Farré et al., 2013; Huang et al., 2021). Nanosensors have the potential to be used as early detection and monitoring systems on-site, allowing for prompt interventions to stop the spread of foodborne infections. Nanomaterials with antibacterial capabilities, such as silver nanoparticles, titanium dioxide nanoparticles, and materials based on graphene, can be used to disinfect food. Nanomaterials with distinctive optical characteristics can be used in sensing applications, such as gold nanoparticles and quantum dots (Jagtiani, 2022; Sahani & Sharma, 2021). Significant information and effort are needed for the journey from food nanotechnology

TABLE 8.1
Applications of nanotechnology in food

Function	Application
Improved food packaging	Enhanced barrier properties
	Active antimicrobial packaging
	Intelligent and responsive packaging
Enhanced nutrient delivery	Nanoencapsulation for improved stability and bioavailability
	Fortification of functional ingredients
Efficient food processing	Nanocatalysis for accelerated reactions
	Nanofiltration for separation and purification
	Nanosensors for real-time process monitoring
Advanced food quality monitoring	Nanosensors for prompt and sensitive recognition
	Nanobiosensors for allergen and biomarker detection
	Portable devices for on-site analysis

Sources: Alfei et al. (2020); Chandra and Panesar (2022).

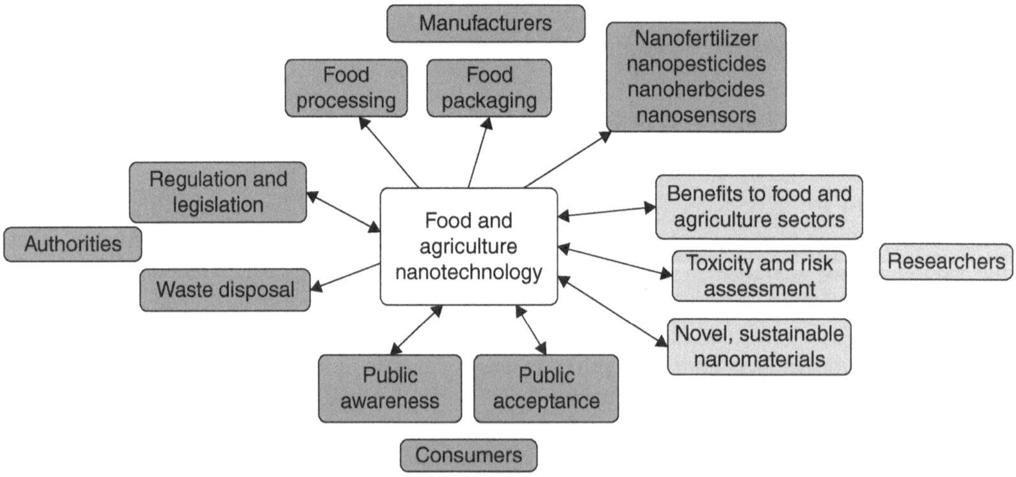

FIGURE 8.2 Food nanotechnology's path from research to market to consumer plates. Reprinted with written permission from He et al. (2019).

development to its market presence. Figure 8.2 encompasses various areas such as nanotoxicity understanding, regulation, public awareness, and their interconnectedness. These systems enable producers of food and regulatory bodies to more effectively and proactively monitor food safety (He et al., 2019; Li et al., 2022).

8.3.1 The Unique Properties of Nanomaterials

Nanomaterials are advantageous instruments in the food sector because they have special qualities that set them apart from their majority equivalents. Because of the in-height surface-to-volume ratio, they interact more constructively with food ingredients like flavorings, nutrients, and pollutants. The chemical and physicalfirmness of materials can be amended by the reduction in particle size at the nanoscale, preventing degradation, aggregation, or spoiling. Encapsulated bioactive substances benefit most from the stability since it shields them from environmental variables including pH, temperature, and oxidation (Qiao et al., 2022; Sahoo et al., 2021; Shawon et al., 2020).

Nanocatalysts such as metal nanoparticles or nanocomposites demonstrate improved catalytic activity. This feature makes it possible for more operative and fixated responses, which enhances food processing effectiveness and uses less energy. Nanocatalysis is used in several food processing steps, including enzymatic hydrolysis, lipid oxidation, and taste creation (Agarwal et al., 2023; Navia-Mendoza et al., 2021; Shah & Mraz, 2020). By adjusting their size, shape, composition, and surface qualities, nanomaterials enable the customization of their physicochemical features. Because of their adaptability, nanoparticles can be modified to satisfy needs in culinary applications (Agarwal et al., 2023; Kumar et al., 2017).

8.3.2 Food Processing and Packaging Strategies Employing Nanotechnology

The processing and packaging setting of food has undergone a profound revolution owing to the combination of nanotechnology. This advancement has introduced innovative substances characterized by enhanced obstructive attributes and dynamic functionalities (Table 8.2). Concurrently, ingenious approaches have emerged to ameliorate the overall caliber, safety, and durability of food products (Emamhadi et al., 2020; Sarfraz et al., 2021). In the field of food processing, nanocomposite materials—which consist of a matrix injected with nanoscale additives—have attracted prominent

TABLE 8.2
Potential uses for nanosensors in the food and agro business

Agriculture	Food Processing	Food Packing	Food Transport	Nutrition
• Nanosensors gauge soil attributes like moisture and pH • Nanosensors detect foodborne pollutants and monitor farm environments • Nanochips track identity and location • Nanocapsules deliver pesticides, fertilizers, and vaccines • Nanosensors and smart delivery systems optimize natural resource use • Nanoparticles deliver growth agents and DNA to plants precisely • Nanoparticles act as early-warning smart sensors for shifting conditions	• Utilizing nanocapsules to enhance flavor • Nanotubes and nanoparticles for gelling and viscosity • Replacing meat cholesterol with plant-based steroids via nanocapsule infusion • Selective binding and removal of food contaminants using nanoparticles • Aptasensors for detecting microbial toxins	• Compact nanosensors for identifying substances like chemicals, pathogens, and toxins in food • Nanosensors integrated into packaging to spot chemical release during food spoilage and act as a digital taste system • Ethylene-detecting nanosensors • Intelligent nanosensor labels on food packaging to enhance microbial safety • Aptasensors for detecting microbial cells, antibiotics, drugs, residues, and heavy metals	• Tiny sensors to oversee storage environment conditions • Nanosensors to track product status during transport and storage • Advanced sensors for overseeing grain, dairy, fruit, and vegetable quality in storage • Aptasensors for identifying microbial cells	• Utilizing nanocapsules in food for nutrient delivery • Using nanocochleates to convey nutrients without impacting food's flavor or appearance

Source: Jafarizadeh-Malmiri et al. (2019).

attention. Notably, the fusion of nanoparticles into polymer matrices serves to develop the structural robustness, gas imperviousness, and thermal resilience of packaging elements (Mohammad et al., 2022; Singh et al., 2023).

Nano-enabled oxygen scavengers typically comprise transition metals, such as iron or cobalt, encapsulated within polymer matrices. These nanocomposites have proven effective in reducing oxygen levels within packages, thereby slowing down oxidative reactions, and preserving product quality. This approach not only extends the shelf life of perishable goods but also maintains the sensory attributes and nutritional content of the food (Emamhadi et al., 2020; Tavker & Sharma, 2022). Silver nanoparticles, recognized for their inherent antimicrobial properties, have garnered attention in the food sector for their potential to inhibit the growth of pathogenic bacteria. Incorporated into packaging materials or coatings, these nanoparticles release controlled amounts of silver ions that interact with the cell membranes of bacteria, disrupting their vital functions and inhibiting reproduction. This strategy lowers the cross-contamination during handling and transportation in addition to limiting bacterial growth on food contact surfaces (Cheng et al., 2022; Saini et al., 2022).

The use of nano-coatings has become a viable strategy for augmenting the quality and preservation of food products. These thin coatings, which are frequently made of nanoscale materials, can be used to protect food surfaces against moisture, air, and microbial contamination. High transparency, controlled release qualities, and improved mechanical strength are just a few benefits that nanocoatings can provide (Chelliah et al., 2021; Patel et al., 2020). Nanosensors built into packing materials can identify volatile chemicals or spoilage indicators emitted by rotting food, giving an early warning of food contamination or deterioration. Nanomaterials with antibacterial capabilities can also be included in intelligent packaging systems, helping to preserve food and eliminating the need for chemical preservatives (Javaid et al., 2021; Singh et al., 2023).

8.3.3 Nanosensors for Food Inspection

Nanosensors have tremendous promise for detecting pollutants and chemical residues in food. These sensors are made to interact with particular target molecules and provide quantifiable signals as a result of their existence. Nanomaterials can be customized to specifically bind and detect pollutants like pesticides, heavy metals, or mycotoxins (Li et al., 2022). Nanosensors can specifically identify the presence of pathogenic bacteria, viruses, or parasites in food samples by employing nanomaterials and bio-recognition components. For example, nanowire, nanofiber, or nanocantilever-based nanobiosensors can be functionalized with particular probes to collect and recognize target pathogens (Bhagat et al., 2022; Shawon et al., 2020). Nanosensors are beneficial in a variety of fields and contribute to methodological breakthroughs by making it possible to explore a broad spectrum of analytes, including macrofoods and compounds like proteins, carbohydrates, lipids, dyes, pigments, and preservatives. A number of instances highlight in Figure 8.3 the development of nanosensors for agricultural and food applications (Shawon et al., 2020; Taj et al., 2022). Nanosensors can be incorporated into portable devices or lab-on-a-chip systems. It offers a practical and affordable platform for food inspection. Nanomaterial-based sensors, like nanowires or nanofilms, can be designed to react to certain changes in these parameters and provide real-time input on food quality (Qiao et al., 2022; Taj et al., 2022).

8.3.4 Making Use of Nanomaterials for Pathogen Detection

To specifically bind and capture target pathogens, certain recognition components, such as antibodies, aptamers, or DNA probes, can be functionalized onto nanomaterials. Using the proper readout techniques, it is possible to identify and quantify changes in electrical conductivity, fluorescence, and electrical impedance (Cardoso et al., 2022). Multiple pathogens can be detected simultaneously in complicated food matrices using nanosensors that can be constructed to contain various

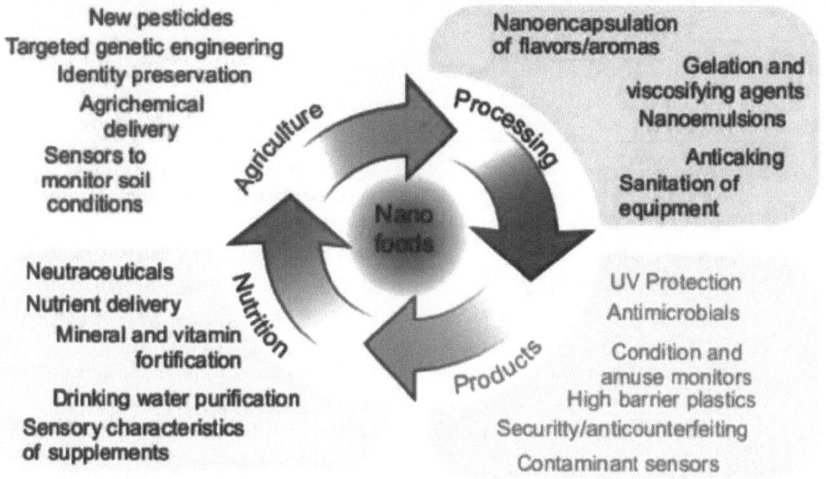

FIGURE 8.3 Benefits of nanotechnology in different sectors of the food sector. Reprinted with written permission from Shawon et al. (2020).

identification features. Both the efficiency and cost-effectiveness of pathogen screening processes are considerably improved by this capacity (Bobrinetskiy et al., 2021).

Gold nanoparticle sensors, specialized in antibodies, enable rapid and sensitive identification of *Salmonella* bacteria. Quantum dot sensors emit fluorescence upon interaction with *Escherichia coli*, facilitating swift bacterial detection. Carbon nanotube sensors, modified with pathogen-specific aptamers, capture norovirus particles selectively. Molecularly imprinted polymer sensors mimic binding sites for aflatoxin, permitting precise toxin detection. Enzyme inhibition sensors detect organophosphate pesticides by gauging enzyme inhibition. These nanosensors epitomize nanotechnology's adaptable role in enhancing food safety through targeted and sensitive pathogen identification. Table 8.3 outlines distinct types of nanosensors employed for pathogen detection, showcasing their diverse applications in ensuring food safety. Also, Table 8.4 illustrates the application of nanosensors in the eradication of pathogens in the food sector (Mishra et al., 2023; Sharifi et al., 2020; Sondhi et al., 2020).

8.3.5 Application of Gold- and Silver-Based Nanosensors

Due to their distinctive optical qualities, which emerge as vivid color changes in response to particular biological interactions, gold and silver nanoparticles have drawn a lot of attention in the creation of nanosensors. These nanoparticles have biomolecules attached to them, such as DNA probes or antibodies, that have an affinity for the infections they are intended to treat. The nanoparticles experience a plasmonic shift that causes a discernible color change when these biomolecules attach to their respective pathogens (Nile et al., 2020; Ravichandran et al., 2021). This phenomenon serves as the foundation for colorimetric assays, which use a change in visual color to determine whether or notinfections are present. Adding specific capture agents along with nanomaterials, like gold nanoparticles or quantum dots, is one technique. The binding event, in accordance with the findings (Ravichandran et al., 2021), results in a discernible signal, such as a color change or fluorescence emission. It can be measured by optical methods (such as fluorescence spectroscopy or spectrophotometry). As a result, dangerous bacteria, viruses, or parasites can be found in food samples and detected with high specificity and sensitivity (Thakur et al., 2022).

Localized surface plasmon resonance (LSPR), which allows gold nanoparticles (AuNPs) to absorb and scatter light at particular wavelengths, is a distinctive optical property of AuNPs. The

TABLE 8.3
Nanosensors for identification of foodborne pathogens

Pathogen	Nanosensor Type	Characteristics and Applications
Bacteria		
*Salmonella*spp.	Gold nanoparticle-based sensors	Functionalized with antibodies for rapid and sensitive detection
E. coli	Quantum dot-based sensors	Emit fluorescent signals upon interaction with bacteria
*Listeria*spp.	Nanowire-based sensors	Detects bacterial binding-induced changes in electrical properties
Viruses		
Norovirus	Carbon nanotube-based sensors	Modified with viral-specific aptamers for selective capture
Hepatitis A	Magnetic nanoparticle-based sensors	Combined with polymer chain reaction (PCR) for rapid virus detection
Mycotoxins		
Aflatoxin	Molecularly imprinted polymer sensors	Designed to mimic aflatoxin binding sites for specific detection
Ochratoxin	Surface plasmonresonance sensors	Functionalized with antibodies for label-free toxin detection
Pesticides		
Organophosphates	Enzyme inhibition-based sensors	Measure pesticide-induced enzyme inhibition
Pyrethroids	Electrochemical nanosensors	Detect pesticide residues through electrochemical changes

Source: Kumar et al. (2021).

development of colorimetric tests for pathogen identification has made use of this property. On binding of the target pathogen, a change in the LSPR wavelength or colorimetric response can be seen, by functionalizing AuNPs with particular antibodies or DNA probes that target pathogen-specific biomarkers (Dester et al., 2022). This colorimetric shift can be quick and on-site pathogen recognized using basic spectrophotometric methods or the naked eye. Silver nanoparticles have shown promise in the development of nanosensors for the detection of infections. AgNPs can be functionalized with certain ligands or probes to selectively bind with target pathogens and change their electrical characteristics. This electrical response can be detected using methods like impedance spectroscopy or field-effect transistors (FETs) (Bruna et al., 2021; Zorraquín-Peña et al., 2020). These nanomaterials are appealing choices for creating reactive and instantaneous detection platforms due to their special characteristics and compatibility with a variety of detection methods (Ramezani et al., 2020; Tavker & Sharma, 2022).

8.3.6 Detection of a Variety of Pathogenic Microorganisms

Nanosensors have transcended the confines of colorimetric assays to detect a broad spectrum of pathogenic microorganisms. From bacteria like *Salmonella* and *E. coli* to viruses such as norovirus and hepatitis A. Quantum dots, semiconductor nanoparticles with tunable optical properties, have been engineered to bind to specific pathogens and emit unique fluorescent signals upon interaction. This fluorescence can be measured and quantified, enabling rapid and accurate identification of pathogens. With the use of proteins, DNA sequences, or cell surface indicators, specific pathogen biomarkers can be recognized and bound by nanosensors. A detectable signal can be produced by these biomarkers through engaging functionalized nanomaterials (Leva-Bueno et al., 2020; Sharifi et al., 2020). Microorganisms such as *Salmonella, E. coli, Listeria monocytogenes*, *Campylobacter*,

TABLE 8.4
Nanosensors for identification of foodborne pathogens

Nanosensors Mechanism	Targeted Pathogen	Relevance to Food Sector
Microchips made of poly (dimethylsiloxane) featuring fluid bilayer membranes and specialized antibodies targeting the toxin.	*Staphylococcus* spp. enterotoxin B	The capacity of this important nanosensor to detect *Staphylococcus* sp. enterotoxin B at concentrations as low as 0.5 ng/mL makes it invaluable to the food industry. Rapid identification of this toxin prevents potential foodborne outbreaks, safeguarding public health and maintaining consumer confidence in food products.
Immune-magnetic beads utilizing G-liposomal nanovesicles.	*E. coli* O157:H7, *Salmonella* spp., *L. monocytogenes*	Regarding food safety, this nanosensor system has shown remarkable sensitivity by identifying *E. coli* O157:H7, *Salmonella*, and *L. monocytogenes* at detection limits as low as 10 CFU/mL.
Detection method utilizing liposomal nanovesicles with universal protein G via immunosorbent assay.	Pathogenic microorganisms	This nanosensor exhibits broad specificity and versatility, capable of identifying multiple pathogenic microorganisms commonly found in foods. Studies have shown successful detection of various pathogens, including *E. coli*, *Salmonella*, and *Campylobacter*.
Utilizing quantum dots modified with high-affinity antibodies for fluorescent sandwich immunoassay	*Clostridium botulinum*, neurotoxin serotype A	Utilizing quantum dots in this nanosensor system offers remarkable sensitivity, allowing the identification of *C.botulinum* neurotoxin type A at extremely low levels. This innovation is crucial for safeguarding packaged foods and preventing instances of botulism.
An immunosensor utilizing quantum dots as fluorescent markers within a waveguide.	*Bacillus anthracis*	The *Bacillus anthracis* spores can be detected in this nanosensor with amazing sensitivity, detecting quantities in the range of as 100 CFU/mL thanks to the employment of quantum dots in combination with a waveguide-based system.
An antigen-targeting peptide is attached to multi-wall carbon nanotubes, enabling detection down to 0.4 picomolar concentration.	*Bacillus anthracis*	With its impressive limit of detection (0.4pM), this nanosensor system offers unparalleled sensitivity in detecting *Bacillus anthracis*. This technology is indispensable for monitoring potential biosecurity threats and ensuring the safety of food production processes.
Immune-sensors utilizing gold nanoparticles modified with antibodies and analyzed through mass spectrometry.	*E. coli*, *Staphylococcus aureus*, *Salmonella enterica*	According to tests with detection limits in the range of 10–100 CFU/mL, this nanosensor uses antibody-modified gold nanoparticles to specifically detect *E. coli*, *S. aureus*, and *S.enterica* with great sensitivity.
Gold nanoparticle-based amperometricimmunosensor exhibiting an extremely low limit of detection (10 CFU/mL).	*Bacillus cereus*	This nanosensor plays a vital role in early identification of *B. cereus* contamination. With its impressive limit of detection (10 CFU/mL), it helps ensure that food products meet stringent quality standards and pose minimal health risks.
Fluorescent silica-coated magnetic nanoparticles containing Fe_3O_4 cores and modified with gentamicin demonstrate a 20-minute detection time for bacterial concentrations as low as 1×10^7 CFU/mL from a 10 mL solution.	*E. coli*	The capacity of this nanosensor device to quickly detect an infection with a limit of detection of 1×10^7 CFU/mL within 20 minutes is crucial for the food industry. Rapid detection of *E. coli* infection provides fast response and possibly outbreak avoidance.

(continued)

TABLE 8.4 (Continued)
Nanosensors for identification of foodborne pathogens

Nanosensors Mechanism	Targeted Pathogen	Relevance to Food Sector
A highly sensitive colorimetric immuno-sensor utilizing cotton swabs and nanoparticles, capable of detecting concentrations ranging from 10 to 10^8 CFU/mL.	*Salmonella typhimurium*, *Salmonella enteritidis*, *S. aureus*, *Campylobacter jejuni*	With its visible detection and sensitivity range of 10–10^8 CFU/mL, this nanosensor technology is an essential tool for foodborne pathogen screening.
Platinum nanoparticles incorporated into pH-sensitive polymer nanobrushes.	*L. monocytogenes*	This nanosensor provides an innovative approach for detecting *L. monocytogenes* contamination in the food sector. Its pH-responsive nature enhances selectivity, and platinum nanoparticle incorporation enhances sensitivity, achieving limits of detection within the range of 10^2–10^3 CFU/mL.

Sources: Annapure and Sathyanarayana (2023); Gao et al. (2020); Kumar et al. (2021); Omar et al. (2020); Shin et al. (2022); Yemets et al. (2022).

and *Norovirus* have been successfully identified using these methods. Nanosensors offer highly precise and reliable detection systems that target specific biomarkers associated with these infections (Bruce-Tagoe & Danquah, 2023; Foddai & Grant, 2020). Here, Table 8.5, demonstrates the list of typical foodborne pathogens, the diseases they cause, and the foods they may contaminate.

8.4 FINDING TOXINS, ADULTERANTS, AND CONTAMINANTS

Nanosensors have shown to be quite successful at detecting poisons, adulterants, and pollutants in food products in addition to pathogens. When present in food, toxins produced by bacteria, fungi, or plants can pose serious health hazards. These harmful substances can pose serious health risks to consumers if present above permissible levels (Riu & Giussani, 2020). Nanosensors constructed on surface-enhanced Raman scattering (SERS) have been advanced to detect mycotoxins, such as aflatoxins, at extremely low concentrations. These nanosensors augment the Raman sign of the mycotoxins, enabling their sensitive and specific identification (Zhang et al., 2022). It can also be utilized to identify food adulteration and contamination. For instance, in the case of honey adulteration with sugar syrups, nanosensors can be functionalized with specific enzymes that react only with authentic honey components. Any deviation from the expected response indicates the presence of adulterants (Kulshreshtha et al., 2017). Similarly, nanosensors can be engineered to perceive the existence of contaminants in food samples. Heavy metals and pesticides, notorious for infiltrating food matrices, face formidable opposition from these nanoscale detectives. Engineered with functionalized nanoparticles, nanosensors exhibit the ability to selectively adsorb these contaminants from food samples. In doing so, they catalyze transformative changes in the nanoparticles' optical or electrical attributes. This metamorphosis, reflective of the extent of contamination, is subsequently quantified, yielding an expedited and precise assessment of food safety (Jeon et al., 2021; Mehrabi et al., 2023; Shin et al., 2022).

Pesticides, veterinary medications, and food additives are examples of chemical contaminants, while toxicants and pathogens are biological ones. As depicted in Figure 8.4, chemical and biological nanosensors are tools for detecting chemical and biological contaminants. When it comes to identifying a variety of pollutants, nanobiosensors are more reliable than chemical nanosensors.

8.5 MONITORING FOOD FRESHNESS WITH TIME-TEMPERATURE AND OXYGEN INDICATORS

In the food industry, it is crucial to ensure the quality and freshness of food goods. By adding time-temperature and oxygen indicators into food packaging materials, nanotechnology has enabled development of creative methods for tracking and evaluating the freshness of food (Jafarizadeh-Malmiri et al., 2019; Li et al., 2023). The temperature history is monitored by time-temperature indicators (TTIs) implanted with nanosensors. These indicators make use of nanomaterials having temperature-sensitive characteristics that vary in observable ways in response to temperature changes. Manufacturers and consumers can quickly identify whether the temperature conditions have risen above the advised limits, thereby jeopardizing the product's quality and safety, by including these indications in the food packing materials (Abedi-Firoozjah et al., 2023; Ma et al., 2022).

Another crucial way that nanosensors are used in food packaging is as oxygen indicators. The presence of oxygen can hasten the deterioration of food products. It causes spoilage by a decline in quality and the expansion of microbes. Continuous monitoring of oxygen levels inside the container is made possible by nanoscale oxygen sensors embedded into packaging materials. These sensors provide quantifiable signals in response to variations in oxygen content. An indication of the oxygen levels is given by the color change or fluorescence intensity (Abedi-Firoozjah et al., 2023; Ashfaq et al., 2022).

TABLE 8.5
A list of the most common bacterial causes of foodborne disease

Bacteria	Food Potentially Contaminated	Main Symptoms
Clostridium botulinum	Inadequately canned or fermented products, homemade preserves, honey	*C. botulinum*, often stemming from improperly canned or fermented goods, including homemade preserves and honey, can manifest as diminished strength, vertigo, impaired vision, difficulties in swallowing, and hindered speech abilities.
*Shigella*spp.	Prepped salads (potato, tuna, chicken, macaroni), raw veggies, dairy, bakery items	*Shigella* spp., typically associated with prepared salads, raw vegetables, dairy products, and bakery items, presents with symptoms such as abdominal discomfort, elevated body temperature, emesis, diarrhea with presence of blood, pus, or mucus in stools, as well as tenesmus.
Bacillus cereus	Various meats, dairy, vegetables, fish, rice dishes	*B. cereus*, commonly encountered in diverse foods encompassing meats, dairy items, vegetables, fish, and rice-based dishes, can incite instances of watery diarrhea, abdominal cramping, and simultaneous feelings of nausea and vomiting.
Campylobacter spp.	Raw chicken, beef, pork, shellfish, raw milk	*Campylobacter* spp., originating from raw meats like chicken, beef, pork, shellfish, and untreated milk, triggers a constellation of symptoms, including fluid-laden diarrhea, heightened body temperature, abdominal discomfort, queasiness, throbbing head pain, and muscular soreness.
Vibrio vulnificus	Raw or undercooked oysters, clams, crabs	*V. vulnificus*, typically stemming from inadequately cooked oysters, clams, and crabs, ushers in symptoms like fever, shivering chills, queasiness, and in individuals with underlying health conditions, it can potentially evolve into bloodstream infection.
Listeria monocytogenes	Raw milk, soft cheeses, raw veggies, meats, smoked fish, fermented sausages	*L. monocytogenes*, often present in raw milk, soft cheeses, untreated vegetables, meats, smoked fish, and fermented sausages, is responsible for eliciting symptoms like nausea, vomiting, diarrhea, and in severe scenarios, could even lead to severe conditions such as meningitis, encephalitis, bloodstream infection, pregnancy-related complications like abortion, and stillbirth.
Clostridium perfringens	Cooked meat, gravies, leftovers	*C. perfringens*, frequently associated with cooked meats, gravies, and leftover meals, brings about instances of severe abdominal cramps and diarrhea.
Yersinia enterocolitica	Meats, oysters, fish, unpasteurized milk	*Y. enterocolitica*, commonly present in meats, oysters, fish, and raw milk, leads to fever, abdominal discomfort, and episodes of diarrhea or vomiting.
Salmonella typhi/ paratyphi	Raw meats, eggs, dairy, seafood, sauces, dressings, tropical foods	*S. typhi* and *S. paratyphi*, typically stemming from raw meats, eggs, dairy, seafood, sauces, dressings, and even tropical foods, give rise to fever resembling typhoid fever, a sense of weariness, headaches, abdominal unease, generalized bodily aches, along with alternating occurrences of diarrhea and constipation.

Source: Zaid et al. (2019).

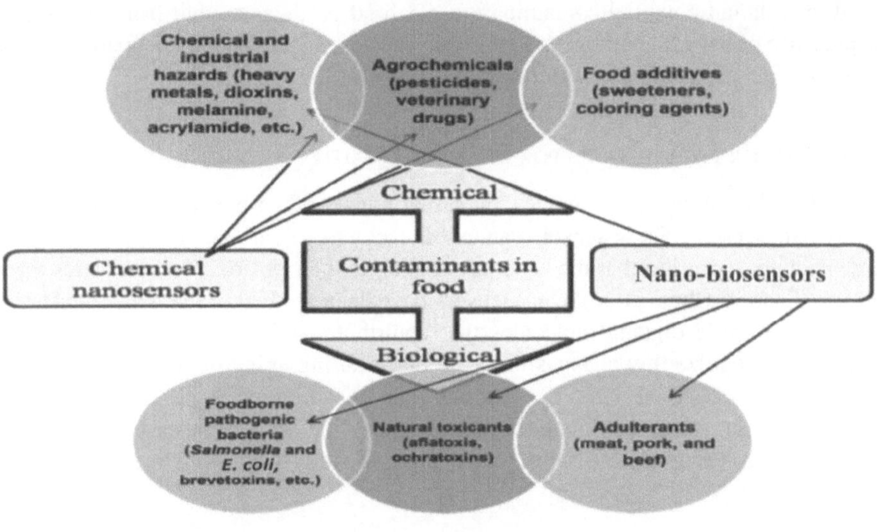

FIGURE 8.4 Nanosensors for the detection of food contaminants. Reprinted with written permission from Patel et al. (2020).

8.6 NANOSENSORSIN FOOD AND BEVERAGE PRODUCTS

An innovative technology that has the potential to completely alter the food and beverage sector is called nanosensors. These tiny, nanomaterial-made gadgets provide unmatched abilities for detecting, monitoring, and guaranteeing the safety and caliber of various food and beverage products. Their unique properties enhance product integrity, minimizing foodborne illnesses, and extending shelf life (Ummi & Siddiquee, 2019). Nanosensors offer the means to address challenges associated with food authenticity and adulteration. With the power to detect minute variations in composition, different nanosensors can distinguish between authentic and adulterated products, ensuring consumers receive what they expect (Castillo-Cambronero et al., 2022; Kannan & Guo, 2020).

8.7 USE OF NANOSENSORS IN NUMEROUS FOOD CATEGORIES

Nanosensors have made significant inroads into various food categories, bringing about transformative changes in assessment and monitoring systems. One prominent application is in dairy products, where nanosensors enable real-time monitoring of milk freshness and quality. For instance, nanosensors embedded in milk packaging can detect changes in pH, temperature, and bacterial activity. Eventually, providing consumers and producers with accurate information about milk's condition (Castillo-Cambronero et al., 2022; Verma et al., 2022). It is also reshaping the landscape of dietary fiber analysis in food products. As a pivotal component of human nutrition, dietary fiber's impact on digestive health and disease prevention cannot be overstated. Nanosensors, designed at the nanoscale, offer unprecedented accuracy and efficiency in quantifying dietary fiber components. These sensors, utilizing optical, electrochemical, and mechanical platforms, detect soluble and insoluble fibers, along with functional compounds like prebiotics (Adam & Gopinath, 2022; Mehadi Galib et al., 2022; Sahani & Sharma, 2021).

Fresh produce benefits from nanosensors integrated into packaging materials, enabling real-time monitoring of environmental conditions to extend shelf life. These nanosensors also find application in baked goods, cereals, and snacks, maintaining texture, taste, and nutritional content. Similarly, it contributes to the safety of meat and seafood products. For instance, in the case of packaged meat, nanosensors can detect gases released by bacterial spoilage, providing an early indication of microbial growth and potential contamination. Seafood products benefit from nanosensors that detect the presence of harmful pathogens like Vibrio species, minimizing the jeopardy of foodborne illnesses (Adetunji et al., 2022; Mohammad et al., 2022).

8.7.1 Nanosensor Technologies for Beverage Quality Assessment

The safety and quality of bottled beverages, including juices and carbonated beverages, are closely monitored by nanosensors. These sensors can assess factors like carbonation levels, pH shifts, and temperature changes, ensuring that the beverages maintain their desired characteristics throughout their shelf life (Saari & Chua, 2020). Nanosensors have been applied to the wine industry to identify and measure volatile organic molecules that contribute to flavor and aroma profiles. These nanosensors may recognize tiny changes that take place during aging or storage as well as distinguish between various kinds of wine. Furthermore, nanosensors in wine corks can detect spoilage due to oxidation, helping maintain quality. In the brewing industry, nanosensors assess fermentation processes, contributing to consistent beer quality (Castillo-Cambronero et al., 2022; Kumar et al., 2017).

8.7.2 Automated Nanotechnology-Driven Packaging for Real-Time Inspection

Automated packaging solutions based on nanotechnology have the advantage of providing feedback and constant monitoring, safeguarding the safety and nutritional value of food goods throughout their shelf life. These advanced packaging systems are embedded with nanosensors capable of detecting various parameters, such as temperature, humidity, gas composition, and even the presence of specific contaminants (Abedi-Firoozjah et al., 2023; He et al., 2019). The data collected by these nanosensors can be wirelessly transmitted to monitoring devices or cloud-based platforms, enabling real-time tracking and analysis. For instance, smart labels equipped with nanosensors can provide instant information about temperature changes, oxygen levels, and other environmental factors that can impact food freshness. This technology enhances consumer confidence and transparency in the food supply chain. Additionally, smart labels employing nanosensors can detect contaminants, such as pathogens or chemical residues, offering a proactive approach to food safety. Nanosensors have thus become indispensable tools in the food and beverage industry, enhancing consumer safety, reducing waste, and optimizing product quality. Their applications extend to a wide range of products, ensuring that the food we consume is of the highest quality and safety standards (Malik et al., 2023; Shawon et al., 2020; Tawade & Wasewar, 2023).

8.8 SYNERGIZING ARTIFICIAL INTELLIGENCE WITH NANOSENSORS

The convergence of nanosensor technology with artificial intelligence (AI) has led to a new era of precision and efficiency in food safety and quality control. AI is capable of processing and interpreting complex datasets with remarkable accuracy. Machine learning algorithms can recognize subtle patterns and trends that might elude human observation, enabling early detection of potential hazards (Camaréna, 2020; Pereira, 2023). AI can optimize nanosensor performance by predicting their behavior under different conditions. This synergy expedites the development of nanosensor-based detection platforms with unprecedented reliability. The application of AI to nanosensors in food safety holds promise not only in detection but also in predictive analytics, risk assessment, and supply chain optimization. By automating decision-making processes, the integration of AI with nanosensors ensures safer and more efficient food production, reducing the potential for contamination and improving overall consumer trust (Camaréna, 2020; Nishant et al., 2020).

8.9 STRATEGIES TO PROMOTE NANOSENSOR ADOPTION IN FOOD

The successful integration of nanosensors into the food industry requires careful consideration to ensure widespread adoption and maximum benefit. To facilitate the incorporation of nanosensors, establishing clear regulatory guidelines is essential. Collaborations between regulatory bodies, research institutions, and industry stakeholders can create standardized protocols for nanosensor safety assessment, deployment, and labeling (Castillo-Cambronero et al., 2022; Yang & Duncan, 2021).

Efficient transfer of nanosensor technology from academia to industry is vital. Scientists, engineers, food technologists, and policymakers facilitate holistic problem-solving. Multidisciplinary teams can address technical challenges, optimize sensor performance, and develop tailored solutions for specific food safety concerns. Developing cost-effective manufacturing processes is essential to ensure affordability for both producers and consumers. Scalable production methods can reduce sensor costs, making them more accessible for widespread use (Rani Sarkar et al., 2022; Ur Rahim et al., 2021). Implementing transparent data handling practices and stringent ethical standards can build trust and encourage responsible nanosensor use. Governments and funding agencies can incentivize innovation in nanosensor technology by providing grants, research funding, and intellectual property protection (Javaid et al., 2021; Patel et al., 2020). These approaches can be combined to help get over the challenges of incorporating nanosensors into the food sector, improving food safety, quality, and consumer satisfaction.

8.10 CONCLUSION

The integration of nanosensors into food safety has introduced a paradigm shift in pathogen detection strategies. The exceptional sensitivity, specificity, and rapidity of nanosensors promise a substantial reduction in foodborne illnesses, instilling greater confidence in consumers.

The utilization of nanomaterials, particularly gold and silver nanoparticles, combined with sophisticated detection techniques, holds the promise of higher precision in detecting diverse pathogenic microorganisms. Additionally, the amalgamation of nanosensors with artificial intelligence not only enhances their analytical capabilities but also facilitates real-time data interpretation, allowing for swift intervention to mitigate potential foodborne outbreaks. Despite these promising prospects, the journey towards widespread nanosensor adoption faces numerous challenges such as regulatory frameworks, ethical considerations, and seamless technology transfer from research to industrial applications. Collaborative efforts between researchers, regulatory bodies, industry stakeholders, and consumers are imperative to address these hurdles effectively.

In conclusion, nanosensors are poised to revolutionize the food industry by ensuring safer and more technologically advanced products. Continuous research, innovation, and responsible implementation are critical to couple the full potential of nanosensors for pathogen detection in food products.

REFERENCES

Abedi-Firoozjah, R., Salim, S. A., Hasanvand, S., Assadpour, E., Azizi-Lalabadi, M., Prieto, M. A., & Jafari, S. M. (2023). Application of smart packaging for seafood: A comprehensive review. *Comprehensive Reviews in Food Science and Food Safety*, *22*(2), 1438–1461. https://doi.org/https://doi.org/10.1111/1541-4337.13117

Adam, T., & Gopinath, S. C. B. (2022). Nanosensors: Recent perspectives on attainments and future promise of downstream applications. *Process Biochemistry*, *117*, 153–173. https://doi.org/https://doi.org/10.1016/j.procbio.2022.03.024

Adetunji, C. O., Ogundolie, F. A., Ajiboye, M. D., Mathew, J. T., Inobeme, A., Titilayo, O., Olaniyan, O. T., Ijabadeniyi, O. A., Ajayi, O. O., & Dauda, W. P. (2022). Chapter 2 – Bio-and nanosensors in the food industry. In A. K. Sukhla (Ed.), *Bio-and Nano-Sensing Technologies for Food Processing and Packaging* (pp. 22–36). Royal Society of Chemistry.

Agarwal, N., Solanki, V. S., Pare, B., Singh, N., & Jonnalagadda, S. B. (2023). Current trends in nanocatalysis for green chemistry and its applications- a mini-review. *Current Opinion in Green and Sustainable Chemistry*, *41*, 100788. https://doi.org/https://doi.org/10.1016/j.cogsc.2023.100788

Agnihotri, A. S., Chungath George, A. M., & Marimuthu, N. (2022). Nanobiosensors: A promising tool for the determination of pathogenic bacteria. In S. Hameed & S. Rehman (Eds.), *Nanotechnology for Infectious Diseases* (pp. 475–495). Springer Singapore. https://doi.org/10.1007/978-981-16-9190-4_21

Alfei, S., Marengo, B., & Zuccari, G. (2020). Nanotechnology application in food packaging: A plethora of opportunities versus pending risks assessment and public concerns. *Food Research International*, *137*, 109664. https://doi.org/https://doi.org/10.1016/j.foodres.2020.109664

Annapure, U. S., & Sathyanarayana, S. R. S. (2023). Chapter 12 – Liposomes as biosensors in the food sector. In C. Anandharamakrishnan & S. Dutta (Eds.), *Liposomal Encapsulation in Food Science and Technology* (pp. 239–254). Academic Press. https://doi.org/https://doi.org/10.1016/B978-0-12-823 935-3.00013-8

Ashfaq, A., Khursheed, N., Fatima, S., Anjum, Z., & Younis, K. (2022). Application of nanotechnology in food packaging: Pros and cons. *Journal of Agriculture and Food Research*, *7*, 100270. https://doi.org/https:// doi.org/10.1016/j.jafr.2022.100270

Bhagat, B., Baruah, P., & Mukherjee, K. (2022). 32 – Application of nanosensors in food inspection. In A. Denizli, T. A. Nguyen, S. Rajendran, G. Yasin, & A. K. Nadda (Eds.), *Nanosensors for Smart Agriculture* (pp. 705–735). Elsevier. https://doi.org/https://doi.org/10.1016/B978-0-12-824554-5.00030-6

Biswas, R., Alam, M., Sarkar, A., Haque, M. I., Hasan, M. M., & Hoque, M. (2022). Application of nanotechnology in food: processing, preservation, packaging and safety assessment. *Heliyon*, *8*(11), e11795. https://doi.org/https://doi.org/10.1016/j.heliyon.2022.e11795

Bobrinetskiy, I., Radovic, M., Rizzotto, F., Vizzini, P., Jaric, S., Pavlovic, Z., Radonic, V., Nikolic, M. V., & Vidic, J. (2021). Advances in nanomaterials-based electrochemical biosensors for foodborne pathogen detection. *Nanomaterials*, *11*(10), 2700. https://doi.org/https://doi.org/10.3390/nano11102700

Bruce-Tagoe, T. A., & Danquah, M. K. (2023). Bioaffinitynanoprobes for foodborne pathogen sensing. *Micromachines*, *14*(6), 1122. https://doi.org/https://www.mdpi.com/2072-666X/14/6/1122

Bruna, T., Maldonado-Bravo, F., Jara, P., & Caro, N. (2021). Silver nanoparticles and their antibacterial applications. *International Journal of Molecular Sciences*, *22*(13), 7202. https://doi.org/https://doi.org/ 10.3390/ijms22137202

Camaréna, S. (2020). Artificial intelligence in the design of the transitions to sustainable food systems. *Journal of Cleaner Production*, *271*, 122574. https://doi.org/https://doi.org/10.1016/j.jclepro.2020.122574

Cardoso, R. M., Pereira, T. S., Facure, M. H. M., dos Santos, D. M., Mercante, L. A., Mattoso, L. H. C., & Correa, D. S. (2022). Current progress in plant pathogen detection enabled by nanomaterials-based (bio)sensors. *Sensors and Actuators Reports*, *4*, 100068. https://doi.org/https://doi.org/10.1016/j.snr.2021.100068

Castillo-Cambronero, G., Cascante-Matarrita, M., Mora-Chacón, A., Castillo-Henríquez, L., & Vega-Baudrit, J. R. (2022). 29 – Use of nanosensor technologies in the food industry. In A. Denizli, T. A. Nguyen, S. Rajendran, G. Yasin, & A. K. Nadda (Eds.), *Nanosensors for Smart Agriculture* (pp. 643–655). Elsevier. https://doi.org/https://doi.org/10.1016/B978-0-12-824554-5.00035-5

Cesewski, E., & Johnson, B. N. (2020). Electrochemical biosensors for pathogen detection. *Biosensors and Bioelectronics*, *159*, 112214. https://doi.org/https://doi.org/10.1016/j.bios.2020.112214

Chandra, P., & Panesar, P. S. (2022). *Nanosensing and Bioanalytical Technologies in Food Quality Control*. Springer. https://doi.org/https://doi.org/10.1007/978-981-16-7029-9

Chelliah, R., Wei, S., Daliri, E. B., Rubab, M., Elahi, F., Yeon, S.-J., Jo, K. H., Yan, P., Liu, S., & Oh, D. H. (2021). Development of nanosensors based intelligent packaging systems: food quality and medicine. *Nanomaterials*, *11*(6), 1515. https://doi.org/https://doi.org/10.3390/nano11061515

Cheng, H., Xu, H., Julian McClements, D., Chen, L., Jiao, A., Tian, Y., Miao, M., & Jin, Z. (2022). Recent advances in intelligent food packaging materials: Principles, preparation and applications. *Food Chemistry*, *375*, 131738. https://doi.org/https://doi.org/10.1016/j.foodchem.2021.131738

Dester, E., Kao, K., & Alocilja, E. C. (2022). Detection of unamplified *E. coli* O157 DNA extracted from large food samples using a gold nanoparticle colorimetric biosensor. *Biosensors*, *12*(5), 274. https://doi.org/ https://doi.org/10.3390/bios12050274

Emamhadi, M. A., Sarafraz, M., Akbari, M., Thai, V. N., Fakhri, Y., Linh, N. T. T., & Mousavi Khaneghah, A. (2020). Nanomaterials for food packaging applications: A systematic review. *Food and Chemical Toxicology*, *146*, 111825. https://doi.org/https://doi.org/10.1016/j.fct.2020.111825

Farré, M., Barceló, D., & Barceló, D. (2013). Analysis of emerging contaminants in food. *TrAC Trends in Analytical Chemistry*, *43*, 240–253. https://doi.org/https://doi.org/10.1016/j.trac.2012.12.003

Foddai, A. C. G., & Grant, I. R. (2020). Methods for detection of viable foodborne pathogens: Current state-of-art and future prospects. *Applied Microbiology and Biotechnology*, *104*(10), 4281–4288. https://doi.org/https://doi.org/10.1007/s00253-020-10542-x

Fung, F., Wang, H.-S., & Menon, S. (2018). Food safety in the 21st century. *Biomedical Journal*, *41*(2), 88–95. https://doi.org/https://doi.org/10.1016/j.bj.2018.03.003

Gao, H., Yan, C., Wu, W., & Li, J. (2020). Application of microfluidic chip technology in food safety sensing. *Sensors*, *20*(6), 1792. https://doi.org/https://doi.org/10.3390/s20061792

He, X., Deng, H., & Hwang, H.-M. (2019). The current application of nanotechnology in food and agriculture. *Journal of Food and Drug Analysis*, *27*(1), 1–21. https://doi.org/https://doi.org/10.1016/j.jfda.2018.12.002

Huang, X., Zhu, Y., & Kianfar, E. (2021). Nano biosensors: Properties, applications and electrochemical techniques. *Journal of Materials Research and Technology*, *12*, 1649–1672. https://doi.org/https://doi.org/10.1016/j.jmrt.2021.03.048

Jafarizadeh-Malmiri, H., Sayyar, Z., Anarjan, N., & Berenjian, A. (2019). Nano-sensors in food nanobiotechnology. In H. Jafarizadeh-Malmiri, Z. Sayyar, N. Anarjan, & A. Berenjian (Eds.), *Nanobiotechnology in Food: Concepts, Applications and Perspectives* (1st ed., pp. 81–94). Springer International Publishing. https://doi.org/https://doi.org/10.1007/978-3-030-05846-3

Jagtiani, E. (2022). Advancements in nanotechnology for food science and industry. *Food Frontiers*, *3*(1), 56–82. https://doi.org/https://doi.org/10.1002/fft2.104

Javaid, M., Haleem, A., Singh, R. P., Rab, S., & Suman, R. (2021). Exploring the potential of nanosensors: A brief overview. *Sensors International*, *2*, 100130. https://doi.org/https://doi.org/10.1016/j.sintl.2021.100130

Jeon, Y., Kim, D., Kwon, G., Lee, K., Oh, C.-S., Kim, U.-J., & You, J. (2021). Detection of nanoplastics based on surface-enhanced Raman scattering with silver nanowire arrays on regenerated cellulose films. *Carbohydrate Polymers*, *272*, 118470. https://doi.org/https://doi.org/10.1016/j.carbpol.2021.118470

Kannan, P., & Guo, L. (2020). Chapter 20 –Nanosensors for food safety. In B. Han, V. K. Tomer, T. A. Nguyen, A. Farmani, & P. Kumar Singh (Eds.), *Nanosensors for SmartCities* (pp. 339–354). Elsevier. https://doi.org/https://doi.org/10.1016/B978-0-12-819870-4.00019-0

Keçili, R., Büyüktiryaki, S., & Hussain, C. M. (2019). Advancement in bioanalytical science through nanotechnology: Past, present and future. *TrAC Trends in Analytical Chemistry*, *110*, 259–276. https://doi.org/https://doi.org/10.1016/j.trac.2018.11.012

Kulshreshtha, N. M., Shrivastava, D., & Bisen, P. S. (2017). Chapter 14 – Contaminant sensors: Nanotechnology-based contaminant sensors. In A. M. Grumezescu (Ed.), *Nanobiosensors* (pp. 573–628). Academic Press. https://doi.org/https://doi.org/10.1016/B978-0-12-804301-1.00014-X

Kumar, V., Guleria, P., & Mehta, S. (2017). Nanosensors for food quality and safety assessment. *Environmental Chemistry Letters*, *15*, 165–177. https://doi.org/10.1007/s10311-017-0616-4

Kumar, V., Guleria, P., Ranjan, S., Dasgupta, N., & Lichtfouse, E. (2021). *Nanotoxicology and Nanoecotoxicology* (1 ed., Vol. 1). Springer Cham. https://doi.org/https://doi.org/10.1007/978-3-030-63241-0

Leva-Bueno, J., Peyman, S. A., & Millner, P. A. (2020). A review on impedimetricimmunosensors for pathogen and biomarker detection. *Medical Microbiology and Immunology*, *209*(3), 343–362. https://doi.org/10.1007/s00430-020-00668-0

Li, X., Liu, D., Pu, Y., & Zhong, Y. (2023). Recent advance of intelligent packaging aided by artificial intelligence for monitoring food freshness. *Foods*, *12*(15), 2976. https://doi.org/https://doi.org/10.3390/foods12152976

Li, Y.-X., Qin, H.-Y., Hu, C., Sun, M.-M., Li, P.-Y., Liu, H., Li, J.-C., Li, Z.-B., Wu, L.-D., & Zhu, J. (2022). Research progress of nanomaterials-based sensors for food safety. *Journal of Analysis and Testing*, *6*(4), 431–440. https://doi.org/10.1007/s41664-022-00235-x

Ma, Y., Yang, W., Xia, Y., Xue, W., Wu, H., Li, Z., Zhang, F., Qiu, B., & Fu, C. (2022). Properties and applications of intelligent packaging indicators for food spoilage. *Membranes*, *12*(5), 477. https://doi.org/https://doi.org/10.3390/membranes12050477

Malhotra, B. D., Srivastava, S., Ali, M. A., & Singh, C. (2014). Nanomaterial-based biosensors for food toxin detection. *Applied Biochemistry and Biotechnology, 174*(3), 880–896. https://doi.org/https://doi.org/10.1007/s12010-014-0993-0

Malik, S., Muhammad, K., & Waheed, Y. (2023). Nanotechnology: A revolution in modern industry. *Molecules, 28*(2), 661. https://doi.org/https://doi.org/10.3390/molecules28020661

MehadiGalib, R., Alam, M., Rana, R., & Ara, R. (2022). Mango (*Mangifera indica* L.) fiber concentrates: Processing, modification and utilization as a food ingredient. *Food Hydrocolloids for Health, 2*, 100096. https://doi.org/https://doi.org/10.1016/j.fhfh.2022.100096

Mehrabi, M. R., Soltani, M., Chiani, M., Raahemifar, K., & Farhangi, A. (2023). Nanomedicine: New frontiers in fighting microbial infections. *Nanomaterials, 13*(3), 483. https://doi.org/https://doi.org/10.3390/nano13030483

Mishra, S., Mishra, S., Dhiman, A., & Singh, R. (2023). Chapter 10 – Nanosensors for the detections of foodborne pathogens and toxins. In A. Sharma, P. S. Vijayakumar, E. P. K. Prabhakar, & R. Kumar (Eds.), *Nanotechnology Applications for Food Safety and Quality Monitoring* (pp. 183–204). Academic Press. https://doi.org/https://doi.org/10.1016/B978-0-323-85791-8.00015-X

Mohammad, Z. H., Ahmad, F., Ibrahim, S. A., & Zaidi, S. (2022). Application of nanotechnology in different aspects of the food industry. *Discover Food, 2*(1), 12. https://doi.org/10.1007/s44187-022-00013-9

Navia-Mendoza, J. M., Filho, O. A., Zambrano-Intriago, L. A., Maddela, N. R., Duarte, M. M., Quiroz-Fernández, L. S., Baquerizo-Crespo, R. J., & Rodríguez-Díaz, J. M. (2021). Advances in the application of nanocatalysts in photocatalytic processes for the treatment of food dyes: A review. *Sustainability, 13*(21), 11676. https://doi.org/https://doi.org/10.3390/su132111676

Nguyen, Q. H., & Kim, M. I. (2020). Nanomaterial-mediated paper-based biosensors for colorimetric pathogen detection. *TrAC Trends in Analytical Chemistry, 132*, 116038. https://doi.org/https://doi.org/10.1016/j.trac.2020.116038

Nile, S. H., Baskar, V., Selvaraj, D., Nile, A., Xiao, J., & Kai, G. (2020). Nanotechnologies in food science: Applications, recent trends, and future perspectives. *Nano-Micro Letters, 12*(1), 45. https://doi.org/10.1007/s40820-020-0383-9

Nishant, R., Kennedy, M., & Corbett, J. (2020). Artificial intelligence for sustainability: Challenges, opportunities, and a research agenda. *International Journal of Information Management, 53*, 102104. https://doi.org/https://doi.org/10.1016/j.ijinfomgt.2020.102104

Omar, S. S., Haddad, M. A., & Parisi, S. (2020). Validation of HPLC and enzyme-linked immunosorbent assay (ELISA) techniques for detection and quantification of aflatoxins in different food samples. *Foods, 9*(5), 661. https://doi.org/https://doi.org/10.3390/foods9050661

Patel, G., Pillai, V., Bhatt, P., & Mohammad, S. (2020). Chapter 21 – Application of nanosensors in the food industry. In B. Han, V. K. Tomer, T. A. Nguyen, A. Farmani, & P. Kumar Singh (Eds.), *Nanosensors for Smart Cities* (pp. 355–368). Elsevier. https://doi.org/https://doi.org/10.1016/B978-0-12-819 870-4.00020-7

Patel, R., Mitra, B., Vinchurkar, M., Adami, A., Patkar, R., Giacomozzi, F., Lorenzelli, L., & Baghini, M. S. (2023). Plant pathogenicity and associated/related detection systems. A review. *Talanta, 251*, 123808. https://doi.org/https://doi.org/10.1016/j.talanta.2022.123808

Pereira, G. C. (2023). Novel nanotechnology-driven prototypes for AI-enriched implanted prosthetics following organ failure. In G. C. Pereira (Ed.), *Gene, Drug, and Tissue Engineering* (pp. 195–237). Springer US. https://doi.org/10.1007/978-1-0716-2716-7_10

Qiao, X., He, J., Yang, R., Li, Y., Chen, G., Xiao, S., Huang, B., Yuan, Y., Sheng, Q., & Yue, T. (2022). Recentadvances in nanomaterial-based sensing for food safety analysis. *Processes, 10*(12), 2576. https://doi.org/https://doi.org/10.3390/pr10122576

Ramezani, M., Esmaelpourfarkhani, M., Taghdisi, S. M., Abnous, K., & Alibolandi, M. (2020). Chapter 22 – Application of nanosensors for food safety. In B. Han, V. K. Tomer, T. A. Nguyen, A. Farmani, & P. Kumar Singh (Eds.), *Nanosensors for Smart Cities* (pp. 369–386). Elsevier. https://doi.org/https://doi.org/10.1016/B978-0-12-819870-4.00021-9

Rani Sarkar, M., Rashid, M. H.-o., Rahman, A., Kafi, M. A., Hosen, M. I., Rahman, M. S., & Khan, M. N. (2022). Recent advances in nanomaterials based sustainable agriculture: An overview. *Environmental Nanotechnology, Monitoring & Management, 18*, 100687. https://doi.org/https://doi.org/10.1016/j.enmm.2022.100687

Ravichandran, M., Kanniah, P., & Kasi, M. (2021). Chapter 5 – Silver nanoparticles as nanomaterial-based nanosensors in agri-food sector. In K. A. Abd-Elsalam (Ed.), *Silver Nanomaterials for Agri-Food Applications* (pp. 103–123). Elsevier. https://doi.org/https://doi.org/10.1016/B978-0-12-823528-7.00023-8

Riu, J., & Giussani, B. (2020). Electrochemical biosensors for the detection of pathogenic bacteria in food. *TrAC Trends in Analytical Chemistry*, *126*, 115863. https://doi.org/https://doi.org/10.1016/j.trac.2020.115863

Saari, N. H. M., & Chua, L. S. (2020). 13 – Nano-based products in beverage industry. In A. M. Grumezescu & A. M. Holban (Eds.), *Nanoengineering in the Beverage Industry* (pp. 405–436). Academic Press. https://doi.org/https://doi.org/10.1016/B978-0-12-816677-2.00014-4

Sahani, S., & Sharma, Y. C. (2021). Advancements in applications of nanotechnology in global food industry. *Food Chemistry*, *342*, 128318. https://doi.org/https://doi.org/10.1016/j.foodchem.2020.128318

Sahoo, M., Vishwakarma, S., Panigrahi, C., & Kumar, J. (2021). Nanotechnology: Current applications and future scope in food. *Food Frontiers*, *2*(1), 3–22. https://doi.org/https://doi.org/10.1002/fft2.58

Saini, A., Panwar, D., Panesar, P. S., & Chandra, P. (2022). Potential of nanotechnology in food analysis and quality improvement.In P. Chandra & P. S. Panesar (Eds.), *Nanosensing and Bioanalytical Technologies in Food Quality Control* (pp. 169–194). Springer Singapore. https://doi.org/10.1007/978-981-16-7029-9_8

Saravanan, A., Kumar, P. S., Hemavathy, R. V., Jeevanantham, S., Kamalesh, R., Sneha, S., & Yaashikaa, P. R. (2021). Methods of detection of food-borne pathogens: A review. *Environmental Chemistry Letters*, *19*(1), 189–207. https://doi.org/10.1007/s10311-020-01072-z

Sarfraz, J., Gulin-Sarfraz, T., Nilsen-Nygaard, J., & Pettersen, M. K. (2021). Nanocomposites for food packaging applications: An overview. *Nanomaterials*, *11*(1), 10. https://doi.org/https://doi.org/10.3390/nano11010010

Shah, B. R., & Mraz, J. (2020). Advances in nanotechnology for sustainable aquaculture and fisheries. *Reviews in Aquaculture*, *12*(2), 925–942. https://doi.org/https://doi.org/10.1111/raq.12356

Sharafeldin, M., & Davis, J. J. (2021). Point of care sensors for infectious pathogens. *Analytical Chemistry*, *93*(1), 184–197. https://doi.org/10.1021/acs.analchem.0c04677

Sharifi, S., Vahed, S. Z., Ahmadian, E., Dizaj, S. M., Eftekhari, A., Khalilov, R., Ahmadi, M., Hamidi-Asl, E., & Labib, M. (2020). Detection of pathogenic bacteria via nanomaterials-modified aptasensors. *Biosensors and Bioelectronics*, *150*, 111933. https://doi.org/https://doi.org/10.1016/j.bios.2019.111933

Shawon, Z. B. Z., Hoque, M. E., & Chowdhury, S. R. (2020). Chapter 6 –Nanosensors and nanobiosensors: Agricultural and food technology aspects. In K. Pal & F. Gomes (Eds.), *Nanofabrication for Smart Nanosensor Applications* (pp. 135–161). Elsevier. https://doi.org/https://doi.org/https://doi.org/10.1016/B978-0-12-820702-4.00006-4

Shin, J. H., Reddy, Y. V. M., Park, T. J., & Park, J. P. (2022). Recent advances in analytical strategies and microsystems for food allergen detection. *Food Chemistry*, *371*, 131120. https://doi.org/https://doi.org/10.1016/j.foodchem.2021.131120

Singh, R., Dutt, S., Sharma, P., Sundramoorthy, A. K., Dubey, A., Singh, A., & Arya, S. (2023). Future of nanotechnology in food industry: Challenges in processing, packaging, and food safety. *Global Challenges*, *7*(4), 2200209. https://doi.org/https://doi.org/10.1002/gch2.202200209

Solís, D., Toro, M., Navarrete, P., Faúndez, P., & Reyes-Jara, A. (2022). Microbiological quality and presence of foodborne pathogens in raw and extruded canine diets and canine fecal samples. *Frontiers in Veterinary Science*, *9*, 799710. https://doi.org/https://doi.org/10.3389/fvets.2022.799710

Sondhi, P., Maruf, M. H., & Stine, K. J. (2020). Nanomaterials for biosensinglipopolysaccharide. *Biosensors*, *10*(1), 2. https://doi.org/https://doi.org/10.3390/bios10010002

Taj, A., Zia, R., Iftikhar, M., Younis, S., & Bajwa, S. Z. (2022). Chapter 31 –Nanosensors for food inspection. In A. Denizli, T. A. Nguyen, S. Rajendran, G. Yasin, & A. K. Nadda (Eds.), *Nanosensors for Smart Agriculture* (pp. 685–703). Elsevier. https://doi.org/https://doi.org/10.1016/B978-0-12-824554-5.00032-X

Tavakoli, H., Mohammadi, S., Li, X., Fu, G., & Li, X. (2022). Microfluidic platforms integrated with nanosensors for point-of-care bioanalysis. *TrAC Trends in Analytical Chemistry*, *157*, 116806. https://doi.org/https://doi.org/10.1016/j.trac.2022.116806

Tavker, N., & Sharma, M. (2022). Chapter 33 –Nanosensors for intelligent food packaging. In A. Denizli, T. A. Nguyen, S. Rajendran, G. Yasin, & A. K. Nadda (Eds.), *Nanosensors for Smart Agriculture* (pp. 737–756). Elsevier. https://doi.org/https://doi.org/10.1016/B978-0-12-824554-5.00014-8

Tawade, P. V., & Wasewar, K. L. (2023). Chapter 4 – Nanotechnology in biological science and engineering. In P. Singh, V. Kumar, M. Bakshi, C. M. Hussain, & M. Sillanpää (Eds.), *Environmental Applications of Microbial Nanotechnology* (pp. 43–64). Elsevier. https://doi.org/https://doi.org/10.1016/B978-0-323-91744-5.00015-1

Thakur, M., Wang, B., & Verma, M. L. (2022). Development and applications of nanobiosensors for sustainable agricultural and food industries: Recent developments, challenges and perspectives. *Environmental Technology & Innovation*, *26*, 102371. https://doi.org/https://doi.org/10.1016/j.eti.2022.102371

Ummi, A. S., & Siddiquee, S. (2019). Nanotechnology applications in food: Opportunities and challenges in food industry. In S. Siddiquee, G. J. H. Melvin, & M. M. Rahman (Eds.), *Nanotechnology: Applications in Energy, Drug and Food* (pp. 295–308). Springer International Publishing. https://doi.org/10.1007/978-3-319-99602-8_15

Ur Rahim, H., Qaswar, M., Uddin, M., Giannini, C., Herrera, M. L., & Rea, G. (2021). Nano-enable materials promoting sustainability and resilience in modern agriculture. *Nanomaterials*, *11*(8), 2068. https://doi.org/https://doi.org/10.3390/nano11082068

Verma, D. K., Thakur, M., Tripathy, S., Mohapatra, B., Singh, S., Patel, A. R., Gupta, A. K., Chávez-González, M. L., Srivastav, P. P., Sandoval-Cortes, J., & Aguilar, C. N. (2022). Chapter 7 – Emerging biosensor technology and its potential application in food. In H. Thatoi, S. Mohapatra, & S. K. Das (Eds.), *Innovations in Fermentation and Phytopharmaceutical Technologies* (pp. 127–163). Elsevier. https://doi.org/https://doi.org/10.1016/B978-0-12-821877-8.00017-8

Yang, T., & Duncan, T. V. (2021). Challenges and potential solutions for nanosensors intended for use with foods. *Nature Nanotechnology*, *16*(3), 251–265. https://doi.org/10.1038/s41565-021-00867-7

Yemets, A., Plokhovska, S., Pushkarova, N., & Blume, Y. (2022). Quantum dot-antibody conjugates for immunofluorescence studies of biomolecules and subcellular structures. *Journal of Fluorescence*, *32*(5), 1713–1723. https://doi.org/10.1007/s10895-022-02968-5

Zaid, M. H. M., Saidykhan, J., Abdullah, J. J. N. A. i. E., Drug, & Food. (2019). Nanosensors based detection of foodborne pathogens. In S. Siddiquee, G. J. H. Melvin, & M. M. Rahman (Eds.), *Nanotechnology: Applications in Energy, Drug and Food* (pp. 377–422). Springer International Publishing. https://doi.org/https://doi.org/10.1007/978-3-319-99602-8_19

Zhang, P., Wang, Y., Zhao, X., Ji, Y., Mei, R., Fu, L., Man, M., Ma, J., Wang, X., & Chen, L. (2022). Surface-enhanced Raman scattering labeled nanoplastic models for reliable bio-nano interaction investigations. *Journal of Hazardous Materials*, *425*, 127959. https://doi.org/https://doi.org/10.1016/j.jhazmat.2021.127959

Zorraquín-Peña, I., Cueva, C., Bartolomé, B., & Moreno-Arribas, M. V. (2020). Silver nanoparticles against foodborne bacteria. Effects at intestinal level and health limitations. *Microorganisms*, *8*(1), 132. https://doi.org/https://doi.org/10.3390/microorganisms8010132

9 Impact of Nanomaterials on Human Health through/via Food Chain

Fahima Siddikey, Md. Istiakh Jahan, Syeda Sabrina Akter, A S M Sayem, and Mahabub Alam

9.1 INTRODUCTION

The start of the first industrial revolution in the mid-1700s stood as a pivotal point in shaping the course of the global economy's history. In recent years, perhaps nothing has had a greater effect on the worldwide economy than the emergence of nanotechnology. The world's nanotechnology market, valued at around US$ 85.39 billion in 2021, is projected to surpass an impressive US$ 288.71 billion by the year 2030. The remarkable chemical and physical properties of nanomaterials, which are intricately tied to their size, have played a significant role in their development. Nanotechnology is playing an important role in the modernization and economic growth of numerous industries, as nanomaterials are distinguished by specific characteristics like surface coating, surface charge, surface area, degree of agglomeration, and particle morphology (Subhan et al., 2021). Nanomaterials are primarily defined based on their size, with materials having diameters ranging from 1 to 100 nm referred to as nanomaterials or nanoparticles (NPs). While researchers in the USA and the European Union (EU) employ various rules and regulations to determine the diameter of NPs and nanomaterials, currently, there exists no universally agreed-upon definition for nanomaterials. Furthermore, different organizations and institutions have their own distinct definitions of nanomaterials (Dekant & Klaunig, 2016).

The application of nanotechnology brings a multitude of benefits across diverse sectors such as medicine, pharmaceuticals, industry, agriculture, and food (Biswas et al., 2022). With its potential to revolutionize food processing, nanotechnologies emerge as a compelling alternative, guaranteeing safety standards and superior quality for food items while fostering a healthier food pattern. Nanomaterials can be comprised of individual components, such as metals or carbon, and also encompass mixtures of elements, like metal oxides or composite arrangements. Nanotechnology is a quickly advancing research field with diverse uses, but its environmental impact remains uncertain. Issues such as ecological, ethical, health and safety, policy, and regulatory concerns arise as NPs are released into the environment, particularly into the soil. NPs can affect soil properties, form toxic compounds, interact with other pollutants, and persist for extended periods, disrupting soil microbial communities, and plant growth, and potentially impacting human health through the food chain. The engineered NPs have the potential to be composed of either metal or carbon materials. Metallic NPs may be classified into three major segments: metals, quantum dots, and metal oxides.

The food chain is an ecological concept that illustrates the flow of energy and nutrients across a succession of organisms within an ecosystem. It depicts the interconnected relationship between decomposers (fungi and bacteria), consumers (carnivores and herbivores), and producers (plants) as they rely on each other for sustenance (Gerardi, 2023). The food chain is of paramount importance in upholding ecological balance and holds direct implications for human well-being. The significance

DOI: 10.1201/9781003438168-9

of the food chain is crucial for human health as it allows the transmission of contaminants like NPs across trophic levels. These particles have the ability to be taken up by plants and form connections with organisms situated at both higher and lower levels of the food chain (Keller & Lazareva, 2014).

The existence of NPs in nature is a subject of increasing concern because of their potential sources, including industrial and consumer products. Upon the integration of NPs into plants, there exists the potential for their transference to higher trophic levels through the consumption of these plants or other organisms within the food chain by animals. Once nanomaterials are internalized by active cells, then they have the potential to accumulate within the membrane of the cell and many regions of the cytoplasm. Alternatively, they may adhere to a diverse range of elements present in the surrounding environment, such as gastrointestinal fluids, microbiota, and food matrix. The process of protein conjugation with engineered nanomaterials has the potential to enhance their cellular uptake (John et al., 2001).

In 2017, a study found that a fish species with behavioral disorders had plastic NPs in its brain, likely due to its position as a top consumer. However, the exact pathways of the nanomaterials within the fish's body remained unclear (Mattsson et al., 2017). Nanoparticles transferring through the food chain raise human health concerns. Their unique physicochemical characteristics allow them to interact with biological systems, potentially disrupting regular cellular processes. Accumulation of these substances in crucial organs and tissues can result in localized toxicity or systemic health consequences. Introducing nanomaterials into aquatic environments, the atmosphere, and the soil can have significant implications for the well-being of animals and plants. This is due to their dispersion within the primary structural elements of plants and the microbial population in the soil (Rico et al., 2011).

9.2 NANOMATERIALS IN THE FOOD CHAIN

9.2.1 Sources of Nanomaterials in the Food Chain

With the progress of research on nanomaterials in the food industry, it is anticipated that the scope of potential uses for nanomaterials in food and its production will expand. This expansion is likely to lead to an increase in human exposure to these substances. The diverse range of human-made or engineered NPs found in food, as depicted in Figure 9.1, contributes to the narrative of human exposure to NPs. The application of engineered nanomaterials in plants and animal husbandry has become a notable factor in the increasing hazards associated with food consumption by consumers. Furthermore, enormous quantities of nanoscale materials, amounting to thousands of tons, are disposed of annually in landfills and water bodies worldwide. As engineered nanomaterials enter the ecosystem, they undergo stochastic transformations and interact with biological entities and inorganic substances, leading to modifications in their properties and affecting their fate (Glenn & Klaine, 2013). Within the food chain, these NPs have the potential to engage with organisms across different feeding levels, extending from lower to upper levels (Keller & Lazareva, 2014). Consequently, concerns have been raised about the gradual increase in human exposure to nanomaterials through diverse pathways, particularly the food chain.

The following are the key points through which nanomaterials can enter the human food chain.

9.2.1.1 Direct Addition

Researchers are exploring the direct addition of nanomaterials to food applications, which can occur during various stages of food production and processing. These additions can enhance food quality, nutritional value, and shelf life, or enable novel functionalities. The process of incorporating NPs into food products serves multiple objectives, including enhancing nutrient delivery, improving sensory attributes, and extending shelf life. However, contamination may arise from nanomaterials originating from food packaging materials or inadvertent contamination during processing or

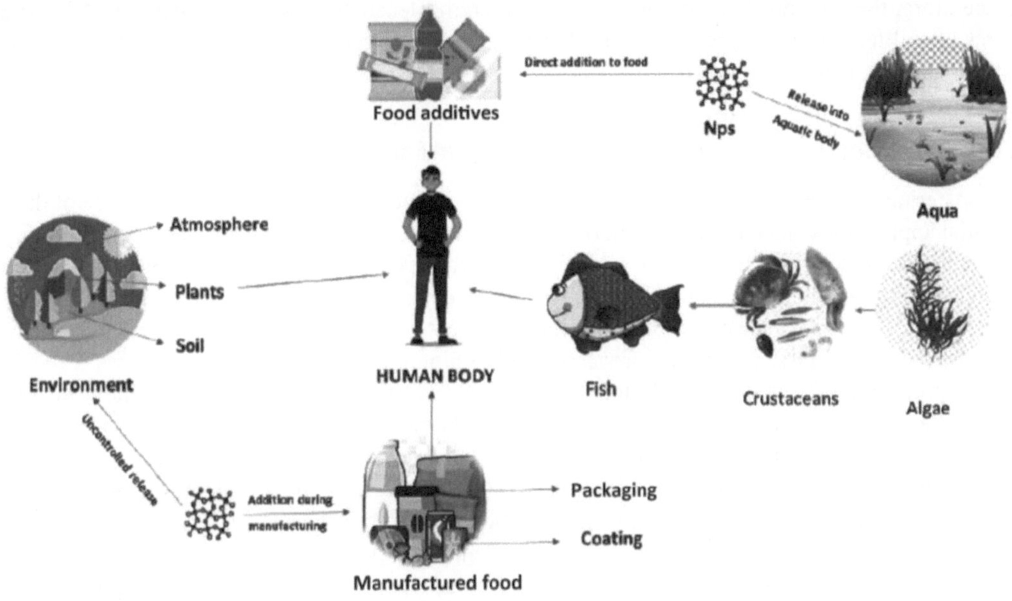

FIGURE 9.1 Sources of nanomaterials in food chain.

production. The direct incorporation of nanomaterials into food has the potential to enter the chain of food, as they may be ingestion of food products containing such materials.

The level of human exposure to food-related nanomaterials is highly dependent on their specific applications in food and industries that are related to food. Incorporating nanomaterials into multi-composite packaging materials, where the other substances are coated with a nanomaterial layer, is unlikely to lead to a significant shift in the food, resulting in minimal exposure for consumers. On the other hand, when nanomaterials are utilized as carriers for bioactive compounds or nutrients directly incorporated into food items, the probability for higher exposure levels exists, dependent on the concentration of nanomaterials in the food and the quantity ingested.

9.2.1.2 Environmental Contamination

Besides being added directly to food items, nanomaterials can also enter the food chain from various sources within the environment (Yah et al., 2012). This means that people are being exposed to nanomaterial pollution through different routes, including direct contamination from water, soil, and air, as well as indirect exposure from consuming plant and animal-based foods (Gothandam et al., 2018).

Plants interact with their environment, allowing engineered nanomaterials to disperse. Nanotechnology is used in agriculture to improve crop productivity by optimizing water and nutrient supply. Human manipulation exposes plants to these materials, potentially causing them to absorb and accumulate within the food chain (Miralles et al., 2012). Nanomaterials that have gotten into the environment in different ways can end up on foliage or other plant components that grow above the ground. They can gather on the surfaces of plant tissues and enter the plant through the stomatal paths (Navarro et al., 2008).

Nanomaterials can infiltrate water reservoirs through various mechanisms, including treatment, agricultural use, and landfill disposal. This contamination has adverse effects on plants, animals, and humans (Gothandam et al., 2018). The inadvertent discharge of waste originating from facilities engaged in the production of NPs has the potential to pollute soil and surface waters, including reservoirs of drinking water, as a result of dispersion via precipitation or wind (Zhu et al., 2012).

Furthermore, the utilization of sewage sludge and landfill leachates that are tainted with NPs as a means of soil fertilization represents a significant pathway for the contamination of soil and underground water (Linbo et al., 2009).

9.2.2 Behavior and Fate of Nanomaterials in the Food Chain

Nanomaterials have attracted attention in various fields, including the food industry, because of their potential applications and unique characteristics. However, concerns about their conduct and destiny within the food chain exist, including their interaction with organisms, potential accumulation within biological systems, and possible negative effects on the environment and human health. Factors like surface chemistry, size, and shape concentration can affect nanomaterial behavior, affecting absorption, distribution, and transformation processes within biological organisms.

This section aims to explore and elucidate the intricate fate of nanomaterials as they traverse through the complex and interconnected food chain of both aquatic and terrestrial ecosystems. The fate of NPs within terrestrial and aquatic food chains has been illustrated in Figure 9.2.

Nanomaterial absorption and accumulation by plants increase worries regarding their potential effect on human health and the food chain. Studies show that factors like nanomaterial characteristics, plant species, and the conditions of the experiment play an important role in affecting how nanomaterials are uptake, accumulation, and transformation within food crops (Rico et al., 2011). Numerous studies have provided evidence supporting the notion that plants possess the ability to uptake nanomaterials through the stomata located on their leaves, as well as through root absorption. The nanomaterials are retained within the plant tissues and subsequently distributed to various plant parts (Ingale & Chaudhari, 2018). Three potential pathways for the transportation of nanomaterials have been identified, namely the apoplastic pathway, cell wall pores, and plasmodesmata channels that establish connections between adjacent cells of plants (Tilney et al., 1991). The translocation of nanomaterials from the plant's roots to the upper plant parts is facilitated by their movement through tissues and entry into the xylem (Chowdary & Sanivada, 2021). Nevertheless, the existence of the Casparian strip within the cell wall of the endodermis in roots may pose challenges to the transportation of nanomaterials (Seeger et al., 2009). This could potentially necessitate the utilization of alternative routes, such as the injured endodermal cells or root apex (Wang et al., 2012). Upon entering the vascular function of plants, nanomaterials have the ability to be transported to the above-ground plant parts through the transpiration stream (Lin et al., 2009). Nanomaterials found in plant tissues can find their way into the human food chain when people consume plants contaminated with NPs or when animals are exposed to plants that have been contaminated with NPs.

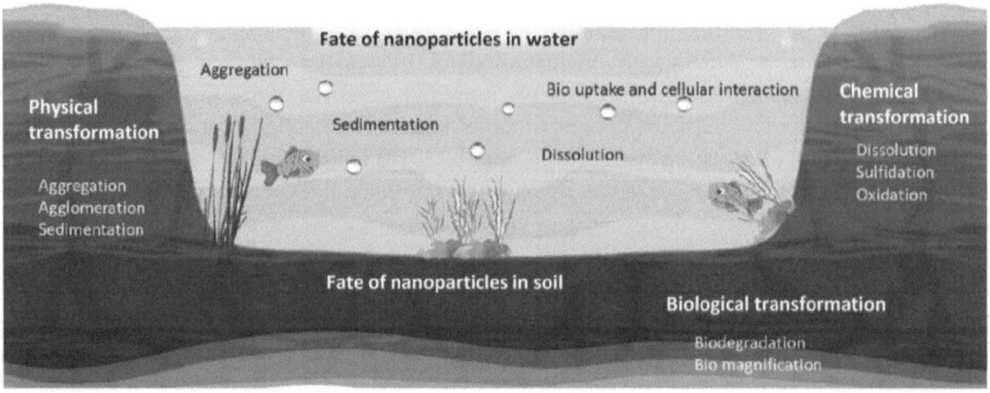

FIGURE 9.2 Fate of NPs within terrestrial and aquatic food chain.

Upon introduction into aquatic environments, engineered nanomaterials engage in interactions with both biotic and abiotic components. The original properties of nanomaterials can be altered by biological entities through various mechanisms. Similarly, abiotic factors such as clay particles can influence the sedimentation and aggregation of engineered nanomaterials in water-based ecosystems (Labille et al., 2015). Biomacromolecules, including carbohydrates, proteins, inorganic acids, and substances like lignite, fulvic acid, hematite, humic acid, sulfates, phosphates, and nitrates, influence the fate of engineered nanomaterials through processes like adsorption, absorption, agglomeration, sedimentation, capping, speciation ionization, and bio-uptake (Glenn & Klaine, 2013). Alterations in engineered nanomaterials can affect their bioavailability and toxicity towards various organisms in the food chain, including producers, consumers, and decomposers. Additionally, lipophilic coatings can impact their bioaccumulation and magnification potential.

9.3 POTENTIAL HEALTH EFFECTS OF NANOMATERIALS IN FOOD

9.3.1 Toxicological Considerations

Nanotechnology finds widespread application in consumer and commercial products like food additives, biocides, soil cleaning, supplements, water purification, agriculture, veterinary drugs, feed, and packaging (Kaphle et al., 2018; Nwani et al., 2010). Numerous advantages, including their high surface area but small size, and improved solubility, have sparked scientific inquiry into their potential toxicity on human health and ecological balance over the past two decades.

Also, the remarkable characteristics that underpin the technical and scientific advantages of nanotechnology give rise to unique biological effects. Nanotechnology's physicochemical properties, for example, size, shape, composition, surface charge, structure, aggregation, concentration, and crystallinity significantly impact the interaction among cells and engineered nanoparticles (ENPs) and pose potential toxicity risks (Figure 9.3). These properties significantly influence the interaction

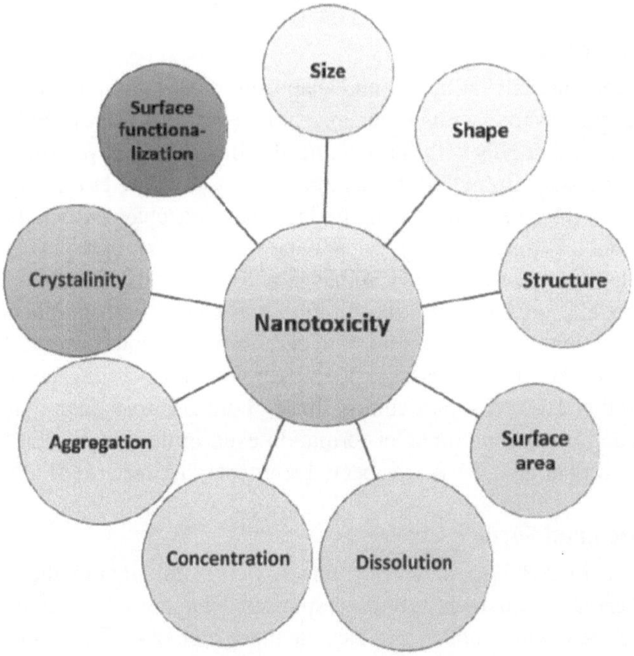

FIGURE 9.3 Factors affecting nanomaterials leading to nanotoxicity.

between ENPs and cells. Therefore, we discuss a list of significant physicochemical properties that contribute to the assessment of nanomaterial toxicity.

9.3.1.1 Particle Size

Nanoparticle size significantly impacts behavior and interactions with biological systems, potentially causing toxicity. The research aims to understand the biological effects of different particle sizes, highlighting their importance in scientific investigations. Nanomaterial toxicity is intricately linked to their size, as NPs (< 1 μm) can enter cells, while larger particles are capable of interacting with cells via the production of particular proteins on their surface. This size-dependent capability to enter biological systems and modify structures and disrupt essential biological processes (Aggarwal et al., 2009; Lovrić et al., 2005). Nanoparticle size is vital for cellular uptake, effective endocytic processes, and eliciting specific physiological responses upon nanomaterial exposure (Li et al., 2015).

One of the studies focused on the cellular toxicity of Ag NPs in relation to their size, finding that smallest (10 nm) NPs induced apoptosis more effectively than larger ones (50 and 100 nm), emphasizing the importance of particle size in biomedical applications (Kim et al., 2012). Furthermore, titanium oxide NPs' toxicity in rat lungs depends on size, with inhalational NPs causing more harm in the alveolar region compared to larger particles. While it is commonly assumed that smaller NP sizes increase the likelihood of cellular entry and potential damage, the mechanisms of toxicity are intricate, indicating that size alone cannot be viewed as the sole influencing factor. Some of the previous studies have reported that smaller particle-size organisms have been linked to increased AgNP accumulation and toxicity (Zhang & Wang, 2018), but contrasting findings have shown that larger particle sizes result in higher accumulation and toxicity in *Gammarusfossarum* (Perde-Schrepler et al., 2019).

Researchers have noted that the *in vivo* toxicity of engineered NPs involves oxidative responses and free radicals, with size playing a crucial role. These free radicals cause DNA damage, lipid peroxidation, and inflammatory responses. A study found that 20 nm Ag NPs had higher toxicity than larger Ag ions but were less toxic (Park et al., 2011).

9.3.1.2 Shape and Structure

In addition to size, nanomaterials' structure and shape significantly impact their toxicity. Different shapes and structures, such as fibers, tubes, spheres, polyhedra, and planes can lead to different toxicity effects. Studies show that NPs with rod- and needle-like shapes have greater uptake compared to cylindrical forms. Rod-shaped ZnONPs have more toxicity compared to spherical ones. The shape of ENPs also influences their behavior during endocytosis and phagocytosis processes. Spherical NPs are more easily and less toxic. Non-spherical nanomaterials can flow via capillaries, leading to other biological results (Gatoo et al., 2014). Akhavan and Ghaderi (2010) found that graphene and carbon nanotubes' toxicity depends on concentration and shape, with graphene causing stronger metabolic activity. Additionally investigating how carbon-based materials are shaped, researchers have also noted diverse toxicity patterns in TiO_2NPs possessing varying crystal structures. Gurr et al. (2005) revealed that TiO_2 NPs, particularly those of anatase form, can induce oxidative DNA damage, lipid peroxidation, and micronuclei formation even in the not present light. Furthermore, the shape-specific toxicity of nickel NPs has been documented by Ispas et al. (2009).

9.3.1.3 Surface Area and Surface Charge

Surface characteristics of NPs significantly affect toxicity, influencing the intake of ions and biomolecules and altering organism or cellular responses. Nanoparticles' inflammatory effects are influenced by surface area, with smaller particles having a higher surface area per unit mass. This increased surface area results in higher reactivity, generation of reactive oxygen species (ROS), and DNA damage at a greater rate than larger-sized particles of the equal mass dose (Vlastou et al., 2017).

In addition, surface charge significantly impacts NP colloidal behavior and organism response by promoting aggregation and agglomeration. It regulates selective adsorption, colloidal behavior, binding to plasma proteins, integrity of the blood–brain barrier, and transmembrane permeability (Kundu et al., 2021).

Calatayud et al. (2014) investigated the effect of functionalized Fe_3O_4NPs on the adsorption of proteins and cell uptake and discovered that cluster formation of protein-NP is influenced by the functional groups, implying the potential of controlling unspecific protein adsorption.

9.3.1.4 Dissolution

Dissolution of NPs significantly impacts their environmental impact, particularly in aquatic environments. They create colloidal dispersions that can either maintain dispersion or aggregate. Factors like shape, temperature, size, surface chemistry, pH, and ionic strength influence the physicochemical properties of silver NPs, leading to precipitation, dissolution, or aggregation. Precipitation or aggregation reduces toxic effects, while dissolution releases silver ions (Ag^+) and increases toxicity (Lee et al., 2018).

9.3.1.5 Aggregation and Concentration

Nanomaterial toxicity is influenced by final vital parameters aggregation and concentration, with factors like surface charge, size, and composition affecting ENPs' toxicity. Carbon nanotubes can cause cytotoxic effects over extended periods (Yang et al., 2008). Concerning the influence of concentration, there is a general trend where higher NP concentrations tend to result in reduced toxicity at upper levels of concentration (Gatoo et al., 2014).

9.3.2 Exposure Pathways of Nanomaterials

With the increasing availability and production of nanotechnology-based products, exposure of nanomaterials to the environment is unavoidable, for both consumers and workers. The circulation and result of NPs within the human body can fluctuate depending on the form of exposure. For the reason of enhancing comprehension of their circulation and subsequent fate within the human body, it is imperative to elucidate these pathways.

9.3.2.1 Inhalation

Nanoparticles enter the lungs through inhaled air, especially in workplaces, posing a significant health risk due to the presence of airborne nanomaterials (Helland et al., 2008). Upon inhalation, NPs can enter the human body through nasal or oral passages and enter the respiratory system. They encounter the pharynx, larynx, and trachea, which are lined with protective cilia. They reach the lungs' lobes via bronchial tubes, where oxygen and carbon dioxide exchange occur in the alveoli. All respiratory system components, including cavities, pharynx, larynx, trachea, bronchial tubes, and alveoli, are susceptible to exposure from inhaled NPs.

9.3.2.2 Ingestion

Nanoparticles can be ingested through the digestive tract through water and food, occurring intentionally or unintentionally. Examples include titanium dioxide NPs in food packaging, food dye E171, candies, gums, dressings, seasonings, and nutritional supplements. Additionally, magnesium oxide and titanium dioxide NPs serve as food preservatives and aid in processing (Ramalingam, 2016). Titanium dioxide NPs' daily intake varies across countries, with the USA consuming 0.2–0.7 mg/kg, while the UK and Germany have 1 mg/kg daily. The potential impact on children is a concern, and further investigation is needed. Silica NPs are present in the processing and storage of processed foods, with 43% of amorphous silica falling within the nanoscale size range (Murugadoss et al., 2017).

9.3.2.3 Dermal Contact

Nanoparticles may enter the body via direct contact with mucous membranes or skin, often in workplaces or consumer products. The dermal pathway is the primary route, with the extensive skin surface area increasing exposure probability. Accurately quantifying NPs permeating the skin presents a challenge due to its extensive surface area.

9.3.2.4 Digestive System

The digestive system can potentially serve as an additional route of entry for NPs. The digestive system refers to a collection of organs that also consists of the liver, gastrointestinal tract, gallbladder, and pancreas. The gastrointestinal tract consists of hollow structures like the mouth, esophagus, stomach, small intestine, large intestine, and anus. Engineered NPs have been widely used in human life, making it plausible that food and beverages may contain these particles. Nanoparticles are used in the food industry for shelf-life extension and flavor enhancement. However, inadvertent NP incorporation can occur during, coating, filtering, and packaging. The increasing use of nanomaterials increases the risk of ingestion through food animals, fish, and water sources.

Individuals can come into contact with NPs through various routes, including surface contamination, direct skin contact, inadvertent ingestion, aquatic organisms' absorption via gills, and targeted exposure. Surface contamination occurs when individuals come into contact with NPs on workplace surfaces, while inadvertent ingestion occurs through direct skin contact or ingestion. Aquatic organisms also experience exposure through their gills, leading to their entry into their circulatory systems. In addition, extensive exposure to nanomaterials in household items can cause skin absorption or ingestion, highlighting potential health and environmental impacts. NPs can enter the body through various sources, including water, air, food, clothing, or drug delivery systems. However, the main exposure routes are the gastrointestinal tract and respiratory tract (Hussain et al., 2005). Note that crystal nanomaterials permeate the body via the respiratory tract, specifically the alveoli, facilitating easy delivery into the organism (Maharramov et al., 2019).

9.3.3 Impact on Human Organs and Systems

The greatest purposes of nanomaterials in food include enhancing health and ensuring consumer safety against pathogens and contaminations. However, concerns about their impact on consumers and food industry workers arise, as well as potential environmental contamination risks. This discussion covers potential risks, migration of nanomaterials, and potential toxicity in food-encapsulation systems.

Nanoparticles may penetrate the human body via various pathways, including inhalation, gastrointestinal uptake, and dermal absorption. They can also penetrate protective barriers like the olfactory mucosa and blood–brain barrier, potentially entering the central nervous system. This attribute is crucial for drug delivery, targeting tissue damage. Nanoparticles migrate quickly due to their size, solubility, and surface charge. However, the correlation between NP characteristics and their distribution and fate within the human body remains an active field of study. Figure 9.4 demonstrates the exposure pathways of NPs in the human body and their health effects.

9.3.3.1 Impact on Respiratory System

The primary route for NPs to enter the human body is through inhalation. Research on animals has demonstrated that specific NPs, based on their size and characteristics, have the potential to reach the alveoli and settle within the lungs (Li & Chen, 2011). Smaller NPs with aerodynamic diameters below 2.5 microns can even penetrate deep into the alveoli (Hoet et al., 2004). Inhalation exposure to NPs in animals can result in acute or long-term lung injury, leading to several lung diseases such as asthma, chronic obstructive pulmonary disease (COPD), mesothelioma, emphysema, and risk of lung cancer (Chen et al., 2011; Han et al., 2011). The respiratory system is vulnerable to NP toxicity,

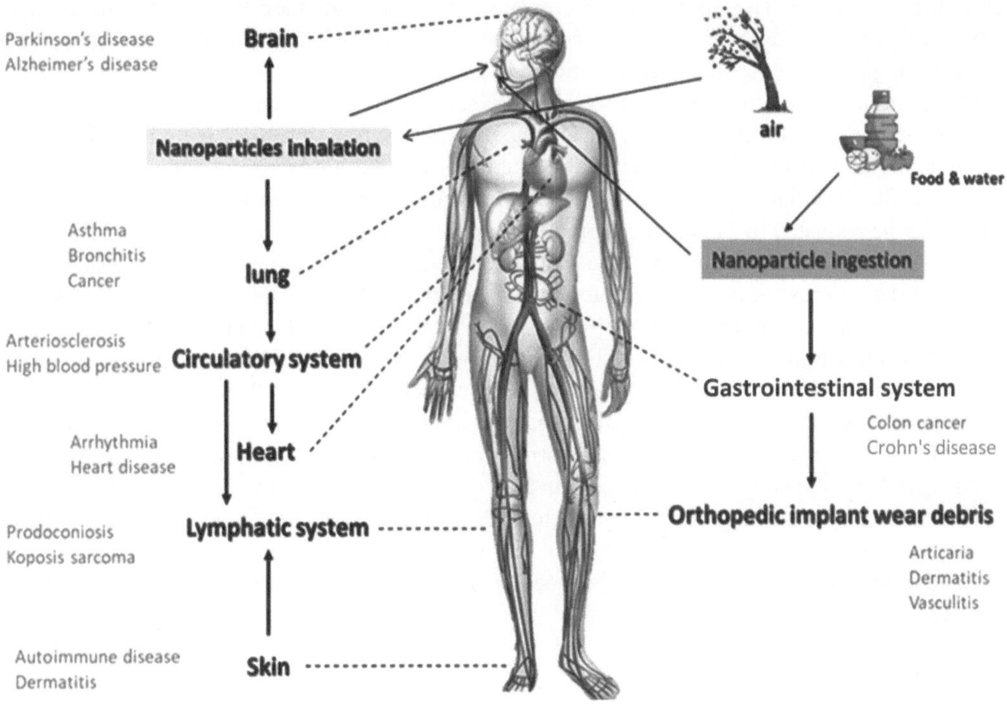

FIGURE 9.4 Exposure pathways of NPs into human body and health effects.

as it receives inhaled particles and cardiac output. Even through other routes, NPs may indirectly affect lung function without direct deposition. Young female workers have reported symptoms like pleural effusions and breathlessness, highlighting the potential adverse effects of NP accumulation in the lungs (Song et al., 2009). Accumulated NPs in the lungs can contribute to the development of asthma, granulomas, and even lung cancer. These findings underscore the significance of understanding the potential hazards related to NP exposure and the necessity to mitigate their harmful effects on the respiratory system.

Inhaled NPs can be eliminated through mucociliary and macrophage clearance mechanisms. Excessive NPs can penetrate the mucous film and multiply within epithelial cells, causing inflammation and increasing the risk of pulmonary infections. Proper elimination is crucial to prevent these particles from accumulating in the lungs (Andujar et al., 2009; Geiser & Kreyling, 2010). When NPs reach the alveoli, they may be engulfed by alveolar macrophages, which then transfer to the alveolar interstitium and approach lymphatic ganglions. Immune cells in this region undergo proliferation and differentiation before entering the bloodstream. Approximately 10–20% of insoluble particles remain in the human lungs due to slow clearance, potentially leading to fibrosis with high exposure. Alveoli provide a favorable entry point for NPs into the systemic circulation due to their short distance between blood and air. Studies on rodents show varying results depending on NP type and administration route. Polyacrylate NPs have been linked to pleural effusion, fatal outcomes, pulmonary fibrosis, and granuloma formation. TiO_2NPs have been demonstrated to instigate pulmonary tumors in rats, but their carcinogenic effects on humans remain inconclusive.

Understanding NP-induced lung diseases involves understanding their mechanisms, including oxidative stress, inflammation, genotoxicity, and fibrosis, which can impair lung function.

Oxidative stress arises when there is some disparity between reactive oxygen production and the system's capability to eliminate them or repair causing damage. Nanoparticles can directly generate ROS or indirectly affect cellular processes, leading to oxidative stress. Studies have shown that NPs, such as titanium oxide (TiO_2) NPs, can accumulate in the lungs, increase ROS levels, induce lipid peroxidation, and decrease antioxidant capacity (Sun et al., 2012). This oxidative stress can trigger harmful biological processes like inflammation and genotoxicity (Sun et al., 2012).

Inflammation is a common response to inhaled NPs and involves inflammatory cell recruitment and cytokine release (Sun et al., 2012). For example, exposure to graphene oxide NPs causes acute lung inflammation in mice, characterized by exudates and elevated pro-inflammatory cytokines (Duch et al., 2011). Inflammatory pathways, including transcription factors (AP-1 NF-κB) and MAP kinases, can be activated by NPs and contribute to the inflammatory response (Li et al., 2010).

Genotoxicity involves NPs damaging effects on cellular DNA and integrity, causing inflammation and oxidative stress through direct interaction or indirect effects (Hubbs et al., 2011). Research has shown that carbon nanotubes and various other NPs can induce DNA damage within lung cells, potentially leading to mutagenic effects. Fibrosis, characterized by the excessive deposition of collagen and tissue scarring, can be triggered by NPs, especially those with high aspect ratios like carbon nanotubes (Cho et al., 2012). Inflammatory effects play a significant role in NP-induced fibrosis. Fibroblast proliferation, granuloma formation, and collagen deposition have been observed in animal studies following exposure to certain NPs.

The severity of lung injury depends on factors like NP physicochemical properties (e.g., shape, size, and surface modification), dose, exposure period, and administration route. Modifications can reduce adverse effects and interactions between NPs and proteins can influence biocompatibility and toxicity.

9.3.3.2 Impact on Gastrointestinal Tract

Nanoparticles may reach the body through the gastrointestinal tract via direct ingestion from food, water, medications, and cosmetics, as well as through the ingestion of NPs cleared from the respiratory tract (Hoet et al., 2004). Recent research has shown that oral administration of Cu NPs, as compared to micro-Cu particles, can cause significant adverse impacts and severe damage in the liver, kidney, and spleen of experimental rodents (Chen et al., 2006). The extensive uses of NPs are linked to various diseases including ulcerative colitis, gastrointestinal cancer, and Crohn's disease, inflammatory bowel disease. They trigger inflammatory responses by absorption and translocation of larger NPs, causing inflammation in the gastrointestinal tract (Ballestri et al., 2001).

Higher NP intake (100 nm to 1 m) from diet contributes to the development of Crohn's disease, but genetic and environmental variables also play a part in its incidence (Lomer et al., 2004).

9.3.3.3 Impact on Central Nervous System

Nanoparticles can bypass the respiratory system's phagocytic defenses and enter the systemic circulation, bypassing the blood–brain barrier. They interact with various organs and target organs. The ability to pass through the blood–brain blockage remains debated, but studies in rats show that a small fraction of inhaled radioactive NPs may reach the brain (Oberdörster et al., 2004). The cross-blood–brain barrier is a potential barrier for NPs to enter the brain, potentially causing diseases like Parkinson's, Alzheimer's, and Huntington's. Conflicting studies suggest radioactivity in the brain may originate from the NP. However, if NPs can cross, they could induce free radical formation and oxidative stress, contributing to these disorders.

9.3.3.4 Effects on the Cardiovascular System

Endothelial dysfunction stands as one potential link connecting ultrafine particles and NPs to cardiovascular ailments. This dysfunction could potentially play a role in triggering conditions like myocardial infarction (heart attack), acute coronary syndrome, and atherosclerosis. Recent studies exposed human endothelial cells from the aorta to iron NPs used in medical imaging, revealing harmful effects that may arise due to the generation of free radicals when NPs directly interact with the endothelium. The presence of these free radicals is believed to contribute to oxidative stress, a significant factor in the advancement of inflammation and the primary phases of atherogenesis. These radicals could trigger an inflammatory reaction, including the release of cytokines, molecules associated with inflammation, and the potential for subsequent cardiovascular complications.

9.3.4 Long-Term Health Risks and Uncertainties

Exposure to NPs occurs daily, as individuals are constantly exposed through various means. For instance, consumers are exposed to NPs by consuming food products that contain NPs for various purposes. Additionally, exposure may occur through the regular use of cosmetics that contain NPs. It is important to consider that the degradation or dissolution of NPs may occur over a prolonged time period, which could be significantly extended than the removal of the therapeutic agents they may carry. Furthermore, the breakdown products resulting from the degradation or dissolution of NPs themselves may possess toxicity. Another factor to consider is that the accumulation and biodistribution of NPs inside the body may change over time (Mohammadpour et al., 2019a,b). These factors collectively raise the potential for long-term health impacts in humans. Long-term health risks associated with NPs include various uncertainties and potential adverse effects, such as carcinogenicity, embryotoxicity, mutagenicity, immunotoxicity, and allergic reactions (Mohammadpour et al., 2019a,b).

Research demonstrated that prolonged exposure to multi-walled carbon nanotubes can enhance the metastasis of breast cancer (Lu et al., 2019). Research shows titanium dioxide NPs induce DNA damage in cell cultures and animal models. Anatase TiO_2, known for its strong photocatalytic activity, is found to be particularly genotoxic among TiO_2NPs based on standard comet assays (Møller et al., 2017).

Extended daily oral intake of aluminum oxide NPs and zinc oxide NPs in rats for 75 days resulted in hepatorenal toxicities and suppressed hepatic expression of mtTFA and PGC-1α proteins (Yousef et al., 2019). Chronic toxicity studies of NPs should assess various toxicology parameters, including hematological, organ, tissue, and genetic analysis. Adequate statistical power analysis is necessary to determine the optimal number of animals, and long-term studies may be necessary. Factors such as route of exposure, duration, dose, frequency, sex, animal age, and consider NPs' physicochemical properties.

9.4 RISK ASSESSMENT AND REGULATION

Risk assessment is crucial for identifying and managing potential risks of NPs, involving hazard identification, risk analysis, and evaluation. It helps understand their effect on the environment and human health, establishes a dose–response relationship, and conducts exposure assessments using various methods. The outcomes guide decision-making, support control measures, and influence safety regulations for NPs.

9.4.1 Current Approaches to Assessing Nanomaterial Risks in Food

Assessing nanomaterials' risk involves considering animal tests, which exhibit greater predictive but show limitations due to physiological differences, and alternatives driven by ethical and legal demands. New experimental approaches using primary human cell cultures and established cell lines are efficient, cost-effective, and reliable risk assessment strategies (Drasler et al., 2017).

For a better understanding of nanomaterial risks in the food chain, improvements and innovations are needed in study methods for assessing toxicity. Traditional methods like MTT, neutral red, calcein AM, and live/dead assays are ineffective due to interactions between carbon nanomaterials and assay markers (Monteiro-Riviere et al., 2009). Therefore, multiple assays are essential for accurately determining nanomaterial toxicity, and comprehensive characterization is crucial. Nanometry, measuring at the nanoscale level, plays an important role in characterizing nanomaterials. Analyzing key parameters like surface charge, surface coating, surface charge, pH and ionic strength solution is essential for stability (Jiang et al., 2009). Furthermore, in response to the swift progress in nanotechnology, new equipment is being developed, and there's a growing emphasis on standardizing techniques and measurements to stay abreast of these developments. Another crucial aspect is the systematic classification of nanomaterials in the food industry is crucial for standardization and determining appropriate dosages. Consumers are exposed to low doses, while workers may encounter higher concentrations. Classifying nanomaterials systematically helps draw accurate conclusions about potential harm, preventing consumer reluctance to consume food containing nanomaterials and impeding food nanotechnology development. The systematic analysis provides a comprehensive understanding of nanomaterial mechanisms and scientifically sound conclusions. Moreover, nanotoxicology assesses long-term human toxicity from NP exposure, substituting impractical human tests with animal model simulations. Toxicology conclusions from animal (*in vivo*) and cell (*in vitro*) models are extendable to humans. Table 9.1 compiles the *in vitro* and *in vivo* toxicity of various NP types.

Lastly, a significant concern is the development of appropriate animal and other models for studying nanomaterials. Sayes et al. (2009) demonstrates that *in vitro* cell culture systems are insufficient for accurately forecasting pulmonary responses to particulates. This highlights the necessity for further development, standardization, and validation of these systems. Animal models provide valuable guidelines for assessing nanomaterial risks to humans, enhancing understanding of potential health impacts and enabling better assessment of exposure conditions.

In summary, nanotoxicity research aims to minimize toxicity and optimize applications of nanomaterials by understanding their underlying mechanisms, toxicity, and potential risks. It is crucial to effectively manage risks and leverage their potential while comprehending their toxicities and underlying mechanisms.

9.4.2 Regulatory Frameworks and Guidelines

The health risks of NPs in food are limited, with the WHO, FAO, and EU not definitively determining their safety. There is no data on genotoxicity, carcinogenicity, or teratogenicity. However, different countries and organizations recognize the need for strengthened safety assessments and prioritization of NPs in food (Magnuson et al., 2013).

Due to the absence of thorough risk assessment models and regulatory frameworks for nanotechnologies, the Woodrow Wilson International Center Project on Emerging Nanotechnologies (PEN) emphasizes the significance of evaluating the end-of-life consequences (Breggin & Pendergrass, 2007). This underscores the necessity to assess and mitigate the possible environmental and health effects of nanomaterials throughout their entire lifecycle, including disposal and waste management. Such considerations are vital for promoting responsible and sustainable utilization of these technologies. The Occupational Safety and Health Administration (OSHA) prioritizes workplace safety by mandating training for laboratory personnel on safety data sheets, labeling, and signage. FDA in the US regulates products utilizing nanotechnology and nanomaterials, including those in biomedical applications.

To ensure the safe use of nanomaterials, comply with federal, state, and local regulations for transporting, handling, storing, disposing, or using these materials. This helps mitigate risks and promotes responsible, safe utilization of nanomaterials.

9.4.3 Challenges and Gaps in Nanomaterial Risk Assessment

Nanomaterials have widespread applications in various fields, but their potential threats to the ecosystem and human health raise concerns for robust risk assessment strategies. Despite progress, challenges and gaps persist in this field.

TABLE 9.1
***In-vivo* and *in-vitro* procedures for risk assessment**

Nanoparticles	Particle Size (nm)	Study Type (*In Vivo/In Vitro*)	Test Organism	Main Results	References
Titanium dioxide	43	*In vivo*	*Pimephalespromelas*(Chordata)	Altered gene expression and fish immunotoxicity	(Jovanović et al., 2011)
Silver	13–17	*In vivo*	*Lymnaeastagnalis*(Mollusca)	Bioaccumulation and growth alteration	(Croteau et al., 2011)
Cupric oxide	55.8	*In vitro*	Cell line of HaCaT(human skin keratinocyte cell line)	DNA impairment, rising lipid peroxidation, glutathione decrease, decline cell workability	(Saud Alarifi et al., 2013)
MSN	50; 500	*In vitro*	IV administration, single dosage in male and female immune-competent inbred BALB/c mice	Both small and large NPs demonstrate cytotoxicity under acute conditions.	(Mohammadpour et al. 2019a)
Silica oxide	43	*In vitro*	Hepatocellular carcinoma cells	Mitochondrial damage, increase in ROS	(Sun et al., 2011)
Iron and iron extracted pure form of MWCNT		*In vitro*	A549 human lung epithelial cells and HepG2 cells	Pure iron-extracted MWCNTs exhibited cytotoxicity.	(Requardt et al., 2019)
Graphene familyNPs		*In vivo*	*Caenorhabditiselegans*	Reduced reproduction rates	(Chatterjee et al., 2008)
Thin bundles (CNT-1) and thick bundles of SWCNTs (CNT-2)		*In vitro* and *vivo*	*In vitro* and *vivo* study	Lung toxicity	(Fujita et al., 2015)
Fullerenes	100–900	*In vivo*	Embryonic zebrafish	Necrotic and apoptotic cell demise, rise in malformations, pericardial edema, etc.	(Usenko et al., 2007)
ZnO Al_2O_3 TiO_2		*In vivo*	*Daniorerio* (Chordata)		(Zhu et al., 2008)
Iron oxide	70–90	*In vivo*	*Eiseniahortensis*	Genotoxicity, DNA lesions, chromosomal damage	(Ciğerci et al., 2018)
Iron oxide	100–150	*In vitro*	Human macrophages	Decrease cell viability	(Jeng & Swanson, 2006)
TiO_2	5, 10, and 32	*In vivo*	*Xenopuslaevis* (Chordata)	High concentrations impacted tadpole growth, mortality, and delayed development	(Zhang et al., 2012)
SilverNPs	30–50	*In vitro*	Human alveolar cell line	ROS increases, cell viability decrease	(Foldbjerg et al., 2011)
SWCNTs			Human embryonic kidney 293 (HEK293) cells	SWCNTs upregulate apoptosis associated genes (Rb, p16 and p53)	(Cui et al., 2005)
Aluminiumoxide	8–12	*In vitro*	HBMVEC	Oxidative stress increase, cell viability decrease	(Chen et al., 2008)
MWCNT	30–50	*In vitro*	Human bronchial epithelial cells (A549)	Viability of cell decreases, DNA damage	(Lindberg et al., 2009)
SWCNT	14	*In vitro*	Mouse lung epithelial cells	Formation of ROS, no damage to DNA	(Jacobsen et al., 2008)

The evaluation of engineered nanomaterials (ENMs) risks is complicated by an insufficient grasp of how nanoscale particles interact with their environment. Risk evaluations must consider both direct effects, the immediate impact of NPs, and indirect effects from transformations and interactions. ENMs can dissolve in water, bind to substances, or degrade into different particle forms, making it difficult to measure exposure levels for organisms in the environment. Quantifying indirect effects in ecotoxicological studies presents challenges, making a comprehensive understanding of both direct and indirect effects essential for accurate risk assessment of ENMs.

Case-by-case risk assessment has been proposed to consider the specific properties of individual nanomaterials (World Health Organization, 2013). However, the complexity of nanomaterials' structures, materials, sizes, and variations makes it challenging to conduct comprehensive risk assessments. A clear determinant of the outcome is needed to assess potential risks. New methods and tools are needed to develop accurate risk assessments. However, detailed information is currently lacking for almost every category of nanomaterial. Enhancing knowledge and gathering comprehensive data is necessary for effective and accurate risk assessment.

Monitoring the health effects and interaction mechanisms of NPs is challenging due to limited data and inefficient characterization techniques. The diversity of nanomaterials complicates risk evaluation and hazard identification. *In vitro,* toxicological studies raise concerns about predicting nanomaterials' effects on human health. The relationship between human health and nanomaterials remains unclear, and there is controversy surrounding which properties determine toxicity. We need further research clarification to better comprehend how NPs affect human health. Comprehensive studies are necessary before universal applications of nanomaterials.

9.4.4 Future Directions and Emerging Research

Advancements in science and technology have led to the need for augmentation or replacement of traditional risk assessment methods. The increasing application of ENMs in goods presents challenges in conventional risk assessment frameworks, causing gaps in understanding toxic mechanisms. Combining AOP and QSAR models offers a promising solution to address these challenges effectively (Cote et al., 2016). Integrating AOP with QSAR models enables connections between ENMs' physicochemical properties and toxicological effects, enabling rapid hazard evaluation based on ENMs' characteristics and potential biological outcomes (Lai et al., 2018).

9.5 MITIGATION STRATEGIES AND SAFETY MEASURES

9.5.1 Nanomaterial Detection and Characterization Techniques

Characterizing nanomaterials is essential for advancing reliable synthesis techniques, encompassing the study of composition, structure, and various properties. A variety of techniques contribute to the understanding of nanomaterials, which is vital for assessing their effects and benefits in food and materials in contact with food. Yet, the characterization of nanomaterials within intricate food matrices presents difficulties, demanding both expertise and advanced tools. Furthermore, the existence of naturally-occurring nanomaterials in food introduces an extra challenge in differentiating between naturally occurring and artificially produced engineered nanomaterials (Tiede et al., 2008).

9.5.1.1 Size and Shape Analysis

The particle size and shape of NPs exert a substantial influence on their behavior, rendering these attributes of utmost importance. Consequently, a variety of methodologies have been devised to quantify and evaluate these characteristics. Comprehending the morphology of nanomaterials is of paramount importance, as it significantly influences their toxicity and various other fundamental characteristics (Zhang et al., 2015). Nanomaterials display a wide array of shapes, encompassing spherical, cubic, tubular, wire-like, polyhedral, and one-dimensional film-like structures. Electron microscopy, with image processing and analysis, is the main method for characterizing NP shape, size, and aggregation state. Raman spectroscopy and X-ray diffraction (XRD), though less common methods, provide valuable shape-related information (Mattarozzi et al., 2017).

9.5.1.2 Crystal Structure

Crystals in nanomaterials have a consistent molecular or ionic arrangement, and multiple crystal phases can coexist. The toxicological impacts and bioavailability of nanomaterials are influenced by their specific crystalline structures, like anatase and rutile in titanium dioxide (Jin et al., 2011). Understanding distinct crystal phases in nanomaterials is vital for evaluating their potential toxic effects and interactions with biological entities. A precise comprehension of nanomaterials' crystalline arrangement plays an important role in predicting their safety and performance in various applications. XRD and electron diffraction are the main techniques used to determine crystal structure. XRD instruments emit a parallel X-ray beam, causing scattering off the crystals in a sample, generating a powder XRD pattern. This pattern helps identify the atomic orientation and, consequently, the crystal structure of NPs in nanomaterials.

9.5.1.3 Characterization of Surface Chemistry of Nanomaterials

Functionalizing the surfaces of nanomaterials by attaching ligands or polymers enhances the stability of colloidal systems and influences the chemical and physical characteristics of NPs like solubility, stability, reactivity, and also the potential reduction of toxicity (Ansar & Kitchens, 2016). Surface characterization techniques are used to understand the surface chemical properties of NPs in detail. The current methods employed for the characterization of the surface chemical properties of NPs include Raman/surface-enhanced Raman spectroscopy (SERS), Zeta potential measurement, thermogravimetric analysis (TGA), and X-ray photoelectron spectroscopy (XPS).

9.5.2 Risk Management and Safety Protocols

The careful consideration of public perceptions regarding the potential threats and benefits related to nanotechnology is crucial for the implementation of risk management strategies in this field. There are three traditional models for managing risks associated with hazardous agents: (a)cost–benefit analysis, (b) the acceptable risk model, and (c) the viability or best technology model that is available (Marchant et al., 2008).

The cost–benefit analysis model for risk management evaluates potential risk options by weighing their costs and benefits, allowing a comprehensive assessment of nanotechnology's implications for public health and the environment (Walsh et al., 2008).

Another one is acceptable risk model strategies rely on conducting risk assessments, in the pursuit of minimizing potential risks associated with nanotechnology. Yet, the lack of established test methods and validated data poses challenges in scientifically generating reliable quantitative estimates for nanotechnology applications (Sweet & Strohm, 2006). Consequently, relying solely on risk-based approaches in nanotechnology is impractical and legally questionable due to insufficient risk information, posing regulatory challenges.

The third traditional risk management principle focuses on the utilization of the best available technology, aiming to reduce risks to feasible levels from technological and economic standpoints. This approach gains attention from policymakers as it provides a pathway to navigate contentious risk discussions by prioritizing risk minimization to the highest feasible extent, regardless of comprehensive knowledge about specific risks or benefits.

Effective control and risk management methods are crucial to protect consumers from potential exposure to engineered nanomaterials. However, their implementation is hindered by insufficient information in the nanomaterials field. To address this, urgent systematic data collection is needed on various aspects, such as exposure, manufacturing, applications, and environmental fate, to inform decision-making and enhance proactive risk management practices.

9.5.3 Responsible Innovation and Sustainable Practices

The responsible advancement and innovation of nanotechnologies have been key areas of focus in both national and international strategies for over 15 years, specifically in the sector of nanoresearch and development (R&D). As per Stilgoe et al. (2013), accountable progress can also be described as "taking care of the future through collective stewardship of science and innovation in the present". Nanoparticles are utilized in the food and farming sectors to provide a wide range of potential advantages to both consumers and society. These applications possess the capacity to make significant contributions towards the advancement of sustainable food and agricultural practices through their ability to enhance nutritional value, improve food characteristics, and facilitate the more efficient distribution of agrochemicals. Acknowledging the paramount significance of responsible innovation in the fields of agriculture and food, it is crucial to recognize and elucidate the potential obstacles that could impede responsible innovation in the nano-agrifood sector. Understanding and addressing these barriers not only secure the nano-agrifood product sustainability and uses but also provides valuable insights and best practices that can be used with other new technologies in the food and agricultural sectors.

9.6 CONCLUSION

On human health, nanomaterials' potential effects on the food chain have emerged as a subject of considerable apprehension. The utilization of nanomaterials in diverse food-related applications is on the rise, owing to their distinctive characteristics. Although the potential benefits of these advancements, such as enhanced food quality and safety, are promising, it is crucial to acknowledge the potential negative impacts they may have on human health. According to existing research, specific nanomaterials can pass through biological obstacles and accumulate in vital organs, posing a potential toxicity risk. Moreover, the knowledge gap on the consequences of prolonged exposure to nanomaterials in the food chain highlights the need for further research. Ensuring safety in food requires rigorous risk assessment, collaboration among governments, scientists, and industries, and protocols covering safe use and disposal. Public awareness and education are crucial for informed decision-making and consumer safety. The continuous investigation, surveillance, and conscientious nanotechnology implantation in the food industry are essential for mitigating adverse effects on human well-being and ensuring sustainable advancement of nanomaterials in food-related contexts.

REFERENCES

Aggarwal, P., Hall, J. B., McLeland, C. B., Dobrovolskaia, M. A., & McNeil, S. E. (2009). Nanoparticle interaction with plasma proteins as it relates to particle biodistribution, biocompatibility and therapeutic efficacy. *Advanced Drug Delivery Reviews*, *61*(6), 428–437. https://doi.org/10.1016/j.addr.2009.03.009

Akhavan, O., & Ghaderi, E. (2010). Toxicity of graphene and graphene oxide nanowalls against bacteria. *ACS Nano*, *4*(10), 5731–5736. https://doi.org/10.1021/nn101390x

Andujar, P., Lanone, S., Brochard, P., & Boczkowski, J. (2009). Effets respiratoires des nanoparticules manufacturées. *Revue des Maladies Respiratoires*, *26*(6), 625–637. https://doi.org/10.1016/S0761-8425(09)74693-5

Ansar, S. M., & Kitchens, C. L. (2016). Impact of gold nanoparticle stabilizing ligands on the colloidal catalytic reduction of 4-nitrophenol. *ACS Catalysis*, *6*(8), 5553–5560. https://doi.org/10.1021/acscatal.6b00635

Ballestri, M., Baraldi, A., Gatti, A. M., Furci, L., Bagni, A., Loria, P., Rapanà, R. M., Carulli, N., & Albertazzi, A. (2001). Liver and kidney foreign bodies granulomatosis in a patient with malocclusion, bruxism, and worn dental prostheses. *Gastroenterology*, *121*(5), 1234–1238. https://doi.org/10.1053/gast.2001.29333

Biswas, R., Alam, M., Sarkar, A., Haque, M. I., Hasan, M. M., & Hoque, M. (2022). Application of nanotechnology in food: processing, preservation, packaging and safetyassessment. *Heliyon*, *8*(11), e11795. https://doi.org/10.1016/j.heliyon.2022.e11795

Breggin, L. K., & Pendergrass, J., 2007. *Where Does the Nano Go? End-of Life Regulation of Nanotechnology*. Environmental Law Institute, USA. Retrieved from https://policycommons.net/artifacts/303726/where-does-the-nano-go-end-of-life-regulation-of-nanotechnology/1220455/on 09 December 2023. CID: 20.500.12592/9crjkg.

Calatayud, M. P., Sanz, B., Raffa, V., Riggio, C., Ibarra, M. R., & Goya, G. F. (2014). The effect of surface charge of functionalized Fe_3O_4 nanoparticles on protein adsorption and cell uptake. *Biomaterials*, *35*(24), 6389–6399. https://doi.org/10.1016/j.biomaterials.2014.04.009

Chatterjee, D. K., Fong, L. S., & Zhang, Y. (2008). Nanoparticles in photodynamic therapy: an emerging paradigm. *Advanced Drug Delivery Reviews*, *60*(15), 1627–1637. https://doi.org/10.1016/j.addr.2008.08.003

Chen, E. Y., Garnica, M., Wang, Y.-C., Chen, C.-S., & Chin, W.-C. (2011). Mucin secretion induced by titanium dioxide nanoparticles. *PLoS One*, 6(1), e16198. https://doi.org/10.1371/journal.pone.0016198

Chen, L., Yokel, R. A., Hennig, B., & Toborek, M. (2008). Manufactured aluminum oxide nanoparticles decrease expression of tight junction proteins in brain vasculature. *Journal of Neuroimmune Pharmacology: The Official Journal of the Society on NeuroImmune Pharmacology*, *3*(4), 286–295. https://doi.org/10.1007/s11481-008-9131-5

Chen, Z., Meng, H., Xing, G., Chen, C., Zhao, Y., Jia, G., Wang, T., Yuan, H., Ye, C., Zhao, F., Chai, Z., Zhu, C., Fang, X., Ma, B., & Wan, L. (2006). Acute toxicological effects of copper nanoparticles in vivo. *Toxicology Letters*, *163*(2), 109–120. https://doi.org/10.1016/j.toxlet.2005.10.003

Cho, W. S., Duffin, R., Poland, C. A., Duschl, A., Oostingh, G. J., Macnee, W., Bradley, M., Megson, I. L., & Donaldson, K. (2012). Differential pro-inflammatory effects of metal oxide nanoparticles and their soluble ions in vitro and in vivo; zinc and copper nanoparticles, but not their ions, recruit eosinophils to the lungs. *Nanotoxicology*, *6*(1), 22–35. https://doi.org/10.3109/17435390.2011.552810

Chowdary, A. R., & Sanivada, S. K. (2021). Impact of nanomaterials on the food chain. In: Kumar, V., Guleria, P., Ranjan, S., Dasgupta, N., & Lichtfouse, E. (Eds.), *Nanotoxicology and Nanoecotoxicology* (vol. 2, pp. 97–117). Springer International Publishing. https://doi.org/10.1007/978-3-030-69492-0_4

Ciğerci, İ. H., Ali, M. M., Kaygısız, Ş. Y., Kaya, B., & Liman, R. (2018). Genotoxic assessment of different sizes of iron oxide nanoparticles and ionic iron in earthworm (*Eisenia hortensis*) coelomocytes by comet assay and micronucleus test. *Bulletin of Environmental Contamination and Toxicology*, *101*(1), 105–109. https://doi.org/10.1007/s00128-018-2364-y

Cote, I., Andersen, M.E., Ankley, G.T., *et al.* (2016). The next generation of risk assessment multi-year study—highlights of findings, applications to risk assessment, and future directions. *Environmental Health Perspectives, 124*(11), 1671–1682. http://dx.doi.org/10.1289/EHP233

Croteau, M. N., Misra, S. K., Luoma, S. N., & Valsami-Jones, E. (2011). Silver bioaccumulation dynamics in a freshwater invertebrate after aqueous and dietary exposures to nanosized and ionic Ag. *Environmental Science & Technology*, *45*(15), 6600–6607. https://doi.org/10.1021/es200880c

Cui, D., Tian, F., Ozkan, C. S., Wang, M., & Gao, H. (2005). Effect of single wall carbon nanotubes on human HEK293 cells. *Toxicology Letters*, *155*(1), 73–85. https://doi.org/10.1016/j.toxlet.2004.08.015

Dekant, W., & Klaunig, J. E. (2016). Toxicology of decamethylcyclopentasiloxane (D5). *Regulatory Toxicology and Pharmacology*, *74*(Suppl), S67–S76. https://doi.org/10.1016/j.yrtph.2015.06.011

Drasler, B., Sayre, P., Steinhäuser, K. G., Petri-Fink, A., & Rothen-Rutishauser, B. (2017). In vitro approaches to assess the hazard of nanomaterials. *NanoImpact*, *8*, 99–116. https://doi.org/10.1016/j.impact.2017.08.002

Duch, M. C., Budinger, G. R., Liang, Y. T., Soberanes, S., Urich, D., Chiarella, S. E., Campochiaro, L. A., Gonzalez, A., Chandel, N. S., Hersam, M. C., & Mutlu, G. M. (2011). Minimizing oxidation and stable nanoscale dispersion improves the biocompatibility of graphene in the lung. *Nano Letters*, *11*(12), 5201–5207. https://doi.org/10.1021/nl202515a

Foldbjerg, R., Dang, D. A., & Autrup, H. (2011). Cytotoxicity and genotoxicity of silver nanoparticles in the human lung cancer cell line, A549. *Archives of Toxicology*, *85*(7), 743–750. https://doi.org/10.1007/s00204-010-0545-5

Fujita, K., Fukuda, M., Endoh, S., Maru, J., Kato, H., Nakamura, A., Shinohara, N., Uchino, K., & Honda, K. (2015). Size effects of single-walled carbon nanotubes on in vivo and in vitro pulmonary toxicity. *Inhalation Toxicology*, *27*(4), 207–223. https://doi.org/10.3109/08958378.2015.1026620

Gatoo, M. A., Naseem, S., Arfat, M. Y., Dar, A. M., Qasim, K., & Zubair, S. (2014). Physicochemical properties of nanomaterials: implication in associated toxic manifestations. *BioMed Research International, 2014*, 498420. https://doi.org/10.1155/2014/498420

Geiser, M., & Kreyling, W. G. (2010). Deposition and biokinetics of inhaled nanoparticles. *Particle and Fibre Toxicology*, *7*(1), 2. https://doi.org/10.1186/1743-8977-7-2

Gerardi, A. (2023). Global Food Safety Initiative (GFSI): underpinning the safety of the global food chain, facilitating regulatory compliance, trade, and consumer trust. In: Knowles, M. E., Anelich, L., Boobis, A., & Popping, B. (Eds.), *Present Knowledge in Food Safety a Risk-Based Approach through the Food Chain* (1st ed., pp. 1089–1098). Academic Press. https://doi.org/10.1016/B978-0-12-819470-6.00058-5

Glenn, J. B., & Klaine, S. J. (2013). Abiotic and biotic factors that influence the bioavailability of gold nanoparticles to aquatic macrophytes. *Environmental Science & Technology*, *47*(18), 10223–10230. https://doi.org/10.1021/es4020508

Gothandam, K., Ranjan, S., Dasgupta, N., Ramalingam, C., & Lichtfouse, E. (2018). *Nanotechnology, Food Security and Water Treatment.* 1st ed. Springer International Publishing, Cham. https://doi.org/10.1007/978-3-319-70166-0

Gurr, J. R., Wang, A. S., Chen, C. H., & Jan, K. Y. (2005). Ultrafine titanium dioxide particles in the absence of photoactivation can induce oxidative damage to human bronchial epithelial cells. *Toxicology*, *213*(1–2), 66–73. https://doi.org/10.1016/j.tox.2005.05.007

Han, B., Guo, J., Abrahaley, T., Qin, L., Wang, L., Zheng, Y., Li, B., Liu, D., Yao, H., Yang, J., Li, C., Xi, Z., & Yang, X. (2011). Adverse effect of nano-silicon dioxide on lung function of rats with or without ovalbumin immunization. *PLoS One*, *6*(2), e17236. https://doi.org/10.1371/journal.pone.0017236

Helland, A., Scheringer, M., Siegrist, M., Kastenholz, H. G., Wiek, A., & Scholz, R. W. (2008). Risk assessment of engineered nanomaterials: a survey of industrial approaches. *Environmental Science & Technology*, *42*(2), 640–646. https://doi.org/10.1021/es062807i

Hoet, P. H., Brüske-Hohlfeld, I., & Salata, O. V. (2004). Nanoparticles—known and unknown health risks. *Journal of Nanobiotechnology*, *2*(1), 12. https://doi.org/10.1186/1477-3155-2-12

Hubbs, A. F., Mercer, R. R., Benkovic, S. A., Harkema, J., Sriram, K., Schwegler-Berry, D., Goravanahally, M. P., Nurkiewicz, T. R., Castranova, V., & Sargent, L. M. (2011). Nanotoxicology—a pathologist's perspective. *Toxicologic Pathology*, *39*(2), 301–324. https://doi.org/10.1177/0192623310390705

Hussain, S. M., Hess, K. L., Gearhart, J. M., Geiss, K. T., & Schlager, J. J. (2005). In vitro toxicity of nanoparticles in BRL 3A rat liver cells. *Toxicology In Vitro: An International Journal Published in Association with BIBRA*, *19*(7), 975–983. https://doi.org/10.1016/j.tiv.2005.06.034

Ingale, A.G., & Chaudhari, A.N. (2018). Nanotechnology in the food industry. In: Gothandam, K., Ranjan, S., Dasgupta, N., Ramalingam, C., & Lichtfouse, E. (Eds.), *Nanotechnology, Food Security and Water Treatment. Environmental Chemistry for a Sustainable World.* Springer, Cham. https://doi.org/10.1007/978-3-319-70166-0_3

Ispas, C., Andreescu, D., Patel, A., Goia, D. V., Andreescu, S., & Wallace, K. N. (2009). Toxicity and developmental defects of different sizes and shape nickel nanoparticles in zebrafish. *Environmental Science & Technology*, *43*(16), 6349–6356. https://doi.org/10.1021/es9010543

Jacobsen, N. R., Pojana, G., White, P., Møller, P., Cohn, C. A., Korsholm, K. S., Vogel, U., Marcomini, A., Loft, S., & Wallin, H. (2008). Genotoxicity, cytotoxicity, and reactive oxygen species induced by single-walled carbon nanotubes and C_{60} fullerenes in the FE1-Muta™ Mouse lung epithelial cells. *Environmental and Molecular Mutagenesis*, *49*(6), 476–487. https://doi.org/10.1002/em.20406

Jeng, H. A., & Swanson, J. (2006). Toxicity of metal oxide nanoparticles in mammalian cells. *Journal of Environmental Science and Health, Part A*, *41*(12), 2699–2711. https://doi.org/10.1080/10934520600966177

Jiang, J., Oberdörster, G., & Biswas, P. (2009). Characterization of size, surface charge, and agglomeration state of nanoparticle dispersions for toxicological studies. *Journal of Nanoparticle Research*, *11*, 77–89. https://doi.org/10.1007/s11051-008-9446-4

Jin, C., Tang, Y., Yang, F. G., Li, X. L., Xu, S., Fan, X. Y., Huang, Y. Y., & Yang, Y. J. (2011). Cellular toxicity of TiO_2 nanoparticles in anatase and rutile crystal phase. *Biological Trace Element Research*, *141*(1–3), 3–15. https://doi.org/10.1007/s12011-010-8707-0

John, T. A., Vogel, S. M., Minshall, R. D., Ridge, K., Tiruppathi, C., & Malik, A. B. (2001). Evidence for the role of alveolar epithelial gp60 in active transalveolar albumin transport in the rat lung. *The Journal of Physiology*, *533*(2), 547–559. https://doi.org/10.1111/j.1469-7793.2001.0547a.x

Jovanović, B., Anastasova, L., Rowe, E. W., Zhang, Y., Clapp, A. R., & Palić, D. (2011). Effects of nanosized titanium dioxide on innate immune system of fathead minnow (*Pimephales promelas* Rafinesque, 1820). *Ecotoxicology and Environmental Safety*, *74*(4), 675–683. https://doi.org/10.1016/j.ecoenv.2010.10.017

Kaphle, A., Navya, P., Umapathi, A., & Daima, H. K. (2018). Nanomaterials for agriculture, food and environment: applications, toxicity and regulation. *Environmental Chemistry Letters*, *16*, 43–58. https://doi.org/10.1007/s10311-017-0662-y

Keller, A. A., & Lazareva, A. (2014). Predicted releases of engineered nanomaterials: from global to regional to local. *Environmental Science & Technology Letters*, *1*(1), 65–70. https://doi.org/10.1021/ez400106t

Kim, T. H., Kim, M., Park, H. S., Shin, U. S., Gong, M. S., & Kim, H. W. (2012). Size-dependent cellular toxicity of silver nanoparticles. *Journal of Biomedical Materials Research. Part A*, *100*(4), 1033–1043. https://doi.org/10.1002/jbm.a.34053

Kundu, D., Khan, M.F., Gogoi, M., & Patra, S. (2021). Environmental impact and econanotoxicity of engineered nanomaterials. In: Kumar, V., Guleria, P., Ranjan, S., Dasgupta, N., & Lichtfouse, E. (Eds), *Nanotoxicology and Nanoecotoxicology Vol. 1. Environmental Chemistry for a Sustainable World* (vol 59). Springer, Cham. https://doi.org/10.1007/978-3-030-63241-0_11

Labille, J., Harns, C., Bottero, J. Y., & Brant, J. (2015). Heteroaggregation of titanium dioxide nanoparticles with natural clay colloids. *Environmental Science & Technology*, *49*(11), 6608–6616. https://doi.org/10.1021/acs.est.5b00357

Lai, R. W. S., Yeung, K. W. Y., Yung, M. M. N., Djurišić, A. B., Giesy, J. P., & Leung, K. M. Y. (2018). Regulation of engineered nanomaterials: current challenges, insights and future directions. *Environmental Science and Pollution Research International*, *25*(4), 3060–3077. https://doi.org/10.1007/s11356-017-9489-0

Lee, W. S., Kim, E., Cho, H. J., Kang, T., Kim, B., Kim, M. Y., Kim, Y. S., Song, N. W., Lee, J. S., & Jeong, J. (2018). The relationship between dissolution behavior and the toxicity of silver nanoparticles on zebrafish embryos in different ionic environments. *Nanomaterials (Basel, Switzerland)*, 8(9), 652. https://doi.org/10.3390/nano8090652

Li, R., Ning, Z., Majumdar, R., Cui, J., Takabe, W., Jen, N., Sioutas, C., & Hsiai, T. (2010). Ultrafine particles from diesel vehicle emissions at different driving cycles induce differential vascular pro-inflammatory responses: implication of chemical components and NFκB signaling. *Particle and Fibre Toxicology*, *7*(1), 6. https://doi.org/10.1186/1743-8977-7-6

Li, X., Liu, W., Sun, L., Aifantis, K. E., Yu, B., Fan, Y., Feng, Q., Cui, F., & Watari, F. (2015). Effects of physicochemical properties of nanomaterials on their toxicity. *Journal of Biomedical Materials Research. Part A*, *103*(7), 2499–2507. https://doi.org/10.1002/jbm.a.35384

Li, Y. F., & Chen, C. (2011). Fate and toxicity of metallic and metal-containing nanoparticles for biomedical applications. *Small (Weinheim an der Bergstrasse, Germany)*, *7*(21), 2965–2980. https://doi.org/10.1002/smll.201101059

Lin, S., Reppert, J., Hu, Q., Hudson, J. S., Reid, M. L., Ratnikova, T. A., Rao, A. M., Luo, H., & Ke, P. C. (2009). Uptake, translocation, and transmission of carbon nanomaterials in rice plants. *Small (Weinheim an der Bergstrasse, Germany)*, *5*(10), 1128–1132. https://doi.org/10.1002/smll.200801556

Linbo, T. L., Baldwin, D. H., McIntyre, J. K., & Scholz, N. L. (2009). Effects of water hardness, alkalinity, and dissolved organic carbon on the toxicity of copper to the lateral line of developing fish. *Environmental Toxicology and Chemistry*, *28*(7), 1455–1461. https://doi.org/10.1897/08-283.1

Lindberg, H. K., Falck, G. C., Suhonen, S., Vippola, M., Vanhala, E., Catalán, J., Savolainen, K., & Norppa, H. (2009). Genotoxicity of nanomaterials: DNA damage and micronuclei induced by carbon nanotubes and graphite nanofibres in human bronchial epithelial cells in vitro. *Toxicology Letters*, *186*(3), 166–173. https://doi.org/10.1016/j.toxlet.2008.11.019

Lomer, M. C., Hutchinson, C., Volkert, S., Greenfield, S. M., Catterall, A., Thompson, R. P., & Powell, J. J. (2004). Dietary sources of inorganic microparticles and their intake in healthy subjects and patients with Crohn's disease. *British Journal of Nutrition*, *92*(6), 947–955. https://doi.org/10.1079/BJN20041276

Lovrić, J., Bazzi, H. S., Cuie, Y., Fortin, G. R., Winnik, F. M., & Maysinger, D. (2005). Differences in subcellular distribution and toxicity of green and red emitting CdTe quantum dots. *Journal of Molecular Medicine (Berlin, Germany)*, *83*(5), 377–385. https://doi.org/10.1007/s00109-004-0629-x

Lu, X., Zhu, Y., Bai, R., Wu, Z., Qian, W., Yang, L., Cai, R., Yan, H., Li, T., Pandey, V., Liu, Y., Lobie, P. E., Chen, C., & Zhu, T. (2019). Long-term pulmonary exposure to multi-walled carbon nanotubes promotes breast cancer metastatic cascades. *Nature Nanotechnology*, *14*(7), 719–727. https://doi.org/10.1038/s41565-019-0472-4

Magnuson, B., Munro, I., Abbot, P., Baldwin, N., Lopez-Garcia, R., Ly, K., McGirr, L., Roberts, A., & Socolovsky, S. (2013). Review of the regulation and safety assessment of food substances in various countries and jurisdictions. *Food Additives & Contaminants. Part A, Chemistry, Analysis, Control, Exposure & Risk Assessment*, *30*(7), 1147–1220. https://doi.org/10.1080/19440049.2013.795293

Maharramov, A. M., Hasanova, U. A., Suleymanova, I. A., Osmanova, G. E., & Hajiyeva, N. E. (2019). The engineered nanoparticles in food chain: potential toxicity and effects. *SN Applied Sciences*, *1*(11), 1362. https://doi.org/10.1007/s42452-019-1412-5

Marchant, G.E., Sylvester, D.J., & Abbott, K.W. (2008). Risk management principles for nanotechnology. *Nanoethics*, *2*, 43–60. https://doi.org/10.1007/s11569-008-0028-9

Mattarozzi, M., Suman, M., Cascio, C., Calestani, D., Weigel, S., Undas, A., & Peters, R. (2017). Analytical approaches for the characterization and quantification of nanoparticles in food and beverages. *Analytical and Bioanalytical Chemistry*, *409*(1), 63–80. https://doi.org/10.1007/s00216-016-9946-5

Mattsson, K., Johnson, E. V., Malmendal, A., Linse, S., Hansson, L. A., & Cedervall, T. (2017). Brain damage and behavioural disorders in fish induced by plastic nanoparticles delivered through the food chain. *Scientific Reports*, *7*(1), 11452. https://doi.org/10.1038/s41598-017-10813-0

Miralles, P., Church, T. L., & Harris, A. T. (2012). Toxicity, uptake, and translocation of engineered nanomaterials in vascular plants. *Environmental Science & Technology*, *46*(17), 9224–9239. https://doi.org/10.1021/es202995d

Mohammadpour, R., Yazdimamaghani, M., Cheney, D. L., Jedrzkiewicz, J., & Ghandehari, H. (2019a). Subchronic toxicity of silica nanoparticles as a function of size and porosity. *Journal of Controlled Release: Official Journal of the Controlled Release Society*, *304*, 216–232. https://doi.org/10.1016/j.jconrel.2019.04.041

Mohammadpour, R., Dobrovolskaia, M. A., Cheney, D. L., Greish, K. F., & Ghandehari, H. (2019b). Subchronic and chronic toxicity evaluation of inorganic nanoparticles for delivery applications. *Advanced Drug Delivery Reviews*, *144*, 112–132. https://doi.org/10.1016/j.addr.2019.07.006

Møller, P., Jensen, D. M., Wils, R. S., Andersen, M. H. G., Danielsen, P. H., & Roursgaard, M. (2017). Assessment of evidence for nanosized titanium dioxide-generated DNA strand breaks and oxidatively damaged DNA in cells and animal models. *Nanotoxicology*, *11*(9–10), 1237–1256. https://doi.org/10.1080/17435390.2017.1406549

Monteiro-Riviere, N. A., Inman, A. O., & Zhang, L. W. (2009). Limitations and relative utility of screening assays to assess engineered nanoparticle toxicity in a human cell line. *Toxicology and Applied Pharmacology*, *234*(2), 222–235. https://doi.org/10.1016/j.taap.2008.09.030

Murugadoss, S., Lison, D., Godderis, L., Van Den Brule, S., Mast, J., Brassinne, F., Sebaihi, N., & Hoet, P. H. (2017). Toxicology of silica nanoparticles: an update. *Archives of Toxicology*, *91*(9), 2967–3010. https://doi.org/10.1007/s00204-017-1993-y

Navarro, E., Baun, A., Behra, R., Hartmann, N. B., Filser, J., Miao, A. J., Quigg, A., Santschi, P. H., & Sigg, L. (2008). Environmental behavior and ecotoxicity of engineered nanoparticles to algae, plants, and fungi. *Ecotoxicology (London, England)*, *17*(5), 372–386. https://doi.org/10.1007/s10646-008-0214-0

Nwani, C. D., Lakra, W. S., Nagpure, N. S., Kumar, R., Kushwaha, B., & Srivastava, S. K. (2010). Toxicity of the herbicide atrazine: effects on lipid peroxidation and activities of antioxidant enzymes in the freshwater fish *Channa punctatus* (Bloch). *International Journal of Environmental Research and Public Health*, *7*(8), 3298–3312. https://doi.org/10.3390/ijerph7083298

Oberdörster, G., Sharp, Z., Atudorei, V., Elder, A., Gelein, R., Kreyling, W., & Cox, C. (2004). Translocation of inhaled ultrafine particles to the brain. *Inhalation Toxicology*, *16*(6–7), 437–445. https://doi.org/10.1080/08958370490439597

Park, M. V., Neigh, A. M., Vermeulen, J. P., de la Fonteyne, L. J., Verharen, H. W., Briedé, J. J., van Loveren, H., & de Jong, W. H. (2011). The effect of particle size on the cytotoxicity, inflammation, developmental toxicity and genotoxicity of silver nanoparticles. *Biomaterials*, *32*(36), 9810–9817. https://doi.org/10.1016/j.biomaterials.2011.08.085

Perde-Schrepler, M., Florea, A., Brie, I., Virag, P., Fischer-Fodor, E., Vâlcan, A., Gurzău, E., Lisencu, C., & Maniu, A. (2019). Size-dependent cytotoxicity and genotoxicity of silver nanoparticles in cochlear cells in vitro. *Journal of Nanomaterials*, *2019*, 6090259. https://doi.org/10.1155/2019/6090259

Ramalingam, S. R. C. (2016). Titanium dioxide nanoparticles induce bacterial membrane rupture by reactive oxygen species generation. *Environmental Chemistry Letters*, *14*, 487–494. https://doi.org/10.1007/s10311-016-0586-y

Requardt, H., Braun, A., Steinberg, P., Hampel, S., & Hansen, T. (2019). Surface defects reduce carbon nanotube toxicity in vitro. *Toxicology In Vitro*, *60*, 12–18. https://doi.org/10.1016/j.tiv.2019.03.028

Rico, C. M., Majumdar, S., Duarte-Gardea, M., Peralta-Videa, J. R., & Gardea-Torresdey, J. L. (2011). Interaction of nanoparticles with edible plants and their possible implications in the food chain. *Journal of Agricultural and Food Chemistry*, *59*(8), 3485–3498. https://doi.org/10.1021/jf104517j

Saud Alarifi, D. A., Verma, A., Alakhtani, S., & Ali, B. A. (2013). Cytotoxicity and genotoxicity of copper oxide nanoparticles in human skin keratinocytes cells. *International Journal of Toxicology*, *32*(4), 296–307. https://doi.org/10.1177/1091581813487563

Sayes, C.M., Reed, K.L., Subramoney, S., Abrams, L., & Warheit, D. B. (2009). Can in vitro assays substitute for in vivo studies in assessing the pulmonary hazards of fine and nanoscale materials? *Journal of Nanoparticle Research*, *11*, 421–431. https://doi.org/10.1007/s11051-008-9471-3

Seeger, E.M., Baun, A., Kästner, M., & Trapp, S. (2009).Insignificant acute toxicity of TiO_2 nanoparticles to willow trees. *Journal of Soils and Sediments*, *9*, 46–53. https://doi.org/10.1007/s11368-008-0034-0

Song, Y., Li, X., & Du, X. (2009). Exposure to nanoparticles is related to pleural effusion, pulmonary fibrosis and granuloma. *European Respiratory Journal*, *34*(3), 559–567. https://doi.org/10.1183/09031936.00178308

Stilgoe, J., Owen, R., & Macnaghten, P. (2013). Developing a framework for responsible innovation. *Research Policy*, *42*(9), 1568–1580. https://doi.org/10.1016/j.respol.2013.05.008

Subhan, M. A., Choudhury, K. P., & Neogi, N. (2021). Advances with molecular nanomaterials in industrial manufacturing applications. *Nanomanufacturing*, *1*(2), 75–97. https://doi.org/10.3390/nanomanufacturing1020008

Sun, L., Li, Y., Liu, X., Jin, M., Zhang, L., Du, Z., Guo, C., Huang, P., & Sun, Z. (2011). Cytotoxicity and mitochondrial damage caused by silica nanoparticles. *Toxicology In Vitro*, *25*(8), 1619–1629. https://doi.org/10.1016/j.tiv.2011.06.012

Sun, Q., Tan, D., Ze, Y., Sang, X., Liu, X., Gui, S., Cheng, Z., Cheng, J., Hu, R., & Gao, G. (2012). Pulmotoxicological effects caused by long-term titanium dioxide nanoparticles exposure in mice. *Journal of Hazardous Materials*, *235*, 47–53. https://doi.org/10.1016/j.jhazmat.2012.05.072

Sweet, L., & Strohm, B. (2006). Nanotechnology—life-cycle risk management. *Human Ecological Risk Assessment*, *12*(3), 528–551. https://doi.org/10.1080/10807030600561691

Tiede, K., Boxall, A. B., Tear, S. P., Lewis, J., David, H., & Hassellöv, M. (2008). Detection and characterization of engineered nanoparticles in food and the environment. *Food Additives and Contaminants*, *25*(7), 795–821. https://doi.org/10.1080/02652030802007553

Tilney, L. G., Cooke, T. J., Connelly, P. S., & Tilney, M. S. (1991). The structure of plasmodesmata as revealed by plasmolysis, detergent extraction, and protease digestion. *The Journal of Cell Biology*, *112*(4), 739–747. https://doi.org/10.1083/jcb.112.4.739

Usenko, C. Y., Harper, S. L., & Tanguay, R. L. (2007). In vivo evaluation of carbon fullerene toxicity using embryonic zebrafish. *Carbon, 45*(9), 1891–1898. https://doi.org/10.1016/j.carbon.2007.04.021

Vlastou, E., Gazouli, M., Ploussi, A., Platoni, K., & Efstathopoulos, E. P. (2017). Nanoparticles: nanotoxicity aspects. *Journal of Physics: Conference Series*, *931*(1), 012020. https://doi.org/10.1088/1742-6596/931/1/012020

Walsh, S., Balbus, J. M., Denison, R., & Florini, K. (2008). Nanotechnology: getting it right the first time. *Journal of Cleaner Production*, *16*(8–9), 1018–1020. https://doi.org/10.1016/j.jclepro.2007.04.015

Wang, Z., Xie, X., Zhao, J., Liu, X., Feng, W., White, J. C., & Xing, B. (2012). Xylem-and phloem-based transport of CuO nanoparticles in maize (*Zea mays* L.). *Environmental Science & Technology*, *46*(8), 4434–4441. https://doi.org/10.1021/es204212z

Yah, C. S., Iyuke, S. E., & Simate, G. S. (2012). A review of nanoparticles toxicity and their routes of exposures. *Iranian Journal of Pharmaceutical Sciences*, *8*(1), 299–314. https://doi.org/10.22037/ijps.v8.41001

Yang, S.-T., Wang, X., Jia, G., Gu, Y., Wang, T., Nie, H., Ge, C., Wang, H., & Liu, Y. (2008). Long-term accumulation and low toxicity of single-walled carbon nanotubes in intravenously exposed mice. *Toxicology Letters*, *181*(3), 182–189. https://doi.org/10.1016/j.toxlet.2008.07.020

Yousef, M. I., Mutar, T. F., & Kamel, M. A. E.-N. (2019). Hepato-renal toxicity of oral sub-chronic exposure to aluminum oxide and/or zinc oxide nanoparticles in rats. *Toxicology Reports*, *6*, 336–346. https://doi.org/10.1016/j.toxrep.2019.04.003

Zhang, J., Wages, M., Cox, S. B., Maul, J. D., Li, Y., Barnes, M., Hope-Weeks, L., & Cobb, G. P. (2012). Effect of titanium dioxide nanomaterials and ultraviolet light coexposure on African clawed frogs (*Xenopus laevis*). *Environmental Toxicology and Chemistry*, *31*(1), 176–183. https://doi.org/10.1002/etc.718

Zhang, K., Zuo, W., Wang, Z., Liu, J., Li, T., Wang, B., & Yang, Z. (2015). A simple route to $CoFe_2O_4$ nanoparticles with shape and size control and their tunable peroxidase-like activity. *RSC Advances*, *5*(14), 10632–10640. https://doi.org/10.1039/C4RA15675G

Zhang, L., & Wang, W.-X. (2018). Dominant role of silver ions in silver nanoparticle toxicity to a unicellular alga: evidence from luminogen imaging. *Environmental Science & Technology*, *53*(1), 494–502. https://doi.org/10.1021/acs.est.8b04918

Zhu, X., Tian, S., & Cai, Z. (2012). Toxicity assessment of iron oxide nanoparticles in zebrafish (*Danio rerio*) early life stages. *PLoS One*, *7*(9), e46286. https://doi.org/10.1371/journal.pone.0046286

Zhu, X., Zhu, L., Duan, Z., Qi, R., Li, Y., & Lang, Y. (2008). Comparative toxicity of several metal oxide nanoparticle aqueous suspensions to zebrafish (*Danio rerio*) early developmental stage. *Journal of Environmental Science and Health, Part A*, *43*(3), 278–284. https://doi.org/10.1080/10934520701792779

10 Nanotechnology Current Scenario and Future Prospects

Ruslan Mehadi Galib, Tanjilla Akter Sumi, and Mahabub Alam

10.1 NANOTECHNOLOGY IN AGRICULTURE AND FOOD: CURRENT APPLICATIONS

10.1.1 Introduction to Nanotechnology

In an era when the world population is expected to exceed 9.7 billion by 2050, the difficulties of ensuring food security, optimizing resource utilization, and limiting environmental impacts have become critical. Although historically successful, traditional agricultural practices are under increased pressure from the demand for effective and sustainable food production. By utilizing the special qualities of nanomaterials and nanodevices, the fusion of nanotechnology and agriculture offers a revolutionary way to address these problems. In-depth analysis of the existing uses and potential future uses of nanotechnology in agriculture and food systems is presented in this article. Furthermore, it emphasizes how important that position has been in redefining numerous aspects of production, resource management, and quality assurance (Malik et al., 2023; Saravanadevi et al., 2022).

The phrase "nanotechnology" describes the handling and engineering of substances at the nanoscale, which typically ranges from 1 to 100 nm. Materials have distinctive physical, chemical, and biological characteristics that set them apart from their bulk counterparts at this scale. Nanotechnology has prepared the path for advancements in agriculture and food that promise previously unheard-of precision and efficiency. The World Health Organization estimates that post-harvest losses harm 25% of the worldwide agricultural market, highlighting the urgent need for novel solutions. To improve crop growth and protection, researchers have achieved considerable advancements in the use of nanotechnology. To increase the effectiveness of agrochemicals like pesticides and fertilizers while reducing their negative effects on the environment, nanoscale delivery devices have been created. Nanoformulations enable controlled and targeted release of these agents, ensuring that they reach their intended targets with minimal wastage. For instance, studies have demonstrated the efficacy of nanoparticles in reducing pesticide usage by up to 70%, thus alleviating concerns about chemical runoff and soil contamination (Faizan et al., 2023; Keçili et al., 2019; Sahoo et al., 2021).

To build barriers against oxygen, moisture, and pathogens, food packaging is increasingly incorporating nanomaterials. Additionally, nanosensor-equipped intelligent packaging systems can track changes in temperature, gas composition, and freshness, giving customers and producers more information with which to make decisions and minimize food waste. Despite the enticing promise of nanotechnology in food and agricultural systems, adoption of this technology is subject to ethical, governmental, and safety concerns. Nanoparticles' interactions with biological systems prompt concerns about their potential long-term effects on the environment and human health. Regulatory

DOI: 10.1201/9781003438168-10

organizations all across the world are attempting to create frameworks that ensure the safe and responsible use of nanomaterials in food and agriculture (Faizan et al., 2023; Jagtiani, 2022).

The road to utilizing nanotechnology is not without obstacles. The literature emphasizes how difficult it is to create nanostructures with the appropriate properties and to comprehend how they behave. Although the science is currently having difficulty with precise control over their properties, biosynthesis techniques employing natural components and the bioinspired approach offer sustainable ways to produce nanomaterials (Malik et al., 2023; Thakur et al., 2022).

10.1.2 Nanomaterials Revolutionizing Agriculture

Nanomaterials, with their amazing nanoscale characteristics, are altering traditional agricultural techniques and addressing important global food system concerns. The unique properties of nanomaterials, such as their increased surface area and higher reactivity, have led to advancements in a variety of agricultural sectors (Biswas et al., 2022). Nanoparticles can be used to deliver nutrients, insecticides, and growth regulators directly to plant cells (Napagoda et al., 2023; Tripathi & Prakash, 2022). Nano-fertilizers with controlled nutrient release mechanisms provide tailored nutrient supply to plants, decreasing waste and improving nutrient uptake. These nanoscale transporters protect nutrients from leaching and volatilization, ensuring their long-term availability to plants (Jakhar et al., 2022). Pesticide nanoformulations have increased efficacy at lower levels, reducing chemical runoff and potential harm to non-target creatures. Pesticides are released via nano-enabled smart delivery systems in response to specified triggers, such as pest presence or environmental circumstances (Kumar et al., 2019; Qiao et al., 2022).

Nanomaterials are being used to create superabsorbent polymers that can hold water and gradually release it to plants. This increases water efficiency and decreases watering frequency. Furthermore, nanosensors can be placed in soil to monitor moisture levels, allowing farmers to optimize irrigation schedules and save water waste. Nanotechnology aids in the repair and remediation of soil health (Oladosu et al., 2022; Rani Sarkar et al., 2022). Through adsorption or chemical interactions, nanoparticles can encapsulate pollutants, lowering their availability to plants and preventing groundwater contamination. Furthermore, nanomaterials improve soil structure, nutrient retention, and microbial activity, creating an environment favorable for plant growth (Bhagat et al., 2022; Jafarizadeh-Malmiri et al., 2019). Plant development and stress tolerance are aided by the use of nanomaterials. Nanoparticles can be used to transport growth factors, hormones, and micronutrients (Bhagat et al., 2022; Singh et al., 2021). As shown in Figure 10.1, the main links between nanotechnology and the food industry include improving food safety, extending shelf life, enhancing taste and nutrient distribution, making it possible to identify pathogens, toxins, and pesticides, and assisting in the creation of functional foods.

Nanotechnology is driving a shift in the food business as well. Food packaging using nanostructures extends the shelf life and improves safety by limiting microbial growth and avoiding spoiling. Nanoscale additions alter flavors and textures to produce new culinary experiences. Additionally, improvements in food quality evaluation, allergen detection, and pathogen monitoring are being driven by bioinspired nanomaterials, supporting food safety (Shawon et al., 2020; Tavker & Sharma, 2022). The fusion of nanotechnology with the agriculture sector is igniting a paradigm shift. It offers a glimpse into a future where food production is not only more efficient and sustainable but also in harmony with environmental and consumer needs (Jafarizadeh-Malmiri et al., 2019; Patel et al., 2020).

10.2 NANOTECHNOLOGY IN FOOD PROCESSING

The substantial impact of nanotechnology on the agriculture and food industries extends to food processing. It has aided in the improvement of several elements of production, preservation, and

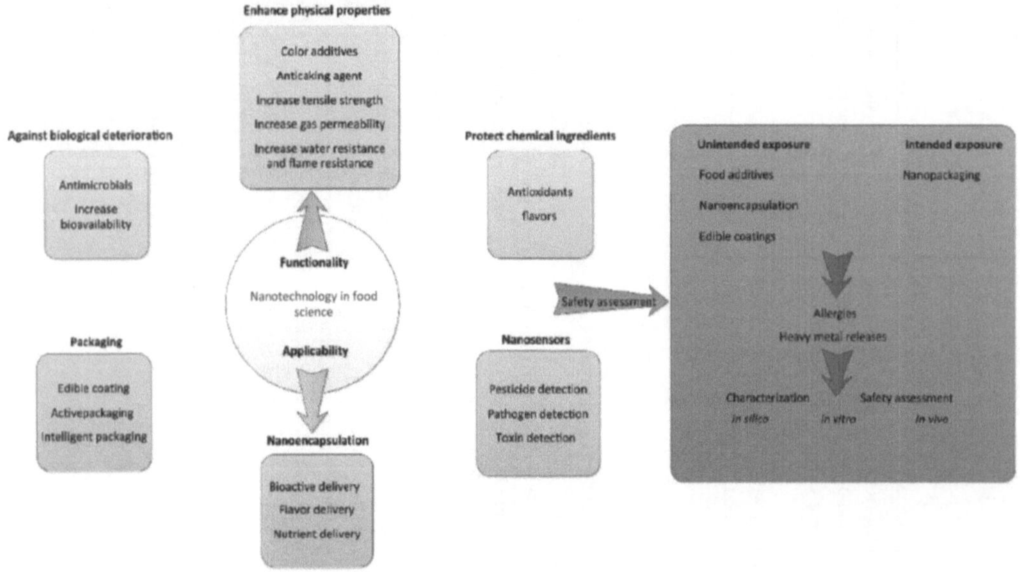

FIGURE 10.1 Development of nanotechnology in food science/industry and its functionality, applicability, and safety assessments. Reprinted with written permission from He and Hwang (2016).

quality improvements. It provides prospects to address food safety, shelf-life extension, nutritional enrichment, and sensory characteristics concerns. Encapsulating beneficial chemicals such as vitamins, antioxidants, and essential oils within nanoscale carriers is possible because to nanotechnology. These nanostructures, which typically consist of lipids or polymers, protect sensitive substances during processing and storage. They also allow for regulated release after eating, which improves bioavailability and nutritional absorption in the human body. Nano-encapsulation could fortify meals with essential nutrients, hence preventing deficiencies and improving general health (Nallamuthu et al., 2017; Tahir et al., 2021).

Nanomaterials are used to provide innovative packaging solutions that enhance shelf life and maintain food safety. Antimicrobial nanoparticles embedded in nanocomposites can limit the growth of spoilage germs and diseases, hence improving food preservation. Nanoemulsions and nanocapsules also improve the solubility and stability of fat-soluble vitamins. Biosensors and nanoprobes allow for the quick identification of hazardous bacteria, allergies, and chemical residues on-site (Ashfaq et al., 2022; Chausali et al., 2022). Using methods like nanoencapsulation and nanoemulsion, engineered nanomaterials are expertly created for use as colorants, flavor enhancers, preservatives, and transporters in food supplements, even extending to animal feed products. The extraordinary qualities of specifically created nanoparticles have a considerable positive impact on food production, acting as both ingredients and supplements. The U.S. FDA has also given the go-ahead for a number of inorganic oxide compounds, including SiO_2 (E551), MgO (E530), and TiO_2 (E171), to perform tasks including reducing clumping, imparting flavor to food, and adding food coloring (Dorier et al., 2017; Weir et al., 2012), details shown in Table 10.1.

The development of intelligent delivery systems that react to environmental cues like pH, temperature, or moisture is made possible by nanotechnology. These systems allow for the regulated release of flavorings, preservatives, or additives, ensuring the best sensory experiences and sustained freshness. Such advancements could cut down on food waste and maximize resource use (Beig et al., 2022). Nanotechnology's application in food processing not only solves pressing problems, but also opens up new avenues for sustainable practices and individualized nutrition. A future with

TABLE 10.1
Present condition of food products enhanced by nanotechnology

Application	Nanomaterials	Current Status	Note	Description
Enhancing Color in Food Processing	TiO_2	Exempt from certification	Allowed up to 1% by weight of the food. Titanium dioxide (TiO_2) is a versatile pigment used to enhance the color and visual appeal of various food products such as candies, coatings, and desserts. Regulatory agencies consider its use safe within specified limits.	Enhances color and visual appeal in food processing.
Red Pigmentation in Pet and Human Foods	Synthetic iron oxide	Exempt from certification	Permitted up to 0.25% (dogs and cats) and 0.1% (human) by weight of the finished food. Synthetic iron oxide contributes red pigmentation to pet food and specific human food items. Controlled limits ensure safe incorporation.	Imparts red pigmentation to pet and human foods.
Polymer Additive and Production Aid	ZnO	Authorized by EC 10/2011	Authorization based on conventional particle size. Zinc oxide (ZnO) serves as a valuable additive and production aid in the food industry. Its authorization hinges on the size of particles used.	Utilized as a polymer additive and production aid.
Diverse Applications of Iron Oxide	Iron oxide	Authorized by EC 10/2011	Authorized as an additive based on specific criteria. Iron oxide finds application as a colorant, enhancing the visual appeal of various food items.	Adds color and aesthetic appeal to food products.
Strengthening PET Packaging	Aluminum oxide	No migration reported	Permitted only in PET bottles up to 20 mg/kg. Aluminum oxide reinforces PET bottles, enhancing structural integrity without causing migration into food contents.	Reinforces PET bottles for improved durability.
Preventing Clumping in Powdered Foods	Silicon dioxide	No migration reported	Permitted only in PET bottles up to 20 mg/kg. Silicon dioxide serves as an effective anti-caking agent, ensuring powdered foods remain free flowing without migration concerns.	Prevents clumping in powdered food products.
Enhancing PET Bottle Color	Cobalt oxide	No migration reported	Permitted only in PET bottles up to 20 mg/kg. Cobalt oxide imparts color to PET bottles without migrating into packaged food, contributing to product aesthetics.	Adds color to PET bottles without migration.
Colorant and Visual Enhancement	Manganese oxide (E530)	No migration reported	Permitted only in PET bottles up to 20 mg/kg. Manganese oxide, acting as a colorant, does not migrate into packaged food, enhancing PET bottle aesthetics.	Adds color to PET bottles without migration.
Aesthetic Enhancement in PET Packaging	Titanium nitride	No migration reported	Permitted only in PET bottles up to 20 mg/kg. Titanium nitride, a colorant, retains its presence in PET bottles without migration into food contents.	Adds color to PET bottles without migration.
Colorant in Polymer-Based Packaging	Carbon black	Authorized by EC 10/2011	Authorized up to $<2.5\%$ w/w in the polymer. Carbon black, a common pigment, is incorporated into polymer-based food packaging to enhance color and appearance within regulated limits.	Enhances color and appearance of polymer-based packaging.

Antimicrobial Enhancement in Food Packaging	Silver-silica Nanox	FCS Inventorya	Permitted at <4 ppm by weight of silver. Silver-silica Nanox is utilized to augment antimicrobial properties in food packaging, prolonging shelf life by inhibiting microbial growth.	Increases shelf life through antimicrobial action.
Flavor Dispersion and Anti-Caking Agent	Silicon dioxide (E551d)	Authorized by EC 1334/2008	Allowed up to <10,000 mg/kg, excluding infant and young children foods. Silicon dioxide serves as an effective anti-caking agent and flavor carrier, enhancing food texture and taste within specified limits.	Prevents caking and enhances flavor dispersion.
Clear Labeling on Produce	Silicon dioxide (E551)	Exempt from certification	Allowed up to <2% of the ink solids. Silicon dioxide aids in clear and legible labeling of fruits and vegetables without requiring certification, improving consumer information.	Facilitates clear and informative labeling on produce.
Preventing Clumping in Powdered Foods	Silicon dioxide (E551)	REGb	Permitted up to <2% by weight of the food. Silicon dioxide functions as an anti-caking agent, maintaining the texture and quality of powdered foods by preventing clumping.	Prevents clumping in powdered food products.

Sources: He et al. (2019); Kearns et al. (2017); Sun et al. (2018); Zambrano-Zaragoza et al. (2014).

safer, healthier, and more ecologically friendly food is possible because to their integration (Chhipa, 2019; He et al., 2019).

10.3 NANOBIOSENSORS FOR FOOD CONTAMINANT DETECTION

Nanoparticles, nanowires, and nanotubes, among other special features, are used in nanobiosensors to increase the detection signal. When these nanomaterials are functionalized with certain biomolecules, like antibodies or aptamers, they display remarkable precision in binding to the intended pollutants. Even minute levels of pollutants can be detected by this binding event, according to Liu et al. (2021). New nanobiosensors are made to identify several pollutants at once. These sensors can identify different pollutants in a single test by integrating a variety of identification elements on a single platform. It also accelerates the detecting process, increasing effectiveness (Bashir et al., 2022; Kaur et al., 2023). Nanobiosensors can identify a wide range of contaminants, including bacterial diseases, allergies, chemical traces, and toxins. Due to their versatility, they are useful for monitoring various locations along the food supply chain, from unprocessed agricultural products to processed foods. The integration of nanobiosensors into food safety procedures enhances the integrity and quality of the food supply (Raghu et al., 2020; Taj et al., 2022).

Common food allergens could be detected by nanobiosensors employing enzyme-linked immunosorbent assays (ELISAs) at a limit of detection as low as 0.1 parts per billion (ppb). The increased surface-to-volume ratio of nanomaterials, which enhances the interaction between analytes and bioreceptors, is the cause of this exceptional sensitivity. Nanobiosensors have high selectivity because biorecognition components like antibodies, aptamers, and enzymes are carefully developed and incorporated into them. They can also identify between certain strains of the bacteria *Escherichia coli*, even in the presence of unharmful bacteria that are typically found in food samples (Bashir et al., 2022; Bruce-Tagoe & Danquah, 2023; Solís et al., 2022). Nanobiosensors make it simple to integrate Internet of Things (IoT) and data analytics systems, enabling continuous monitoring, data sharing, and predictive modeling. Through this connection, the infrastructure for overall food safety is improved, and proactive monitoring of dangerous pollutants is made possible (Nagarajan et al., 2022; Taj et al., 2022).

10.4 NANOTECHNOLOGY IN FOOD PACKAGING: SAFETY AND SHELF LIFE

10.4.1 Overview and Importance of Nanotechnology in Food Packaging

Food packaging materials are essential in the modern food industry. Aside from containment, they protect food, extend its shelf life, and provide essential information to consumers (Figure 10.2) (Cheng et al., 2022). In recent times, nanotechnology has transformed food packaging by innovative approaches to enhancing safety, extending shelf life, and resolving environmental issues are provided. As the world searches for improved ways to preserve food quality and battles food waste. The use of nanotechnology in packaging materials has drawn a lot of attention. Numerous complex materials made possible by nanotechnology, such as nano-clays, nanocomposites, and nanolaminates, give packaging materials exceptional barrier properties. Perishable food is shielded from environmental factors like oxygen, moisture, and ultraviolet (UV) radiation by these nanoparticles (Figure 10.3), which collectively create an impenetrable screen (Mu et al., 2019; Zabihzadeh Khajavi et al., 2020). According to Martins et al. (2022), applying nano-clay to packaging sheets reduces oxygen permeability by 75%, hence increasing the shelf life of packaged foods.

To successfully stop the development of dangerous germs and oxidative degradation, nanoparticles can be created to release antimicrobial agents, antioxidants, and oxygen scavengers. Packaging with

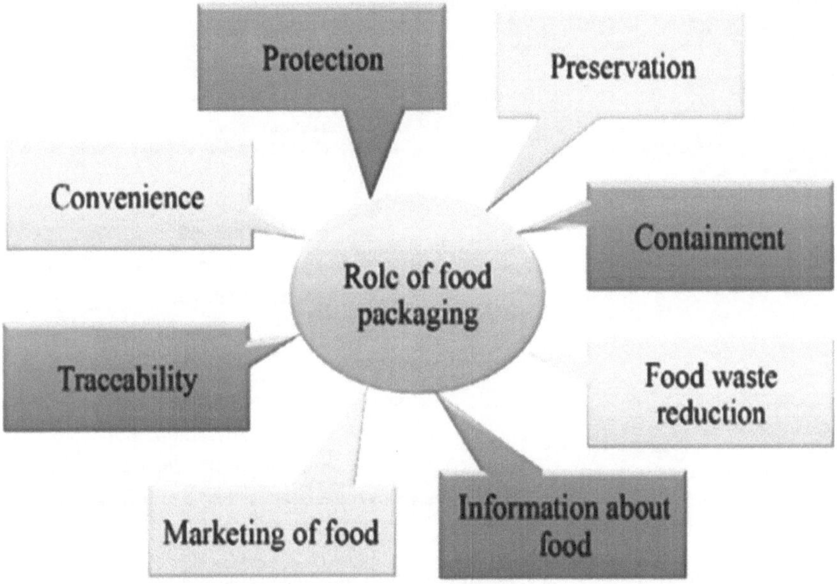

FIGURE 10.2 The role of food packaging materials. Reprinted with written permission from Jaiswal et al. (2019).

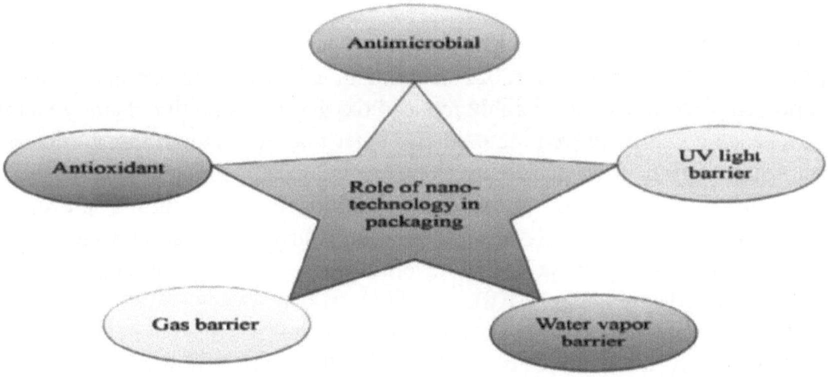

FIGURE 10.3 The role of nanotechnology in food packaging. Reprinted with written permission from Jaiswal et al. (2019).

nanomaterial enhancement reduces early deterioration and the need for frequent replacement. By reducing the needless disposal of edible food, this not only results in financial savings for both producers and consumers but also promotes sustainable resource utilization (Huang et al., 2020; Kong et al., 2023). To ensure consumer protection, the deployment of nanotechnology necessitates thorough safety analyses. Regulatory organizations evaluate the possibility of nanoparticles migrating into food, including the Food and Drug Administration (FDA) and the European Food Safety Authority (EFSA). The incorporation of nanotechnology into food packaging ushers in a new era of improved sustainability, decreased food waste, and increased safety (Amin et al., 2022; Owusu-Apenten & Vieira, 2023).

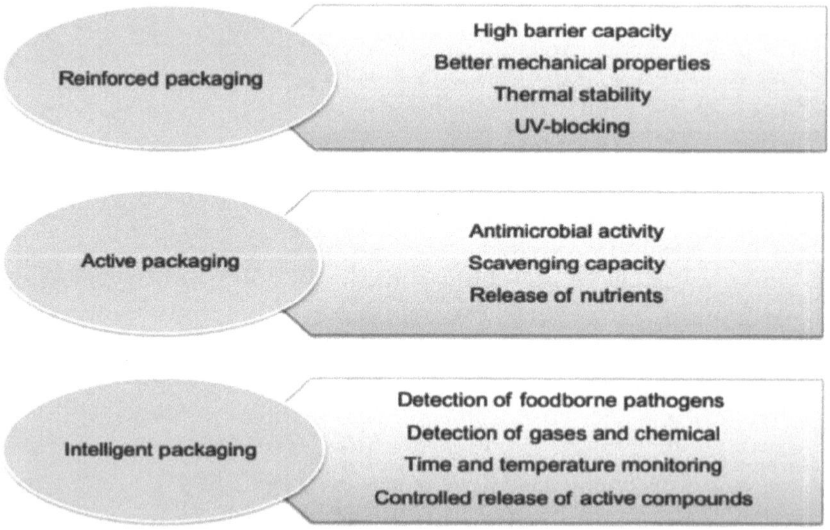

FIGURE 10.4 Possible applications of nanotechnology in food packaging, not only the actual applications but also the ones that in a near future will be available on the shelves. Reprinted with written permission from Cerqueira et al. (2018).

10.5 ADVANCED NANOMATERIALS FOR FOOD PACKAGING

Graphene nanoplatelets have demonstrated significant gas and moisture barrier properties. In accordance with research by Amin et al. (2022) and Emamhadi et al. (2020), graphene nanoplatelets inserted into polyethylene films considerably reduced oxygen permeability, thereby increasing the shelf life of packaged goods by preventing oxidative deterioration. Another transformational application of nanotechnology in food packaging is nanocoating. Numerous studies have been done on silver nanoparticles, which are recognized for their antibacterial qualities, to prevent microbial development on packaging surfaces. It demonstrated how silver nanoparticles can stop common foodborne germs from growing on packaging materials, improving the overall safety of packaged goods (Azeredo et al., 2019; Ma et al., 2022).

As nanomaterials have developed, they have exhibited capabilities that have changed the way that conventional packaging functions into a dynamic and intelligent system, illustrated in Figure 10.4. The addition of nanoparticles, such as graphene oxide, nanosilver, and montmorillonite nano-clays, results in a remarkable impermeability to gases and moisture. Additionally, it emphasizes how water vapor transmission rates are dramatically reduced by 50% with nano-clay-enhanced coatings, extending the freshness of packaged goods (Kim et al., 2022; Maresca & Mauriello, 2022). The properties of polymers and nanoparticles are combined in nanocomposites, which have higher mechanical strength, thermal stability, and gas barrier performance. Packaging films with improved mechanical integrity and gas barrier functions have been produced by incorporating nanoparticles like graphene and carbon nanotubes into polymer matrices (Chaudhary et al., 2020; Nile et al., 2020). The surface characteristics of packing materials can be changed using a variety of methods, including nanocoatings. Nanoparticles with antibacterial properties, including zinc oxide and silver, can be added. It prevents pathogenic bacteria from growing on the package surface. Due to their high surface area-to-volume ratio and porous nature, electrospun nanofibers have become more popular. With the addition of antimicrobials, enzymes, and sensing components, these nanofibers can be used in active and intelligent packaging. Enhanced safety and environmental sustainability are made possible by nanostructured coatings, active nanocomposites, nanobiosensors, and eco-friendly nanomaterials (Ashfaq et al., 2022; Barage et al., 2022; Dogra et al., 2023).

10.6 EXTENDING SHELF LIFE AND ENSURING FRESHNESS

The preservation of perishable commodities has been transformed by the inclusion of nanotechnology into food packaging, which considerably increases their shelf life and preserves their freshness. By actively interacting with the food goods they contain, these innovative packaging solutions ensure their quality from manufacturing to consumption (Babu, 2022; Nile et al., 2020). Packaging materials contain nanobarcodes that act as data carriers for delivering real-time data on temperature, humidity, and other environmental factors. TTIs, powered by nanotechnology, provide observable signs of changes in food quality brought on by variations in temperature. Based on the real freshness status of the products, these indicators enable consumers to make educated decisions (Pirsa et al., 2022).

Nanosensors are made to recognize volatile substances that food releases as it deteriorates. As a visual cue for the freshness of the packaged goods, nanosensors' color change or signal emission in response to particular gases. Controlled environments can be created inside the package because of nanocoatings that have been impregnated with nanoparticles. To optimize the environment and stop microbial development, oxidative reactions, and moisture ingress, these nanoparticles either release or absorb gases (Ghosh et al., 2022; Janjarasskul & Suppakul, 2018). In addition to forming a barrier of protection, edible nanocomposites also release bioactive substances that maintain freshness. These edible packaging materials protect products from contamination while boosting their general quality by incorporating antioxidants or antibacterial agents (Singha et al., 2022). These developments show how nanotechnology has the potential to reduce food waste, promote sustainability, and ensure the supply of fresh, healthy products to consumers (Ghosh et al., 2022; Singha et al., 2022).

10.7 HEALTH AND SAFETY CONSIDERATIONS OF NANOMATERIALS IN FOOD PACKAGING

Analyzing the possible risks caused by nanoparticles in food packaging requires thorough risk assessment processes. These methods take a wide range of variables into account, such as the size of the nanoparticles, their surface area, their chemical makeup, and potential migration routes. The identification of potential negative effects and the establishment of safety thresholds are aided by thorough analysis of these factors (Sharma et al., 2023; Singha et al., 2022). A major worry is the propensity of nanoparticles to transfer from packaging materials into food. It is essential to understand the scope and methods of migration as well as the ensuing bioavailability within the human body to assess any potential health dangers. To determine the safety of nanoparticles used in food packaging, comprehensive toxicological investigations are essential. These investigations look at how nanoparticles affect cells and molecules, identifying any cytotoxicity, genotoxicity, or unfavorable immunological reactions (Chinchkar et al., 2023; Mironava et al., 2014).

Building trust and enabling informed decision-making requires educating the public about the use of nanotechnology in food packaging, addressing concerns, and dispelling myths. To establish policies, uphold safety requirements, and conduct post-market surveillance, regulators, researchers, and industry must work together. Increased safety and longer shelf life may result from the use of nanotechnology in food packaging. However, proactive steps are needed to protect consumer health, such as detailed risk evaluations, thorough toxicological investigations, open communication, and efficient regulatory control (Chen et al., 2020; Dainelli et al., 2008; Sharma et al., 2021).

10.8 FUTURE PERSPECTIVES AND RESEARCH DIRECTIONS

10.8.1 Prospects of Nanotechnology

Future uses of nanotechnology in agriculture and food show great promise for addressing problems associated with a growing global population, resource sustainability, and food security. Emerging research directions and potential applications are reshaping the environment in which these businesses will operate (Gorczyca et al., 2021). The system of food distribution is well-positioned to gain from nanotechnology. Nanosensor-equipped foods can provide real-time updates on their condition. It can also be used to help functional meals include adequate amounts of vital nutrients such vitamins, minerals, probiotics, and dietary fibers (Mehadi Galib et al., 2022; Tawade & Wasewar, 2023). Nanoencapsulating stress-tolerant genes or chemicals can increase a plant's tolerance to drought, salt, and extreme temperatures. This innovation could significantly assist in ensuring food security in the face of changing climate trends (Awuchi et al., 2022; Sahoo et al., 2021).

Sophisticated nanobiosensors can be included into packing materials to swiftly detect pollutants, diseases, and spoilage symptoms. By giving customers instant access to information on the caliber and safety of their food, this promotes decision-making. Agrochemicals may be administered precisely where they are required using nano-enabled delivery systems, reducing their negative environmental effects, and maximizing resource efficiency. The introduction of more ecologically friendly farming practices could result from this transition, which could improve crop nutrition and pest management. Because of nanoemulsions and nanocomposites, food can have a varied texture and appearance, resulting in new flavor experiences and appealing aesthetics. This opens possibilities for creating novel culinary experiences and increasing the potential of gastronomy (Jagtiani, 2022; Pandey et al., 2023; Sahoo et al., 2021).

Nanotechnology enhances food systems by controlling nutrient release, using biodegradable packaging, and enhancing component stability. Methods like nanoemulsions and nanocomposites are used to create novel formulations with improved properties and health advantages. Additionally, by accurately providing fertilizers and insecticides while avoiding environmental damage, nanotechnology promotes environmentally friendly farming (Jampílek et al., 2019). Due to safety concerns regarding the possible toxicity of nanoparticles in food products, thorough risk assessments are necessary. Additional concerns including consumer knowledge, labeling, and regulatory frameworks must be taken into consideration to ensure the ethical development and deployment of nanotechnology in the food industry. As the food industry uses nanotechnology, these potential are expected to revolutionize how we process, package, and preserve the quality of our food supply, resulting in safer, more nutrient-dense, and sustainable food items (Guleria et al., 2023; Kiss, 2020).

10.9 NOVEL NANOMATERIALS AND OPTIMIZATION FOR FOOD APPLICATIONS

The fundamental goal of using innovative nanomaterials in food is to increase the bioavailability of essential nutrients. Liposomes and solid lipid nanoparticles are examples of nanoencapsulation techniques that enable better stability and controlled release of bioactive substances. For instance, beta-carotene in nanocapsules showed greater stability over time. Additionally, it encapsulates fat-soluble vitamins, greatly improving their bioaccessibility and solubility (Campelo et al., 2020). Nanoemulsions and nanogels are essential for changing the appearance and texture of food products. Nanostructured components are used to improve the texture of gluten-free food, make dairy products creamier, and produce eye-catching visual effects in beverages. Scientists can precisely construct food matrices at the nanoscale using nanotechnology, obtaining desirable sensory qualities, and improving the entire customer experience (Harish et al., 2022; Nile et al., 2020). Utilizing nanoencapsulation technology, specific dosages of savory and chemical components can be provided, minimizing the reliance on excessive additive consumption in food products. This strategy guarantees that additives are only released under certain consumption circumstances, thereby providing customers with a more wholesome and controlled food experience (Assadpour & Jafari, 2019; Tahir et al., 2021).

The ability to specifically create nanomaterials with certain capabilities is made possible by advances in computational modeling. The development of nanomaterials specialized for certain uses, such as controlled release, flavor modification, or antimicrobial activity, is driven by simulation-driven methodologies. By reducing trial-and-error iterations, this approach expedites material creation. It is

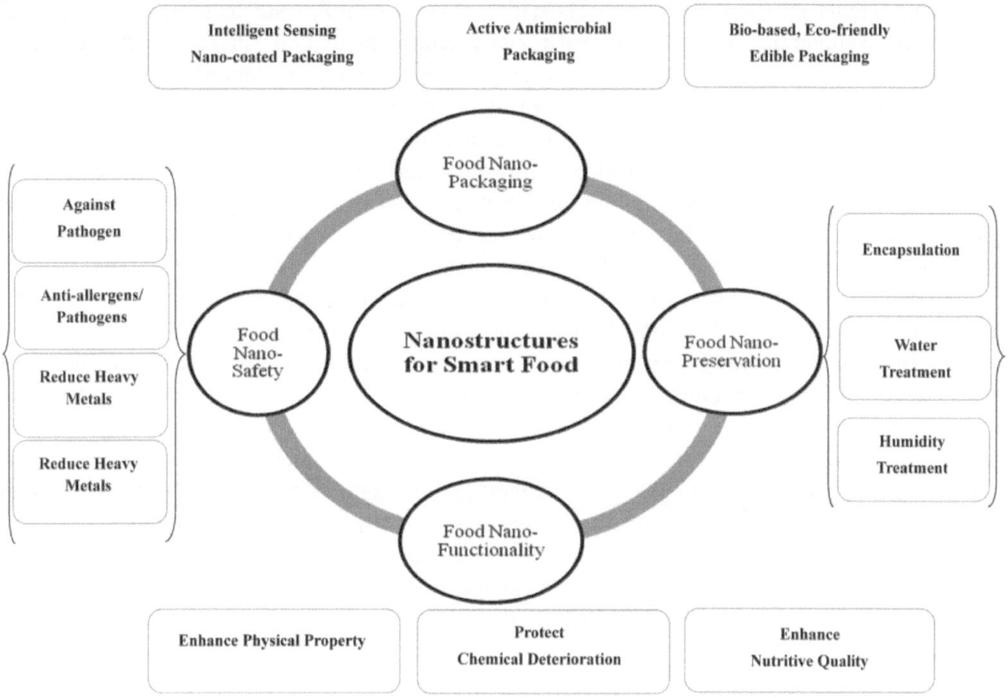

FIGURE 10.5 Systematic representation of application of nanoparticles in various areas of food industry. Reprinted with written permission from Bajpai et al. (2018).

crucial to strike a balance between the advantages of integrating nanomaterials and security issues. The World Health Organization (WHO) highlights the value of extensive risk analyses prior to the commercialization of food items based on nanotechnology. Optimization procedures must avoid undesirable interactions that can degrade nutritional value or sensory qualities. Exploring innovative nanomaterials and strategically optimizing them for use in culinary applications have the potential to fundamentally alter how we prepare, package, and consume food (Emamhadi et al., 2020; Tawade & Wasewar, 2023; Ummi & Siddiquee, 2019). The main applications of nanoparticles, nanomaterials, and nanostructures in the food business are outlined in Figure 10.5 (Bajpai et al., 2018).

10.10 ENVIRONMENTAL AND HEALTH IMPLICATIONS OF NANOTECHNOLOGY IN FOOD

Conducting long-term ecological studies to assess the persistence and possible accumulation of nanomaterials in various environmental compartments is one of the most difficult tasks. These investigations are essential for identifying possible bioaccumulation, trophic transfer, and general ecological perturbations that may occur over lengthy periods of time. Specific aspects of nanoparticles, such as size, surface characteristics, and potential interactions with biological systems, should be considered while developing standardized testing procedures. Researchers and regulatory authorities can get accurate and comparable data by developing standardized processes (Kabir et al., 2018; Thakur et al., 2022).

The use of life cycle assessments (LCAs) can give insight into how food products utilizing nanotechnology will affect the environment. Stakeholders can find possibilities to reduce environmental burdens and improve sustainability by examining the full life cycle, from raw material extraction to end-of-life disposal. Molecular simulations and other forms of computational modeling provide prescriptive insights into how nanomaterials will behave in complex surroundings. This method facilitates risk evaluations, predicts potential toxicity, and helps us understand how nanoparticles interact with biological systems. For the future perspective, it is crucial to comprehend nanomaterial fate in various environmental compartments, assess effects on creatures that are not the intended targets, and design efficient mitigation techniques. Stakeholder engagement fosters a balanced perspective on the implications of nanotechnology in food and ensures that concerns are addressed transparently and responsibly. By collaborating on regulations and guidelines, international regulatory agencies can create comprehensive frameworks for risk assessment and management (Jabeen et al., 2020; Martínez et al., 2021; Scott-Fordsmand et al., 2017).

The environment may become contaminated by the disposal, leaching, or runoff of nanomaterials employed in food-related applications. Water quality, aquatic habitats, and soil health may all be negatively impacted by these particles. For instance, silver nanoparticles, which are frequently used in food packaging for their antibacterial qualities, may have an effect on aquatic creatures and upset the equilibrium of ecosystems (Li et al., 2022; Wigger et al., 2020). The creation of safe exposure limits and regulatory standards will be influenced by an evaluation of the potential health concerns related to ingestion, inhalation, and cutaneous exposure to nanomaterials. While certain nanoparticles may increase the accessibility of nutrients, others may cause immunological reactions or even damage human cells (Alfei et al., 2020; Rani Sarkar et al., 2022). By uniting scientific studies, standardized tests, stakeholder engagement, and robust regulations, we can responsibly integrate nanotech in production and packaging while preserving well-being and sustainability (Loyal et al., 2023).

10.11 PUBLIC PERCEPTION AND ADOPTION OF NANOTECHNOLOGY

The promise for radical change offered by nanotechnology is evident in the rapidly changing landscape of sectors like food and agriculture. Public acceptance and perception, however, are crucial for its successful integration. According to studies, people are more cautious when using new technologies. Customized communication strategies can be created by determining the variables that affect risk perception. Seminars, conferences, and online tools that offer thorough yet understandable explanations of nanotechnology's advantages and safety in the context of food production are examples of effective public engagement. It is crucial to have open lines of communication between scientists, regulators, industrial stakeholders, and the general public. This helps dispel myths, disseminate factual information, and foster trust in relation to food-related nanotechnology (Mukherjee et al., 2019; Rathore & Mahesh, 2021).

Extended shelf life, lower food waste, and features like higher nutritional content are sought. Research that quantifies these benefits may have a substantial impact on consumer views. By using participatory tools like citizen panels, the public makes sure that their opinions are heard and taken into account when making policy. Complex scientific concepts can be made simpler for the general public by working with social media influencers, science communicators, and traditional media outlets. It is essential to do research on consumer preferences and opinions regarding food made with nanotechnology. Products that are in line with consumer values and preferences can be developed with the help of data-driven insights (Handford et al., 2015; Handy et al., 2008).

10.12 CONCLUSION

The incorporation of nanotechnology into the existing setting of innovation is advantageous for food and agriculture. This technology helps to modify processes, address problems, and lead the way for sustainability and efficiency. Precision tools for increasing yields of crops and resource efficiency are provided by agricultural nanotechnology. Nanoparticles distribute nutrients more efficiently, lessening their negative effects on the environment. Nanomaterials help in pest control and soil health restoration. The effects of nanotechnology on the food industry are also significant. By using advanced packing materials that guard against pollutants and spoiling factors, it increases the shelf life of food. With quick, on-site contamination and allergen detection, nanobiosensors change food safety. The usage of nanomaterials improves the texture, taste, and nutritional value of food products. However, this integration brings up important issues. Impacts on the environment and public health require careful consideration. Regulations must guarantee security and ethical innovation. Adoption is influenced by ethical issues and public perception. The fusion of biotechnology, data analytics, and nanotechnology offers possibilities for a robust food system. Precision farming will change farming methods and reduce its negative environmental effects. We must combine studies on the environmental impact, health, ethics, and public perception to assure nanotechnology's future in food. We can appropriately integrate nanotechnology in sustaining well-being and sustainability by fusing extensive research, standardized testing, stakeholder participation, and strong laws.

REFERENCES

Alfei, S., Marengo, B., & Zuccari, G. (2020). Nanotechnology application in food packaging: A plethora of opportunities versus pending risks assessment and public concerns. *Food Research International, 137*, 109664. https://doi.org/10.1016/j.foodres.2020.109664

Amin, U., Khan, M. K. I., Maan, A. A., Nazir, A., Riaz, S., Khan, M. U., Sultan, M., Munekata, P. E. S., & Lorenzo, J. M. (2022). Biodegradable active, intelligent, and smart packaging materials for food applications. *Food Packaging and Shelf Life, 33*, 100903. https://doi.org/10.1016/j.fpsl.2022.100903

Ashfaq, A., Khursheed, N., Fatima, S., Anjum, Z., & Younis, K. (2022). Application of nanotechnology in food packaging: Pros and cons. *Journal of Agriculture and Food Research, 7*, 100270. https://doi.org/10.1016/j.jafr.2022.100270

Assadpour, E., & Jafari, S. M. (2019). Chapter 3 – Nanoencapsulation: Techniques and developments for food applications. In A. López Rubio, M. J. FabraRovira, M. MartínezSanz, & L. G. Gómez-Mascaraque (Eds.), *Nanomaterials for Food Applications* (pp. 35–61). Elsevier. https://doi.org/10.1016/B978-0-12-814130-4.00003-8

Awuchi, C. G., Morya, S., Dendegh, T. A., Okpala, C. O. R., & Korzeniowska, M. (2022). Nanoencapsulation of food bioactive constituents and its associated processes: A revisit. *Bioresource Technology Reports, 19*, 101088. https://doi.org/10.1016/j.biteb.2022.101088

Azeredo, H. M. C., Otoni, C. G., Corrêa, D. S., Assis, O. B. G., de Moura, M. R., & Mattoso, L. H. C. (2019). Nanostructured antimicrobials in food packaging – recent advances. *Biotechnology Journal, 14*(12), 1900068. https://doi.org/10.1002/biot.201900068

Babu, P. J. (2022). Nanotechnology mediated intelligent and improved food packaging. *International Nano Letters, 12*(1), 1–14. https://doi.org/10.1007/s40089-021-00348-8

Bajpai, V. K., Kamle, M., Shukla, S., Mahato, D. K., Chandra, P., Hwang, S. K., Kumar, P., Huh, Y. S., & Han, Y.-K. (2018). Prospects of using nanotechnology for food preservation, safety, and security. *Journal of Food and Drug Analysis, 26*(4), 1201–1214. https://doi.org/10.1016/j.jfda.2018.06.011

Barage, S., Lakkakula, J., Sharma, A., Roy, A., Alghamdi, S., Almehmadi, M., Hossain, M. J., Allahyani, M., & Abdulaziz, O. (2022). Nanomaterial in food packaging: A comprehensive review. *Journal of Nanomaterials, 2022*, 6053922. https://doi.org/10.1155/2022/6053922

Bashir, O., Bhat, S. A., Basharat, A., Qamar, M., Qamar, S. A., Bilal, M., & Iqbal, H. M. N. (2022). Nano-engineered materials for sensing food pollutants: Technological advancements and safety issues. *Chemosphere, 292*, 133320. https://doi.org/10.1016/j.chemosphere.2021.133320

Beig, B., Niazi, M. B. K., Sher, F., Jahan, Z., Malik, U. S., Khan, M. D., Américo-Pinheiro, J. H. P., & Vo, D.-V. N. (2022). Nanotechnology-based controlled release of sustainable fertilizers. A review. *Environmental Chemistry Letters, 20*(4), 2709–2726. https://doi.org/10.1007/s10311-022-01409-w

Bhagat, B., Baruah, P., & Mukherjee, K. (2022). Chapter 32 – Application of nanosensors in food inspection. In A. Denizli, T. A. Nguyen, S. Rajendran, G. Yasin, & A. K. Nadda (Eds.), *Nanosensors for Smart Agriculture* (pp. 705–735). Elsevier. https://doi.org/10.1016/B978-0-12-824554-5.00030-6

Biswas, R., Alam, M., Sarkar, A., Haque, M. I., Hasan, M. M., & Hoque, M. (2022). Application of nanotechnology in food: Processing, preservation, packaging and safety assessment. *Heliyon, 8*(11), e11795. https://doi.org/10.1016/j.heliyon.2022.e11795

Bruce-Tagoe, T. A., & Danquah, M. K. (2023). Bioaffinity nanoprobes for foodborne pathogen sensing. *Micromachines*, *14*(6), 1122. https://doi.org/10.3390/mi14061122

Campelo, P. H., Sant'Ana, A. S., & Pedrosa Silva Clerici, M. T. (2020). Starch nanoparticles: Production methods, structure, and properties for food applications. *Current Opinion in Food Science*, *33*, 136–140. https://doi.org/10.1016/j.cofs.2020.04.007

Cerqueira, M. A., Vicente, A. A., & Pastrana, L. M. (2018). Chapter 1 – Nanotechnology in food packaging: Opportunities and challenges. In M. Â. P. R. Cerqueira, J. M. Lagaron, L. M. Pastrana Castro, & A. A. M. de Oliveira Soares Vicente (Eds.), *Nanomaterials for Food Packaging* (pp. 1–11). Elsevier. https://doi.org/10.1016/B978-0-323-51271-8.00001-2

Chaudhary, P., Fatima, F., & Kumar, A. (2020). Relevance of nanomaterials in food packaging and its advanced future prospects. *Journal of Inorganic and Organometallic Polymers and Materials*, *30*(12), 5180–5192. https://doi.org/10.1007/s10904-020-01674-8

Chausali, N., Saxena, J., & Prasad, R. (2022). Recent trends in nanotechnology applications of bio-based packaging. *Journal of Agriculture and Food Research*, *7*, 100257. https://doi.org/10.1016/j.jafr.2021.100257

Chen, Z., Han, S., Zhou, S., Feng, H., Liu, Y., & Jia, G. (2020). Review of health safety aspects of titanium dioxide nanoparticles in food application. *NanoImpact*, *18*, 100224. https://doi.org/10.1016/j.impact.2020.100224

Cheng, H., Xu, H., Julian McClements, D., Chen, L., Jiao, A., Tian, Y., Miao, M., & Jin, Z. (2022). Recent advances in intelligent food packaging materials: Principles, preparation and applications. *Food Chemistry*, *375*, 131738. https://doi.org/10.1016/j.foodchem.2021.131738

Chhipa, H. (2019). Chapter 6 – Applications of nanotechnology in agriculture. In V. Gurtler, A. S. Ball, & S. Soni (Eds.), *Methods in Microbiology* (Vol. 46, pp. 115–142). Academic Press. https://doi.org/10.1016/bs.mim.2019.01.002

Chinchkar, A. V., Singh, A., Kamble, M. G., Prabhakar, P. K., Meghwal, M., & Sharma, A. (2023). Chapter 14 – Nanotechnology applications for quality determination of RTE and packaged food. In A. Sharma, P. S. Vijayakumar, E. P. K. Prabhakar, & R. Kumar (Eds.), *Nanotechnology Applications for Food Safety and Quality Monitoring* (pp. 265–288). Academic Press. https://doi.org/10.1016/B978-0-323-85791-8.00002-1

Dainelli, D., Gontard, N., Spyropoulos, D., Zondervan-van den Beuken, E., & Tobback, P. (2008). Active and intelligent food packaging: Legal aspects and safety concerns. *Trends in Food Science & Technology*, *19*, S103–S112. https://doi.org/10.1016/j.tifs.2008.09.011

Dogra, V., Kishore, C., Mishra, A., Verma, A., & Gaur, A. (2023). Coatings: Types and synthesis techniques. In A. Verma, S. K. Sethi, & S. Ogata (Eds.), *Coating Materials: Computational Aspects, Applications and Challenges* (pp. 17–31). Springer Nature Singapore. https://doi.org/10.1007/978-981-99-3549-9_2

Dorier, M., Béal, D., Marie-Desvergne, C., Dubosson, M., Barreau, F., Houdeau, E., Herlin-Boime, N., & Carriere, M. (2017). Continuous in vitro exposure of intestinal epithelial cells to E171 food additive causes oxidative stress, inducing oxidation of DNA bases but no endoplasmic reticulum stress. *Nanotoxicology*, *11*(6), 751–761. https://doi.org/10.1080/17435390.2017.1349203

Emamhadi, M. A., Sarafraz, M., Akbari, M., Thai, V. N., Fakhri, Y., Linh, N. T. T., & Mousavi Khaneghah, A. (2020). Nanomaterials for food packaging applications: A systematic review. *Food and Chemical Toxicology*, *146*, 111825. https://doi.org/10.1016/j.fct.2020.111825

Faizan, M., Ahmad, S. M., Ahamad, L., Chen, C., & Yu, F. (2023). Nanotechnology in agriculture. In A. Raina, M. R. Wani, R. A. Laskar, N. Tomlekova, & S. Khan (Eds.), *Advanced Crop Improvement, Volume 1: Theory and Practice* (pp. 33–46). Springer International Publishing. https://doi.org/10.1007/978-3-031-28146-4_2

Ghosh, T., Raj, G. V. S. B., & Dash, K. K. (2022). A comprehensive review on nanotechnology based sensors for monitoring quality and shelf life of food products. *Measurement: Food*, *7*, 100049. https://doi.org/10.1016/j.meafoo.2022.100049

Gorczyca, A., Pociecha, E., & Matras, E. (2021). Nanotechnology in agriculture, the food sector, and remediation: prospects, relations, and constraints. In R. Prasad (Ed.), *Environmental Pollution and Remediation* (pp. 1–34). Springer Singapore. https://doi.org/10.1007/978-981-15-5499-5_1

Guleria, G., Thakur, S., Shandilya, M., Sharma, S., Thakur, S., & Kalia, S. (2023). Nanotechnology for sustainable agro-food systems: The need and role of nanoparticles in protecting plants and improving crop productivity. *Plant Physiology and Biochemistry*, *194*, 533–549. https://doi.org/10.1016/j.plaphy.2022.12.004

Handford, C. E., Dean, M., Spence, M., Henchion, M., Elliott, C. T., & Campbell, K. (2015). Awareness and attitudes towards the emerging use of nanotechnology in the agri-food sector. *Food Control*, *57*, 24–34. https://doi.org/10.1016/j.foodcont.2015.03.033

Handy, R. D., Owen, R., & Valsami-Jones, E. (2008). The ecotoxicology of nanoparticles and nanomaterials: Current status, knowledge gaps, challenges, and future needs. *Ecotoxicology*, *17*(5), 315–325. https://doi.org/10.1007/s10646-008-0206-0

Harish, V., Tewari, D., Gaur, M., Yadav, A. B., Swaroop, S., Bechelany, M., & Barhoum, A. (2022). Review on nanoparticles and nanostructured materials: Bioimaging, biosensing, drug delivery, tissue engineering, antimicrobial, and agro-food applications. *Nanomaterials*, *12*(3), 457. https://doi.org/10.3390/nano12030457

He, X., Deng, H., & Hwang, H.-m. (2019). The current application of nanotechnology in food and agriculture. *Journal of Food and Drug Analysis*, *27*(1), 1–21. https://doi.org/10.1016/j.jfda.2018.12.002

He, X., & Hwang, H.-M. (2016). Nanotechnology in food science: Functionality, applicability, and safety assessment. *Journal of Food and Drug Analysis*, *24*(4), 671–681. https://doi.org/10.1016/j.jfda.2016.06.001

Huang, S., Liu, X., Chang, C., & Wang, Y. (2020). Recent developments and prospective food-related applications of cellulose nanocrystals: A review. *Cellulose*, *27*(6), 2991–3011. https://doi.org/10.1007/s10570-020-02984-3

Jabeen, F., Arshad, M. I., Mahmood Zia, K., Ul Hasan, M. S., Younas, M., Akhtar, M., & Rehman, A. U. (2020). Chapter 15 – Computational modeling for bionanocomposites. In K. Mahmood Zia, F. Jabeen, M. N. Anjum, & S. Ikram (Eds.), *Bionanocomposites* (pp. 367–420). Elsevier. https://doi.org/10.1016/B978-0-12-816751-9.00015-5

Jafarizadeh-Malmiri, H., Sayyar, Z., Anarjan, N., & Berenjian, A. (2019). Nano-sensors in food nanobiotechnology. In H. Jafarizadeh-Malmiri, Z. Sayyar, N. Anarjan, & A. Berenjian (Eds.), *Nanobiotechnology in Food: Concepts, Applications and Perspectives* (pp. 81–94). Springer International Publishing. https://doi.org/10.1007/978-3-030-05846-3_6

Jagtiani, E. (2022). Advancements in nanotechnology for food science and industry. *Food Frontiers*, *3*(1), 56–82. https://doi.org/10.1002/fft2.104

Jaiswal, L., Shankar, S., & Rhim, J.-W. (2019). Chapter 3 – Applications of nanotechnology in food microbiology. In V. Gurtler, A. S. Ball, & S. Soni (Eds.), *Methods in Microbiology* (Vol. 46, pp. 43–60). Academic Press. https://doi.org/10.1016/bs.mim.2019.03.002

Jakhar, A. M., Aziz, I., Kaleri, A. R., Hasnain, M., Haider, G., Ma, J., & Abideen, Z. (2022). Nano-fertilizers: A sustainable technology for improving crop nutrition and food security. *NanoImpact*, *27*, 100411. https://doi.org/10.1016/j.impact.2022.100411

Jampílek, J., Kráľová, K., Campos, E. V. R., & Fraceto, L. F. (2019). Bio-based nanoemulsion formulations applicable in agriculture, medicine, and food industry. In R. Prasad, V. Kumar, M. Kumar, & D. Choudhary (Eds.), *Nanobiotechnology in Bioformulations* (pp. 33–84). Springer International Publishing. https://doi.org/10.1007/978-3-030-17061-5_2

Janjarasskul, T., & Suppakul, P. (2018). Active and intelligent packaging: The indication of quality and safety. *Critical Reviews in Food Science and Nutrition*, *58*(5), 808–831. https://doi.org/10.1080/10408398.2016.1225278

Kabir, E., Kumar, V., Kim, K.-H., Yip, A. C. K., & Sohn, J. R. (2018). Environmental impacts of nanomaterials. *Journal of Environmental Management*, *225*, 261–271. https://doi.org/10.1016/j.jenvman.2018.07.087

Kaur, G., Bhari, R., & Kumar, K. (2023). Nanobiosensors and their role in detection of adulterants and contaminants in food products. *Critical Reviews in Biotechnology*, 1–15. https://doi.org/10.1080/07388551.2023.2175196

Kearns, H., Goodacre, R., Jamieson, L. E., Graham, D., & Faulds, K. (2017). SERS detection of multiple antimicrobial-resistant pathogens using nanosensors. *Analytical Chemistry*, *89*(23), 12666–12673. https://doi.org/10.1021/acs.analchem.7b02653

Keçili, R., Büyüktiryaki, S., & Hussain, C. M. (2019). Advancement in bioanalytical science through nanotechnology: Past, present and future. *TrAC Trends in Analytical Chemistry*, *110*, 259–276. https://doi.org/10.1016/j.trac.2018.11.012

Kim, I., Viswanathan, K., Kasi, G., Thanakkasaranee, S., Sadeghi, K., & Seo, J. (2022). ZnO nanostructures in active antibacterial food packaging: preparation methods, antimicrobial mechanisms, safety issues, future prospects, and challenges. *Food Reviews International*, *38*(4), 537–565. https://doi.org/10.1080/87559129.2020.1737709

Kiss, É. (2020). Nanotechnology in food systems: A review. *Acta Alimentaria AAlim*, *49*(4), 460–474. https://doi.org/10.1556/066.2020.49.4.12

Kong, J., Ge, X., Sun, Y., Mao, M., Yu, H., Chu, R., & Wang, Y. (2023). Multi-functional pH-sensitive active and intelligent packaging based on highly cross-linked zein for the monitoring of pork freshness. *Food Chemistry*, *404*, 134754. https://doi.org/10.1016/j.foodchem.2022.134754

Kumar, S., Nehra, M., Dilbaghi, N., Marrazza, G., Hassan, A. A., & Kim, K.-H. (2019). Nano-based smart pesticide formulations: Emerging opportunities for agriculture. *Journal of Controlled Release*, *294*, 131–153. https://doi.org/10.1016/j.jconrel.2018.12.012

Li, Y.-X., Qin, H.-Y., Hu, C., Sun, M.-M., Li, P.-Y., Liu, H., Li, J.-C., Li, Z.-B., Wu, L.-D., & Zhu, J. (2022). Research progress of nanomaterials-based sensors for food safety. *Journal of Analysis and Testing*, *6*(4), 431–440. https://doi.org/10.1007/s41664-022-00235-x

Liu, X., Huang, L., & Qian, K. (2021). Nanomaterial-based electrochemical sensors: Mechanism, preparation, and application in biomedicine. *Advanced NanoBiomed Research*, *1*(6), 2000104. https://doi.org/10.1002/anbr.202000104

Loyal, A., Pahuja, S. K., Sharma, P., Malik, A., Srivastava, R. K., & Mehta, S. (2023). Chapter 17 – Potential environmental and human health implications of nanomaterials used in sustainable agriculture and soil improvement. In A. Husen (Ed.), *Engineered Nanomaterials for Sustainable Agricultural Production, Soil Improvement and Stress Management* (pp. 387–412). Academic Press. https://doi.org/10.1016/B978-0-323-91933-3.00017-9

Ma, Y., Yang, W., Xia, Y., Xue, W., Wu, H., Li, Z., Zhang, F., Qiu, B., & Fu, C. (2022). Properties and applications of intelligent packaging indicators for food spoilage. *Membranes*, *12*(5), 477. https://doi.org/10.3390/membranes12050477

Malik, S., Muhammad, K., & Waheed, Y. (2023). Nanotechnology: A revolution in modern industry. *Molecules*, *28*(2), 661. https://doi.org/10.3390/molecules28020661

Maresca, D., & Mauriello, G. (2022). Development of antimicrobial cellulose nanofiber-based films activated with nisin for food packaging applications. *Foods*, *11*(19), 3051. https://doi.org/10.3390/foods11193051

Martínez, G., Merinero, M., Pérez-Aranda, M., Pérez-Soriano, E. M., Ortiz, T., Villamor, E., Begines, B., & Alcudia, A. (2021). Environmental impact of nanoparticles' application as an emerging technology: A review. *Materials*, *14*(1), 166. https://doi.org/10.3390/ma14010166

Martins, P. C., Latorres, J. M., & Martins, V. G. (2022). Impact of starch nanocrystals on the physicochemical, thermal and structural characteristics of starch-based films. *LWT*, *156*, 113041. https://doi.org/10.1016/j.lwt.2021.113041

MehadiGalib, R., Alam, M., Rana, R., & Ara, R. (2022). Mango (*Mangifera indica* L.) fiber concentrates: Processing, modification and utilization as a food ingredient. *Food Hydrocolloids for Health*, *2*, 100096. https://doi.org/10.1016/j.fhfh.2022.100096

Mironava, T., Hadjiargyrou, M., Simon, M., & Rafailovich, M. H. (2014). Gold nanoparticles cellular toxicity and recovery: Adipose derived stromal cells. *Nanotoxicology*, *8*(2), 189–201. https://doi.org/10.3109/17435390.2013.769128

Mu, R., Hong, X., Ni, Y., Li, Y., Pang, J., Wang, Q., Xiao, J., & Zheng, Y. (2019). Recent trends and applications of cellulose nanocrystals in food industry. *Trends in Food Science & Technology*, *93*, 136–144. https://doi.org/10.1016/j.tifs.2019.09.013

Mukherjee, A., Maity, A., Pramanik, P., Shubha, K., Joshi, D. C., & Wani, S. H. (2019). Chapter 18 – Public perception about use of nanotechnology in agriculture. In M. Ghorbanpour & S. H. Wani (Eds.), *Advances in Phytonanotechnology* (pp. 405–418). Academic Press. https://doi.org/10.1016/B978-0-12-815322-2.00019-5

Nagarajan, S. M., Deverajan, G. G., Chatterjee, P., Alnumay, W., & Muthukumaran, V. (2022). Integration of IoT based routing process for food supply chain management in sustainable smart cities. *Sustainable Cities and Society*, *76*, 103448. https://doi.org/10.1016/j.scs.2021.103448

Nallamuthu, I., Khanum, F., Fathima, S. J., Patil, M. M., & Anand, T. (2017). Chapter 16 – Enhanced nutrient delivery through nanoencapsulation techniques: The current trend in food industry. In A. M. Grumezescu (Ed.), *Nutrient Delivery* (pp. 619–651). Academic Press. https://doi.org/10.1016/B978-0-12-804304-2.00016-0

Napagoda, M., Jayathunga, D., & Witharana, S. (2023). Introduction to nanotechnology. In S. Witharana & M. T. Napagoda (Eds.), *Nanotechnology in Modern Medicine* (pp. 1–17). Springer Nature Singapore. https://doi.org/10.1007/978-981-19-8050-3_1

Nile, S. H., Baskar, V., Selvaraj, D., Nile, A., Xiao, J., & Kai, G. (2020). Nanotechnologies in food science: applications, recent trends, and future perspectives. *Nano-Micro Letters*, *12*(1), 45. https://doi.org/10.1007/s40820-020-0383-9

Oladosu, Y., Rafii, M. Y., Arolu, F., Chukwu, S. C., Salisu, M. A., Fagbohun, I. K., Muftaudeen, T. K., Swaray, S., & Haliru, B. S. (2022). Superabsorbent polymer hydrogels for sustainable agriculture: A review. *Horticulturae*, *8*(7), 605. https://doi.org/10.3390/horticulturae8070605

Owusu-Apenten, R., & Vieira, E. (2023). Food regulatory agencies. In R. Owusu-Apenten & E. R. Vieira (Eds.), *Elementary Food Science* (pp. 57–79). Springer International Publishing. https://doi.org/10.1007/978-3-030-65433-7_3

Pandey, G., Tripathi, S., Bajpai, S., & Kamboj, M. (2023). Approaches, challenges, and prospects of nanotechnology for sustainable agriculture. In F. Fernandez-Luqueno & J. K. Patra (Eds.), *Agricultural and Environmental Nanotechnology: Novel Technologies and Their Ecological Impact* (pp. 83–103). Springer Nature Singapore. https://doi.org/10.1007/978-981-19-5454-2_3

Patel, G., Pillai, V., Bhatt, P., & Mohammad, S. (2020). Chapter 21 – Application of nanosensors in the food industry. In B. Han, V. K. Tomer, T. A. Nguyen, A. Farmani, & P. Kumar Singh (Eds.), *Nanosensors for Smart Cities* (pp. 355–368). Elsevier. https://doi.org/10.1016/B978-0-12-819870-4.00020-7

Pirsa, S., Sani, I. K., & Mirtalebi, S. S. (2022). Nano-biocomposite based color sensors: Investigation of structure, function, and applications in intelligent food packaging. *Food Packaging and Shelf Life*, *31*, 100789. https://doi.org/10.1016/j.fpsl.2021.100789

Qiao, X., He, J., Yang, R., Li, Y., Chen, G., Xiao, S., Huang, B., Yuan, Y., Sheng, Q., & Yue, T. (2022). Recent advances in nanomaterial-based sensing for food safety analysis. *Processes*, *10*(12), 2576. https://doi.org/10.3390/pr10122576

Raghu, H. V., Parkunan, T., & Kumar, N. (2020). Application of nanobiosensors for food safety monitoring. In N. Dasgupta, S. Ranjan, & E. Lichtfousc (Eds.), *Environmental Nanotechnology* (Vol. 4, pp. 93 129). Springer International Publishing. https://doi.org/10.1007/978-3-030-26668-4_3

Rani Sarkar, M., Rashid, M. H.-o., Rahman, A., Kafi, M. A., Hosen, M. I., Rahman, M. S., & Khan, M. N. (2022). Recent advances in nanomaterials based sustainable agriculture: An overview. *Environmental Nanotechnology, Monitoring & Management*, *18*, 100687. https://doi.org/10.1016/j.enmm.2022.100687

Rathore, A., & Mahesh, G. (2021). Public perception of nanotechnology: A contrast between developed and developing countries. *Technology in Society*, *67*, 101751. https://doi.org/10.1016/j.techsoc.2021.101751

Sahoo, M., Vishwakarma, S., Panigrahi, C., & Kumar, J. (2021). Nanotechnology: Current applications and future scope in food. *Food Frontiers*, *2*(1), 3–22. https://doi.org/10.1002/fft2.58

Saravanadevi, K., Renuga Devi, N., Dorothy, R., Joany, R. M., Rajendran, S., & Nguyen, T. A. (2022). Chapter 1 – Nanotechnology for agriculture: An introduction. In A. Denizli, T. A. Nguyen, S. Rajendran, G. Yasin, & A. K. Nadda (Eds.), *Nanosensors for Smart Agriculture* (pp. 3–23). Elsevier. https://doi.org/10.1016/B978-0-12-824554-5.00013-6

Scott-Fordsmand, J. J., Peijnenburg, W. J. G. M., Semenzin, E., Nowack, B., Hunt, N., Hristozov, D., Marcomini, A., Irfan, M. A., Jiménez, A. S., Landsiedel, R., Tran, L., Oomen, A. G., Bos, P. M. J., & Hund-Rinke, K. (2017). Environmental risk assessment strategy for nanomaterials. *International Journal of Environmental Research and Public Health*, *14*(10), 1251. https://doi.org/10.3390/ijerph14101251

Sharma, A., Ranjit, R., Kumar, N., Kumar, M., & Giri, B. S. J. B. E. J. (2023). Nanoparticles based nanosensors: Principles and their applications in active packaging for food quality and safety detection. *Biochemical Engineering Journal*, *193*, 108861. https://doi.org/10.1016/j.bej.2023.108861

Sharma, P. K., Dorlikar, S., Rawat, P., Malik, V., Vats, N., Sharma, M., Rhyee, J. S., & Kaushik, A. K. (2021). Chapter 1 – Nanotechnology and its application: A review. In K. R. Khondakar & A. K. Kaushik (Eds.), *Nanotechnology in Cancer Management* (pp. 1–33). Elsevier. https://doi.org/10.1016/B978-0-12-818154-6.00010-X

Shawon, Z. B. Z., Hoque, M. E., & Chowdhury, S. R. (2020). Chapter 6 –Nanosensors and nanobiosensors: Agricultural and food technology aspects. In K. Pal & F. Gomes (Eds.), *Nanofabrication for Smart Nanosensor Applications* (pp. 135–161). Elsevier. https://doi.org/10.1016/B978-0-12-820702-4.00006-4

Singh, N. A., Rai, N., & Marwal, A. (2021). Nanosensors for the detection of chemical food adulterants. In V. Kumar, P. Guleria, S. Ranjan, N. Dasgupta, & E. Lichtfouse (Eds.), *Nanotoxicology and Nanoecotoxicology* (Vol. 2, pp. 25–53). Springer International Publishing. https://doi.org/10.1007/978-3-030-69492-0_2

Singha, K., Regubalan, B., Pandit, P., Maity, S., & Ahmed, S. (2022). Chapter 1 – Introduction to nanotechnology-enhanced food packaging industry. In J. Parameswaranpillai, R. E. Krishnankutty, A. Jayakumar, S.M. Rangappa, S. Siengchin (Eds.), *Nanotechnology-Enhanced Food Packaging* (pp. 1–17). Wiley Online Library. https://doi.org/10.1002/9783527827718.ch1

Solís, D., Toro, M., Navarrete, P., Faúndez, P., & Reyes-Jara, A. (2022). Microbiological quality and presence of foodborne pathogens in raw and extruded canine diets and canine fecal samples. *Frontiers in Veterinary Science*, *9*, 799710. https://doi.org/10.3389/fvets.2022.799710

Sun, Y., Fang, L., Wan, Y., & Gu, Z. (2018). Pathogenic detection and phenotype using magnetic nanoparticle-urease nanosensor. *Sensors and Actuators B: Chemical*, *259*, 428–432. https://doi.org/10.1016/j.snb.2017.12.095

Tahir, A., Shabir Ahmad, R., Imran, M., Ahmad, M. H., Kamran Khan, M., Muhammad, N., Nisa, M. U., Tahir Nadeem, M., Yasmin, A., Tahir, H. S., Zulifqar, A., & Javed, M. (2021). Recent approaches for utilization of food components as nano-encapsulation: A review. *International Journal of Food Properties*, *24*(1), 1074–1096. https://doi.org/10.1080/10942912.2021.1953067

Taj, A., Zia, R., Iftikhar, M., Younis, S., & Bajwa, S. Z. (2022). Chapter 31 –Nanosensors for food inspection. In A. Denizli, T. A. Nguyen, S. Rajendran, G. Yasin, & A. K. Nadda (Eds.), *Nanosensors for Smart Agriculture* (pp. 685–703). Elsevier. https://doi.org/10.1016/B978-0-12-824554-5.00032-X

Tavker, N., & Sharma, M. (2022). Chapter 33 –Nanosensors for intelligent food packaging. In A. Denizli, T. A. Nguyen, S. Rajendran, G. Yasin, & A. K. Nadda (Eds.), *Nanosensors for Smart Agriculture* (pp. 737–756). Elsevier. https://doi.org/10.1016/B978-0-12-824554-5.00014-8

Tawade, P. V., & Wasewar, K. L. (2023). Chapter 4 – Nanotechnology in biological science and engineering. In P. Singh, V. Kumar, M. Bakshi, C. M. Hussain, & M. Sillanpää (Eds.), *Environmental Applications of Microbial Nanotechnology* (pp. 43–64). Elsevier. https://doi.org/10.1016/B978-0-323-91744-5.00015-1

Thakur, R. K., Prasad, P., Savadi, S., Bhardwaj, S. C., Gangwar, O. P., & Kumar, S. (2022). Nanotechnology for agricultural and environmental sustainability. In R. Goel, R. Soni, D. C. Suyal, & M. Khan (Eds.), *Survival Strategies in Cold-Adapted Microorganisms* (pp. 413–424). Springer Singapore. https://doi.org/10.1007/978-981-16-2625-8_18

Tripathi, A., & Prakash, S. (2022). Chapter 1 –Nanobiotechnology: Emerging trends, prospects, and challenges. In S. Ghosh, S. Thongmee, & A. Kumar (Eds.), *Agricultural Nanobiotechnology* (pp. 1–21). Woodhead Publishing. https://doi.org/10.1016/B978-0-323-91908-1.00006-7

Ummi, A. S., & Siddiquee, S. (2019). Nanotechnology applications in food: Opportunities and challenges in food industry. In S. Siddiquee, G. J. H. Melvin, & M. M. Rahman (Eds.), *Nanotechnology: Applications in Energy, Drug and Food* (pp. 295–308). Springer International Publishing. https://doi.org/10.1007/978-3-319-99602-8_15

Weir, A., Westerhoff, P., Fabricius, L., Hristovski, K., & von Goetz, N. (2012). Titanium dioxide nanoparticles in food and personal care products. *Environmental Science & Technology*, *46*(4), 2242–2250. https://doi.org/10.1021/es204168d

Wigger, H., Kägi, R., Wiesner, M., & Nowack, B. (2020). Exposure and possible risks of engineered nanomaterials in the environment – current knowledge and directions for the future. *Reviews of Geophysics*, *58*(4), e2020RG000710. https://doi.org/10.1029/2020RG000710

Zabihzadeh Khajavi, M., Ebrahimi, A., Yousefi, M., Ahmadi, S., Farhoodi, M., Mirza Alizadeh, A., & Taslikh, M. (2020). Strategies for producing improved oxygen barrier materials appropriate for the food packaging sector. *Food Engineering Reviews*, *12*(3), 346–363. https://doi.org/10.1007/s12393-020-09235-y

Zambrano-Zaragoza, M. L., Mercado-Silva, E., Del Real L, A., Gutiérrez-Cortez, E., Cornejo-Villegas, M. A., & Quintanar-Guerrero, D. (2014). The effect of nano-coatings with α-tocopherol and xanthan gum on shelf-life and browning index of fresh-cut "Red Delicious" apples. *Innovative Food Science & Emerging Technologies*, *22*, 188–196. https://doi.org/10.1016/j.ifset.2013.09.008

11 Market Potential of Innovations Based on Food Nanotechnology

Mahabub Alam, Tanjilla Akter Sumi, Raju Mia, and Md. Jamil Imtiaj

11.1 INTRODUCTION

The emergence of nanoscience as a transformative discipline has catalyzed advancements across diverse domains, permeating fields as disparate as electronics, medicine, and materials science. Within the realm of food science, the integration of nanotechnology, termed "food nanotechnology," stands poised to engender a paradigm shift by introducing novel possibilities for innovation and enhancement within the culinary landscape (Biswas et al., 2022). The amalgamation of nanoscience's precision and food science's culinary artistry unveils a rich tapestry of opportunities wherein the manipulation of matter at the nanoscale enables the tailoring of food properties, flavors, textures, and nutritional profiles with unparalleled precision.

11.1.1 Brief Overview of Food Nanotechnology

At its core, food nanotechnology harnesses the principles of nanoscience to engineer, modify, and manipulate the structure and properties of food materials at nanometer dimensions. This manipulation encompasses the creation of nanomaterials, nanoscale additives, and nanostructured delivery systems to optimize sensory attributes, nutritional bioavailability, and shelf life (Sahoo et al., 2021). The underpinning philosophy of food nanotechnology revolves around the utilization of nanoparticles and nanoscale phenomena to confer beneficial effects on food products (Cummins et al., 2020). These effects extend beyond the mere enhancement of sensory experiences, encompassing improvements in functional aspects such as nutrient delivery, bioactive encapsulation, and targeted release mechanisms.

Within the realm of food nanotechnology, the notion of encapsulating bioactive compounds, modulating textures at the nanoscale, and engineering nanoscale delivery systems has been demonstrated as a means to extend the realm of culinary experiences (Cummins et al., 2020). These advancements embody a fusion of science and art, where culinary prowess is augmented by scientific understanding, enabling chefs and food technologists to push the boundaries of gastronomic creativity while enhancing consumer well-being.

11.1.2 Importance of Assessing Market Potential for Food Nanotechnology Innovations

Innovation, although fueled by scientific ingenuity, must be guided by market realities to navigate the path from laboratory discovery to consumer acceptance. Within the context of food nanotechnology, market potential assessment emerges as an indispensable compass, steering

DOI: 10.1201/9781003438168-11

researchers, industry professionals, and policymakers toward aligning their efforts with the ever-evolving landscape of consumer preferences, regulatory frameworks, and economic viability (Read et al., 2016).

The multifaceted nature of market potential encompasses not only the assessment of demand but also the exploration of factors that govern adoption, acceptance, and scalability. Market potential assessment, therefore, transcends being a mere numeric exercise and assumes the role of a strategic guidepost that bridges the chasm between innovation and commercial success. By anchoring innovation within the currents of market dynamics, stakeholders can cultivate a synergy that capitalizes on consumer aspirations while adhering to ethical considerations and regulatory imperatives (Ramakrishna et al., 2020).

In this era of heightened consumer awareness, the marriage of scientific prowess with market insights stands as a pivotal strategy for innovation's triumph. The ensuing exploration traverses the landscape of food nanotechnology's market dynamics, from trends and growth drivers to challenges and opportunities, enabling stakeholders to navigate the nuanced interplay between technological innovation and the ever-shifting currents of consumer demands and societal imperatives.

11.2 MARKET ANALYSIS

11.2.1 Current Trends in the Food Industry

The contemporary food industry operates within a dynamic milieu shaped by evolving consumer preferences, societal concerns, and technological breakthroughs. Health and wellness have emerged as cardinal tenets driving consumer choices, with an increasing emphasis on functional foods that offer physiological benefits beyond basic nutrition. The inclination toward sustainability, propelled by mounting environmental awareness, has engendered a demand for eco-friendly practices throughout the food supply chain. Concurrently, the pursuit of culinary individuality is fostering an appetite for personalized gastronomic experiences, driven by a diverse consumer base with distinct cultural backgrounds and dietary preferences. Additionally, the demand for transparency in food sourcing and production is surging, underpinned by a desire for authenticity and ethical considerations.

The intrinsic synergy between food nanotechnology and these trends is evident. The comprehensive Table 11.1 highlights a wider range of nanomaterials and their applications in both food production and packaging. The ability to enhance nutrient delivery, extend shelf life, and manipulate textures at the nanoscale aligns seamlessly with the quest for functional foods. The development of sustainable packaging materials leveraging nanotechnology addresses the sustainability imperative, aligning innovation with environmental consciousness (Tiwari, 2023). Furthermore, nanotechnology's precision in flavor delivery and sensory enhancement harmonizes with the demand for personalized culinary experiences. The capacity to trace and authenticate food origins, augmented by nanoscale identification mechanisms, resonates with the call for transparency in food supply chains (Nizar et al., 2018).

11.2.2 Growth Drivers for Food Nanotechnology

The ascent of food nanotechnology is propelled by a constellation of growth drivers that converge at the crossroads of technological advancement, consumer preferences, and global imperatives. At the crux of these drivers lies nanotechnology's evolution itself, as advancements in materials science, fabrication techniques, and analytical tools facilitate the creation of novel nanomaterials and nanostructures. This technological trajectory not only empowers the design of innovative food products but also engenders consumer curiosity and receptiveness to these novel creations (Valkenburg et al., 2020).

TABLE 11.1
Different nanomaterials and their applications in food production and packaging

Nanomaterial	Applications in Food Production and Packaging	References
Titanium oxide (TiO_2)	Enhances brightness in food products, acts as a food additive.	(Peters et al., 2014)
Silver (Ag)	Utilized for foliar applications and exhibits potent antibacterial properties.	(Ren et al., 2016)
Gold (Au)	Used in the meat industry for various applications.	(Rastogi et al., 2022)
Iron oxide (Fe_2O_3)	Employed in nutrient fertilizers, water treatment, and enhances food bioavailability.	(Cao et al., 2022)
Silicon dioxide (SiO_2)	Acts as an anticaking agent in food products.	(Bashir et al., 2023)
Zinc oxide (ZnO)	Used for plant nutrient supplementation and as a fruit spray.	(Semida et al., 2021)
Copper (Cu)	Promotes germination, exhibits antimicrobial activity.	(Liu et al., 2022)
Nano-clay	Enhances the shelflife and durability of food packaging.	(Nath et al., 2022)
Micelles	Improves liquid solubility, especially in food products.	(Zheng & McClements, 2020)
Lipid nanoparticles	Serve as emulsifiers and antioxidants in food formulations.	(Oehlke et al., 2017)
Starch nanospheres	Employed for physical or covalent interactions in food systems.	(Kumari et al., 2020)
Liposomes	Facilitate nano-encapsulation, enhance solubility, improve bioavailability, and enable cell-specific targeting in food applications.	(Subramani & Ganapathyswamy, 2020)
Multi-walled carbon nanotubes (MWCNTs)	Used in seed priming, the food industry, and phosphate-buffered solutions.	(Cai et al., 2022)
Single-wall carbon nanotubes (SWCNTs)	Integrated with biomolecules for various food-related applications.	(Lee, 2023)
chitosan	Promotes root growth, acts as an anti-fungicide, and facilitates nutrient delivery in agriculture and food production.	(Alam et al., 2020; Haris et al., 2023)
Nanomaterials in packaging	Enhance the performance and safety of food packaging.	(Bradley et al., 2011)
Nano-clay	Improves barrier properties, extending the shelflife of packaged food.	(Zehra et al., 2022)
Nanoscale enzyme immobilization	Increases surface area and stabilizes enzymes in packaging for temperature and pH control.	(Mohammadi et al., 2022)
Oxygen sensors	Used to detect oxygen levels in modified atmosphere food packaging.	(Zhu et al., 2017)
Oxygen scavengers	Extend the shelflife of food products by removing oxygen from packaging.	(Rodríguez-Núñez et al., 2018)
Titanium nitride (TiN)	Enhances thermal properties and acts as a UV filter in packaging materials.	(Vidakis et al., 2022)
Carbon black	Functions as a food additive and can be incorporated into packaging materials.	(Bott et al., 2014)
Aluminum	Serves as a filler in polymers and enhances scratch and abrasion resistance in coatings for packaging.	(Hsissou et al., 2021)
Nanosilver	Exhibits antimicrobial, antibiotic, and antistatic properties in packaging materials.	(Deshmukh et al., 2019)
Copper nanoparticles	Demonstrates antibacterial effects against microorganisms in food packaging.	(Shankar & Rhim, 2014)

(continued)

TABLE 11.1 (Continued)
Different nanomaterials and their applications in food production and packaging

Nanomaterial	Applications in Food Production and Packaging	References
UV-blocking nanomaterials (TiO_2, MgO, SiO_2, ZnO)	Act as UV blockers and photo-catalytic disinfecting agents in food packaging.	(Nawab et al., 2023)
Nano-sensors	Include oxygen sensors, time-temperature indicators, and freshness indicators for detecting spoilage due to contaminants.	(Shruti et al., 2024)
Fumed silica	Used as a food additive and can also enhance the properties of food packaging materials.	(Videira-Quintela et al., 2022)
Nanofibers	Used in food packaging to enhance mechanical strength and barrier properties.	(Bikiaris & Triantafyllidis, 2013)

Parallel to technological advancement, health-conscious consumers are catalyzing the demand for nutritional precision offered by nanoscale nutrient delivery systems (Shunin et al., 2018). The convergence of aging populations and escalating health concerns amplifies this demand, heralding opportunities for precision nutrition tailored to specific health needs. Moreover, in an era where food security and sustainability are pressing global challenges, nanotechnology's potential to optimize resource utilization, minimize waste, and enhance food preservation resonates with initiatives aimed at ensuring the nourishment of a burgeoning world population (Maluin et al., 2021).

11.2.3 Key Market Segments and Their Characteristics

The application of food nanotechnology spans a spectrum of key market segments, each characterized by distinct consumer needs and industry imperatives (Achrol & Kotler, 2012). Functional foods, imbued with health-enhancing attributes, exemplify the convergence of technological innovation and consumer health consciousness. The packaging industry, grappling with sustainability concerns, finds promise in nanotechnology-enabled biodegradable materials and active packaging systems (Jagtiani, 2022). Nutraceuticals, poised at the intersection of food and pharmaceuticals, capitalize on nanotechnology's potential for targeted and controlled nutrient release. Lastly, the realm of convenience foods seeks to harmonize modern lifestyles with nutritional quality, where nanotechnology offers avenues to enhance nutritional content without sacrificing convenience (Kumar et al., 2019). Figure 11.1 depicts the progress and utilization of nanotechnology in the agricultural and food production sectors, as well as the resulting consequences and impacts. The interplay between food nanotechnology and these market segments encapsulates both consumer aspirations and industry objectives. By aligning innovation with the unique demands of each segment, stakeholders stand to unlock the transformative potential of nanotechnology, realizing a convergence of enhanced consumer experiences and sustainable industry growth. In this expansive realm of market analysis, the synergistic relationship between evolving consumer preferences and technological capabilities crystallizes. The subsequent sections delve deeper into the methodologies employed to assess market potential, the intricate factors influencing it, and the real-world case studies that embody the harmonious fusion of innovation and market resonance.

11.3 MARKET POTENTIAL ASSESSMENT METHODS

11.3.1 Market Size Estimation

The endeavor to quantify the potential of food nanotechnology within the market landscape necessitates the deployment of meticulous methodologies for market size estimation. These

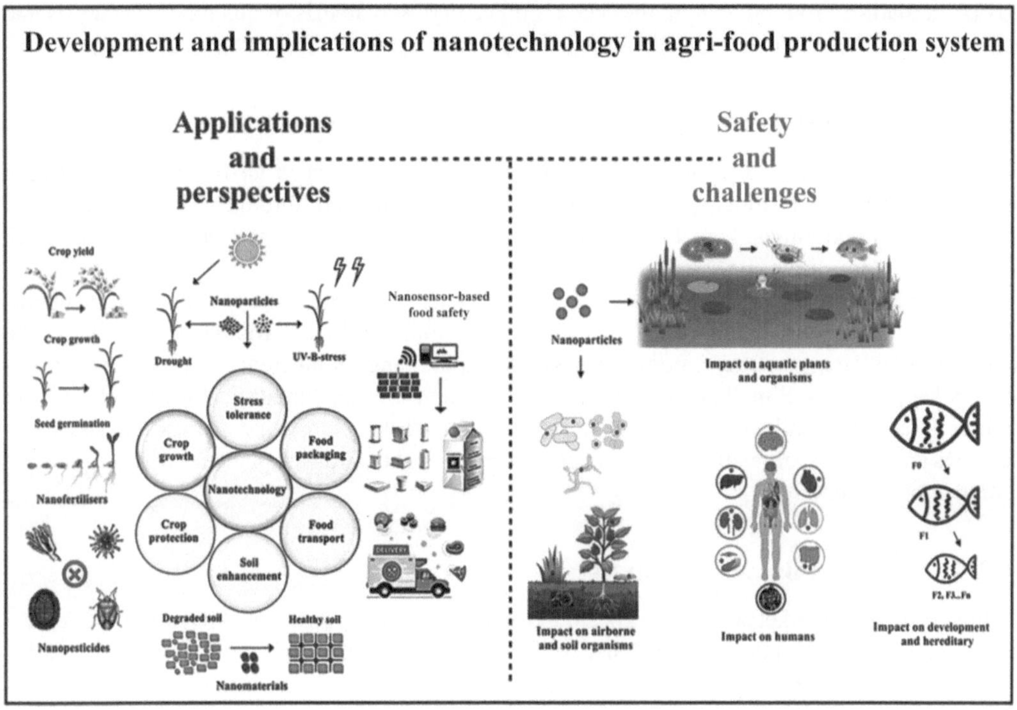

FIGURE 11.1 The advancement and application of nanotechnology in the agricultural and food production system and its consequences. Reprinted with written permission from Elsevier of Ashraf et al. (2021).

methodologies manifest as analytical tools that synthesize macroeconomic data, consumer behavior patterns, and industry trends to project the extent of market demand for innovative food nanotechnology products.

Top-down and bottom-up approaches constitute the pillars of market size estimation. The top-down approach involves extrapolating data from larger economic indicators to discern the potential market share of food nanotechnology within the broader food industry (Chau et al., 2007). This method capitalizes on macro-level statistics, encompassing factors such as national food consumption trends, economic indicators, and demographic profiles. In contrast, the bottom-up approach zooms into the minutiae of consumer behaviors, scrutinizing purchase patterns, preferences, and demographics to ascertain the potential consumer base for innovative nanotechnology-infused food products (Chau et al., 2007). Combining these methodologies provides a comprehensive perspective that illuminates the magnitude of the market canvas.

11.3.2 Market Growth Projections

As innovation forges ahead, projecting the growth trajectory of the food nanotechnology sector becomes paramount. Herein, Compound Annual Growth Rate (CAGR) analysis and trend extrapolation emerge as potent tools for forecasting future expansion (Korotky, 2013; Tsilingeridis et al., 2023). CAGR encapsulates the annual growth rate of a market over a specified period, offering insights into the sector's velocity and development pace. This metric, amalgamating historical data with future predictions, underscores the sector's anticipated maturation.

Trend extrapolation, on the other hand, encapsulates the act of extending historical market trends into the future. By analyzing past growth patterns and their underlying drivers, extrapolation provides a roadmap for anticipating the sector's trajectory. A synthesis of quantitative analysis and qualitative intuition empowers stakeholders to infer future dynamics from historical evolution.

11.3.3 Competitive Analysis

Navigating the complex labyrinth of the market ecosystem mandates an in-depth examination of competitors and positioning. Competitive analysis facilitated through SWOT (Strengths, Weaknesses, Opportunities, Threats) assessments and benchmarking, offers a structured framework to dissect the competitive landscape.

SWOT analyses dissect internal attributes (strengths and weaknesses) and external factors (opportunities and threats) that influence a firm's market positioning (Hunt & Madhavaram, 2012). The identification of strengths and opportunities aids in capitalizing on inherent advantages and emerging market prospects. Simultaneously, scrutinizing weaknesses and threats enables stakeholders to proactively address challenges and mitigate risks. Moreover, benchmarking against industry peers facilitates an appraisal of relative performance, offering insights into best practices and areas necessitating improvement.

11.3.4 Consumer Behavior Analysis

Central to assessing market potential is a profound understanding of consumer behaviors, preferences, and sentiments. In this endeavor, consumer behavior analysis emerges as a vital instrument, encapsulating the exploration of consumers' psychological and social underpinnings that guide their purchasing decisions (Pecoraro & Uusitalo, 2014).

Surveys, focus groups, and data analytics constitute the trinity of consumer behavior analysis (Chinman et al., 2006). Surveys yield quantitative insights by posing structured queries to a representative sample, enabling the quantification of preferences and tendencies. Focus groups, on the other hand, delve into qualitative exploration, delving into the nuances of consumer motivations, opinions, and perceptions. Complementing these approaches, data analytics harnesses the vast expanse of digital footprints to extract behavioral patterns and trends from a deluge of consumer data.

11.4 FACTORS INFLUENCING MARKET POTENTIAL

11.4.1 Regulatory Landscape and Safety Considerations

Navigating the intricate nexus of food nanotechnology and the regulatory landscape necessitates a meticulous understanding of prevailing regulations and safety considerations. As innovation ushers novel ingredients and processes, the regulatory framework serves as both a protector of public health and a gatekeeper to market entry (Chawla et al., 2021).

The deployment of nanomaterials in food products triggers concerns that transcend traditional ingredient assessments. Their unique physicochemical properties raise questions about potential toxicity, bioavailability, and unintended interactions within the human body (Mehta & Rietjens, 2023). Consequently, regulatory bodies are tasked with crafting frameworks that balance innovation's potential benefits with the prudence of safety evaluations. Stakeholders must engage in transparent dialogue with regulatory agencies, furnishing data that substantiates both the efficacy and safety of nanotechnology-enabled food products. The harmonization of innovation and safety ensures consumer trust, cultivates industry credibility, and paves the path for regulatory compliance.

11.4.2 Consumer Acceptance and Perception

Innovation's trajectory is intricately intertwined with consumer sentiment and perception. Within the realm of food nanotechnology, consumer acceptance stands as an overarching determinant of market success. The amalgamation of novelty, technological complexity, and potential health implications engenders a delicate dance between consumer curiosity and apprehension (Dahabieh et al., 2018).

Understanding consumer risk perceptions, ethical considerations, and cultural dimensions is instrumental. Transparent communication, educational campaigns, and ethical discussions play pivotal roles in aligning innovation with consumer values. Beyond this, the psychological factors that govern consumer choices, such as familiarity, trust, and sensory appeal, underscore the necessity of harmonizing innovation with consumer expectations.

11.4.3 Economic Factors and Cost-Effectiveness

While innovation fuels scientific progress, economic viability remains an inescapable cornerstone. Within the scope of food nanotechnology, the exploration of cost-effectiveness permeates discussions of market potential. Innovations that enhance flavors, nutritional delivery, and safety must also be positioned within a framework of affordability and economic sustainability (Malik et al., 2023).

Economic assessments encompass considerations beyond production costs, embracing consumer willingness to pay and price elasticity (Bernheim & Taubinsky, 2018). Stakeholders must weigh innovation's potential benefits against costs, ensuring that enhanced attributes translate into perceived value for consumers. The economic landscape extends beyond individual products, encompassing supply chain intricacies, infrastructure investments, and competitive pricing strategies.

11.4.4 Technological Feasibility and Scalability

Innovation's translation from laboratory to marketplace pivots on the axis of technological feasibility and scalability. While nanotechnology excels in the laboratory, its journey toward commercialization mandates meticulous assessments of manufacturability, reliability, and scalability (Tsung et al., 2023).

Innovation's success hinges on bridging the gap between scientific ingenuity and industrial practicality. Ensuring that nanoscale innovations can be produced consistently, at scale, and without compromising safety is pivotal (Sanguansri & Augustin, 2006). Moreover, the assessment of scalability extends to upstream considerations such as raw material sourcing and downstream aspects like waste management.

11.5 CASE STUDIES OF SUCCESSFUL INNOVATIONS

11.5.1 Nano-Emulsions for Enhanced Flavor Delivery

One compelling case study within the realm of food nanotechnology is the utilization of nano-emulsions to revolutionize flavor delivery. Nano-emulsions are colloidal dispersions with droplet sizes in the nanometer range, conferring attributes of enhanced stability, improved solubility, and amplified bioavailability (Abbas et al., 2013). In the context of flavor enhancement, nano-emulsions excel as vehicles for encapsulating volatile flavor compounds.

This innovation finds resonance with the current trend of personalized gastronomic experiences. The encapsulation of delicate volatile compounds safeguards them from premature degradation, facilitating a prolonged and controlled release upon consumption. This not only elevates sensory attributes but also offers the potential for novel culinary experiences through the manipulation of release kinetics. Furthermore, by enhancing the solubility of hydrophobic flavor compounds, nano-emulsions expand the possibilities of flavor incorporation into various food matrices, transcending traditional formulation constraints (Islam et al., 2023).

11.5.2 Nanoscale Nutrient Delivery for Functional Foods

Nanoscale nutrient delivery systems represent another success story in the realm of food nanotechnology. The convergence of health-conscious consumer preferences and precision nutrition has prompted the development of nanotechnology-driven nutrient delivery platforms. These systems encapsulate vitamins, minerals, and bioactive compounds within nanoscale carriers, facilitating targeted delivery to specific physiological sites (Chai et al., 2018).

This case study exemplifies the harmonious coalescence of innovation and health trends. By enhancing bioavailability and controlled release, nanoscale nutrient delivery amplifies the health impact of functional foods. Consumers seeking tailored nutrition can benefit from optimized nutrient delivery, ensuring efficient absorption and utilization. The potential to target specific health concerns, such as nutrient deficiencies, underscores the capacity of food nanotechnology to address individual health needs while aligning with broader health and wellness trends (Block et al., 2011).

11.5.3 Nanosensors for Food Safety Assurance

The incorporation of nanosensors to enhance food safety assurance stands as a triumphant example of the fusion of innovation and consumer concerns. Nanosensors, with their exquisite sensitivity and real-time detection capabilities, offer a transformative solution to address concerns related to foodborne pathogens, contaminants, and spoilage (Patel et al., 2020).

In an era marked by increasing awareness of foodborne illnesses and supply chain vulnerabilities, nanosensors introduce an avenue for real-time monitoring and early detection. By offering rapid and accurate detection of pathogens and spoilage indicators, nanosensors bolster consumer confidence in food safety. This case study underscores the pivotal role of innovation in aligning with societal concerns while simultaneously cultivating market potential through enhanced food quality and safety.

11.6 CHALLENGES AND RISKS

11.6.1 Potential Barriers to Market Adoption

While the realm of food nanotechnology brims with transformative potential, it is not devoid of challenges that could impede seamless market adoption. Regulatory uncertainties stand as a formidable barrier, as the evolving landscape of nanotechnology necessitates continuous dialogue between innovators and regulatory bodies to ensure compliance without stifling innovation. Ethical considerations, particularly surrounding the potential health impacts of nanoparticles, require comprehensive risk assessment and transparent communication to alleviate consumer apprehensions Srivastava, 2023).

Consumer perception of risk, fueled by concerns over unknown long-term effects, underscores the necessity of comprehensive safety assessments and transparent communication strategies (Singh & Kumar, 2023). Technical hurdles, encompassing issues like scale-up feasibility and quality control, pose challenges that demand scientific rigor and cross-disciplinary collaboration to surmount.

11.6.2 Risk Assessment and Mitigation Strategies

Mitigating the challenges that accompany innovation's journey entails a judicious blend of risk assessment and strategic planning. Regulatory engagement, spanning from early dialogue to compliance initiatives, ensures that innovations adhere to evolving safety standards while allowing for flexibility to accommodate novel technologies. Transparent communication campaigns that elucidate both benefits and potential risks foster consumer trust, emphasizing ethical responsibility and aligning innovation with societal values.

The global adoption and utilization of nanotechnology in the food and agriculture sectors are generally regarded as a promising approach to enhancing productivity, albeit with certain health and safety concerns (Fortunati et al., 2019). The interdisciplinary utilization of nanoparticles, their potential accumulation, and ingestion through everyday products have raised concerns among health scientists and toxicologists. The adverse effects primarily hinge on the nanoscale dimensions and interactions at the cellular level (Bauer et al., 2017). Such interactions may lead to genomic alterations and influence the proteomic structure.

According to published reports, prolonged consumption of ready-to-eat foods may have detrimental effects on gut microbiota, potentially causing dysbiosis (Gangadoo et al., 2021). Other notable effects include the production of reactive oxygen species (ROS), DNA damage, genotoxic effects, and damage to vital organs in humans. The generation of ROS can result in oxidative stress and cytotoxicity, interacting with extracellular matrices. The impact of various ingested engineered nanomaterials on cell proliferation, initiation of apoptosis, and release of pro-inflammatory and inflammatory cytokines has been reported in *invivo* studies, indicating cytotoxic effects on the liver, kidneys, stomach, and spleen (Kermanizadeh et al., 2014).

Furthermore, nano-emulsions with lipophilic cores exhibit different rates and extents of digestion and absorption in the gastrointestinal tract, potentially leading to adverse effects due to their chemical nature as surfactants (Qian et al., 2012). Several *invitro* and *invivo* investigations have demonstrated their cytotoxic effects on various human organs, but clinical trial data regarding safety and efficacy are still pending. Therefore, additional research is required to establish clinical dosing efficacy and identify cytotoxic biomarkers to develop comprehensive guidelines.

Advisory guidelines could be developed for commercially available food and agricultural products to address potential long-term exposure-related adverse effects. The evolution of nanotechnology and its applications in the food and agriculture industries emphasizes the importance of judiciously selecting nanoparticles, considering both their advantages and disadvantages. A visual representation of various nanoparticles employed in the food and agriculture sectors, along with their potential toxicity on different human organs, is shown in Figure 11.2. This visual aid aims to assist the scientific community in evaluating cytotoxicity, ethical dosing, and future safety considerations for the listed nanoparticles used in the agri-food production system. Research and development investments aimed at addressing technological hurdles are pivotal. Collaborative efforts that span academia, industry, and regulatory bodies offer diverse perspectives, aiding in identifying and surmounting technical roadblocks. The orchestration of multidisciplinary expertise, along with continuous monitoring and adaptation, fosters innovation resilience and positions stakeholders to navigate barriers effectively.

11.7 FUTURE OUTLOOK AND OPPORTUNITIES

11.7.1 Emerging Trends in Food Nanotechnology

The trajectory of food nanotechnology unfolds against a backdrop of evolving scientific frontiers and dynamic consumer landscapes. Emerging trends herald a future brimming with potential for innovation and market evolution. The refinement of nano-delivery systems stands as a testament to nanotechnology's journey toward precision, enabling targeted nutrient delivery and personalized health solutions (Şenel, 2021). The integration of nanosensors within the food supply chain augurs enhanced transparency and traceability, assuaging consumer concerns and cultivating trust. Furthermore, sustainable practices are poised to blossom with the advent of nanotechnology-driven biodegradable packaging materials and energy-efficient processes. The seamless fusion of innovation and sustainability resonates with the growing call for responsible consumption.

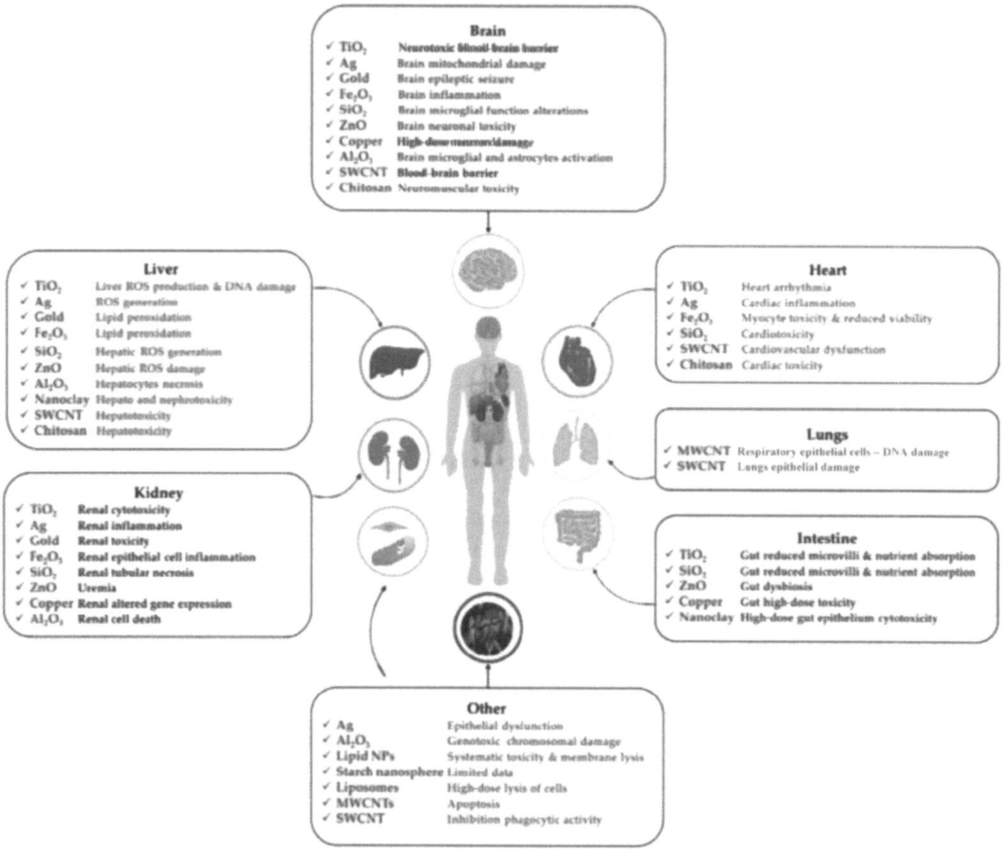

FIGURE 11.2 Illustrative depiction of the impact of different nanoparticles found in food and agricultural products on various human organs (brain, heart, lungs, intestines, liver, kidneys, and more) and their associated adverse effects. Reprinted with written permission from Elsevier of Ashraf et al. (2021).

11.7.2 Untapped Market Segments and Potential Applications

As innovation accelerates, hitherto untapped market segments beckon explorations into novel applications. The aging global population catalyzes the demand for precision nutrition tailored to age-related health concerns, underscoring the potential for innovation in geriatric health. The burgeoning trend of alternative proteins, addressing environmental and ethical considerations, aligns with nanotechnology's capacity to create textured and flavorful plant-based products (Kumar et al., 2023). The world of pet foods offers another fertile ground, where personalized nutrition and functional enhancements align with consumers' dedication to the well-being of their companions. The orchestration of nanotechnology's precision and these burgeoning market niches stands as an exciting avenue for future exploration.

11.7.3 Collaboration and Partnership Opportunities

The future of food nanotechnology transcends individual efforts, beckoning collaboration across diverse sectors. Collaborative partnerships between academia, industry, and governmental bodies offer a conduit for cross-pollination of knowledge, resources, and expertise. The convergence of disciplines fosters a holistic approach that tackles challenges from multiple angles, propelling innovation and market integration. Collaborative endeavors also offer avenues for navigating the intricate regulatory landscape, leveraging collective insights to ensure compliance while driving forward.

Furthermore, partnerships between industry players and consumer advocacy groups bridge the chasm between consumer expectations and industry initiatives. Transparent dialogues, educational campaigns, and shared insights foster alignment between innovation trajectories and societal values, ultimately engendering consumer trust and brand loyalty.

11.8 CONCLUSION

The voyage through the realm of food nanotechnology unveils a confluence of scientific innovation, consumer preferences, and market dynamics. This synthesis underscores the transformative potential of nanoscience within the culinary landscape, where precision engineering enhances flavors, textures, and nutritional profiles. The meticulous assessment of market potential acts as a compass, guiding stakeholders in navigating the intricate interplay between innovation and market resonance.

Market analysis illuminates the trends, growth drivers, and market segments that converge with nanotechnology's capabilities. The assessment of market potential methods, encompassing size estimation, growth projections, competitive analysis, and consumer behavior scrutiny, provides a nuanced perspective of innovation's reception. The factors governing market potential – regulatory landscapes, consumer perceptions, economic viability, and technological scalability – underscore the complex tapestry that innovation weaves within the market ecosystem.

Real-world case studies underscore the symbiotic dance between innovation and market needs, exemplifying the transformative potential of nano-emulsions, nutrient delivery systems, and nanosensors. Challenges and risks underscore the delicate balance between innovation and adoption, prompting a strategic interplay of risk assessment and mitigation strategies.

As the horizon of food nanotechnology unfurls, emerging trends, untapped markets, and collaborative partnerships beckon. The future resonates with precision nutrition, sustainable practices, and interdisciplinary collaborations that marry scientific acumen with societal aspirations. This holistic journey through the landscape of food nanotechnology unearths its potential to shape the culinary experience, enhance consumer well-being, and mold the contours of the food industry's evolution. The symphony of innovation, market dynamics, and consumer preferences orchestrates a harmonious narrative that propels food nanotechnology toward a future teeming with possibilities.

In the intersection of scientific rigor, market acumen, and consumer resonance lies the transformative power that can reshape the culinary landscape and nourish a future where innovation and sustenance intertwine.

REFERENCES

Abbas, S., Hayat, K., Karangwa, E., Bashari, M., & Zhang, X. (2013). An overview of ultrasound-assisted food-grade nanoemulsions. *Food Engineering Reviews*, *5*, 139–157. https://doi.org/10.1007/s12393-013-9066-3

Achrol, R. S., & Kotler, P. (2012). Frontiers of the marketing paradigm in the third millennium. *Journal of the Academy of Marketing Science*, *40*(1), 35–52. https://doi.org/10.1007/s11747-011-0255-4

Alam, M., Hossain, M. A., & Sarkar, A. (2020). Effect of edible coating on functional properties and nutritional compounds retention of air dried green banana (*Musa sapientum* L.). *IOSR Journal of Environmental Science, Toxicology and Food Technology (IOSR-JESTFT)*, *14*(2), 51–58.

Ashraf, S. A., Siddiqui, A. J., Elkhalifa, A. E. O., Khan, M. I., Patel, M., Alreshidi, M., Moin, A., Singh, R., Snoussi, M., & Adnan, M. (2021). Innovations in nanoscience for the sustainable development of food and agriculture with implications on health and environment. *Science of the Total Environment*, *768*, 144990. https://doi.org/10.1016/j.scitotenv.2021.144990

Bashir, O., Hussain, S. Z., Ameer, K., Amin, T., Beenish, Ahmed, I. A. M., Aljobair, M. O., Gani, G., Mir, S. A., Ayaz, Q., & Nazir, N. (2023). Influence of anticaking agents and storage conditions on quality characteristics of spray dried apricot powder: shelf life prediction studies using Guggenheim–Anderson–de Boer (GAB) model. *Foods*, 12(1), 171. https://doi.org/10.3390/foods12010171

Bauer, J., Meza, L. R., Schaedler, T. A., Schwaiger, R., Zheng, X., & Valdevit, L. (2017). Nanolattices: an emerging class of mechanical metamaterials. *Advanced Materials*, *29*(40), 1701850. https://doi.org/10.1002/adma.201701850

Bernheim, B. D., & Taubinsky, D. (2018). Chapter 5 – Behavioral public economics. In B. D. Bernheim, S. DellaVigna, & D. Laibson (Eds.), *Handbook of Behavioral Economics: Applications and Foundations 1* (Vol. 1, pp. 381–516). North-Holland. https://doi.org/10.1016/bs.hesbe.2018.07.002

Bikiaris, D. N., & Triantafyllidis, K. S. (2013). HDPE/Cu-nanofiber nanocomposites with enhanced antibacterial and oxygen barrier properties appropriate for food packaging applications. *Materials Letters*, *93*, 1–4. https://doi.org/10.1016/j.matlet.2012.10.128

Biswas, R., Alam, M., Sarkar, A., Haque, M. I., Hasan, M. M., & Hoque, M. (2022). Application of nanotechnology in food: processing, preservation, packaging and safety assessment. *Heliyon*, *8*(11), e11795. https://doi.org/10.1016/j.heliyon.2022.e11795

Block, L. G., Grier, S. A., Childers, T. L., Davis, B., Ebert, J. E., Kumanyika, S., Laczniak, R. N., Machin, J. E., Motley, C. M., & Peracchio, L. (2011). From nutrients to nurturance: a conceptual introduction to food well-being. *Journal of Public Policy & Marketing*, *30*(1), 5–13.

Bott, J., Störmer, A., & Franz, R. (2014). Migration of nanoparticles from plastic packaging materials containing carbon black into foodstuffs. *Food Additives & Contaminants: Part A*, *31*(10), 1769–1782. https://doi.org/10.1080/19440049.2014.952786

Bradley, E. L., Castle, L., & Chaudhry, Q. (2011). Applications of nanomaterials in food packaging with a consideration of opportunities for developing countries. *Trends in Food Science & Technology*, *22*(11), 604–610. https://doi.org/10.1016/j.tifs.2011.01.002

Cai, S., Zhang, P., Guo, Z., Jin, F., Wang, J., Song, Z., Nadezhda, T., Lynch, I., & Dang, X. (2022). Multi-walled carbon nanotubes improve nitrogen use efficiency and nutritional quality in *Brassica campestris*. *Environmental Science: Nano*, *9*(4), 1315–1329. https://doi.org/10.1039/D1EN01211H

Cao, X., Yue, L., Wang, C., Luo, X., Zhang, C., Zhao, X., Wu, F., White, J. C., Wang, Z., & Xing, B. (2022). Foliar application with iron oxide nanomaterials stimulate nitrogen fixation, yield, and nutritional quality of soybean. *ACS Nano*, *16*(1), 1170–1181. https://doi.org/10.1021/acsnano.1c08977

Chai, J., Jiang, P., Wang, P., Jiang, Y., Li, D., Bao, W., Liu, B., Liu, B., Zhao, L., & Norde, W. (2018). The intelligent delivery systems for bioactive compounds in foods: physicochemical and physiological conditions, absorption mechanisms, obstacles and responsive strategies. *Trends in Food Science & Technology*, *78*, 144–154. https://doi.org/10.1016/j.tifs.2018.06.003

Chau, C.-F., Wu, S.-H., & Yen, G.-C. (2007). The development of regulations for food nanotechnology. *Trends in Food Science & Technology*, *18*(5), 269–280. https://doi.org/10.1016/j.tifs.2007.01.007

Chawla, R., Sivakumar, S., & Kaur, H. (2021). Antimicrobial edible films in food packaging: current scenario and recent nanotechnological advancements – a review. *Carbohydrate Polymer Technologies and Applications*, *2*, 100024. https://doi.org/10.1016/j.carpta.2020.100024

Chinman, M., Young, A. S., Hassell, J., & Davidson, L. (2006). Toward the implementation of mental health consumer provider services. *The Journal of Behavioral Health Services & Research*, *33*(2), 176–195. https://doi.org/10.1007/s11414-006-9009-3

Cummins, C., Lundy, R., Walsh, J. J., Ponsinet, V., Fleury, G., & Morris, M. A. (2020). Enabling future nanomanufacturing through block copolymer self-assembly: a review. *Nano Today*, *35*, 100936. https://doi.org/10.1016/j.nantod.2020.100936

Dahabieh, M. S., Bröring, S., & Maine, E. (2018). Overcoming barriers to innovation in food and agricultural biotechnology. *Trends in Food Science & Technology*, *79*, 204–213. https://doi.org/10.1016/j.tifs.2018.07.004

Deshmukh, S. P., Patil, S. M., Mullani, S. B., & Delekar, S. D. (2019). Silver nanoparticles as an effective disinfectant: a review. *Materials Science and Engineering: C*, *97*, 954–965. https://doi.org/10.1016/j.msec.2018.12.102

Fortunati, E., Mazzaglia, A., & Balestra, G. M. (2019). Sustainable control strategies for plant protection and food packaging sectors by natural substances and novel nanotechnological approaches. *Journal of the Science of Food and Agriculture*, *99*(3), 986–1000. https://doi.org/10.1002/jsfa.9341

Gangadoo, S., Nguyen, H., Rajapaksha, P., Zreiqat, H., Latham, K., Cozzolino, D., Chapman, J., & Truong, V. K. (2021). Inorganic nanoparticles as food additives and their influence on the human gut microbiota. *Environmental Science: Nano*, *8*(6), 1500–1518. https://doi.org/10.1039/D1EN00025J

Haris, M., Hussain, T., Mohamed, H. I., Khan, A., Ansari, M. S., Tauseef, A., Khan, A. A., & Akhtar, N. (2023). Nanotechnology – a new frontier of nano-farming in agricultural and food production and its development. *Science of the Total Environment*, *857*, 159639. https://doi.org/10.1016/j.scitotenv.2022.159639

Hsissou, R., Seghiri, R., Benzekri, Z., Hilali, M., Rafik, M., & Elharfi, A. (2021). Polymer composite materials: a comprehensive review. *Composite Structures*, *262*, 113640. https://doi.org/10.1016/j.compstruct.2021.113640

Hunt, S. D., & Madhavaram, S. (2012). Managerial action and resource-advantage theory: conceptual frameworks emanating from a positive theory of competition. *Journal of Business & Industrial Marketing*, *27*(7), 582–591. https://doi.org/10.1108/08858621211257356

Islam, F., Saeed, F., Afzaal, M., Hussain, M., Ikram, A., & Khalid, M. A. (2023). Food grade nanoemulsions: promising delivery systems for functional ingredients. *Journal of Food Science and Technology*, *60*(5), 1461–1471. https://doi.org/10.1007/s13197-022-05387-3

Jagtiani, E. (2022). Advancements in nanotechnology for food science and industry. *Food Frontiers*, *3*(1), 56–82. https://doi.org/10.1002/fft2.104

Kermanizadeh, A., Gaiser, B., Johnston, H., Brown, D., & Stone, V. (2014). Toxicological effect of engineered nanomaterials on the liver. *British Journal of Pharmacology*, *171*(17), 3980–3987. https://doi.org/10.1111/bph.12421

Korotky, S. K. (2013). Semi-empirical description and projections of Internet traffic trends using a hyperbolic compound annual growth rate. *Bell Labs Technical Journal*, *18*(3), 5–21. https://doi.org/10.1002/bltj.21625

Kumar, P., Mehta, N., Abubakar, A. A., Verma, A. K., Kaka, U., Sharma, N., Sazili, A. Q., Pateiro, M., Kumar, M., & Lorenzo, J. M. (2023). Potential alternatives of animal proteins for sustainability in the food sector. *Food Reviews International*, *39*(8), 5703–5728. https://doi.org/10.1080/87559129.2022.2094403

Kumar, S., Nehra, M., Dilbaghi, N., Marrazza, G., Hassan, A. A., & Kim, K.-H. (2019). Nano-based smart pesticide formulations: emerging opportunities for agriculture. *Journal of Controlled Release*, *294*, 131–153. https://doi.org/10.1016/j.jconrel.2018.12.012

Kumari, S., Yadav, B. S., & Yadav, R. B. (2020). Synthesis and modification approaches for starch nanoparticles for their emerging food industrial applications: a review. *Food Research International*, *128*, 108765. https://doi.org/10.1016/j.foodres.2019.108765

Lee, J. (2023). Carbon nanotube-based biosensors using fusion technologies with biologicals & chemicals for food assessment. *Biosensors*, *13*(2), 183. https://doi.org/10.3390/bios13020183

Liu, Z., Guo, S., Fang, X., Shao, X., & Zhao, Z. (2022). Antibacterial and plant growth-promoting properties of novel Fe_3O_4/Cu/CuO magnetic nanoparticles. *RSC Advances*, *12*(31), 19856–19867. https://doi.org/10.1039/D2RA03114K

Malik, S., Muhammad, K., & Waheed, Y. (2023). Nanotechnology: a revolution in modern industry. *Molecules*, *28*(2), 661. https://doi.org/10.3390/molecules28020661

Maluin, F. N., Hussein, M. Z., Nik Ibrahim, N. N., Wayayok, A., & Hashim, N. (2021). Some emerging opportunities of nanotechnology development for soilless and microgreen farming. *Agronomy*, *11*(6), 1213. https://doi.org/10.3390/agronomy11061213

Mehta, J. M., & Rietjens, I. M. (2023). Chapter 69 – Pros and cons of hazard-versus risk-based approaches to food safety regulation. In M. E. Knowles, L. E.Anelich, A. R. Boobis, & B. Popping (Eds.), *Present Knowledge in Food Safety* (pp. 1068–1087). Academic Press. https://doi.org/10.1016/B978-0-12-819470-6.00024-X

Mohammadi, Z. B., Khoshtinat, K., Ghasemi, S., & Ahmadi, Z. (2022). Nanotechnology and food grade enzymes. In A. Dutt Tripathi, K. K. Darani, & S. K. Srivastava (Eds.), *Novel Food Grade Enzymes: Applications in Food Processing and Preservation Industries* (pp. 455–487). Springer Nature Singapore. https://doi.org/10.1007/978-981-19-1288-7_17

Nath, D., Santhosh, R., Pal, K., & Sarkar, P. (2022). Nanoclay-based active food packaging systems: a review. *Food Packaging and Shelf Life*, *31*, 100803. https://doi.org/10.1016/j.fpsl.2021.100803

Nawab, R., Iqbal, A., Niazi, F., Iqbal, G., Khurshid, A., Saleem, A., & Munis, M. F. H. (2023). Review featuring the use of inorganic nano-structured material for anti-microbial properties in textile. *Polymer Bulletin*, *80*(7), 7221–7245. https://doi.org/10.1007/s00289-022-04418-5

Nizar, N. N. A., Zainal, I. H., Bonny, S. Q., Pulingam, T., Vythalingam, L. M., & Ali, M. E. (2018). 22 – DNA and nanobiosensor technology for the detection of adulteration and microbial contamination in religious food. In M. E. Ali & N. N. A. Nizar (Eds.), *Preparation and Processing of Religious and Cultural Foods* (pp. 409–431). Woodhead Publishing. https://doi.org/10.1016/B978-0-08-101892-7.00022-5

Oehlke, K., Behsnilian, D., Mayer-Miebach, E., Weidler, P. G., & Greiner, R. (2017). Edible solid lipid nanoparticles (SLN) as carrier system for antioxidants of different lipophilicity. *PLoS One*, *12*(2), e0171662. https://doi.org/10.1371/journal.pone.0171662

Patel, G., Pillai, V., Bhatt, P., & Mohammad, S. (2020). Chapter 21 – Application of nanosensors in the food industry. In B. Han, V. K. Tomer, T. A. Nguyen, A. Farmani, & P. Kumar Singh (Eds.), *Nanosensors for Smart Cities* (pp. 355–368). Elsevier. https://doi.org/10.1016/B978-0-12-819870-4.00020-7

Pecoraro, M., & Uusitalo, O. (2014). Exploring the everyday retail experience: the discourses of style and design. *Journal of Consumer Behaviour*, *13*(6), 429–441. https://doi.org/10.1002/cb.1492

Peters, R. J. B., van Bemmel, G., Herrera-Rivera, Z., Helsper, H. P. F. G., Marvin, H. J. P., Weigel, S., Tromp, P. C., Oomen, A. G., Rietveld, A. G., & Bouwmeester, H. (2014). Characterization of titanium dioxide nanoparticles in food products: analytical methods to define nanoparticles. *Journal of Agricultural and Food Chemistry*, *62*(27), 6285–6293. https://doi.org/10.1021/jf5011885

Qian, C., Decker, E. A., Xiao, H., & McClements, D. J. (2012). Nanoemulsion delivery systems: influence of carrier oil on β-carotene bioaccessibility. *Food Chemistry*, *135*(3), 1440–1447. https://doi.org/10.1016/j.foodchem.2012.06.047

Ramakrishna, S., Ngowi, A., Jager, H. D., & Awuzie, B. O. (2020). Emerging industrial revolution: symbiosis of Industry 4.0 and circular economy: the role of universities. *Science, Technology and Society*, *25*(3), 505–525. https://doi.org/10.1177/0971721820912918

Rastogi, S., Kumari, V., Sharma, V., & Ahmad, F. J. (2022). Gold nanoparticle-based sensors in food safety applications. *Food Analytical Methods*, *15*(2), 468–484. https://doi.org/10.1007/s12161-021-02131-z

Read, S. A. K., Kass, G. S., Sutcliffe, H. R., & Hankin, S. M. (2016). Foresight study on the risk governance of new technologies: the case of nanotechnology. *Risk Analysis*, *36*(5), 1006–1024. https://doi.org/10.1111/risa.12470

Ren, Y.-y., Yang, H., Wang, T., & Wang, C. (2016). Green synthesis and antimicrobial activity of monodisperse silver nanoparticles synthesized using Ginkgo Biloba leaf extract. *Physics Letters A*, *380*(45), 3773–3777. https://doi.org/10.1016/j.physleta.2016.09.029

Rodríguez-Núñez, J. R., Madera-Santana, T. J., Burrola-Núñez, H., & Martínez-Encinas, E. G. (2018). Composite materials based on PLA and its applications in food packaging. In *Composites Materials for Food Packaging* (pp. 355–399). https://doi.org/10.1002/9781119160243.ch12

Sahoo, M., Vishwakarma, S., Panigrahi, C., & Kumar, J. (2021). Nanotechnology: current applications and future scope in food. *Food Frontiers*, *2*(1), 3–22. https://doi.org/10.1002/fft2.58

Sanguansri, P., & Augustin, M. A. (2006). Nanoscale materials development – a food industry perspective. *Trends in Food Science & Technology*, *17*(10), 547–556. https://doi.org/10.1016/j.tifs.2006.04.010

Semida, W. M., Abdelkhalik, A., Mohamed, G. F., Abd El-Mageed, T. A., Abd El-Mageed, S. A., Rady, M. M., & Ali, E. F. (2021). Foliar application of zinc oxide nanoparticles promotes drought stress tolerance in eggplant (*Solanum melongena* L.). *Plants*, *10*(2), 421. https://doi.org/10.3390/plants10020421

Şenel, S. (2021). Nanotechnology and animal health. *Pharmaceutical Nanotechnology*, *9*(1), 26–35. https://doi.org/10.2174/2211738508666200910101504

Shankar, S., & Rhim, J.-W. (2014). Effect of copper salts and reducing agents on characteristics and antimicrobial activity of copper nanoparticles. *Materials Letters*, *132*, 307–311. https://doi.org/10.1016/j.matlet.2014.06.014

Shruti, A., Bage, N., & Kar, P. (2024). Nanomaterials based sensors for analysis of food safety. *Food Chemistry*, *433*, 137284. https://doi.org/10.1016/j.foodchem.2023.137284

Shunin, Y., Bellucci, S., Gruodis, A., & Lobanova-Shunina, T. (2018). Nanotechnology application challenges: nanomanagement, nanorisks and consumer behaviour. In Y. Shunin, S. Bellucci, A. Gruodis, & T. Lobanova-Shunina (Eds.), *Nonregular Nanosystems: Theory and Applications* (pp. 337–395). Springer International Publishing. https://doi.org/10.1007/978-3-319-69167-1_11

Singh, R., & Kumar, S. (2023). Regulatory and safety concerns regarding the use of active nanomaterials in food industry. In R. Singh & S. Kumar (Eds.), *Nanotechnology Advancement in Agro-Food Industry* (pp. 269–306). Springer, Singapore. https://doi.org/10.1007/978-981-99-5045-4_8

Srivastava, S. (2023). Chapter 22 – Ethical issues regarding the use of nanobiotechnology-based products. In R. Pratap Singh, C. O. Adetunji, R. L. Singh, J. Singh, P. R. Solanki, & K. R. B. Singh (Eds.), *Nanobiotechnology for the Livestock Industry* (pp. 435–473). Elsevier. https://doi.org/10.1016/B978-0-323-98387-7.00005-7

Subramani, T., & Ganapathyswamy, H. (2020). An overview of liposomal nano-encapsulation techniques and its applications in food and nutraceutical. *Journal of Food Science and Technology*, *57*(10), 3545–3555. https://doi.org/10.1007/s13197-020-04360-2

Tiwari, A. (2023). Advancement of materials to sustainable & green world. *Advanced Materials Letters*, *14*(3), 2303–1724. https://doi.org/10.5185/amlett.2023.031724

Tsilingeridis, O., Moustaka, V., & Vakali, A. (2023). Design and development of a forecasting tool for the identification of new target markets by open time-series data and deep learning methods. *Applied Soft Computing*, *132*, 109843. https://doi.org/10.1016/j.asoc.2022.109843

Tsung, T.-H., Tsai, Y.-C., Lee, H.-P., Chen, Y.-H., & Lu, D.-W. (2023). Biodegradable polymer-based drug-delivery systems for ocular diseases. *International Journal of Molecular Sciences*, *24*(16), 12976. https://doi.org/10.3390/ijms241612976

Valkenburg, G., Mamidipudi, A., Pandey, P., & Bijker, W. E. (2020). Responsible innovation as empowering ways of knowing. *Journal of Responsible Innovation*, *7*(1), 6–25. https://doi.org/10.1080/23299460.2019.1647087

Vidakis, N., Petousis, M., Mountakis, N., Korlos, A., Papadakis, V., & Moutsopoulou, A. (2022). Trilateral multi-functional polyamide 12 nanocomposites with binary inclusions for medical grade material extrusion 3D printing: the effect of titanium nitride in mechanical reinforcement and copper/cuprous oxide as antibacterial agents. *Journal of Functional Biomaterials*, *13*(3), 115. https://doi.org/10.3390/jfb13030115

Videira-Quintela, D., Guillen, F., Martin, O., Cumbal, L., & Montalvo, G. (2022). Antibacterial and antioxidant triple-side filler composed of fumed silica, iron, and tea polyphenols for active food packaging. *Food Control*, *138*, 109036. https://doi.org/10.1016/j.foodcont.2022.109036

Zehra, A., Wani, S. M., Bhat, T. A., Jan, N., Hussain, S. Z., & Naik, H. R. (2022). Preparation of a biodegradable chitosan packaging film based on zinc oxide, calcium chloride, nano clay and poly ethylene glycol incorporated with thyme oil for shelf-life prolongation of sweet cherry. *International Journal of Biological Macromolecules*, *217*, 572–582. https://doi.org/10.1016/j.ijbiomac.2022.07.013

Zheng, B., & McClements, D. J. (2020). Formulation of more efficacious curcumin delivery systems using colloid science: enhanced solubility, stability, and bioavailability. *Molecules*, *25*(12), 2791. https://doi.org/10.3390/molecules25122791

Zhu, R., Desroches, M., Yoon, B., & Swager, T. M. (2017). Wireless oxygen sensors enabled by Fe (II)-polymer wrapped carbon nanotubes. *ACS Sensors*, *2*(7), 1044–1050. https://doi.org/10.1021/acssensors.7b00327

Index

Note: Page numbers in **bold** indicate tables, *italic* numbers indicate figures, on the corresponding pages.

A

Acceptable risk model 217
Acoustic cavitation 9
Active packaging **94**, 122
Active substances, contamination by 134
Additive manufacturing 11–12
Adoption of nanosensors, strategies for 197
Adulterants, detection of 193
Advantages and disadvantages of nanotechnology **105**, 106
Aflatoxins **90**, 133, 146, 193
Agriculture sector, advancements in 225, 226
Alcalases **33**
Altinkaynak, C. 38
Amylases 21, **22**
  α-Amylase 21, **30**, **31**, **32**, **34**
  β-Amylase 21, **22**, **32**
Amyloglucosidase **29**
Anethole 72
Anodic stripping voltammetry (ASV) 165
Antibacterial capabilities 71–72, *73*, **79**, 97–98
Antibiotics, contamination by 134, 136
Anticancer capabilities 67–69, *70*, **76–77**
Antidiabetic capabilities 66–67, *68*, **76**
Anti-inflammatory capabilities 69–71
Antiviral capabilities 72–74, *74*, **81–82**
Apple juice clarification 36
Arsenic (As) 157
Asastaxanthin 67–68
Aspartic proteases **22**, 24
Assistive methods of synthesis 9–10
Astaxanthin 66

B

Bacteria **190**
Bacteriocins 72
Beverages, safety and quality of 196
Bioactive compounds, clinical trials of 74–75, **76–82**
Biochemical techniques 163–165
Biofilms, inhibition of **94**
Biofuel cells 165
Biogenic synthesis 4–5, *5*
Biological contaminations 133
Biological methods of synthesis 4–6, *5*
Biomass, production of NPs and 6
Biomolecules, templates based on 6
Bio-nanocomposites 101
Bio nanosensors **105**
Biopolymers in food packaging 118
Biosensors 170–171
Bulk encapsulation systems 99

C

Cancer, capabilities against 67–69, *70*
  clinical trials of bioactive compounds 74, **76–77**, **78**, **81**
  vitamin A 75
Carbohydrases 21, **22–23**, 24
Carbon dots (CDs) 147
Carbon nanomaterials, food analysis and 135
Carbon quantum dots (CQDs) 163, 166, 168
Carboxyl proteases **22**, 24
Cardiovascular system 212–213
Carotenoids 62
Carotenuto, P. 68
Carvone 72
Cellulases **22**, 24, **29**, **31**, **33**, **34**
Cellulose encapsulated nano-emulsion 71
Central nervous system 212
Cereals, heavy metals contamination in 158, *159*
Characterization techniques 216
Chemical contaminations 133–134
Chemical methods of synthesis 1–4, *2*, *4*
Chemical vapor deposition (CVD) 3
Chemical vapor synthesis (CVS) 3
Chemotherapeutic drugs 12
Chitinases **23**, 25, **29**, **30**
Chitosan in food packaging **113–114**
Chromogenic substances 162
Chronoamperometry 136
Clinical trials of bioactive compounds 74–75, **76–82**
Coatings, nano- 188, 232
Co-immobilization 36, 43
CO_2LAP process 8
Colds **80**
Colorimetric analysis 162–163, 190
Colorimetric sensors 169–170
Common cold **80**
Conductimetry 165
Consumer perception and adoption 236, 248–249
Controllable hydrogel 12
Coupling agents 26
Covalent bonding 21, 27, **28–29**, **31–35**
COVID-19 74–75, **76**, **77**, **78–79**, **80**
Crocin-gold nanoparticles (AuNPs) 67
Crystal structure 217
Curcumin 73
Cyclic voltammetry (CV) 135, 136
Cysteine proteases **22**, 24

D

Dendrimers 60–61
Dermal contact 210
Dextran-aldehyde 27

Diabetes, capabilities against 66–67, *68*, **76**
Dietary fiber 62–63, 195
Differential pulse voltammetry (DPV) 135, 136
Digestive system 210
Dimetridazole (DMZ) 136, *141*
Disadvantages and advantages of nanotechnology **105**, 106
Diseases, foodborn 103
Dissolution of nanoparticles 209
DNA as biomoleculer template 6
DNA tetrahedron nanostructures (DTNs) 146, *147*
DNAzymes 167, 172
Downstream processes **94**
Drug release, controllable hydrogel 12

E

EC Regulation No 258/97 104
Edible films 101
Egg, heavy metals contamination of *159*, 160
Elderly pneumonia **81**
Electrochemical nanosensors 135–136, **137–139**, *140*, 140–142, *141*
Electrochemical techniques 163–165
Electrochemiluminescence (ECL) 164–165
Electrode modifiers 164
Electrospraying technique 8, 12
Essential fatty acids 63
Essential oils 97–98
Extracellular production of NPs 4–5

F

Fernandez Caresani, J.R. 40
Fiber, dietary 62–63, 101
Fish and sea foods, heavy metals contamination of 161
Flame spray pyroysis (FSP) 8
Flavonoids 63–64
Flavor delivery, nano-emulsions and 249
Flavorzyme immobilization 37
Fluorescence technique 165–167, 173
Fluorescent sensors 142, **143–145**, 146–147, *148*, 168–169
Fluorinated graphene/gold nanocage (FGP/ AuNC) 172
Food and beverage sector 195
Food bundling, nanostructures in 93–94
Food-grade enzymes
  amylases 21, **22**
  carbohydrases 21, 24
  cellulases **22**, 24
  chitinases **23**, 25
  classes (summary) 21, **22–23**
  β-Galactosidase **23**, **30**
  glucose oxidase **23**, 26
  lactoperoxidase **23**, 26
  lipases **23**, 25
  lysozymes **23**, 25
  oxioreductases **23**, 26
  pectinases **22**, 24
  proteases **22**, 24
Food inspection, nanosensors in 188
Food poverty 99
Food preparation, nanostructures in 93–94
Forster resonance energy transfer (FRET) 166, 168
Fragmentation, particle 6–7, *7*
Fruits, heavy metals contamination in *159*, 159–160

G

β-Galactosidase **23**, 24, **28**, **30–31**, **32**, **33**, 37, 39
Gastrointestinal tract 212
Genetically modified crops 104
Glucoamylase 21, **22**, **28**
Glucose oxidase **23**, 26, **29**, **30**, **32**, **35**
α-Glucosidase **33**
Glutaraldehyde 27
Glutathione 6
Gold nanoparticles 36, 140, 189–190
Graphene 42–43, 146
  heavy metals 173
Graphene-based nanotechnology 166
Graphene nanoplatelets 232
Graphene oxide-based microbots (GOx- microbots) 172
Graphene oxide (GO) 168–169
Green tea supplementation 75

H

Heavy metals
  arsenic (As) 157
  health concerns 157–158
Hesperidin 75
Hesperidin-loaded gold nanoparticles (Hsp-AuNPs) 71
High-energy ball milling (HEBM) 6–7, *7*
HIV-1
  clinical trials of bioactive compounds **77**
  selenium supplementation 75, 83
Hydrocolloid systems 99
Hydrothermal synthesis 3

I

Immobilization
  co-immobilization 36, 43
  irreversible 27
  methods for 20–21
Immunomodulation 69–71
Inert gas condensation (IGC) 7
Inflammation, capabilities against 69–71
Ingestion of nanoparticles 209
Inhalation of nanoparticles 209
Innovation, responsible 218
Inspection, food, nanosensors in 188
Intracellular production of NPs 4–5
Iron fortification of foods **79**
Irreversible immobilization 27

J

Juice clarification 36

L

Lactic acid bacteria 98–99
Lactoperoxidase **23**, 26, **29**, **31**, **35**
Laser-induced breakdown spectroscopy (LIBS) 172

Laser pyroysis (LAP) 8
Layer-by-layer embodiment 99
Lipases **23**, 25, **28–29**, **30**, **31**, **32**, **33–34**
Lipid-based nanoparticles 60
Localized surface plasmon resonance (LSPR) 169
Luminescent metal-organic frameworks (LMOFs) 166
Lutein 66
Lycopene 66
Lysozymes **23**, 25, **29**, **30**, **32**, **34**, 39

M

Magnetic nanoparticles (MNPs) 27
Magnetic silica nanocomposites 36
Medication, controllable hydrogel 12
Melt mixing 9
Metal-based QDs 166
Metallo proteases **22**, 24
Metal nanomaterials 162
Metal-organic frameworks (MOFs) 142, 165, 171
Microbial lipases 25
Microemulsion technique 2–3
Microorganisms, production of NPs and 4–5, *5*
Microwave-assisted production techniques 9–10
Milling 6–7, *7*
Minerals 65
Molecularly imprinted polymers (MIPs) 136
Monodispersed gold nanoparticles (AuNPs) 162
Mukhopadhyay, A. 41
Multi-walled carbon nanotube (MWCNTs) **33**, **34**, 41–42
MWCNT *see* multi-walled carbon nanotube (MWCNTs)
Mycotoxins **90**, 133, 146, 193

N

Nano additives **94**, **105**
Nano-capsules 59
Nanocarriers 104
Nanocatalysts 186
Nanocellulose in food packaging **113**
Nano-coatings 188, 232
Nanocomposite materials 186, 188
Nanoemulsions 58–59, 249
Nanoflowers **30–31**, 38–39
Nanoliposomes 71
Nanosheets **34–35**, 42–43
Nanosized hydroxyapatite (NHAP) 167
Nanospheres 61
Nanostructured approach for immobilization 26–27, **28–35**, 36–37
Nanostructured lipid carriers (NLCs) 60
Nanowires 6
Natural microbial agents 97–98

O

Optical chemical sensors (OCSs) 173
Optical nanosensors 142, **143–145**, 146–147, *147*, *148*
Organic-inorganic hybrid nanoflowers **30–31**, 38–39
Oxidative stress 66, **77**
Oxioreductases **23**, 26
Oxygen scavengers, nano-enabled 188

P

Paraquat (PQ) 146
Parasites *see* pathogen detection
Particle fragmentation 6–7, *7*
Particle size 208, 216–217
Patulin 136, *140*
Pectinases **22**, 24, 27, **34**, 36
Pectinex **22**
Pesticides, contamination by 134, **190**, 193
Pharmaceuticals, contamination by 134, 136
Phenolic acids 64
Phenolic compounds 72
Physical adsorption 27
Physical methods of synthesis 6–9, *7*
Physical pollutants, contamination by 134
Physical vapor deposition (PVD) 8
Phyto-assisted production techniques 10
Plant biometabolites 6
Plant-extract mediated NPs 10
Plasma-enhanced chemical vapor deposition (PECVD) 3–4
Plasmonic sensors 169
Plastic waste 118
Pneumonia, elderly **81**
Poly aldehyde kefiran 27
Polymeric micells 57
Polymeric nanomaterials, food analysis and 135
Polyol synthesis 3, *4*
Porphyrin-functionalized metal-organic framework (porph@MOF) 72
Probiotics 98–99
Production techniques
  physical methods 6–9, *7*
  physical vapor deposition (PVD) 8
  phyto-assisted 10
  plant biometabolites 6
  polyol synthesis 3, *4*
Proteases **22**, 24, **29**, **31**, **32**, **34**, **35**
Protein-based nanocarriers 60
Public perception and adoption 236, 248–249
PVP-taxifolin liposomes (PVP TAX-Lips) 70

Q

Quantum dots (QDs) 168
Quercetin loaded silk fibroin nanoparticles (QSFNs) 70
Quercetin nanoparticles (QT-NPs) 70–71

R

Responsible innovation 218
Reverse micelles 2, 116
Reversible physical adsorption 27
Rutin 72

S

Salmonella 136, 140, *141*
Secondary metabolites 61
Selenium nanoparticles (SeNPs) 67, 71
Selenium supplementation 75, 83
Sensor systems for contaminant detection 132

Serine proteases **22**, 24
SERS sensors 169
Shape of particles 216–217
Signal tags 164
Silver in food packaging **114**
Silver nanoparticles 188, 189–190
Single-cell encapsulation systems 99
Single layer carbon nanotubes (SWCNTs) **34**, 41
Size of particles 208, 216–217
Skin contact 210
Smart labels 196
Smart packaging **94**, 100, 122–123
Sol-gel process 1–2, *2*
Solid lipid nanoparticles (SLNs) 60
Sonochemistry 9
Square wave voltammetry (SWV) 135–136
Starch in food packaging **113**
Superparamagnetic Fe_3O_4 nanoparticles 27
Surface chemistry of nanomaterials 217
Surface-enhanced Raman scattering (SERS) sensors 169
Sustainable practices 218
SWCNT *see* single layer carbon nanotubes (SWCNTs)

T

Templates based on biomolecules 6
Thiol proteases **22**, 24
Toxins, detection of 193
Transportation **187**
Trends in nanotechnology **94**
Trypsin **30**, **35**

U

Ultrasound-assisted production techniques 9
Upconversion nanoparticles (UCNPs) 167

V

Vegetables, heavy metals contamination in 158–160, *159*
Viruses **190**
Vitamins 64–65
Voltammetric techniques 135–136

W

Wine industry 196

X

Xylanase 36

Z

Zinc oxide production technique 96, *96*
Zinc supplementation 75